Composite Materials, Volume I: Properties, Nondestructive Testing, and Repair

MEL M. SCHWARTZ

To join a Prentice Hall PTR mailing list, point to:
http://www.prenhall.com/register

Prentice Hall PTR, Upper Saddle River, New Jersey 07458
http://www.prenhall.com

Library of Congress Cataloging-in-Publication Data

Schwartz, Mel M.
Composite materials / Mel Schwartz.
p. cm.
Includes bibliographical references and index.
Contents: v. 1. Properties, nondestructive testing, and repair.
ISBN 0-13-300047-8
1. Composite materials. I. Title.
TA418.9.C6S37 1996
620.1′18—dc20 96-28559
CIP

Acquisitions editor: *Bernard M. Goodwin*
Editorial/production supervision
and interior design: *bookworks*
Manufacturing manager: *Alexis Heydt*
Cover design: *Scott Weiss*
Cover design director: *Jerry Votta*

The background material on the cover is a reproduction of a photograph of Kevlar, a fiber produced by DuPont which typifies raw materials used to fabricate composite parts. The Kevlar displayed here is used by Sikorsky Aircraft in the manufacturing of helicopter parts. Photograph courtesy of Sikorsky Aircraft.

Prentice-Hall, Inc.
A Pearson Education Company
Upper Saddle River, NJ 07458

The publisher offers discounts on this book when ordered in bulk quantities. For more information, contact:

Corporate Sales Department
PTR Prentice Hall
One Lake St.
Upper Saddle River, NJ 07458

Phone: 800-382-3419
FAX: 201-236-7141
E-mail: corpsales@prenhall.com

Printed in the United States of America
10 9 8 7 6 5 4 3 2 1

ISBN 0-13-300047-8

Prentice-Hall International (UK) Limited, London
Prentice-Hall of Australia Pty. Limited, Sydney
Prentice-Hall Canada Inc., Toronto
Prentice-Hall Hispanoamericana, S.A., Mexico
Prentice-Hall of India Private Limited, New Delhi
Prentice-Hall of Japan, Inc., Tokyo
Pearson Education Asia Pte. Ltd., Singapore
Editoria Prentice-Hall do Brasil, Ltda., Rio De Janeiro

**To Mother and Dad, whose emotional strength,
free-flowing love, and caring concern
have helped shape my maturation.**

Contents

2 Fibers and Matrices 18

Preface

It has been recognized for many years that heterophase boundaries—interfaces between dissimilar materials—play a key role in determining the properties and performance of modern materials systems. The influence of heterophase boundaries in composites is widely recognized, and Chapter 2 discusses this matter.

The field of material technology will require scientific and industrial leadership in the next decade because this is one of several areas expected to experience unbelievable growth. To be involved in this explosion, highly specialized individuals with in-depth knowledge of materials design, process design, and product design and corporations must develop new materials and new processes to point to new pathfinder directions, routes, and roadmaps, and to stimulate thinking.

As has happened in the past, new scientific endeavors can be expected to create a demand for new and different materials. New ideas being explored include intermetallics, metal-ceramics, metal-carbons or graphites, ceramic-carbons, and plastic-ceramics. In the near future, hypersonic flight will create a need for many different and new combinations of these materials at material temperatures of 982–1482°C for up to 1 h for the trip from atmosphere to orbit and back again. High-modulus materials will also be required for parts of the structure. Other needs will require composites for liquid pressure tanks, high-conductivity structures, oxidation- and temperature-resistant ceramic matrices, diesel engines, and brake motors, as well as for a broad spectrum of uses in other industries such as electronics and computers.

It has been shown that in the modern world, advances in materials not only expand our technological capabilities but also become catalysts that drive capitalistic entrepreneurship and in turn allow society to maintain economic growth. Continuing advances in technology are absolutely critical to sustaining and improving the quality of life.

The "engineered materials age" is *now*, and we are able to put together, in some instances atom by atom, materials that are lighter, stronger, more conductive, more resilient, and smarter, and that have "memories."

Many of the materials we produce today are considered technologically mature. But the processes used to manufacture these metal, ceramic, and polymer composites are going through a revolution.

In the future smart software will design both the products and the production method. Concurrent engineering of composite materials is the technique needed to meet and keep a generation ahead of competition and stay focused on better manufacturability. This is significantly important because difficult-to-manufacture parts are universally difficult to manufacture, especially when the manufacturing processes employed have been developed for moderate to high volumes and requirements have actually fallen to low-volume levels in response to the worldwide recession or because of competitive pressures.

The early chapters of Volume I cover a general introduction to the major composite families, the fibers and matrix combinations and their properties, and how these materials have been strengthened by various mechanisms while overcoming, for example, incompatibility problems. The remaining chapters cover composite design, modeling, mechanical properties, nondestructive evaluation and repair methods, and the major uses and applications in the various industries.

Mel M. Schwartz

1

General Introduction to Advanced Composites

1.0 INTRODUCTION

It is evident that material advances have been the key to significant technology breakthroughs throughout history. The Stone Age, the Iron Age, the Industrial Revolution, the nuclear age, the electronic revolution, the aerospace era of today—all have critically depended on, or resulted from, breakthroughs in material technology.

Today we are in the midst of a new revolution triggered by the onset of advanced composites. This radically new class of materials is characterized by the marriage of quite diverse individual components that work together to produce capabilities that far exceed those of their separate elements. Their unique properties make them the enabling materials for major technological advances. Industry representatives believe these materials will be critical to the economic trade picture well into the twenty-first century.

Typically, advanced materials have been characterized by a lengthy development cycle (20 years).[1] Today, the use of composite materials in structures of all kinds is accelerating rapidly, with the major impact already being felt in the aerospace industry where the use of composites has directly enhanced the capability of fuel-efficient aircraft in the commercial arena and new-generation aircraft in the military sphere. The increasing usage of these materials is spreading worldwide, capitalizing on developments that were the direct result of a large investment in the technology over the last two or more decades.

1.1 WHAT ARE COMPOSITES?

In their broadest form, composites are the result of embedding high-strength, high-stiffness fibers of one material in a surrounding matrix of another material. The fibers of interest for composites are generally in the form of either single fibers about the thickness of a human hair or multiple fibers twisted together in the form of a yarn or tow. When properly produced, these fibers—usually of a nonmetallic material such as carbon, silicon carbide, boron, or alumina—can have very high values of strength and stiffness. As a result of work that started in the United States in the early 1950s, there are available to us thin, continuous fibers of a variety of materials, together with the manufacturing capability to produce them on a continuous basis. In addition to continuous fibers, there are also varieties of short fibers, whiskers, platelets, and particulates intended for use in discontinuous reinforced composites.

Fiber-reinforced composite materials consist of fibers of high strength and modulus embedded in or bonded to a matrix with distinct interfaces (boundaries) between them. In this form, both fibers and matrix retain their physical and chemical identities, yet they produce a combination of properties that cannot be achieved with either of the constituents acting alone. In general, the fibers are the principal load-carrying members, whereas the surrounding matrix keeps them in the desired location and orientation, acts as a load transfer medium between them, and protects them from environmental damages due to elevated temperature or humidity, for example. Thus, even though the fibers provide reinforcement for the matrix, the latter also serves a number of useful functions in a fiber-reinforced composite material.

The principal fibers in commercial use are various types of glass and carbon, as well as Kevlar and those mentioned previously. All these fibers can be incorporated into a matrix either in continuous lengths or in discontinuous (chopped) lengths. The matrix material may be a polymer, a metal, or a ceramic. Various chemical combinations, compositions, and microstructural arrangements are possible in each matrix category.

The most common form in which fiber-reinforced composites are used in structural applications is called a laminate. Laminates are obtained by stacking a number of thin layers of fibers and matrix and consolidating them into the desired thickness. Fiber orientation in each layer, as well as the stacking sequence of various layers, can be controlled to generate a wide range of physical and mechanical properties for the composite laminate.

1.1.1 Advanced Composites

Whereas the high properties of the fibers are in part a result of their being in fiber form, as fibers they are not useful from a practical point of view. The key to taking advantage of their uniquely high properties is to embed them in a surrounding matrix of another material. The matrix acts as a support for the fibers, transports applied loads to the fibers, and is capable of being formed into useful structural shapes. The right kind of matrix can also provide ductility and toughness properties that the much more brittle fibers do not possess. The term *advanced composites* is used to differentiate those with high-performance characteristics, generally strength and stiffness, from the simpler forms like reinforced plastics.

Historically, the term advanced composites has been taken to mean "high-performance" composites. Many believe that this definition is far too restrictive and eliminates many of the applications and materials with the most potential for future growth. The development of any composite requires balancing many factors, including performance, fabrication speed, and total cost. With high-performance materials, the desire for improved properties is the dominant requirement. For many applications, however, better performance, although desirable, is not the primary need. In fact, materials may already be available with properties that meet or even exceed the performance requirements. Instead, the problem is to produce parts at sufficient speeds and low enough costs to obtain them cost-effectively. For lack of a better term, such composites can be called *cost-performance materials.*

Processing methods such as liquid molding (resin transfer molding and structural reaction injection molding), pressure molding, filament winding and tow placement, thermal forming, and pultrusion offer the most potential for reduced cost and increased speed and will be discussed in Volume II. Industry representatives believe that they must harness the chemical and physical changes that occur during fabrication to the extent that is required for the processes to be optimized and controlled. Consequently, processing science and on-line process control are key issues for the future.

1.2 COMPOSITES AND THEIR HISTORY

Modern structural composites, frequently referred to as advanced composites, are blends of two or more materials, one of which is composed of stiff, long fibers and, for polymeric composites, a resinous binder or matrix that holds the fibers in place. The fiber is strong and stiff relative to the matrix, and generally it is orthotropic; that is, it has different properties in two different directions. For advanced structural composites, the fiber is long, with a length-to-diameter ratio of over 100. The strength and stiffness of the fiber are much greater, perhaps multiples of those of the matrix material. When the fiber and the matrix are joined to form a composite, they both retain their individual identities and both influence the composite's final properties directly. The resulting composite consists of layers, or laminas, of fibers and matrix (Figure 1.1) stacked in such a way as to achieve the desired properties in one or more directions.

Modern composite materials evolved from the simplest mixtures of two or more materials to obtain a property that was not there before. The Bible mentions the combining of straw with mud to make bricks. The three key historical steps leading to modern composites were

1. The commercial availability of fiberglass filaments in 1935. This work led to the main commercial use of fiber-reinforced plastics (FRP) in the construction of aircraft radomes in the United States beginning in about 1942. Eventually, the right reinforcement and the right resin at the right price saw the production of fiber-reinforced plastic translucent sheet in 1949, followed rapidly in the early 1950s by developments in fiber-reinforced plastic boat hulls, car bodies, and truck cabs.
2. The development of strong aramid, glass, and carbon fibers in the late 1960s and early 1970s. These developments parallel the development of resins dating back to

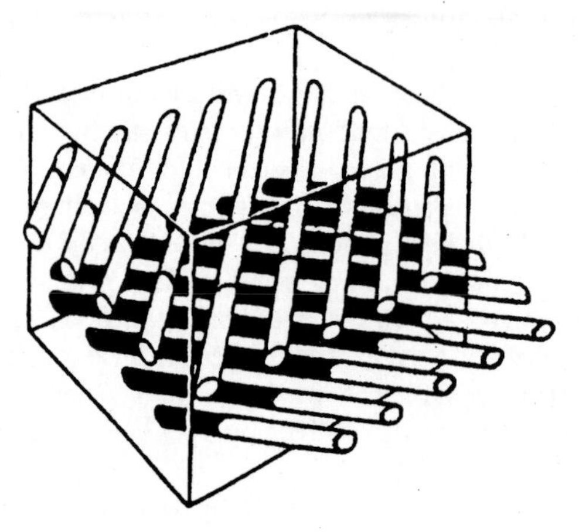

Figure 1.1. Schematic of an advanced composite laminate.

1969 (phenolics) and 1937 (epoxies) and of many other important thermosetting resins available today, which can be used to form special fiber-reinforced components (e.g., polyimides, phenolics, vinyl esters, furanes, silicones, polyurethanes, and urethane acrylates).

The foundation for the development of the other major class of plastics—thermoplastics—was laid down before that of thermosets, and thermoplastics development has been rapid. Note the recent introduction of polyether ether ketone (PEEK), which although expensive is gaining acceptance in high-performance applications where lightweight, temperature resistance, and high mechanical properties are essential.

As in the case of resin development, the pace of fiber development has accelerated since the 1950s with the introduction of E-glass, R-glass, S-glass, and special acid- and alkali-resistant glasses. In addition, textile processing technology has enabled glass rovings to be processed into an enormous variety of fabrics which have been brought to the marketplace in the form of chopped strand mat (CSM), woven roving (WR), stitched cross-plied rovings, combination fabrics (CSM/WR), unidirectional tapes and fabrics, woven cloth, and multidirectional fabrics.

Glass fiber has been the major reinforcement in the FRP industry, but the desire for lighter, stronger, stiffer structures has seen the introduction of carbon and polyaramid fibers.

Carbon fiber was first used about 100 years ago as filaments in electric lamps; these early fibers were relatively weak and of little use as reinforcements. By comparison, today's carbon fibers are miles apart from their predecessors and have exceptionally high strength and stiffness. However, since the development of these second-generation carbon fibers in 1963, their growth in FRP has been limited to specialized applications where cost is relatively unimportant, for example, aircraft and sporting goods.

Polyaramid fibers were first developed in 1965 and although not as strong or stiff as carbon fibers, they offer considerable performance advantages over glass fibers. They are lighter than carbon fiber, are less expensive, and exhibit exceptional impact resistance. In the FRP industry polyaramid fibers have found acceptance in gas pressure bottle construction, aerospace applications, sporting goods, and marine applications.

3. The promulgation of analytical methods for structures made from these fibers.[2–5]

The increases in consumption of composite materials were primarily due to the need for nonconductive electrical components, noncorroding and noncorrosive storage containers and transfer lines, and sporting goods. The technologies for matrices and for the fabrication of useful structures with stronger fiber reinforcement were commercialized in the two decades after 1970. Along with the new fibers, new matrices were developed, and new commercial fabrication techniques were introduced. These developments were due in part to military aircraft designers who were quick to realize that these materials could increase the speed, maneuverability, or range of an aircraft by lowering the weight of its substructures.

1.3 ADVANTAGES OF COMPOSITES

Designers of structures have been quick to capitalize on the high strength-to-weight or modulus-to-weight ratios of composites. But there are other advantages, as well as some disadvantages. The advantages include

Weight reduction (high strength- or stiffness-to-weight ratio)
Tailorable properties (strength of stiffness can be tailored to be in the load direction)
Redundant load paths (fiber to fiber)
Longer life (no corrosion)
Lower manufacturing costs because of lower part count
Inherent damping
Increased (or decreased) thermal or electric conductivity.

The disadvantages include

Cost of raw materials and fabrication
Possible weakness of transverse properties
Weak matrix and low toughness
Environmental degradation of matrix
Difficulty in attaching
Difficulty with analysis.

The advantages extend not only to aircraft but to everyday activities also, such as longer drives with a graphite-shafted golf club (because more of the mass is concentrated in the club head) and less fatigue and pain because a graphite composite tennis racquet has inherent damping. Generally the advantages accrue for any fiber composite combination, whereas the disadvantages are more obvious with certain combinations. Proper design and material selection can avoid many of the disadvantages.

1.4 GENERAL CHARACTERISTICS OF COMPOSITES

Many fiber-reinforced composite materials offer a combination of strength and modulus that is either comparable to or better than that of many traditional metallic materials. Because of their low specific gravities, the strength-to-weight ratios and modulus-to-weight ratios of these composite materials are markedly superior to those of metallic materials. In addition, fatigue strength-to-weight ratios, as well as fatigue damage tolerances, of many composite laminates are excellent (Table 1.1).

Traditional structural materials, such as steel and aluminum alloys, are considered isotropic because they exhibit nearly equal properties irrespective of the direction of measurement. In general, the properties of a fiber-reinforced composite depend strongly on the direction of measurement. For example, the tensile strength and modulus of a unidirectionally oriented fiber-reinforced laminate are maximum when these properties are measured in the longitudinal direction of the fibers. At any other angle of measurement, these properties are lower. The minimum value is observed at 90° to the longitudinal direction. Similar angular dependence is observed for other physical and mechanical properties, such as coefficient of thermal expansion (CTE), thermal conductivity, and impact strength. Bi- or multidirectional reinforcement, either in the planar form or in the lami-

TABLE 1.1. Tensile Properties of Some Metallic and Structural Composite Materials

Material[a]	Specific Gravity	Modulus (GPa)	Tensile Strength (MPa)	Yield Strength (MPa)	Ratio of Modulus to Weight (10^6 m)[b]	Ratio of Tensile Strength to Weight (10^3 m)[b]
SAE 1010 steel (cold-worked)	7.87	207	365	303	2.68	4.72
AISI 4340 steel (quenched and tempered)	7.87	207	1722	1515	2.68	22.3
Al 6061-T6 aluminum alloy	2.70	68.9	310	275	2.60	11.7
Al 7178-T6 aluminum alloy	2.70	68.9	606	537	2.60	22.9
Ti-6A1-4V titanium alloy (aged)	4.43	110	1171	1068	2.53	26.9
17-7 PH stainless steel (aged)	7.87	196	1619	1515	2.54	21.0
INCO 718 nickel alloy (aged)	8.2	207	1399	1247	2.57	17.4
High-strength carbon fiber-epoxy (unidirectional)	1.55	137.8	1550		9.06	101.9
High-modulus carbon fiber-epoxy (unidirectional)	1.63	215	1240		13.44	77.5
E-glass fiber-epoxy (unidirectional)	1.85	39.3	965		2.16	53.2
Kevlar 49 fiber-epoxy (unidirectional)	1.38	75.8	1378		5.60	101.8
Boron fiber-6061 A1 alloy (annealed)	2.35	220	1109		9.54	48.1
Carbon fiber-epoxy (quasi-isotropic)	1.55	45.5	579		2.99	38

[a] For unidirectional composites, the reported modulus and tensile strength values are measured in the direction of the fibers.
[b] The modulus-to-weight ratio and the strength-to-weight ratios are obtained by dividing the absolute values by the specific weight of the respective material. Specific weight is defined as weight per unit volume. It is obtained by multiplying density by acceleration due to gravity.

nated construction, yields a more balanced set of properties. Although these properties are lower than the longitudinal properties of a unidirectional composite, they still represent a considerable advantage over common structural materials on a unit weight basis.

The design of a fiber-reinforced structure is considerably more difficult than that of a metal structure, principally because of the difference in its properties in different directions. However, the anisotropic nature of a fiber-reinforced composite material creates a unique opportunity for tailoring its properties according to the design requirements. This design flexibility can be utilized to selectively reinforce a structure in the directions of major stresses, increase its stiffness in a preferred direction, fabricate curved panels without any secondary forming operation, or produce structures with zero CTEs.

In addition to the directional dependence of properties, there are a number of other differences between structural metals and fiber-reinforced composites. For example, metals in general exhibit yielding and plastic deformation. Most fiber-reinforced composites are elastic in their tensile stress-strain characteristics. However, the heterogeneous nature of these materials provides mechanisms for high-energy absorption on a microscopic scale comparable to the yielding process. Depending on the type and severity of the external loads, a composite laminate may exhibit gradual deterioration in properties but usually does not fail in a catastrophic manner. Mechanisms of damage development and growth in metal and composite structures are also quite different.

CTEs for many fiber-reinforced composites are much lower than those of metals. As a result, composite structures may exhibit a better dimensional stability over a wide temperature range. However, the differences in thermal expansion between metals and composite materials may create undue thermal stresses when the materials are used in conjunction, for example, near an attachment.

Another unique characteristic of many fiber-reinforced composites is their high internal damping. This leads to better vibrational energy absorption within the material and results in reduced transmission of noise and vibrations to neighboring structures. The high damping capacity of composite materials can be beneficial in many automotive applications in which noise, vibration, and harshness are critical issues for passenger comfort.

An advantage attributed to fiber-reinforced polymers is their noncorroding behavior. However, many polymeric matrix composites are capable of absorbing moisture from the surrounding environment, which creates dimensional changes as well as adverse internal stresses within the material. If such behavior is undesirable in an application, the composite surface must be protected from moisture diffusion by appropriate paints or coatings. Among the other environmental factors that may cause degradation in the mechanical properties of some polymer matrix composites are elevated temperatures, corrosive fluids, and ultraviolet rays. In metal matrix composites, oxidation of the matrix as well as adverse chemical reactions between the fibers and matrix are of great concern at high-temperature applications.[6]

1.4.1 Unreinforced Composites Versus Reinforced Composites

Most materials used in structural applications are polymers, metals, or ceramics, and in many present applications these materials perform satisfactorily in their unmodified or unreinforced form. When the thermal stability and strength of the material are not critical, low-cost polymeric materials such as acrylates, epoxies, and polycarbonates can perform acceptably. Likewise, metals such as aluminum, steel, copper, and tungsten are adequate for lightweight structural components, tooling, electric conductors, and lamp filaments, respectively. Because cost is the controlling factor, the present performance of many ceramic structural products such as window glass, structural bricks, and cement blocks is considered satisfactory. However, in many applications where performance is the controlling factor (e.g., aerospace, transportation, underwater vessels), advanced structural materials are needed that are stronger, stiffer, lighter in weight, and more resistant to hostile environments. Unreinforced, the polymer, metal, and ceramic materials available today cannot meet many of these requirements. This is especially true if the structural component must be exposed to extremely high temperatures for extended periods of time. The graph in Figure 1.2 shows the approximate temperature limits for the use of various structural materials.[7]

1.4.2 Composite Material Characteristics in Fiber Form

Natural fibers such as cotton and wool are some of the oldest materials. These fibers were used by early humans when strength and light weight were critical. However, only in the past 50 years, with the development of analytical techniques such as x-ray diffraction, has

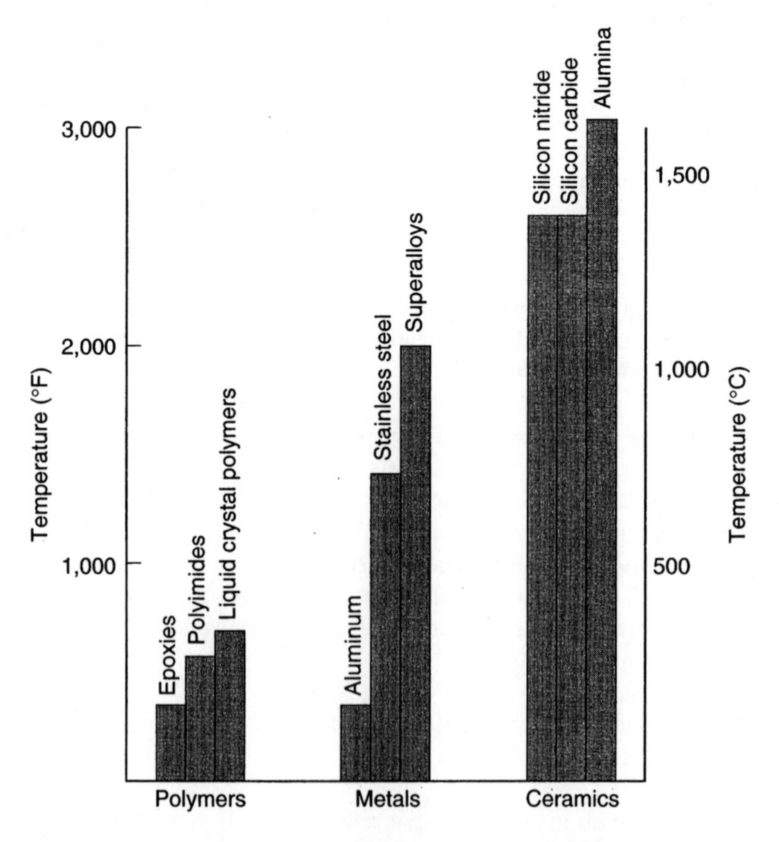

Figure 1.2. Maximum-use temperatures of various structural materials.

the reason for the unusual properties of materials in fiber form been understood. Scientists now know that the molecules within fibers tend to align along the fiber axis. This preferred alignment makes the strength and modulus (stiffness) of both natural fibers and synthetic fibers superior to those of the same material in a randomly oriented bulk form. As an example, Table 1.2 lists the strength and modulus of a typical polymer in various forms. Whereas the strength of an injection-molded polyamide plate is only 0.08 GPa, the tensile strength of the same polymer is over 5 times greater when it is extruded into a textile-grade fiber. If this same textile-grade fiber is stretched in an extensive drawing process, an industrial-grade tire cord fiber can be produced that is 10 times stronger and nearly twice as stiff as the injection-molded polymer. Chemically, all these materials are identical, differing only in the orientation and structure of the solid polymer. When both natural and synthetic polymers are extruded and/or drawn into fiber form, the processes of extrusion and extension orient the structure along the fiber axis. This results in high strength and increased stiffness for much the same reason that an oriented mass of strings (a rope) is stronger and stiffer than the same mass of strings with no orientation.

Rigid, liquid-crystal-forming polymers (e.g., aramid fibers) can develop nearly perfect orientation and alignment during fiber formation. This allows a kilogram of fibers formed from this rigid polyaramid molecule to be five times stronger than a kilogram of steel and still be five times as stiff. Because the density of aramid fiber is only one-fifth that of steel, this new class of synthetic high-performance fibers already is an obvious replacement for metal in many applications.

Brittle materials, like carbon, also have a higher strength and stiffness when formed into fibers. High-performance carbon fibers formed from pitch are now available commercially with a tensile strength of 3.9 GPa. This is approximately 1000 times greater than the strength of unoriented carbon in bulk form. In the case of brittle materials, the higher strength of fibers is caused by two factors. First, like polymeric fibers, the molecular structure and orientation are improved by the fiber formation process. Second, because the failure of brittle objects is caused mainly by flaws, the small size of fibers limits the size of the flaws that can exist. Thus, in addition to forming a more perfect structure, brittle materials in fiber form contain smaller flaws, further enhancing the tensile strength.[8]

Unfortunately, the increased tensile strength of fibers does not come without a

TABLE 1.2. Properties of Polyamid in Various Forms

Form	Tensile Strength (GPa)	Tensile Modulus (GPa)	Orientation
Injection-molded	0.08	2.5	Random
Textile-grade fiber	0.43	2.5	Medium
Industrial-grade fiber	0.92	4.5	High
Kevlar	3.50	186.0	Near perfect

penalty. Fibers, like rope, display this increased strength only when the load is applied parallel to the fiber axis. Even though the tensile strength parallel to the fiber axis increases as the orientation and structure become more perfect in the fiber direction, this same increase causes a decrease in strength perpendicular to the fiber axis. For example, the strength of a carbon fiber perpendicular to the fiber axis is 10 times less than the strength parallel to the axis. Also, as the orientation of a fiber increases, it often becomes brittle, making it more susceptible to damage by abrasion. Thus, to take advantage of the high strength of fibrous materials in a structure, the fibers must be oriented in the direction of the applied load and separated to prevent damage by abrasion.

Mechanical reinforcement of matrices can also be accomplished by using short, randomly oriented fibers, crystal whiskers, or particulates. These types of reinforcement offer directionally independent (isotropic) reinforcement, but the degree of reinforcement is not as great as that obtainable from longer continuous-filament fibers.

1.5 MAJOR COMPOSITE CLASSES

The major classes of structural composites that exist today can be categorized as polymer matrix composites (PMCs), metal matrix composites (MMCs), ceramic matrix composites (CMCs), carbon-carbon composites (CCCs), intermetallic composites (IMCs), or hybrid composites.

1.5.1 PMCs

PMCs are the most developed class of composite materials in that they have found widespread application, can be fabricated into large, complex shapes, and have been accepted in a variety of aerospace and commercial applications. They are constructed of components such as carbon or boron fibers bound together by an organic polymer matrix. These reinforced plastics are a synergistic combination of high-performance fibers and matrices. The fiber provides the high strength and modulus, whereas the matrix spreads the load as well as offering resistance to weathering and corrosion. Composite strength is almost directly proportional to the basic fiber strength and can be improved at the expense of stiffness. High-modulus organic fibers have been made with simple polymers by arranging the molecules during processing, which results in a straightened molecular structure. Optimization of stiffness and fiber strength remain a fundamental objective of fiber manufacture. In addition, because of differences in flexibility between fiber intra- and interfibrillar amorphous zones, shear stresses can result and eventually lead to a fatigue crack.

1.5.2 MMCs

These composites consist of metal alloys reinforced with continuous fibers, whiskers (a version of short fibers that are in the form of single crystals), or particulates (fine particles, as distinct from fibers). Because of their use of metals as matrix materials, they have a higher temperature resistance than PMCs but in general are heavier. They are not as widely used as PMCs but are finding increasing application in many areas. Further development of manufacturing and processing techniques are essential to bringing down product costs and accelerating the use of MMCs.

Research continues on particulate- and fiber-reinforced MMCs because of substantial improvements in their strength and stiffness as compared to those of unreinforced metal alloys. Also, the advantages of metals as matrices, compared to polymer matrices, are their high tensile strength and shear modulus, high melting point, small coefficient of expansion, resistance to moisture, dimensional stability, ease of joining, high ductility, and toughness. Efforts are directed at light-alloy composites capable of use in low- to medium-temperature applications for space structures and for high-temperature applications such as engines and airframe components. The extreme low density of graphite used as a reinforcement, coupled with its very high modulus, makes it a highly desirable material. However, one obstacle is the poor interfacial bond between graphite and metals such as aluminum and magnesium. The interfacial bond can be strengthened by using a metal carbide coating. MMCs exhibit high performance as a result of their good strength-to-weight ratios. The development of directionally solidified alloys has increased the high-temperature capability further by aligning grain boundaries away from the principal stress direction. The extreme low density of graphite, coupled with its very high modulus, makes it a highly desirable reinforcement.

1.5.3 CMCs

Monolithic ceramic materials have a natural high-temperature resistance but also have fundamental limitations in structural applications owing to their propensity for brittle fracture. The incorporation of a reinforcement, for example, ceramic fiber reinforcement, into the ceramic matrix can improve the forgivability of the material by allowing cracking to be retarded by the fiber-matrix interfaces. CMCs are a class of structural materials with reinforcements such as SiC fibers embedded in a ceramic matrix such as Al_2O_3, Si_3N_4 or SiC. The reinforcements can be continuous fibers, chopped fibers, small discontinuous whisker platelets, or particulates. They have the potential for high-temperature application above 1649°C.

The brittleness mentioned earlier arises because ceramics cannot deform by shear as metals do to relieve stress. The average fracture toughness, the major weakness of ceramics, is doubled with CMCs. They have enabled fabricators to advance technology to being able to fabricate structures capable of withstanding a variety of extreme environments. CMCs also provide a competitive advantage over superalloys with up to 70% lower density and a 500°C higher maximum-use temperature. Continuous-fiber materials can be used as reinforcements for CMCs and MMCs. However, no ceramic reinforcement performs well above 1371°C, and at present only graphite has the potential for use. The thermal stability of whisker and fibrous reinforcement limits their use to temperatures below about 1371°C. Many components in turbine engines and hypervelocity flight structures operate above 1649°C. Matrices such as ceramic, refractory metal, and intermetallic materials are available for operational temperatures above 1649°C, but newer reinforcements are needed. A major problem facing the aerospace industry is the lack of high-temperature reinforcements that do not oxidize; however, high-temperature fiber coatings are one solution currently under investigation.

1.5.4 CCCs

CCCs consist of carbon fiber reinforcements embedded in a carbonaceous matrix. Preliminary processing is very much like that for PMCs, but the organic matrix is subsequently heated up to the point where it is converted to carbon. Carbon-carbon is a superior structural material for applications where resistance to very high temperatures and thermal shock is required. No other material has higher specific strength properties (strength-to-density ratio) at temperatures in excess of 1371°C. Oxidation protection systems, low-cost manufacturing, and scale-up of C/C structures are needed to effect more widespread use and subsequent flow-down to industrial applications. In the 30 years or so that this class of materials has been under development, it has found the applications shown in Figure 1.3.

CCC material systems can be generally classified into two usage categories: nonstructural composites and structural composites. The maturity of these two categories is different. The nonstructural class of materials is in production for commercial and military use and is relatively mature. The structural class of materials, which have very high payoff applications, is not in production and will require considerable development. The main reasons for the considerable difference in development of the two classes of C/C composites are the significantly higher requirements involving temperature, service life, and load-carrying capability of structural applications.

1.5.5 IMCs

IMCs are currently of extreme interest for use in future high-temperature, high-efficiency, high-performance gas turbine engines for both civil and military applications.

By going the composite route, the low density of the intermetallic compounds can be utilized to good advantage, and if low-density, high-strength fibers are available, the

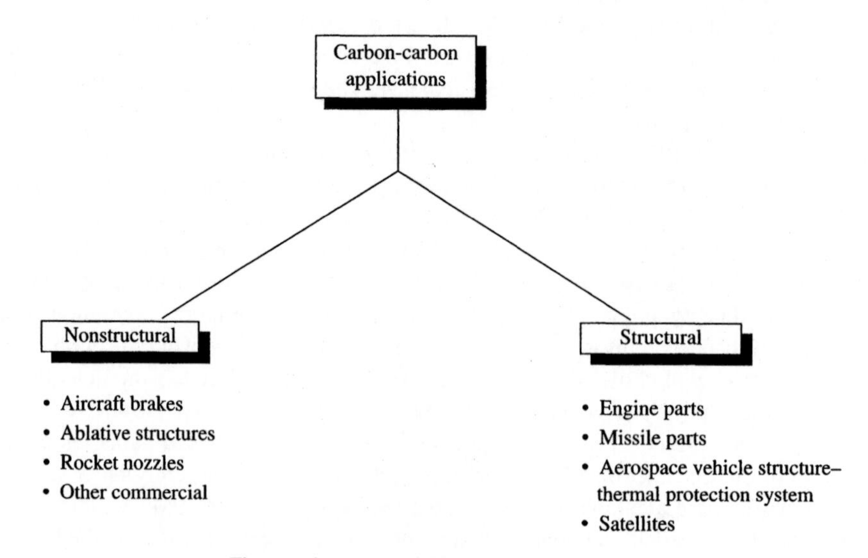

Figure 1.3. Types of CCC applications.

low strength of intermetallic matrices becomes less of an issue. Thus, the matrix can be optimized for other properties, most importantly ductility, oxidation resistance, and density. The influence of the fiber on strength properties of a composite has been discussed by McDanels and Stephens[9] regarding the use of SiC in aluminide matrices; the predicted strength-to-density ratio for SiC-reinforced aluminide composites was shown to be essentially independent of matrix strength. An example of the experimental results of this concept is illustrated by the results of Brindley.[10] The tensile behavior was shown to achieve a predicted rule of mixture behavior. And on a density-corrected basis, the composite was shown to have superior tensile properties compared to wrought nickel base and cobalt base alloys and a single-crystal superalloy, NASAIR 100.

One of the major problems facing the successful development of IMCs is the compatibility between fiber and matrix both from a chemical viewpoint and from the mismatch in CTE. An example of the extreme complexity of fiber-matrix interaction occurs in the SCS-6 fiber-Ti_3Al + Nb matrix system where the SiC fiber has a two-layer carbon zone on the surface which further contributes to the chemical reactions involved.

Ti_3Al (or alpha) and TiAl (or gamma) are the two main titanium aluminide IMCs being investigated by the National Aeronautics and Space Administration (NASA) and the U.S. Air Force.[9,11–13]

All the problems associated with IMCs provide challenges and opportunities for materials and structures researchers. To overcome the compatibility problems of the fibers and the matrices in IMCs, studies on the kinetics of reaction types and methodology have been undertaken to identify the most appropriate fibers for the intermetallic matrices.[14] IMCs are further discussed in the last chapter of Volume II.

1.5.6 Hybrid Composites

Hybrid composite materials represent the newest of the various composite materials currently under development. The hybrid composite category covers both the hybridizing of a composite material with other materials (either other composites or base unreinforced materials) and composites using multiple reinforcements. Further, this category covers the use of multiple materials (at least one of which is a composite) in structural applications and highlights the multiple uses and advantages of composite materials.

Hybrid composites can be divided into five major subcategories: (1) hybrid composite materials, (2) selective reinforcements, (3) thermal management, (4) smart skins and structures, and (5) ultralightweight materials.

1.5.6.1 Hybrid Composite Materials (HCMs). HCM is defined as a composite material system derived from the integrating of dissimilar materials at least one of which is a basic composite material. A typical example of a hybrid composite material is a reinforced polymer composite combined with a conventional unreinforced homogeneous metal. The hybrid composite material blends the desirable properties of two or more types of materials into a single material system which displays the beneficial characteristics of the separate constituents. An existing example of a hybrid composite is aramid-reinforced aluminum laminate (ARALL), which consists of high-strength aluminum alloy sheets interleafed with layers of aramid fiber-reinforced adhesive as illustrated in Figure 1.4. The ARALL hybrid composite is baselined on several secondary structural compo-

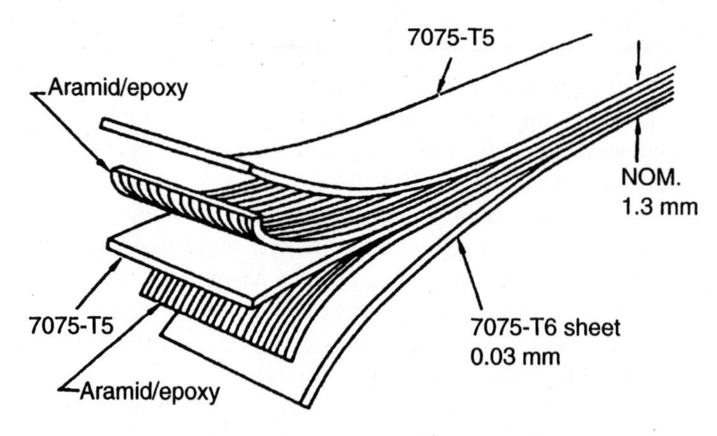

Figure 1.4. ARALL is an example of a hybrid composite material in production.

nents of fixed-wing subsonic aircraft. A second example of a hybrid composite is a CCC with a single side application of the refractory metal rhenium. This carbon-carbon-rhenium material is being developed for thermal management heat pipes on space-based radiator systems. Other examples include interpenetrating polymer networks (IPNs), which are hybrid resin matrices consisting of thermoset and thermoplastic resin combinations. Still another hybrid composite material concept involves multiple reinforcement types within a common matrix such as chopped fibers and continuous fibers within a polymer matrix.

Other HCMs include new composite materials such as nanocomposites,[15–21] functionally gradient materials (FGMs),[22–24] hybrid materials (hymats),[25] interpenetrating polymer networks (IPNs),[26] microinfiltrated macrolaminated composites (MIMLCs),[27] and liquid crystal polymers (LCPs),[28] which may force development of previously uneconomical process routes if they offer the path to a technical solution for advanced system capability. These materials present opportunities for reducing the number of stages in turbine engines and in so doing may be economically beneficial even at a higher materials cost because the smaller number of stages leads to greater economy of use.

HCMs technology is in its infancy in comparison to that of the other types of composite materials. Whereas ARALL and IPNs have been used for the past decade, the other types of hybrid composite materials are truly embryonic. From nearly all aspects of research, these hybrid composite materials offer great potential for structural applications, however, their widespread use remains a decade or more in the future.

1.5.6.2 Selective Reinforcement. Selective reinforcement is the category of hybrid composites that provide reinforcement to a structural component in a local area or areas by means of adding a composite material. An example of this is the use of superplastic forming–diffusion bonding (SPF/DB) as a means of integrating a titanium-reinforced MMC into a base titanium structure. As part of the initial design approach, consideration must be given to the tooling required for placement of the reinforcing material within the structure. This is accomplished by building into the form tooling areas in which the

MMC material is placed that permit SPF expansion of the base titanium to the MMC during processing and allowing diffusion bonding to occur, thus accomplishing integral reinforcing of the final structure.

The selective reinforcement design approach allows the aerospace designer to utilize the more costly, higher-performance materials only where they are required and the less expensive materials in areas where they can perform the job. This approach leads to an optimization of both cost and performance in the most ideal case. Similarly, problems associated with use of the reinforcing composite, such as low mechanical joint strength, can be eliminated by not reinforcing the area where the mechanical joining occurs. In reality, while this approach can be very efficient, it requires the introduction of multiple materials to solve the design problem and may in turn result in increased fabrication costs and risks.

1.5.6.3 Thermal Management. In the field of thermal management, composite materials and hybrid composite materials can be innovatively constructed to effectively limit the maximum temperature of structural hardware and rapidly transfer heat from hot areas to cool areas. This capability arises from the unmatched thermal conductivity of graphite fiber which is higher (in the fiber direction) than that of oxygen-free high-conductivity copper (OFHC). The graphite fibers act as heat paths and, by suitable arrangement in the structure of interest, can remove heat by transmitting it along its length. In contrast, the matrix materials can be a thermal insulator so that thermal conduction through the thickness is lower by orders of magnitude than in-plane. This allows designers to develop a structure that is a thermal insulator in some directions but a thermal conductor in others.

By combining graphite and glass fibers in a phenolic resin, a hybrid is developed that exhibits high fire and flame resistance while still being easily formable for contoured shapes. Hybrids such as this are being baselined for aircraft interiors to carry structural loads, reduce weight, and provide safety for passengers in case of an accident. The combination of graphite and glass fibers reduces structural weight and costs, and the phenolic resin increases flame resistance and reduces smoke in case of a fire. These materials are being used for aircraft hard cabin furnishings such as walls, window frames, bins, ceilings, and bulkheads.

1.5.6.4 Smart Skins and Smart Structures. Smart skins and smart structures are related in that each contains embedded, nonstructural elements. A smart structure contains sensors that monitor the health of the structure itself, such as fiber optics to determine temperature and structural deformations or cracks. A smart skin contains circuitry and electronic components that enable the skin to double as part of the electronic system of the parent vehicle, be it an aircraft or a missile.

Smart structure technology, like that of smart skins, is still evolving. The present technology consists of the incorporation of sensors into structural elements in the material processing stage to better control cure (or consolidation), and so on.

1.5.6.5 Ultralightweight Materials. The category of ultralightweight materials includes the emerging family of liquid crystal ordered polymers, which by virtue of their molecular structure exhibit extremely high specific strength and specific stiffness. These highly directional materials are similar to composite materials in that the long, ordered molecular chains within the polymer act very much like reinforcing fibers in a composite material. Examples of these materials are poly-*p*-phenylene benzobisthiazole (PBZT) and poly-*p*-phenylene benzobisoxazole (PBO). Another example is gel-spun polyethylene. These polymers exhibit extremely high specific strength and specific stiffness and can potentially be used as reinforcing fibers in composite laminates, as rope or cable, or as a self-reinforced thin film structure. These ordered polymers, in the form of thin films, find use in shear webs and skin applications for aircraft. These thin films can also be processed into honeycomb for lightweight structural applications. It has been estimated that a shear web made of PBZT would be one-eighth the weight of an aluminum web and one-sixth the weight of a graphite epoxy web.

REFERENCES

1. Knauer, B. 1993. Construction and analysis of high performance composites. *Compos. Struct.* 24(3):181–91.
2. Jones, R. M. 1975. *Mechanics of Composite Materials.* Scripta Book, Washington, DC.
3. Tsai, S. W., and H. T. Hahn. 1980. *Introduction to Composite Materials.* Technomic, Westport, CT.
4. Agarwal, B. D., and L. J. Broutman. 1980. *Analysis and Performance of Fiber Composites.* Wiley-Interscience, New York.
5. Peters, S. T. 1992. Advanced composite materials and processes. In *Handbook of Plastics, Elastomers, and Composites*, 2nd ed., ed. C. Harper, 5.1–5.73, McGraw-Hill, New York.
6. Bittence, J. 1987. Guide to selected engineering materials. *Adv. Mater. Process.* 2(1).
7. National Materials Advisory Board. 1992. High performance synthetic fibers for composites. NMAB-458.
8. Reinhart, T. J. 1987. Introduction to composites. In *Composites Engineered Materials Handbook,* ed. T. J. Reinhart, 27–39. ASM International, Metals Park, OH.
9. McDanels, D. L., and J. R. Stephens. 1988. High temperature engine materials technology—Intermetallics and metal matrix composites. NASA TM-100844.
10. Brindley, P. K., P. A. Bartolotta, and S. J. Klima. 1988. Investigation of a SiC/Ti-24Al-11 Nb composite. NASA TM-100956.
11. Smith, C. J., and J. E. Johnson. 1989. Advanced materials for 21st century UBE. *34th ASME Int. Gas Turbine Conf.,* June 1989, Toronto.
12. Klein, H. 1989. NASA high temperature advanced propulsion study. Interim report, NASA Lewis Research Center, Cleveland, OH.
13. Petrasek, D. W. 1988. High temperature intermetallic matrix composites. *HITEMP Review—1988,* Advanced High Temperature Engine Materials Technology Program, pp. 67–82. NASA CP-10025.
14. Stephens, J. R. 1990. Intermetallic and ceramic matrix composites for 815 to 1370°C gas turbine engine applications. *Int. Conf. Met. Ceram. Matrix Compos. Process. Modeling Mech. Behav.,* ed. R. B. Bhagat, A. H. Clauer, et al., pp. 3–12, February 1990, Anaheim, CA.
15. Averback, R. S. 1992. Nanophase ceramics. University of Illinois at Urbana-Champaign. AR0-25526, 10-MS.

16. Messersmith, P. B., and S. I. Stupp. 1992. Synthesis of nanocomposites: Organoceramics. *J. Mater. Res.* 7(9):2599–2611.

17. Komarneni, S. 1992. Nanocomposites. *J. Mater. Chem.* 2(12):1219–30.

18. Drexler, K. E. 1992. *Nanosystems: Molecular Machinery, Manufacturing, and Computation.* John Wiley, New York.

19. Joint Publications Research Service Report. JST-92-033, December 1992.

20. Drexler, K. E. 1981. Molecular engineering: An approach to the development of general capabilities for molecular manipulation. *Proc. Nati. Acad. Sci. USA* 78(9):5275–78.

21. Hillig, W. B. 1991. Ceramic matrix composites—Research and development in Japan. JTEC Panel Report on Advanced Composites in Japan, pp. 61–85. FR 3/91, ECS 8902528.

22. Sasaki, M., and T. Hirai. 1990. Fabrication and thermal barrier characteristics of CVD SiC/C functionally gradient material. *Proc. First Int. Symp. FGM,* ed. M. Yamanouchi, M. Koizumi, T. Hirai, et al., p. 83, Sendai.

23. Moya, J. S., A. J. Sanchez-Herencia, J. Requena, et al. 1992. Functionally gradient ceramics by sequential slip casting, *Mater. Lett.* 14(5/6):333–35.

24. Takahashi, M., Y. Itoh, and H. Kashiwaya. 1992. New ways to make functionally gradient materials. *Weld J.* March:93.

25. Frazier, W. E., M. E. Donnellan, P. Architetto, et al. 1991. The status of HYMATs—A new category of hybrid materials. *J. Met.* 43(5):10–15.

26. Clarke, D. R. 1992. Interpenetrating phase composites. *J. Am. Ceram. Soc.* 75(4):739–59.

27. Bose, A., and J. Lankford. 1991. MIMLCs: New composite architecture. *Adv. Mater. Process.* 140(1):18–22.

28. Hunt, M. 1993. Star Wars materials launch commercial products. *Mach. Des.* March 26:52–58.

BIBLIOGRAPHY

NATIONAL CENTER FOR ADVANCED TECHNOLOGIES. National advanced composites strategic plan. Symposium draft, 1990.

2

Fibers and Matrices

2.0 INTRODUCTION

The major constituents of a fiber-reinforced composite material are the reinforcing fibers and a matrix that acts as a binder for the fibers. Other constituents that may also be found are coupling agents, coatings, and fillers. Coupling agents and coatings are applied on the fibers to improve their wetting with the matrix as well as to promote bonding across the fiber-matrix interface. Both in turn promote a better load transfer between the fibers and the matrix. Fillers are used with some polymeric matrices primarily to reduce cost and improve their dimensional stability.

Manufacture of a composite structure starts with the incorporation of a large number of fibers into a thin layer of matrix to form a *lamina* (ply). The thickness of a lamina is usually in the range 0.1–1 mm. In a continuous unidirectional orientation all the fibers are in one direction while in a bidirectional orientation the same amount of fiber is divided equally into the longitudinal and transverse directions. For a lamina containing unidirectional fibers, the composite material has the highest strength and modulus in the longitudinal direction of the fibers. However, in the transverse direction, its strength and modulus are very low. For a lamina containing bidirectional fibers, the strength and modulus can be varied by employing different amounts and different types of fibers in the longitudinal and transverse directions. For a balanced lamina, these properties are the same in both directions.

A lamina can also be constructed using discontinuous (short) fibers in a matrix. The fibers can be arranged either in a unidirectional orientation or in a random orientation.

Discontinuous-fiber-reinforced composites have lower strength and modulus than continuous-fiber-reinforced composites. However, with random orientation of fibers, it is possible to obtain nearly equal mechanical and physical properties in all directions in the plane of the lamina.

The thickness required to support a given load or maintain a given deflection in a fiber-reinforced composite structure is obtained by stacking several laminas in a specified sequence to form a *laminate.* Various laminas in a laminate may contain fibers either all in one direction or all in different directions. It is also possible to combine different kinds of fibers to form either an interply or an intraply hybrid laminate. An interply hybrid laminate consists of different kinds of fibers in different laminas, whereas an intraply hybrid laminate consists of two or more different kinds of fibers interspersed in the same lamina. Generally, the same matrix is used throughout the laminate so that a coherent interlaminar bond is formed between the laminas. An exception to this is hybrid laminates, for example, ARALL, which utilize a sandwich construction of a fiber-reinforced polymer and thin aluminum sheets with adhesive layers between them.

2.1 ADVANCES IN POLYMER FIBER TECHNOLOGY

Advances in the performance of fibers have come about because of a continuity of effort in fiber materials and fiber-processing research and development. This is illustrated in Figure 2.1 for the case of organic polymer fibers, and similar illustrations apply to other types of fibers. In the early 1920s the first synthetic fibers were produced from cellulose. Because this natural polymer degrades before it melts, this early synthetic fiber was precipitated from a concentrated polymer solution. After precipitation the cellulose fibers

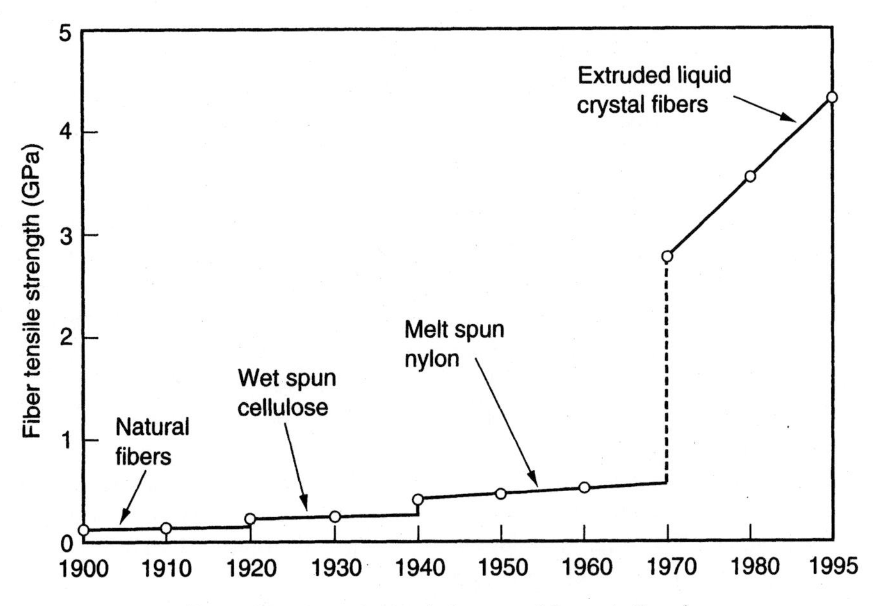

Figure 2.1. Strength of typical commercial organic fibers.[1]

had to be drawn out in order to orient the polymer molecules and improve the mechanical properties. Nylon, one of the earliest synthetically produced polymers, was introduced as a fiber in the 1940s. Because the polymer melted before it degraded, a melt spinning process was developed. With the recent discovery of LCPs, fully oriented fibers can now be spun. Thus, as Figure 2.1 indicates, the mechanical properties of fibers have improved dramatically over the past 55 years, and the major breakthroughs have been due to the development of new materials and processing techniques.

High-performance fibers represent a major area of growth for the synthetic fibers industry, and a number of these fibers are already available commercially. Projections of world demand for advanced composites indicate that it will reach approximately 500 million metric tons by the early decades of the twenty-first century, of which industrial and other applications will grow to 55% of market share and the aircraft and aerospace market share will drop to 45%. It is important to note that this is a conservative estimate based on the current price of advanced fibers: $15–$20 per pound. It is forecast that if the cost of these fibers could be reduced to a few dollars per pound, the demand would be a factor of 10 higher.[1]

2.2 FIBERS

Fibers are the principal constituent in a fiber-reinforced composite material. They occupy the largest volume fraction in a composite laminate and share the major portion of the load acting on a composite structure. Proper selection of the type, amount, and orientation of fibers is very important, because it influences the following characteristics of a composite laminate:

1. Specific gravity
2. Tensile strength and modulus
3. Compressive strength and modulus
4. Fatigue strength and fatigue failure mechanisms
5. Electric and thermal conductivities
6. Cost.

A number of commercially available fibers and their properties are listed in Table 2.1. The first point to note in this table is the extremely small filament diameter of the fibers. Because such small sizes are difficult to handle, the useful form of commercial fibers is a *strand,* which is produced by gathering a large number of filaments into a bundle.

In a composite matrix the fibers are surrounded by a thin layer of matrix material that holds the fibers permanently in the desired orientation and distributes an applied load among all the fibers. The matrix also plays a strong role in determining the environmental stability of the composite article as well as mechanical factors such as toughness and shear strength. Because the reinforcing fibers can be oriented during fabrication of an item, composites can be tailored to meet increased load demands in specific directions. The combined fiber-matrix system is an engineered material designed to maximize mechanical and environmental performance.

TABLE 2.1. Properties of Selected Commercial Reinforcing Fibers

Fiber	Typical Diameter (μm)[a]	Specific Gravity	Tensile Modulus (GPa)	Tensile Strength (GPa)	Strain to Failure (%)	Coefficient of Thermal Expansion (10^{-6}/°C)[b]	Poisson's Ratio
Glass							
E-glass	10	2.54	72.4	3.45	4.8	5	0.2
S-glass	10	2.49	86.9	4.30	5.0	2.9	0.22
PAN carbon							
T-300[c]	7	1.76	231	3.65	1.4	–0.6, 7–12	0.2
AS-1[d]	8	1.80	228	3.10	1.32		
AS-4[d]	7	1.80	248	4.07	1.65		
T-40[c]	5.1	1.81	290	5.65	1.8	–0.75	
IM-7[d]	5	1.78	301	5.31	1.81		
HMS-4[d]	8	1.80	345	2.48	0.7		
GY-70[e]	8.4	1.96	483	1.52	0.38		
Pitch carbon							
P-55[c]	10	2.0	380	1.90	0.5	–1.3	
P-100[c]	10	2.15	758	2.41	0.32	–1.45	
Aramid							
Kevlar 49[f]	11.9	1.45	131	3.62	2.8	–2, 59	0.35
Kevlar 149[f]		1.47	179	3.45	1.9		
Technora[g]		1.39	70	3.0	4.4	–6	
Extended-chain polyethylene							
Spectra 900	38	0.97	117	2.59	3.5		
Spectra 1000	27	0.97	172	3.0	2.7		
Boron	140	2.7	393	3.1	0.79	5	0.2
SiC							
Monofilament	140	3.08	400	3.44	0.86	1.5	
Nicalon (multi-filament)[i]	14.5	2.55	196	2.75	1.4		
Al_2O_3							
FiberFP[f]	20	3.95	379	1.90	0.4	8.3	
Al_2O_3-SiO_2[j]							
Fiberfrax (discon-tinuous)	2–12	2.73	103	1.03–1.72			

[a] 1 μm = 0.0000393 in.
[b] 1 m/m per °C = 0.556 in./in. per °F.
[c] Amoco.
[d] Hercules.
[e] BASF.
[f] DuPont.
[g] Teijin.
[h] Allied-Signal.
[i] Nippon Carbon.
[j] Carborundum.

There is an important, but not generally well understood, difference between the development time for traditional materials compared to that for high-performance fibers. Because a composite material is a complex system of two components coupled at an interface, the time required to develop and optimize new high-performance fibers for a particular application is much longer than that needed for the development of traditional materials. For composite applications it normally takes 5–10 years to develop a new high-performance reinforcing fiber.

Composite materials containing fibers (whether they are short staple fibers, whiskers, or fibers in continuous-filament form such as roving or textiles) provide considerable flexibility in the design of structures. Because of this, composites of inexpensive glass fibers embedded in a plastic matrix material have been used extensively in medium-to-high-volume applications by the transportation, construction, and recreation industries for over 40 years in applications such as auto body panels, boat hulls, and chemical tanks. However, high-performance fibers dramatically expand the opportunities for composite materials. When high-performance fibers such as carbon and Kevlar, at fiber loadings typically greater than 45%, are surrounded by the same plastic matrix, the material becomes an advanced composite. Thus, in many ways, advanced composite materials represent a major breakthrough for a composite material technology that has been extensively utilized for many years. It is the added strength and stiffness of these new reinforcing fibers that allows the new advanced composites to outperform current metal and metal alloy structures.

By dispersing fibers or particles of one material in a matrix of another material, today's designer can obtain structural properties that neither material exhibits on its own.[2] For example, a metal alloy selected for its resistance to high temperature but having low resistance to creep at use temperature can be reinforced with fibrous inorganic oxide fibers to provide enhanced creep resistance and still be stable at high temperatures. A ceramic matrix, brittle and sensitive to impact or fracture induced by thermal stresses, may be reinforced with ceramic fibers to increase its resistance to crack propagation, providing greater toughness and protecting against catastrophic failure. The addition of reinforcing fibers to provide equal mechanical properties at a greatly reduced weight is often an important reason for choosing composites over traditional structural materials. Another vital consideration is the substitution of readily available materials for critical elements in short supply or those available only from foreign sources. Composite materials made from abundant, domestically available materials such as carbon, polymers, ceramics, and common metals often outperform these imported strategic materials.

2.2.1 Projections for High-Performance Fibers

The fiber research needed depends on both the application and the fiber in question. For example, even though carbon fibers are commercially available, research directed toward product improvement and cost reduction still yields significant payback. Research funded by the federal government is already yielding progress in ceramic fibers and whisker technology, but new applications, such as the hypersonic transport vehicle, may require significant advances in this art. Present metals and superalloys cannot withstand the operating temperatures predicted for the national aerospace plane (NASP), and the use of fiber-reinforced composites is the likely solution. As existing composite materials are exposed to ever-increasing temperatures, it appears that present fiber reinforcement materials do not meet many of these requirements and that new fibers and whiskers are probably needed. At elevated temperatures, present high-performance fibers exhibit excessive grain growth or oxidize, resulting in deficiencies such as low modulus or strength properties, excessive creep rates, thermal expansion mismatch, or reaction with matrices. Other potential applications for high-performance fibers include electronic and weapons systems, where the ability to match the thermal expansion coefficients of adjacent components and to dissipate heat is critical. This may be the major future market for pitch-based carbon fibers, which can develop a thermal conductivity at least three times greater than that of copper. The application of these fibers in dissipating heat could revolutionize both the size and operating speed of computer and electronic systems.

In all these projected areas of high-performance fiber development, it is critical that support be continuous and that the longer time required to develop these fibers for composite applications be recognized. Unfortunately, development of new fibers with promising properties is costly. Thus, federal funding may be necessary both to ensure a domestic source for these fibers and to support the research and development needed to improve manufacturability and reduce costs.

Present performance-driven applications provide the opportunity and the need to develop a strong domestic technology base for the high-performance fibers required for composite materials. Future high-volume markets such as automotive and construction applications will be cost-driven, and it is vital that the domestic fiber industry be prepared to aggressively compete in these markets. It is this potential for tremendous future growth, coupled with the fact that high-performance fibers are critical for many present high-technology products, that makes basic fiber research and the health of the domestic fiber industry vital to both the U.S. economy and national security.

Because the composites industry is highly international, extensive fiber science and technology bases also exist in Western Europe and Japan, where there are strong commitments to support the development of high-performance fibers. Unless steps are taken to strengthen the domestic fiber science and technology base, to facilitate its industrial application and to broaden it for high-performance fibers, the United States might lose its present competitive position in this key industry.[3]

2.2.2 High-Performance Fibers for PMCs

High-performance fibers are often classified by the matrix used to support them. This is a convenient classification because it effectively specifies the temperature rating of the composite structure: polymeric, ambient to 427°C; metal matrix, 816–1371°C; and ceramic, 1093–1649°C.

Figure 2.2 provides comparative specific strength versus specific modulus data for various reinforcing fibers. Continuous reinforcement is prevalent in structural applications where efficient load-carrying capability is the primary consideration, such as aircraft primary and secondary structures and aerodynamic fairings. Typical uses of short fibers are for reinforcing thermoplastic matrices for injection molding and thermoset matrices for compression and sheet molding processes. The major fibers in use today are glass, carbon, and aramid.

A number of new fibers are becoming available that offer potential for advances in

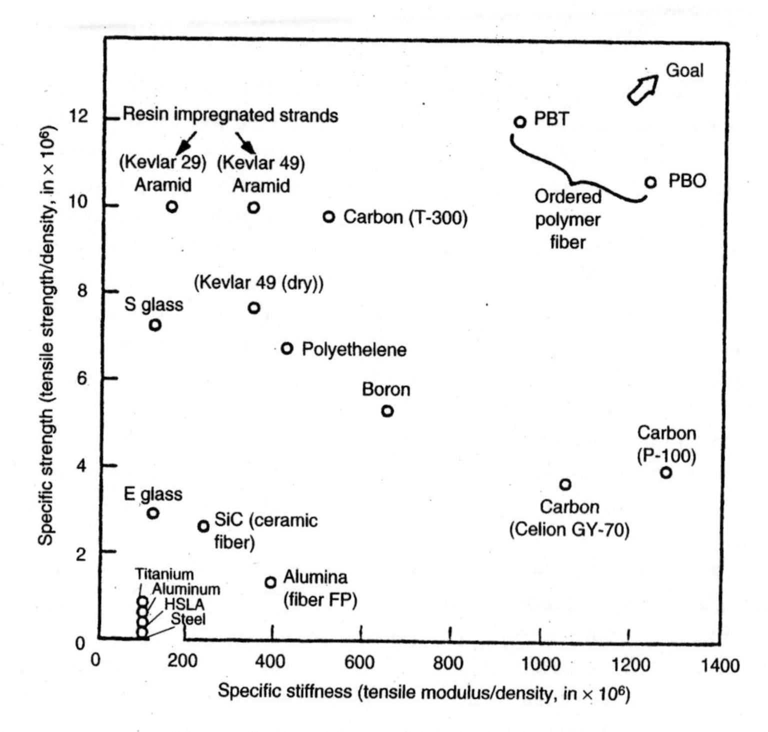

Figure 2.2. Specific properties of advanced fibers.

stiffness with accompanying increases in strength. New, ultrahigh-stiffness carbon fibers offer significant weight savings improvement as well as high thermal conductivity for thermal management applications. New carbon fibers are being fabricated with noncircular cross sections to enhance resin bonding and mechanical properties such as compression and shear strength. Polyethylene, with directional properties and extreme light weight, is another reinforcement that offers considerable advantages. It is ideal for applications where weight is critical and upper use temperature is not high. Liquid crystal polymers are another approach to enhancing composite properties. Research is being directed toward using LCPs as a reinforcement in relatively tough matrix materials. The result could be an in situ composite that avoids prepregging operations and still yields good composite properties.

Additionally, fiber materials can be produced as semifinished products that include

- Chopped fibers
- Strand mats
- Woven fabrics
- Multiaxial layers
- Rovings and yarns.

Additionally, in recent years there have been dramatic innovations in developing hybrid yarns.

2.2.2.1 Hybrid yarns. The area of textile precursors (or what might notionally be termed textile prepregs) has been a new and innovative development in thermoplastic composites (TPCs). To overcome the problem of poor drapability with melt-impregnated prepregs, new types of hybrid yarns have been developed. A hybrid yarn consists of a combination of a reinforcing fiber yarn and a yarn spun from a thermoplastic resin. Different approaches have been adopted to blend the yarns.

Coweaving. On a relatively coarse scale the two yarns may be cowoven; that is, the two fibers are fed into a textile-weaving operation to produce a desired weave pattern that places the fibers in a particular pattern. Coweaving involves bundles of reinforcement filaments in combination with bundles of thermoplastic matrix filaments. Subsequent fabrication involves the assembly of a laminated structure and consolidation by the application of heat and pressure. It must be appreciated that in the cowoven structure the fibers of the matrix and the reinforcement are discrete and that during consolidation time must be allowed for the molten thermoplastic to penetrate and wet the reinforcing filaments. There is some concern that at this coarse level of predispersion of the matrix resin there is a risk of producing nonuniformity in the composite. The advantages of coweaving are the production of a pliable fabric, the ability to produce broad goods, the variety of matrix fibers and reinforcement fibers that can be combined, and the ease of varying the fiber ratios. Cowoven fabrics have been produced from multifilament and slit film thermoplastic matrix components with a variety of reinforcements such as carbon-graphite, S-2 glass, and quartz.

Comingled Yarns and Fabrics. On a fine scale, yarns can be comingled such that the reinforcement and the matrix are interspersed at the individual filament level. The fabrics produced from comingled yarns exhibit excellent handling characteristics. The use of a special fiber sizing agent permits the fabric to be made tacky simply by spraying with a fine mist of water. As in the case of composites based on cowoven materials, a potential limitation of this approach appears to be the risk of incomplete wetting of the reinforcement fibers during the consolidation process. Nevertheless, the more intimate predispersion of the matrix and the reinforcement in the comingled systems might be expected to produce fewer defects in terms of resin-rich regions and nonwetted fibers.

The comingled yarns can be converted to various forms such as woven fabric, unidirectional tape, and braided sleeving. Other products have been produced as a spun staple yarn. The filaments are discontinuous and are twisted in the same way as the textile yarn. According to users, this approach offers the following advantages over comingled continuous-filament materials.

- A greater degree of dispersion is achieved and maintained during processing, such as weaving.
- A high degree of flexibility in composition and the fiber-to-resin ratio is possible.
- Extremely lightweight hybrid yarns can be produced at a lower cost than that of the continuous-filament analogs.
- The yarn has textilelike qualities, which facilitates weaving, knitting, and braiding.

In France, a stretch broken yarn spinning process is being developed for composite applications. The process is elegantly simple and involves the stretching and breaking of continuous filaments into long fibers. It is claimed that this treatment effectively removes the weak points in the fibers and thereby improves tenacity and processability. An attractive feature of the process is its versatility: fiber options include carbon, aramid, glass, and thermoplastic fibers. After the stretch treatment the fibers are spun into yarns. The system permits close control of the relative proportions of the reinforcement and matrix fibers and ensures that the matrix polymer is correctly positioned to achieve excellent consolidation. Intimately blended spun yarns are available based on PEEK, polyetherimide (PEI), polyphenylene sulfide (PPS), and polyamide (PA) matrix materials with high-modulus (HM) carbon, E-glass, and aramid as reinforcement materials.

The hybrid yarn approach offers the following benefits when compared with the traditional prepreg form of TPC.

- Fabrics made from hybrid yarns can be produced in a number of weave styles that provide conformability to complex mold shapes.
- The composition of composites is broadened because, in principle, any weavable fiber and any spinnable polymer can be combined.
- The textile technology base is well established and highly automated and permits precise location of the reinforcement fiber.

2.2.3 Glass Fibers

Glass fibers are the most common of all reinforcing fibers for PMCs. Glass fibers account for the major share of the market for reinforcement fibers for use in the PMC fiber-reinforced plastics industry. E-glass is available as continuous filament, chopped staple, and random fiber mats suitable for most methods of resin impregnation and composite fabrication. S-glass, originally developed for aircraft components and missile casings, has the highest tensile strength of all fibers in use. However, the compositional difference and higher manufacturing cost make it more expensive than E-glass. A lower-cost version of S-glass, called S-2 glass, has been made available in recent years. Although S-2 glass is manufactured with less stringent nonmilitary specifications, its tensile strength and modulus are similar to those of S-glass.

S-glass is primarily available as rovings and yarn, and with a limited range of surface treatments. S-glass fibers are being used in hybrid reinforcement systems in combination with graphite fibers and aramid fibers. R-glass is a similar high-strength, high-modulus fiber developed in France.

The chemical composition of E- and S-glass fibers is just like that of common soda-lime glass (window and container glasses), the principal ingredient in all glass fibers being silica (SiO_2). Other oxides, such as B_2O_3 and Al_2O_3, are added to modify the network structure of SiO_2 as well as to improve its workability. Unlike soda-lime glass, the Na_2O and K_2O content of E- and S-glass fibers is quite low, which gives them a better corrosion resistance to water as well as higher surface resistivity. The internal structure of glass fibers is a three-dimensional, long network of silicon, oxygen, and other atoms arranged in a random fashion. Thus, glass fibers are amorphous (noncrystalline) and isotropic (equal properties in all directions).

The basic commercial form of continuous glass fibers is a strand, which is a collection of parallel filaments numbering 204 or more. A roving is a group of untwisted parallel strands (*end*) wound in a cylindrical *forming package*. Rovings are used in continuous molding operations such as filament winding and pultrusion. They can also be preimpregnated with a thin layer of polymeric resin matrix to form *prepregs*. Prepregs are subsequently cut into required dimensions, stacked, and cured into the final shape in batch molding operations such as compression molding and hand lay-up molding.

Chopped strands are produced by cutting continuous strands into short lengths. The ability of the individual filaments to hold together during or after the chopping process depends largely on the type and amount of the size applied during fiber manufacturing operation. Strands of high integrity are called "hard," and those that separate more readily are called "soft." Chopped strands ranging in length from 3.2 to 12.7 mm are used in injection molding operations. Longer strands, up to 50.8 mm in length, are mixed with a resinous binder and spread in a two-dimensional random fashion to form chopped strand mats. These mats are used mostly for hand lay-up moldings and provide equal properties in all directions in the plane of the structure.

Glass fibers are also available in woven form, such as woven roving or woven cloth. Woven roving is a coarse, drapable fabric in which continuous rovings are woven in two mutually perpendicular directions. Woven cloth is made from twisted continuous strands called yards. Both woven roving and cloth provide bidirectional properties that depend on the style of weaving, as well as relative fiber counts in the length (*warp*) and crosswise (*fill*) directions. A layer of woven roving is sometimes bonded with a layer of chopped strand mat to produce a woven roving mat. All these forms of glass fibers are suitable for hand lay-up moldings.

The average tensile strength of freshly drawn glass fibers may exceed 3.45 GPa. However, surface damage (flaws) produced by abrasion, either by rubbing against each other or by contact with the processing equipment, tends to reduce it to values in the range of 1.72–2.07 GPa. Strength degradation is increased as the surface flaws grow under cyclic loads, which is one of the major disadvantages of using glass fibers in fatigue applications. Surface compressive stresses obtained by alkali ion exchange or elimination of surface flaws by chemical etching may reduce the problem; however, commercial glass fibers are not available with such surface modifications.

The tensile strength of glass fibers is also reduced in the presence of water or under sustained loads (static fatigue). Water bleaches out the alkalis from the surface and deepens the surface flaws already present in fibers. Under sustained loads, the growth of surface flaws is accelerated owing to stress corrosion by atmospheric moisture. As a result, the tensile strength of glass fibers is decreased with increasing time of load duration (Figure 2.3).

2.2.4 Aramid Fibers

Aramid fibers, first introduced on a commercial basis by DuPont in the early 1970s, are based on linear, rigid, rodlike polymer chains comprising para-linked aromatic amides. During the spinning process involved in fiber production, the molecular chains become highly oriented along the fiber axis, and this produces markedly anisotropic properties.

Kevlar 49, a DuPont aramid fiber product, belongs to a group of highly crystalline

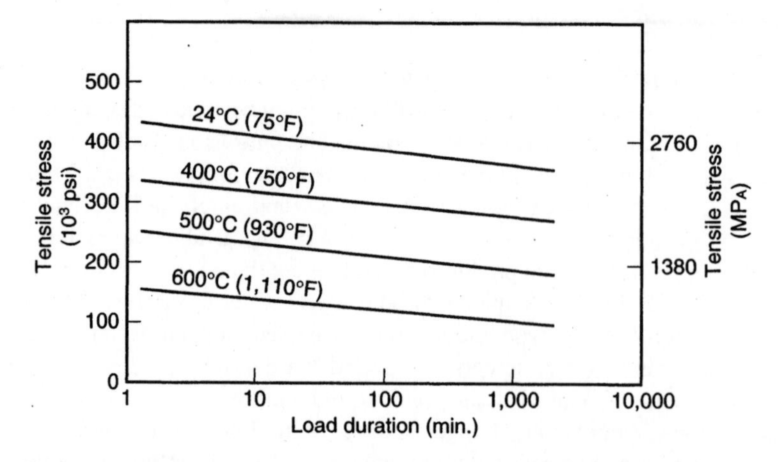

Figure 2.3. Reduction in tensile stress in E-glass fibers as a function of time at various temperatures.[3]

aramid (aromatic polyamide) fibers that have the lowest specific gravity and the highest tensile strength-to-weight ratio of current reinforcing fibers. They possess a density, typically 1.44 g/cm^3, which is lower than that of both glass fiber (by about 40%) and carbon fiber (by about 20%), and this results in the fibers having high specific strength values. The fibers also exhibit reasonably high levels of elongation to break. However, the compressive strength of aramid fibers is only about 20% of the tensile strength, and composites containing aramid fiber reinforcement are not recommended for structural applications involving high compressive loads. Aramid composites exhibit ductile behavior in compression and bending with considerable energy absorption. The continuous-use temperature of aramid fibers ranges from 160 to 200°C, and other advantageous properties are excellent fatigue and wear resistance, good electrical properties, high toughness, and good chemical resistance.

There is a range of fiber types. In the case of DuPont, the Kevlar range covers a spectrum of strength and modulus values, from low-modulus Kevlar 29, through high-modulus Kevlar 49, to very-high-modulus Kevlar 149 and very-high-strength Kevlar 129. Kevlar 49 is the grade most widely used in the fabrication of nonstructural and semistructural components in commercial aircraft and helicopters, rocket motor cases, pressure vessels, and boats. Kevlar 149, offering low moisture pickup, is being evaluated for aircraft, helicopter, and sporting goods applications. The latest addition to the product range, designated Kevlar Hp, is reported to be characterized by a smoother surface finish with virtually no fibrils. Although the properties are comparable to those of Kevlar 29 and 49, it is claimed that the new fiber offers better impact resistance and damage tolerance in composites. Major applications for Kevlar Hp are envisaged in the sporting goods and marine sectors of the high-performance composites market.

Other aramid fibers include Technora HM 50 from Teijin and Unitika in Japan, and Twaron and Twaron HM from Akzo-Enka in the Netherlands and West Germany (Table 2.1).

Like carbon fibers, aramid fibers also have a negative CTE in the longitudinal direction, which is utilized in designing low-thermal-expansion composite printed circuit boards. The major disadvantages of aramid fiber-reinforced composites are their low compressive strengths and difficulty in cutting and machining.

Kevlar 49 fibers are commercially available as untwisted yarns (with 134, 267, 768, and 1000 filaments per yarn), roving (3072 and 5000 filaments per roving), and fabrics.

Although the tensile stress-strain behavior of Kevlar 49 is linear, fiber fracture is usually preceded by longitudinal fragmentation, splintering, and even localized drawing. In bending, Kevlar 49 fibers exhihit a high degree of yielding on the compression side. Such a noncatastrophic failure mode is not observed in glass or carbon fibers and gives Kevlar 49 composites superior damage tolerance against impact or other dynamic loading. One interesting application of this characteristic of Kevlar 49 fibers is found in the soft, lightweight body armor and helmets used for protecting police officers and military personnel.

Kevlar 49 fibers do not melt or support combustion but start to carbonize at about 427°C. The maximum long-term use temperature recommended for Kevlar 49 is 160°C. Fibers are quite sensitive to ultraviolet light. Prolonged direct exposure to sunlight causes discoloration and significant loss in tensile strength. The problem is less pronounced in composite laminates in which the fibers are covered with a matrix. Ultraviolet light-absorbing fillers can be added to the matrix to further reduce the problem.

Kevlar 49 fibers are hygroscopic and can absorb up to 6% moisture content (i.e., maximum moisture absorption is directly proportional to the relative humidity and is attained in 16–36 h). Absorbed moisture seems to have very little effect on the tensile properties of Kevlar 49 fibers.

A second-generation Kevlar fiber is Kevlar 149, which has the highest tensile modulus of all commercially available aramid fibers. The tensile modulus of Kevlar 149 is 40% higher than that of Kevlar 49; however, its strain to failure is lower. Kevlar149 has an equilibrium moisture content of 1.2% at 65% relative humidity and 22°C, which is nearly 70% lower than that of Kevlar 49 under similar conditions. Kevlar 149 also has a lower creep rate than Kevlar 49.[4]

2.2.5 Carbon Fibers (Graphite Fibers)

Carbon fibers, more than all other fibrous reinforcements, have provided the basis for the development of PMCs as advanced structural engineering materials.

Carbon fibers are commercially available with a variety of tensile moduli ranging from 207 GPa on the low side to 1035 GPa on the high side. In general, low-modulus fibers have lower specific gravities, lower cost, higher tensile and compressive strengths, and higher tensile strain to failure than high-modulus fibers. Among the advantages of carbon fibers are their exceptionally high tensile strength-to-weight ratios and tensile modulus-to-weight ratios, very low CTEs (which provides dimensional stability in such applications as space antennas), and high fatigue strengths. The disadvantages are their low impact resistance and high electric conductivity, which may cause "shorting" in unprotected electrical machinery. Their high cost has so far excluded them from widespread commercial application. They are used mostly in the aerospace industry, where weight savings is considered more critical than cost.

Structurally, carbon fibers contain a blend of amorphous carbon and graphitic carbon. Their high tensile modulus results from the graphitic form, in which carbon atoms are arranged in crystallographically parallel planes of regular hexagons. The planes of the carbon atoms are held together by weak van der Waals-type forces, and a strong covalent bond exists between the carbon atoms in a plane. This results in highly anisotropic physical and mechanical properties for the fiber.[4]

The structure and properties of carbon fibers are dependent on the raw material used and the process conditions of manufacture. The manufacturing process involves the oxidation, pyrolysis, and graphitization of a carbon-containing precursor fiber, typically a polyacrylonitrile (PAN) fiber or a mesophase pitch fiber.

Carbon fibers are manufactured from two types of precursors (starting materials), namely, textile precursors and pitch precursors. The most common textile precursor is PAN.

Pitch, a by-product of petroleum refining or coal coking, is a lower-cost precursor than PAN. The carbon atoms in pitch are arranged in low-molecular-weight aromatic ring patterns. Heating to temperatures above 300°C polymerizes (joins) these molecules into long, two-dimensional sheetlike structures. The highly viscous state of pitch at this stage is referred to as mesophase. Pitch filaments are produced by melt spinning mesophase through a spinneret.

It has been clearly demonstrated that the modulus of carbon fibers derived from PAN precursors increases continuously with final heat treatment temperature, an effect attributable to an increase in the preferred orientation of the layer planes in the fiber (Figure 2.4).

PAN carbon fibers are generally categorized into high-strength, high-modulus, and ultrahigh-modulus types. The high-strength PAN carbon fibers, such as T-300 and AS-4 in Tables 2.1 and 2.2, have the lowest modulus, whereas the ultrahigh-modulus PAN carbon fibers, such as GY-70, have the lowest tensile strength as well as the lowest tensile

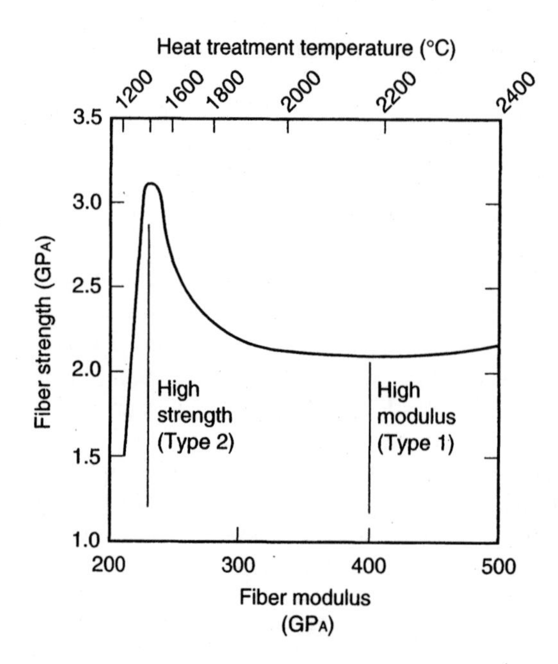

Figure 2.4. Influence of heat treatment temperature on the strength and modulus of carbon fibers.[3]

TABLE 2.2. Classification of Carbon Fibers

Fiber Classification	Young's Modulus (GPa)	Tensile Strength (MPa)	Precursor Material
Low modulus	35–70	350–1000	Pitch
Intermediate modulus	200–300	2700–5500	PAN
High modulus	350–420	2000–2750	PAN, pitch
Very high modulus	450–800	1700–2600	PAN, pitch

strain to failure. Recently, a number of intermediate-modulus, high-strength PAN carbon fibers, such as T-40 and IM-7, have been developed that also possess the highest strain to failure. Another point to note in Table 2.1 is that the pitch carbon fibers have very high modulus values, but their tensile strength and strain to failure are lower than those of PAN carbon fibers.

Carbon fibers are commercially available in three basic forms, namely, long, continuous tow, chopped (6–50 mm long), and milled (30–3000 μm long). The long, continuous tow, which is simply a bundle of 1000–160,000 parallel filaments, is used for high-performance applications.

Carbon fiber tows can also be woven into two-dimensional (2-D) fabrics of various styles. Hybrid fabrics containing commingled or cowoven carbon and other fibers, such as E-glass, Kevlar, PEEK, and PPS, are also available. Techniques of forming three-dimensional (3-D) weaves with fibers running in the thickness direction have also been developed.[3]

2.2.6 Polyethylene Extended-Chain Fibers

Polyethylene (PE) fibers, commercially available under the trade name Spectra, are produced by solution-spinning a high-molecular-weight PE. These fibers outperform all others with respect to microwave transparency in radome applications.

Spectra PE fibers have the highest strength-to-weight ratio of all commercial fibers available to date. Two other outstanding features of Spectra fibers are their low moisture absorption (1% compared to 5–6% for Kevlar 49) and high abrasion resistance, which makes them very useful in marine composites, such as boat hulls and water skis.

The major limitation of these materials is the lower upper-temperature limit (about 120°C) beyond which the fibers lose their strength. Another problem with Spectra fibers is their poor adhesion with resin matrices, which can be partially improved by surface modification with gas plasma treatment.

Spectra fibers provide high impact resistance for composite laminates even at low temperatures and are finding growing applications in ballistic composites such as armor and helmets. However, their use in high-temperature aerospace composites is limited unless they are used in conjunction with stiffer carbon fibers to produce hybrid laminates with more impact damage tolerance than all-carbon laminates.

2.2.7 Boron Fibers

Boron fiber used as a reinforcement for polymeric and metallic materials is available in many forms, several diameters, and on substrates of tungsten or carbon. Because vapor deposition of boron on a carbon monofilament substrate is still relatively new, when the material is serving as a reinforcement, "boron" refers to boron deposited on tungsten.

The most common method for producing continuous boron filaments is a chemical vapor plating process in which the reduction of boron trichloride by hydrogen gas takes place on a moving incandescent tungsten filament.

Boron-fiber composites cost more than carbon-fiber composites and have superior mechanical properties. The 0.10-, 0.14-, and 0.20-mm boron fibers have tensile strengths of 2758, 3516, and 3654 MPa, respectively. All three fibers have a modulus of approximately 400-GPa tensile strength, and PAN carbon fibers have a 228-GPa modulus. Coupled with their relatively large diameter, boron fibers offer excellent resistance to buckling, which in turn contributes to high compressive strength for boron fiber-reinforced composites. The principal disadvantage of boron is its high cost, which is even higher than that of many forms of carbon fibers. For this reason, its use is at present restricted to aerospace applications.

The most common substrate used in the production of boron fibers is tungsten wire, typically 0.0127 mm in diameter. It is continuously pulled through a reaction chamber in which boron is deposited on its surface at 1100–1300°C. The resulting fiber diameter is controlled by varying the speed of pulling as well as the deposition temperature. Currently, commercial boron fibers are produced in diameters of 0.1, 0.142, and 0.203 mm, which are much larger than those of other reinforcing fibers.

During boron deposition, the tungsten substrate is converted to tungsten boride by the diffusion and reaction of boron with tungsten. The core diameter increases from 0.0127 mm to 0.0165 mm, placing boron near the core in tension. However, near the outer surface of the boron layer a state of biaxial compression exists, which makes the boron fiber less sensitive to mechanical damage. The adverse reactivity of boron fibers with metals is reduced by chemical vapor deposition (CVD) of SiC on boron fibers, which produce borsic fibers.

Tungsten is an expensive, dense substrate used for vapor deposition of boron; substituting a carbon monofilament for the tungsten is both practical and economical. Manufacturers have successfully vapor-deposited boron onto a carbon monofilament, which differs from the carbon fiber used in C/epoxy (Ep) composites. The carbon monofilament has a diameter of 0.03 mm, and the mechanical properties of boron fiber made from it are somewhat less than those of boron on tungsten, though still superior to those of regular carbon fiber. Because the carbon monofilament substrate has a larger diameter than the tungsten filament substrate, a boron-on-carbon fiber contains less boron than a boron-on-tungsten fiber for a given final diameter.

A 0.107-mm boron-on-carbon fiber has been rated at a tensile strength of 3275 MPa and a modulus of 365.4 GPa. A 0.14-mm fiber has the same tensile strength and a 379-GPa modulus. The 0.107-mm fiber can now be directly substituted for the 0.10-mm boron-on-tungsten fiber in a B/Ep prepreg tape.

2.2.8 Other Organic Fibers

Akzo (Enka) has recently introduced a polyetherimide fiber in filament and staple form. The material is reported to offer high temperature resistance ($T_g = 215$°C) and good environmental resistance. Filament diameters are in the range 15–36 μm. Illustrative fiber properties are, for the undrawn fiber elongation 80% and hot air shrinkage 1.7%, and for the drawn fiber elongation 38% and hot air shrinkage 9%.

Dow Chemical produces paraphenylene polybenzobisoxazole fibers. The rigid-rod, liquid crystal polymer is claimed to have a unique combination of high strength, stiffness, and environmental resistance. When compared with Kevlar aramid fibers, PBO fibers are reported to offer higher tensile strength and modulus and equal compressive strength. Experimental PBO fibers have been produced with a tensile strength of 5700 MPa, tensile modulus of 360 GPa, compressive strength of 200 MPa, and density of 1.58 g/cm^3. The low compressive strength limits the applications of these fibers in unhybridized form. There is also a family of polybenzazole (PBZ) fibers with claims of typical properties: tensile strength 3450–5510 MPa, tensile modulus 345–485 GPa, and compressive strength 690–825 MPa.

2.2.9 MMC Fibers

The family of materials classified as MMCs comprises a very broad range of advanced composites of great importance to both industrial and aerospace applications. However, the development and use of MMCs are still in their infancy when compared to monolithic materials or even PMC systems. Therefore, only a handful of applications have been designed and produced, but these are illustrative of the potential of MMCs. As in the case of other composites, the family of MMCs is made up of many varieties of materials which can be categorized based on their matrix composition, fabrication process, or reinforcement type. The generic listing in Table 2.3 illustrates the range of possibilities, which have all been fabricated and investigated with varying degrees of success.

A few MMCs are composed of a metal matrix and metal reinforcement,[5] however, most MMCs are reinforced with nonmetallic fibers, particles, or whiskers.[5]

Currently the reinforcing constituents of greatest interest include alumina, boron, graphite, and silicon carbide. All these reinforcements are available as continuous fibers, whereas the main whisker reinforcements are silicon carbide and alumina. Those used in particulate form include silicon carbide, boron carbide, alumina, and titanium carbide.[6]

2.2.10 MMC Fiber-Matrix Interface

In order for the metal matrix to effectively distribute the load borne by the composite, the matrix must adhere well to the fibers (or other forms of reinforcement) and the matrix must have good shear strength. If these conditions are met, when the strong, stiff (but usually brittle) reinforcement fibers break, plastic flow occurs at the tip of any crack in the matrix and absorbs the energy, thus reducing the stress concentration. For a fiber-reinforced MMC, if the matrix-fiber interface is sufficiently strong and the matrix is ductile, any cracks occurring will tend to be deflected parallel to the fibers rather than propagating across the fibers.[7,8] For whiskers or for short, discontinuous fibers, the amount of load that can be transferred from the matrix to the reinforcement is less than in the case of continuous fibers because the ends of the fibers cannot contribute to the support.[9] For particle-reinforced composites, strengthening results from the particles mechanically restraining matrix deformation.[10,11]

TABLE 2.3. Advantages of MMCs[1]

Matrix (Alloy Class)	Reinforcement Fiber	Whisker and Particle	Process
Lead	Glass Boron Carbon		Cast Diffusion bond
Magnesium	Boron Carbon Alumina	Silicon carbide Boron carbide	Cast Diffusion bond Extrude
Aluminum	Glass Boron Steel Silicon carbide Carbon Alumina	Silicon carbide Boron carbide Glass	Cast, diffusion bond, extrude
Copper	Carbon Tungsten	Silicon carbide	Cast, diffusion bond, electroplate
Titanium	Boron Silicon carbide Alumina	Beryllium	Diffusion bond
Nickel	Boron Alumina Tungsten Carbon	Alumina	Cast, diffusion bond, electroplate

2.2.10.1 Adherence. Adherence of the matrix to the reinforcement is an important factor in determining composite properties.[12] Optimum adherence of the matrix to the reinforcement may involve some degree of chemical reaction at the interface. However, if the reaction becomes too extensive, the matrix-reinforcement bond may weaken because of the properties of any compounds formed, and it may also degrade the reinforcement. As examples, boron fiber-reinforced aluminum was observed to fail during fracture through debonding at the matrix-fiber interface. However, when this composite was exposed to a 500°C treatment in an argon atmosphere, the room-temperature composite strength was found to increase, with the failures then occurring in the fiber itself. Analytical studies of the interface showed that reaction products had formed during the high-temperature exposure. These products served to strengthen the matrix-fiber interface.[13] A more detailed study of the as-prepared B/Al composite found that the interface between the B and the Al was covered by Al_2O_3 and that this oxide decreased the interfacial bonding. The 500°C heat treatment caused the oxide to be replaced by several B/Al compounds that provided much better interfacial strength.[14]

The other extreme of reactivity between matrix and reinforcement is that B, graphite (Gr), and SiC fibers dissolve in many superalloy matrices at high temperature and are therefore useless as reinforcements in this application. As a result, tungsten fibers may be the only reinforcement in this case.[15]

In general, to obtain an optimum matrix-reinforcement interface, it is necessary that the matrix at least wet the reinforcement. If the work of adhesion between the matrix and the reinforcement exceeds the surface tension of the matrix, wetting is assumed to occur. Experimentally, wetting is said to have occurred when a liquid spreads spontaneously on a solid surface.[16] Three approaches have been taken to improve wetting: pretreatment of the reinforcement, modification of the matrix through alloying, and coating the reinforcement.[17] Each of these approaches has met with some success. For example, Gr fibers must be cleaned chemically to remove contamination that interferes with wetting by metals.[16] Lithium additions apparently improve the wetting of aluminum to its reinforcement by weakening the Al_2O_3 film covering the aluminum.[16] Coatings on the reinforcement that enhance wetting apparently also work by disrupting the oxide on the molten metal matrix during fabrication of the composite.[17]

2.2.11 MMC Advantages

The use of metals as matrices imparts important properties to the resultant composites. High matrix strength and elastic modulus impart high composite shear and transverse strength and stiffness. As an example, a boron fiber-reinforced aluminum system exhibits transverse tensile strength equal to that of the unreinforced matrix and transverse stiffness twice that of the matrix. Similarly, the matrix can impart significant toughness and resistance to the operating environment resembling the characteristics of the parent metal. Metallike thermal and electrical properties also are of importance.

It is obvious that MMCs are used only where their superior properties are required. The most notable of these are high strength-to-weight and stiffness-to-weight ratios. In addition, MMCs generally continue to exhibit these improved properties up to much higher temperatures than the acceptable usage temperature range for the unreinforced materials.

In general, it can be seen that the strength of the composite is usually better than for unreinforced aluminum (improvements for a given aluminum MMC generally range from a factor of 3 to a factor of 10[10]), and the modulus (a measure of the stiffness) is improved significantly.

2.2.12 MMC Orientation

The degree of ordering and orientation of the reinforcement in a composite plays the major role in determining whether or not the composite has isotropic or anisotropic (directionally dependent) properties (mechanical, electrical, thermal, magnetic, or chemical reactivity) within the material. As would be expected, composites reinforced with aligned, continuous fibers show much greater strength in the direction of the fibers. Whisker-reinforced composites may or may not exhibit isotropic behavior, depending on the fabrication and forming techniques used, whereas particle-reinforced composites should exhibit no directional dependence on properties unless the metal matrix itself is anisotropic.

The ability to predict the strength and other properties of MMCs is improving because of several factors, including improvements in the uniformity of the reinforcement materials, improvements in fabrication, and an increasing database for MMC properties.

Also, an improved understanding of the behavior of some of the more complex composites has resulted from improved models and computational methods.

2.2.13 Continuous MMC Reinforcement

The main continuous-fiber reinforcements currently in use are B, Gr, Al_2O_3, and SiC. The main characteristics are given in Table 2.4. This table shows the multifilament family, which includes Gr, SiC, and Al_2O_3 fibers, whereas the monofilament family is based on B only. The multifilaments are available in the form of single yarns or two- or three-dimensional weaves.[18]

2.2.13.1 Metal oxide fibers. There is much incentive to make aluminosilicate fibers with a high Al_2O_3 content or 100% Al_2O_3 fibers because of the higher temperature resistance and higher moduli of these fibers, which makes them attractive for use in composites. However, the high melting point and low viscosity of molten Al_2O_3 preclude melt spinning, and so other methods of forming fibers have been developed.

The properties of Al_2O_3 fibers have led to their use in high-performance composites. A 20-μm-diameter 99% α-alumina fiber had a tensile strength of 1400 MPa when measured at a gauge length of 10 mm. When coated with silica to heal any surface flaws, the same fiber had a tensile strength of 1900 MPa. This high tensile strength and modulus in alumina composites are retained after heating in air at 1000°C.

Boron fibers made by CVD have received, in some cases, fiber coatings of SiC or B_4C to retard reactions between the B and the metal matrix at high temperatures.

Carbon fibers are not suitable for forming Al-based MMC because of the fiber degradation during processing, but T-300 especially is used successfully to form the cheapest magnesium composite. Sometimes CVD coating of carbon fiber, using either nickel or silicon, has been employed to improve the wettability.

Continuous fibers based on Al_2O_3 (Fiber FP) have a high purity and a large grain size, which means that the fiber is extremely brittle and difficult to handle but may offer better MMC properties.

Nicalon and Tyranno SiC fibers are used widely because of their attractive combination of strength, stiffness, and handling characteristics. MMC materials reinforced with Tyranno exhibit high transverse strength and are used mainly in the aerospace industry (see Tables 2.1 and 2.4).

Nicalon fibers (40% Si, 50% C, 8% O, 2% N) have a tensile strength of 2413–3790 MPa and moduli ranging from 190 to 410 GPa. Although the Si_3N_4 fibers are still experimental, their potential looks promising because they are simpler and less expensive to fabricate than SiC. Continuous filaments of 5- to 10-μm diameter have been produced, with a typical 10-μm-diameter fiber having a tensile strength of 2447 MPa and a modulus of >200 GPa and able to withstand continuous use at 1200°C. The raw material is a by-product of the production of silane, which contributes to the processing route being cheaper than that used to produce an equivalent SiC fiber.

In regard to glass fibers, the Si_3N_4 continuous fiber, which has been also produced in the United States as HPZ, has a two and one-half to three times greater tensile modulus, up to 2.45×10^5 MPa, whereas HPZ has a tensile strength measured as 3312 MPa, which is also higher than that of glass. HPZ is said to have outstanding compression

TABLE 2.4. Properties of Reinforcing Fibers for MMCs[18–20]

Fiber	Protection System	Trade Name	Manufacturer	Diameter (μm)	Type	Elastic Modulus (GN/m^2)	Tensile Strength (GN/m^2)	Density (g/cm^3)	Coefficient of Thermal Expansion ($10^{-6}/K$)
Boron (CVD)									
B on W			Textron, CTI, VMC	100, 140, 200	Single	400	3.5–4.1	2.5–2.6	5.0
B on C			Textron	100, 140	Single	370	3.3	2.3	5.0
SiC coated B	SiC	Borsic	CTI	100,145	Single	400	3.0	2.6	5.0
B_4C coated B	B_4C		Textron	145	Single	370	4.0	2.3	5.0
Carborundum									
SiC on W (CVD)	C + TiB_2	SM1240	BP	100	Single	400	3.75	3.4	
SiC on C (CVD)	C	SCS6	Textron	145	Single	427	4.0	3.0	4.8
SiC + O (precursor)		Nicalon	Nippon Carbon	15	Tow	196	2.75	2.55	
SiC + O + Ti (precursor)		Tyranno	Ube	8–10	Tow	210	2.9	2.3	
Carbon									
PAN high strength			Various	7–9	Tow	240–300	3.5–5.0	1.7–1.8	–0.75
PAN high modulus			Various	7–9	Two	350–450	2.3–3.0	1.8–1.9	–1.15
Pitch system		P100	UCC	11	Tow	700	2.1	2.1	–1.45
Alumina		Fiber FP	DuPont	20	Tow	380	1.5	3.9	8.0

strength, which improves its strength-to-weight properties. This in turn broadens the material's design flexibility within a part compared to that of glass fiber and, depending on the resin system, to that of carbon fiber as well.[19]

2.2.14 Short MMC Fibers

When considering aligned fibers, short fibers show high strength in composites. These fibers include the oxides, Saffil (Al_2O_3) and Kaowool, which have been used mainly for the reinforcement of automobile engine components. Short fibers of carbon and several ceramics show the characteristics in the accompanying table.

Characteristics of Ceramic Short Fibers

Fiber	Size		Density	Ultimate Thermal Strength	
	Length, l (mm)	Diameter, d (μm)	(g/cm^3)	(GPa)	E (GPa)
Carbon T-300	2.5	7.8	1.75	3.45	230
SiC Nicalon	1–6	10–15	2.55	3	195
Al_2O_3 FP	3–6	15–25	3.96	1.7	380
Al_2O_3 Saffil	0.1–1	1–5	3.30	2	300

2.3 WHISKERS

Fiber diameters classify the various fiber morphologies, which consist of whiskers (<1 μm), continuous multifilament yarn (5–25 μm), and continuous monofilament (>100 μm). These include aggregate forms such as wool and rigid preforms (whiskers, staple) and yarns and wovens (continuous fibers).[20–30]

Whiskers are characterized by their fibrous, single-crystal structures, which have almost no crystalline defects. Numerous materials, including metals, oxides, carbides, halides, and organic compounds, have been prepared under controlled conditions in the form of whiskers. Generally, a whisker has a single dislocation running along the central axis. This relative freedom from discations means that the yield strength of a whisker is close to the theoretical strength of the material.[21]

The method of vapor deposition is widely used in whisker preparation. The performance of whiskers at elevated temperatures is far better than that of any other fibers.[21] Because of their outstanding specific mechanical characteristics, much work has been carried out in fabricating MMCs using whiskers. The small diameter of the whiskers ($d = 0.1$–2 μm) means that their lengths (l) are sufficient ($l/d \sim 50$–100) to permit load transfer.[31]

They are a discontinuous reinforcement with a high aspect ratio. Because of their small diameters, whiskers have very few defects to initiate fracture and as a result have much higher strengths than discontinuous fibers.

Silicon carbide, silicon nitride, carbon and potassium titanate whiskers are currently available. Among these, silicon carbide whiskers seem to offer the best opportunities for MMC reinforcement. Presently, silicon carbide whisker reinforcement is produced from rice husk, which is a low-cost material.

TABLE 2.5. Properties of Some Whiskers[21]

Material	Tensile Strength (kN/mm^2)	Density (g/cm^3)	Young's Modulus (kN/mm^{-2})	Specific Strength (kN/mm^{-2})	Specific Modulus (kN/mm^{-2})
Alumina	21	3.96	430	5.3	110
Silicon carbide	21	3.21	490	6.5	150
Graphite	20	1.66	710	12.0	430
Boron carbide	14	2.52	490	5.6	190
Silicon nitride	14	3.18	380	4.4	120

The physical characteristics of whiskers are responsible for different chemical reactivity with the matrix alloy.[31] For instance, high-strength carbon fibers exhibit a much higher chemical reactivity toward liquid aluminum than do high-modulus carbon fibers owing to their different states of crystallization. Table 2.5 shows the properties of some whiskers.

Besides those just mentioned, many other materials including metal oxides and halides have been produced as whiskers. The metal oxides include Al_2O_3, MgO, MgO-Al_2O_3, Fe_2O_3, BeO, MoO, NiO, Cr_2O_3, and ZnO. Whiskers grow by two different mechanisms: tip growth and basal growth. Tip growth occurs in the formation of SiC whiskers as well as Si_3N_4, Al_2O_3, and AlN whiskers. In basal growth, the whisker material migrates to the base of the whisker and extrudes the whisker from the substrate.

2.3.1 SiC Whiskers

Ceramic whiskers are in demand for application as reinforcing fillers in both MMC and CMC materials. At present, SiC whiskers are most widely used for this purpose because of their high strength. However, the application of other ceramic whiskers such as Si_3N_4 is expected to increase with increasing and diversifying utilization of composite materials.

β′-Sialon whiskers are another prospective material for high-temperature composite filler. β′-Sialon is a solid solution of β-Si_3N_4 in which Si and N are substituted by Al and O.[32]

Ceramic whisker growth has been studied mainly for SiC. SiC whiskers have been produced by various methods, and these are well reviewed by Milewski et al.[33]

One method is the production of SiC whiskers from rice hulls which contain 10–20 wt% SiO_2. Rice hulls are first heated in an oxygen-free atmosphere to 700–900°C to remove the volatiles and then to 1500–1600°C for 1 h to produce SiC whiskers. The final heat treatment is at 800°C in air, which removes free carbon. The resulting SiC whiskers contain 10 wt% of SiO_2 and up to 10 wt% Si_3N_4. The tensile strength and modulus of these whiskers are reported as 13 and 700 GPa, respectively.

An exhaustive study and evaluation were conducted and reported by Niwano,[34] in which he outlined the manufacturing process and properties of SiC whiskers and the properties of SiC whisker-reinforced ceramics, metals, and plastics. He showed that when SiC whiskers were added to an Al_2O_3 matrix, significant improvements in both fracture toughness and strength were achieved over monolithic Al_2O_3. Increases in fracture toughness are achieved in mullite, zirconia, and Si_3N_4 matrix composites as well. SiC whiskers can also effect remarkable improvements in the strength and elastic modulus of aluminum alloys.[34]

Akiyama and Yamamoto[35] also conducted an extensive investigation of SiC whiskers (Tokawhiskers), and Table 2.6 shows a summary of currently available whiskers.

By combining Nicalon SiC fibers and 10% SiC whiskers and the glass matrix, the following values were obtained.

Hybrid elastic limit 650 MPa, from 330 MPa for SiC fibers

Transverse strength 134 MPa, from 47 MPa

Interlaminar shear strength 51 MPa, from 12 MPa for SiC fibers.

Finally, there are single-crystal α-SiC platelets, which have several advantages over whiskers. They do not agglomerate like whiskers, and so they are easier to disperse; furthermore, platelets do not retard sintering or interfere with densification, which makes them ideal for pressureless sintering. Platelets are grown to size by a diffusion process that uses growth dopants to promote surface diffusion above 2100–2200°C. The platelets are grown as α-phase whiskers, which withstand higher temperatures than β-phase whiskers. The final product is a smooth, perfect hexagon, unlike that resulting from other processes in which the whiskers appear to have been ground. Mechanical studies must be performed to learn how to process the platelets into a composite.[36,37]

2.3.2 Si_3N_4 Whiskers

Si_3N_4 whiskers, with their superb heat resistance, may one day be the reinforcement of choice for high-strength, high-temperature ceramic composites. Scientists have not yet given Si_3N_4 whiskers the same scrutiny brought to the more common SiC whiskers, but this is changing. Researchers are trying to determine how processing affects whisker crystallinity and the extent of polymorphic transformation of the Si_3N_4, and how to control whisker segregation during processing.

2.3.3 Al_2O_3 Whiskers

Single-crystal α-Al_2O_3 whiskers have been found to be easier to disperse than SiC whiskers, and so higher reinforcement loadings can be achieved.[36] The diameters of aluminum oxide whiskers (4–7 μm versus ~ 0.5 μm for SiC) make them less prone to agglomeration and therefore easier to disperse. (Their aspect ratio is 10:13.) Their larger size also makes them less likely to float in the air and become a health problem, a potential threat to those who work with SiC whiskers.

A new Al_2O_3-based ceramic whisker, called Alborex, has been developed for use as a reinforcing material in fiber-reinforced plastics and metals.[37,38] The whisker has a composition of $9Al_2O_3 \bullet 2B_2O_3$. It has a melting point of 1440°C and excellent thermal resistance; its tensile elasticity is 40 tons/mm^2, close to that of a SiC whisker, and its strength is greater than that of a potassium hexatitanate ($K_2Ti_6O_{13}$) whisker. Because its Al_2O_3 content is 86%, it has great stability with respect to metals such as aluminum.

TABLE 2.6. Summary of Commercially Available Whiskers[35]

	Whisker	Manufacturer[a]	Product Name	Specific Gravity	Diameter (μm)	Length (μm)	Main Application
SiC	β-SiC	Tokai Carbon	Tokawhiskers	3.19	0.1–1.6	50–200	
	β-SiC	Tateho Chemical	SCW	3.18	0.05–1.5	20–200	FRP
	α-SiC	ACMC (USA)	SILAR	3.2	av 0.6	10–80	FRM
	β-SiC	American Matrix, Mitsui Toatsu Chemicals, Shin-Etsu Chemical, Nippon Light Metal, Kobe Steel		3.2	1–3	100–500	FRC
Si_3N_4	α-Si_3N_4	Tateho Chemical	SNW	3.18	0.1–0.6	20–200	FRP
	α-Si_3N_4	Ube Industries		3.18	0.1–0.4	5–20	FRC
	β-Si_3N_4	Ube Industries		3.19	0.1–1.5	10–50	
Al_2O_3	α-Al_2O_3						
C	Polycrystalline graphite	Nikkiso (Asahi Chemical Industry)	Graskers (graphite whiskers)	2.25	0.1–1.0	10–200	FRP
Potassium titanate	$K_2O \cdot 6TiO_2$	Ohtsuka Chemical, (Kubota)	Tismo	<3.3	0.2–0.5	10–20	FRP

[a] Companies indicated in parentheses are those with plans to manufacture whiskers.

2.3.4 Miscellaneous Ceramic Whiskers

In reinforcing aluminum, TiB_2 whiskers have many advantages over SiC. TiB_2 is more thermodynamically stable than SiC and less reactive. SiC cannot be used in aluminum above its 650°C melting point, but TiB_2 is stable and wets out all the way up to 900°C. It can also be a stable reinforcement for Ti, Si_3N_4, and SiC.

NiO, MgO, and MgO-NiO whiskers have been grown from their molten salts by a variant of the vapor-liquid-solid (VLS) process (used to grow SiC whiskers). Crystals up to 5 mm long and from 1 to 10 μm in diameter have been obtained.

A new process has been developed to produce primarily synthesized sialon material consisting of mainly silicon oxynitride (SiO_2-N_2) and aluminum oxynitride (Al-O-N) in solid solution. The resulting material is a homogeneous Si-Al-O-N whisker-powder blend. The process involves a specially designed combustion synthesis technique said to yield at least 20 wt% whisker in the composite. Such a composite is expected to have enhanced fracture toughness and improved creep properties compared with conventional sialons. These benefits are in addition to already good sialon properties, including high strength, wear resistance, low coefficient of expansion, and high resiliency in corrosive environments.

2.4 PARTICULATES

Particulates are the most common and cheapest reinforcement materials. They produce the isotropic property of MMCs, which show a promising application in structural fields. Presently, higher volume fractions of reinforcements have been achieved for various kinds of ceramic particles (oxides, carbides, nitrides).[31]

It is observed from research that 20% SiC in particulate form shows improvements in yield and tensile strength of a similar percentage. There is especially no change in density, but stiffness has improved by up to 50%,[39] which breaks the general "rule" that the specific thickness of all engineering metals, regardless of their density, is roughly the same. Kohara[40] has worked on SiC_p/Al and SiC_w/Al composite materials, showing that the SiC particulate-reinforced aluminum matrix composites are not as strong as SiC-whisker-reinforced composites.

However, SiC particulate-reinforced aluminum matrix composites have a good potential for use as wear-resistant materials. Actually, particulates lead to a favorable effect on properties such as hardness, wear resistance, and compressive strength. Additionally, a tight particle size and control of purity are the main additional requirements for composites. Table 2.7 lists the particulates currently available.

A number of particulate-reinforced systems have been in use industrially for many years, including cermets, employed in the electronics industry for the tracks of precision variable resistors, and high-speed cutting tool tips.[41]

Most frequently there are two types of discontinuous reinforcements: whiskers and particulates. The most common types of particulates are Al_2O_3, B_4C, SiC, TiC, and WC. Whiskers generally cost more than particulates. For instance, SiC whiskers cost $95 per pound, whereas SiC particulates costs $3 per pound. In terms of tailorability, a very important advantage in MMC applications, particulate reinforcement possess various desir-

TABLE 2.7. A Selection of Reinforcements for CMCs[41]

Form	Name and Chemical Formula		Dimensions
Particulate (powder)	Silicon carbide	SiC	All are in the range 0.5–5 μm
	Silicon nitride	Si_3N_4	
	Titanium diboride	TiB_2	
	Titanium carbide	TiC	
	Zirconia	ZrO_2	
Whiskers	β-Silicon carbide	β-SiC	All have dimensions in the ranges 0.1–3 μm diameter and 5–20 μm length
	α-Silicon nitride	α-Si_3N_4	
	β-Silicon nitride	β-Si_3N_4	
Platelets	α-Silicon carbide	α-SiC	Hexagonal platelets 3–10 μm thick and 30–100 μm wide
Continuous fibers	Carbon	C	~7 μm diameter
	Silicon carbide (nonstoichiometric fibers with excess oxygen)	Si-C-O	~10–15 μm diameter
	Silicon carbide monofilaments (chemically vapor-deposited silicon carbide on either a tungsten (W) or carbon (C) core)	SiC	~100 to 150-μm-diameter fibers on either a W or a C core of typically ~7- to 10-μm diameter

able properties. B_4C and SiC, for instance, are widely used, inexpensive commercial abrasives that can offer good wear resistance as well as high specific stiffness. TiC has a high melting point and chemical inertness, which are desirable properties for processing and stability in use. WC displays high strength and hardness at high temperatures.

In composites, a general rule is that mechanical properties such as strength and stiffness tend to increase as the reinforcement length increases. Particulates can be considered the limit of short fibers. Particulate-reinforced composites are isotropic; that is, they have the same mechanical properties in all directions.[42–47]

2.5 WIRE

Metallic filaments are called wire, and they are characterized by their high elastic moduli. Among them, molybdenum and tungsten are the most outstanding. Presently, stainless steel wire is creating interest. However, the main disadvantage of metal filaments with the probable exception of beryllium, is that their density is higher than that of ceramic whiskers.

These fibers are generally more ductile than any other fibers, and so now the effort is toward using metallic filaments to fabricate composites for carrying tensile load with high toughness.

2.6 MATRICES

The role of the matrix in a fiber-reinforced composite is (1) to transfer stresses between the fibers, (2) to provide a barrier against an adverse environment, and (3) to protect the surface of the fibers from mechanical abrasion. The matrix plays a minor role in the tensile load-carrying capacity of a composite structure. However, selection of a matrix has a major influence on the interlaminar shear as well as on in-plane shear properties of the

TABLE 2.8. Matrix Materials

Polymeric
- Thermoset polymers (resins)
  - Epoxies: principally used in aerospace and aircraft applications
  - Polyester, vinyl esters: commonly used in automotive, marine, chemical, and electrical applications
  - Phenolics: used in bulk molding compounds
  - Polyimides, polybenzimidazoles (PBI), polyphenylquinoxaline (PPQ): for high-temperature aerospace applications (temperature range 250–400°C)
- Thermoplastic polymers
  - Nylons (such as nylon 6, nylon 6,6), thermoplastic polyesters [such as polyethylene terephthalate (PET), PBT], polycarbonate (PC), polyacetals: used with discontinuous fibers in injection-molded articles
  - Polyamide-imide (PAI), polyether ether ketone (PEEK), polysulfone (PSUL), polyphenylene sulfide (PPS), polyether imide (PEI): suitable for moderately high-temperature applications with continuous fibers

Metallic
- Aluminum and its alloys, titanium alloys, magnesium alloys, copper-based alloys, nickel-based superalloys, stainless steel: suitable for high-temperature applications (temperature range 300–500°C)

Ceramic
- Aluminum oxide (Al_2O_3), carbon, silicon carbide (SiC), silicon nitride (Si_3N_4): suitable for high-temperature applications

composite material. The interlaminar shear strength is an important design consideration for structures under bending loads, whereas the in-plane shear strength is important for structures under torsional loads. The matrix provides lateral support against the possibility of fiber buckling under compression loading, thus influencing to some extent the compressive strength of the composite material. The interaction between fibers and matrix is also important in designing damage-tolerant structures. Finally, the processability of and defects in a composite material depend strongly on the physical and thermal characteristics, such as viscosity, melting point, and curing temperature of the matrix.

Table 2.8 lists various matrix materials that have been used either commercially or in research. Among these, thermoset polymers, such as epoxies and polyesters, are in greatest commercial use, mainly because of the ease of processing with these materials. Metallic and ceramic matrices are considered primarily for high-temperature applications.

Additional functions of a matrix are to

Keep the fibers in place in the structure
Help distribute or transfer load
Protect the filaments, both in the structure and before fabrication
Control the electrical and chemical properties
Carry interlaminar shear.

The needs or desired properties of a matrix, depending on the purpose of the structure, are to

Minimize moisture absorption
Wet and bond to fiber
Flow to penetrate completely and eliminate voids during the compacting and curing process

Be elastic in order to transfer load to fibers
Have strength at elevated temperatures (depending on the application)
Have a low-temperature capability (depending on the application)
Have excellent chemical resistance (depending on the application)
Have low shrinkage
Have a low coefficient of thermal expansion
Have reasonable strength, modulus, and elongation (elongation must be greater than that of fiber)
Be easily processible
Have dimensional stability (maintain its shape).

There are several matrix choices available. Each type has an impact on the processing technique and the environmental properties of the finished composite. The first choice is among PMC, MMC, CMC, IMC, CCC, and so on. If this first key selection is PMC, then the next decision is between a thermoplastic and a thermoset composite matrix. Thermoplastic matrices (polymers that can be repeatedly softened by increasing the temperature and hardened by decreasing the temperature) have been developed to increase the hot-wet use temperature and the fracture toughness of composites. Although TPCs are not in general use, their properties are well documented because of the sponsorship of development programs by the U.S. Air Force. Table 2.9 shows the relative advantage because their large-scale use is still in the future. The following special considerations are in order for TPCs.

Because high temperatures (up to 300°C) are required for processing, special autoclaves, processes, ovens, and bagging materials may be needed.
The fiber finishes used for thermosetting resins are not compatible with thermoplastic matrices, necessitating alternative treatment.
TPCs can have either greater or much lower solvent resistance than a thermoset material. If the stressed matrix of the composite is not resistant to the solvent, the attack and destruction of the composite may be nearly instantaneous.

The properties of some thermoplastic matrices are given in references 1, 3, 4, 5, 11, and 32. A thermosetting matrix, in contrast to thermoplastic resins, sets at some temperature (room temperature or above) and cannot be reshaped by subsequent heating. In general, thermosetting polymers contain two or more ingredients, namely, a resinous matrix and a curing agent. Solidification of the composite matrix starts either when the resin and the curing agent are mixed or when the matrix is heated, which causes a reaction between the resin and its curing agent.

The following are common thermoset matrices for composites:

- Polyesters and vinyl esters
- Epoxy (Ep)
- Polyimide (PI)
- Bismaleimide (BMI).

TABLE 2.9. Composite Matrix Tradeoffs[4]

Property	Thermoset	Thermoplastic	Notes
Resin cost	Low to medium high, based on resin requirements	Low to high; premium thermoplastic prepregs cost more than thermoset prepregs	Decreases for thermoplastics as volume increases
Formulation	Complex	Simple	
Melt viscosity	Very low	High	High melt viscosity interferes with fiber impregnation
Fiber impregnation	Easy	Difficult	
Prepreg tack or drape	Good	None	Simplified by comingled fibers
Prepreg stability	Poor	Good	
Composite voids	Good	Good to excellent	
Processing cycles	Long	Short to long (long processing degrades polymer)	
Fabrication costs	High for aerospace, low for pipes and tanks with glass fibers	Low (potentially)	
Composite mechanical properties	Fair to good	Good	
Interlaminar fracture toughness	Low	High	
Resistance to fluids or solvents	Good	Poor to excellent (choose matrix well)	Thermoplastics stress craze
Damage tolerance	Poor to excellent	Fair to good	
Resistance to creep	Good	Not known	
Database	Very large	Small	
Crystallinity problems	None	Possible	Crystallinity affects solvent resistance

Each of these resin systems has some drawbacks which must be accounted for in design and manufacturing plans.

2.6.1 Disadvantages of TPCs

In the present state of the art the following disadvantages of TPCs constrain their acceptance in the marketplace.

- Prepreg tapes of TPC matrices tend to be stiff and boardy and to lack the attractive drape and tack of their thermoset competitors (this makes handling difficult both in terms of conformability to complex mold shapes and in bonding and tacking together laminated structures).
- Semicrystalling thermoplastic materials require a carefully controlled cooling program after consolidation of the composite in order to develop an appropriate degree

of crystallinity (this extends the processing time and adds to the process control requirements).

- The newness of the TPC technology is a problem because there is no established long-term performance database to guide composite design and materials selection. Similarly, fabrication techniques are in an early stage of development, and many need further refinements.
- Thermoplastic prepregs and hybrid yarns and fabrics are high-cost materials.

The thrust of the development of TPCs is dominated by the demands of the aerospace sector, and this limits the opportunities for application of lower-performance systems in other market sectors.[48]

2.6.2 Advantages of TPCs

The key advantages of TPCs are

- Enhanced processability with respect to
  - reduced cycle times, with increased production rates and major cost savings
  - elimination of the low-temperature storage requirement associated with thermoset resins
  - maintenance of consistent properties on extended storage
  - being less energy-intensive than thermosets (no cure required)
  - absence of solvents (in most systems) eliminating the requirement for solvent removal and associated health concerns
  - recycling of scrap
- High-temperature stability
- High toughness and damage tolerance
- Improved microcrack resistance
- Higher strains to failure
- Excellent damping characteristics
- Greatly reduced moisture absorption
- Low outgassing and better radiation tolerance (of importance in space applications)
- Improved solvent resistance
- Lower capital investment requirements.

2.6.3 Problems Associated with Thermoset Matrices

There are several problems that consistently arise with thermoset matrices and prepregs that do not apply to TPC starting materials.

Frequent variation from batch to batch involving

- Effects of small amounts of impurities
- Effects of small changes in chemistry
- Changes in matrix component vendor or manufacturing location.

Void generation, caused by

- Premature gelation
- Premature pressure application.

Because of these problems, if raw material and processing costs are comparable for the two matrices, the choice should always be TPCs without regard to the other advantages resulting in the composite. These problems lead to a great increase in quality control efforts, which may represent the bulk of the final costs of the composite structure.

The following general notes are more-or-less applicable to all thermoset matrices.

The higher the service temperature limitation, the lower the strain to failure.

The greater the service temperature, the more difficult the processing, which may be due to volatiles in the matrix, higher melt viscosity, or longer heating curing cycles.

The greater the service temperature or the greater the curing temperature, the greater the chance for development of color in the matrix.

Higher service temperatures and higher curing temperatures may result in better flame resistance (although this is not evident for epoxies with curing temperatures between 121 and 177°C.

2.6.4 Thermoplastic Matrix Polymers

High-performance TPCs have been developed with glass, aramid, and carbon fiber reinforcements and with all of the high-temperature-resistant engineering thermoplastic resins as matrix materials. A listing of the major thermoplastic resin types of current interest is presented in Table 2.10 which indicates the thermal transition temperature (T_g and T_m). The list is not comprehensive but serves to emphasize the particular interest in PPS, polyarylsulfone (PAS), polyether sulfide (PES), PIs, and polyamide-imide. Of all the resin types, the leading candidate for aerospace applications appears to be polyketone. A family of resins is being developed that exhibit differences in the thermal transition temperatures, as illustrated in Table 2.10. Illustrative of the variants are polyether ether ketone, polyether ketone (PEK and PEKK), polyarylether ketone (PAEK), and polyether ketone ether ketone ketone (PEKEKK).

The advanced polymer composite APC-2 is the most well-established polyketone-based material. PESs are a family of amorphous materials and exhibit high glass transition temperatures. ICI-Fiberite has developed a biphenyl-modified PES, Victrex HTA/IM8 carbon fiber unidirectional tape, which is attracting attention for use in components in an advanced tactical fighter (ATF) development program. Udel polysulfone is also being used in solution to form thermoplastic prepreg tapes. Quadrax has offered biaxial tapes incorporating Radel, a polyarylether sulfone (PAES).

PIs represent a diverse class of materials, some of which are truly thermoplastic and others of which have some thermoplastic character but are strictly thermosets. These include polyetherimide (PEI), used with carbon prepregs, and Cypac 7005, used for structural applications in the previously mentioned ATF program. A number of suppliers are

TABLE 2.10. High-Performance Thermoplastic Matrix Polymers[48]

Polymer Composition	Acronym	Morphology	Glass Transition	Melting Temperature	Processing Temperature (°C)	Manufacturer	Trade Name
Polyether ether ketone	PEEK	Semicrystalline	143	343	360–400	ICI	Victrex PEEK
Polyphenylene sulfide	PPS	Semicrystalline	138	288	340–370	Phillips 66	Ryton PPS
Polyether ketone	PEK	Semicrystalline	165	365	400–450	BASF	Ultrapek PEK
Polyether ketone ether ketone ketone	PEKEKK	Semicrystalline	173	370	420–450	BASF	Ultrapek
Polyether sulfone	PES	Amorphous	260	—	400–450	ICI	Victrex HTA
Polyether imide	PEI	Amorphous	217	—	350–400	GE Plastics	Ultem PEI
		Semicrystalline	270	380	380–420	Mitsui Toatsu	P-IP
Pseudothermoplastics							
Polyimide	LARC-TPI	Amorphous	250	—	330–350	Rogers, Mitsui Toatsu	
	PI	Amorphous	265	—	350	CIBA-Geigy	Matrimide 9725

offering amorphous thermoplastic fully imidized and polymerized PI resins that are suitable for solution impregnation of reinforcement fibers, and finally, there is a thermoplastic prepreg based on J-2 and Avimid thermoplastic PIs with carbon fiber and aramid fibers as reinforcement.

The polyquinoxaline (PQ), polyphenylquinoxaline (PPQ), and polybenzimidazole (PBI) systems are representative of candidate materials having the greatest potential for thermal and oxidative stability that are still in the experimental stage. Processing of these materials is difficult, and the PQ/PPQ systems will require extensive development work in the future to make their properties attractive. A description of the characteristics of some of the leading TPCs follows.

2.6.4.1 PEEK. PEEK is a linear aromatic thermoplastic based on the following repeating unit in its molecules.

Ketone Ether Ether

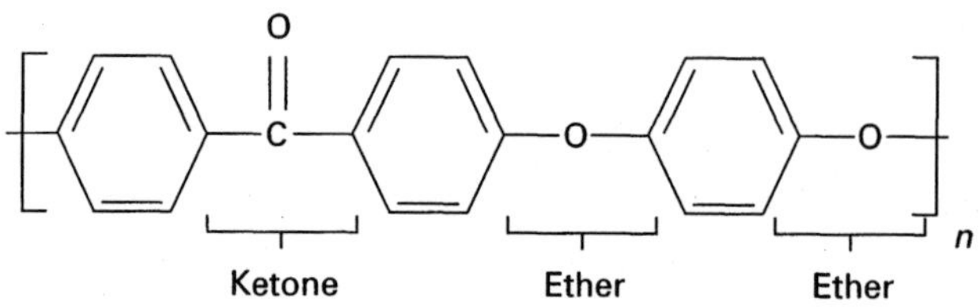

Polyether ether ketone

Continuous-carbon fiber-reinforced PEEK composites are known in the industry as aromatic polymer composite (APC).

PEEK is a semicrystalline polymer, and amorphous PEEK is produced when the melt is quenched. The presence of fibers in PEEK composites tends to increase the crystallinity to a higher level because the fibers act as nucleation sites for crystal formation. Increasing crystallinity increases both the modulus and the yield strength of PEEK but reduces its strain to failure.[49]

PEEK's maximum continuous-use temperature is 250°C. PEEK is the foremost thermoplastic matrix that may replace epoxies in many aerospace structures. Its outstanding property is its high fracture toughness, which is 50–100 times higher than that of epoxies. Another important advantage of PEEK is its low water absorption, which is less than 0.5% at 23°C compared to 4–5% for conventional aerospace epoxies. Being semicrystalline, it does not dissolve in common solvents.

2.6.4.2 Polyphenylene sulfide (PPS). PPS is a linear semicrystalline polymer with the following repeating unit in its molecules.

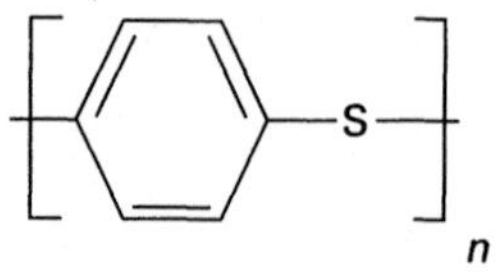

Polyphenylene sulfide

PPS is normally 65% crystalline. It has a glass transition temperature of 85°C and a crystalline melting point of 285°C. Melt processing of PPS requires heating the polymer in the temperature range 300–345°C. The continuous-use temperature is 240°.

This thermoplastic material has the ability to cross-link by air-oxidation at elevated temperatures, resulting in an irreversible cure. PPS has outstanding resistance to heat, flame, and chemicals and has excellent electrical insulation characteristics. Its extreme inertness to organic solvents and inorganic salts and bases makes it an exceptional corrosion-resistant coating for use in contact with food. Being crystalline in form, PPS does not have the toughness of amorphous plastics such as polycarbonate and polysulfone. Though it tends to be brittle, it is resistant to environmental stress cracking. The mechanical properties can be significantly improved by adding glass fiber reinforcement. PPS is one of the most expensive commercial moldable thermoplastics, and so it is greatly dependent on its unique combination of properties for selection.

2.6.4.3 Polysulfone. Polysulfone is an amorphous thermoplastic with the following repeating unit.

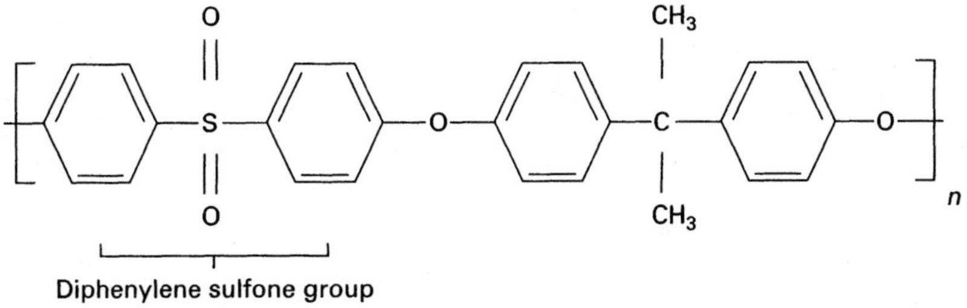

Diphenylene sulfone group

Polysulfone

Polysulfone has a glass transition temperature of 185°C and a continuous-use temperature of 160°C. The melt processing temperature is between 310 and 410°C. Although it has good resistance to mineral acids, alkalies, and salt solutions, it swells, stress-cracks, or dissolves in polar organic solvents.

The family of polysulfones also includes polyaryl sulfones and polyarylether sulfones. The principal characteristics of polysulfones include their outstanding heat resistance, exceptional resistance to creep, rigidity, transparency, and resistance to greases, solvents, and chemicals. Additionally, they are self-extinguishing. They are among the higher-priced engineering thermoplastics and so are used only in situations where polycarbonates or other cheaper materials are not suitable.

2.6.4.4 Thermoplastic PIs. Thermoplastic PIs are linear polymers derived by condensation polymerization of a polyamic acid and an alcohol. Depending on the types of polyamic acid and alcohol, various thermoplastic PIs can be produced. The chemical reaction takes place in the presence of a solvent and produces water as a by-product. The resulting polymer has a high melt viscosity and must be processed at relatively high temperatures. Unlike thermosetting PIs, thermoplastic PIs can be reprocessed by the application of heat and pressure.

PEI and PAI are melt-processable thermoplastic PIs. Both are amorphous polymers with high glass transition temperatures, 217°C for PEI and 280°C for PAI. The processing temperature is 350°C or above. Their chemical structures are shown here.

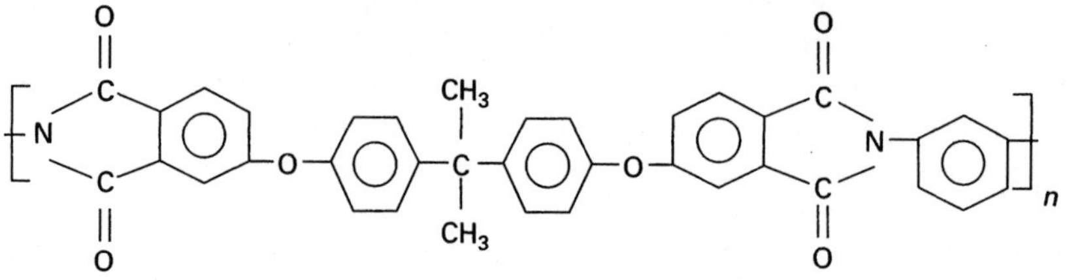

Polyetherimide

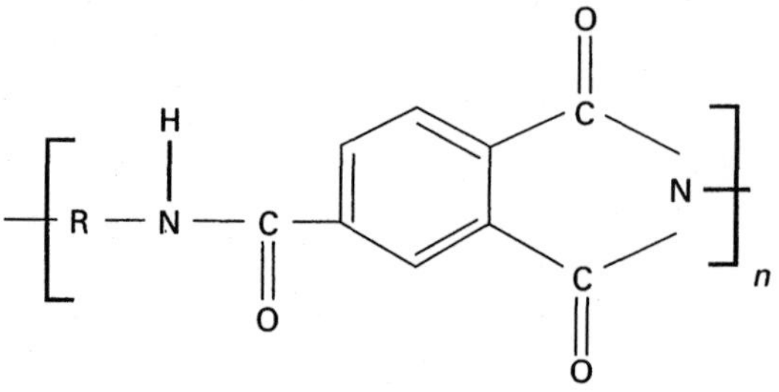

Polyamide-imide

Two other thermoplastic PIs, known as K polymers and LARC-TPI (for Langley Research Center thermoplastic imide), are generally available as prepolymers dissolved in suitable solvents. In this form, they have low viscosities, and so the fibers can be coated with their prepolymers to produce flexible prepregs. Curing, which for these polymers means imidization or imide ring formation, requires heating up to 300°C or more.

The glass transition temperature for K polymers and LARC-TPI are 250 and 265°C, respectively. Both are amorphous polymers and offer excellent heat and solvent resistance. Because their molecules are not cross-linked, they are not as brittle as thermosetting polymers. They are processed with fibers from low-viscosity solutions much like thermosetting resins; yet, after imidization, they can be made to flow and be shape-formed like conventional thermoplastics by heating above their T_g.[3,50]

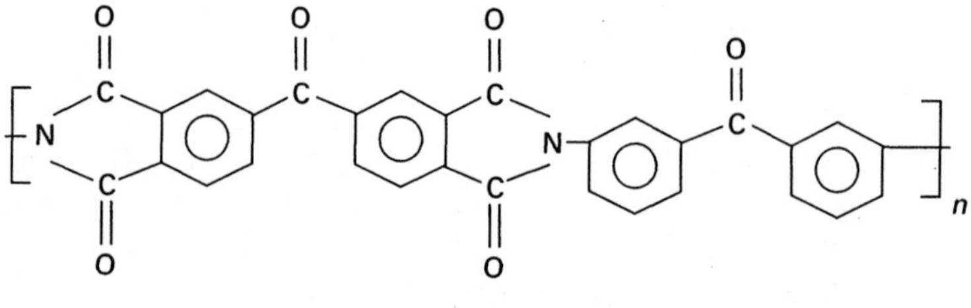

LARC-TPI

2.6.5 Thermoset Resins (Polymers)

The thermoset resins form the most widely used class of matrix materials in polymer-based composites. However, although well established, they suffer from a number of well-known limitations and it is to address these problems that current research efforts are largely devoted.

Because major consideration in the application of composite materials in primary aircraft structures is their damage tolerance, considerable attention has been given to the development of techniques for toughening thermoset matrix resins. The following approaches have been adopted in the last few years.

- A reduction in the cross-link density of the matrix, (either by adjustment of the amine-to-epoxy ratio or by the introduction of high-molecular-weight chain-extended epoxy resins)
- Development of a two-phase morphology in which there is a finely dispersed, discrete second phase comprising a low-molecular-weight, functional liquid rubber such as carboxy-terminated butadiene acrylonitrile rubber (CTBN)
- Development of a two-phase morphology in which the dispersed second phase comprises a functionally terminated thermoplastic
- Development of a two-phase morphology in a thermoset-thermoplastic combination in which the phases are cocontinuous
- Introduction of energy-absorbing interleaves, comprising rubber particles or films of high-melting thermoplastics, in laminated composite constructions.

The major resin developments have been aimed at the toughened high-performance thermoset systems (epoxy, bismaleimide, and cyanate ester), and several proprietary technologies have been developed. Where appropriate, the same basic toughening technique is being adopted across the range of thermoset resins as developers seek to gain a competitive advantage in performance in leading-edge aircraft applications.

Another major consideration of current interest is the development of systems with an improved hot-wet performance. Because moisture uptake is often associated with a loss in mechanical properties under hot or wet conditions, a goal of resin formulators has been the investigation of systems with low moisture uptake.

Another general feature of recent developments in thermoset resins has been widespread introduction of lower-viscosity systems designed to facilitate processing and fabrication.

The constraints imposed by regulatory bodies on the use of hazardous materials in the workplace has focused attention on the need to eliminate chemicals that are suspected of or known to pose toxic or carcinogenic threats to health. Considerable effort has been devoted to the development of thermoset systems that are free of methylene diamine (MDA).

2.6.5.1 Epoxy. Epoxy resins will retain their dominant position in the thermoset resin sector over the next 5 years despite competition from other resins that offer superior high-temperature performance. Although a number of factors serve to inhibit the development of radically new chemical systems, epoxy resin suppliers will continue to introduce innovations in response to the demand for improved processability and performance. Concerns regarding health and the costs involved in obtaining the necessary approval and registration of truly new chemical systems are major barriers to radical departures from established chemistry.

Starting materials for epoxy matrix are low-molecular-weight organic liquid resins containing a number of epoxide groups, which are three-membered rings consisting of one oxygen atom and two carbon atoms.

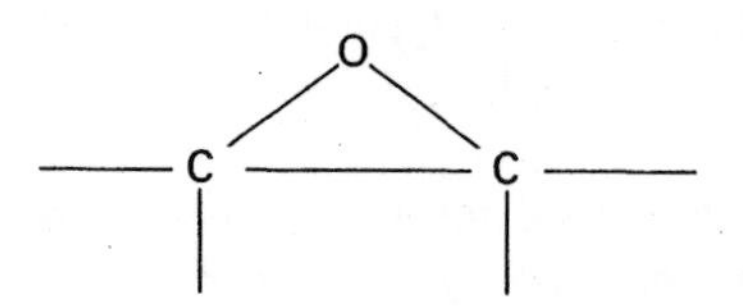

Epoxy matrix, as a class, has the following advantages over other thermoset matrices:

- A wide variety of properties because a large number of starting materials, curing agents, and modifiers are available
- An absence of volatile matters during cure
- Low shrinkage during cure
- Excellent resistance to chemicals and solvents
- Excellent adhesion to a wide variety of fillers, fibers, and other substrates
- High or low strength and flexibility
- Resistance to creep and fatigue
- Good electrical properties
- Solid or liquid resins in the uncured state
- A wide range of curative options.

The principal disadvantages are its relatively high cost and long cure time, as well as

- Resins and curative somewhat toxic in uncured form
- Heat distortion point lowered by moisture absorption
- A change in dimensions and physical properties as a result of moisture absorption
- Use (dry) limited to about 200°C upper temperature
- Difficult to combine toughness and high-temperature resistance
- A high thermal coefficient of expansion
- A high degree of smoke liberation in a fire
- May be sensitive to ultraviolet light degradation.

These resins cost more than polyesters and do not have the high-temperature capability of BMIs or PIs; but because of their advantages they are used widely.

The epoxy resins are capable of upper service temperatures in the range 125–175°C, depending on the composition. The toughened versions of epoxy resins are tailored for use up to 125°C, whereas the more rigid epoxy compositions are capable of service temperatures up to 175°C.

The significant differences in the properties of the various epoxy materials are in the moduli, strains to failure, and glass transition temperatures. The glass transition temperature, which controls the use temperature of the various epoxy materials, is high (247°C) for the brittle epoxies and much lower for the toughened epoxies (76 and 185°C). Therefore, it is clear that one compromises the use of temperature to gain toughness in these resin materials.[51–59]

2.6.5.2 Phenolic. Phenolic resins, developed over 80 years ago, are beginning to find many nonstructural uses in the automotive industry, especially for highly stressed, elevated-temperature (150–200°C) applications.

For example, a glass-reinforced phenolic is being investigated for use as an intake manifold, as a 10–15% phenolic part, and in advanced composite application in secondary structures.

2.6.5.3 Polyester. The starting material for a thermoset polyester matrix is an unsaturated polyester resin, and the curing time for polyester resins depends on the decomposition rate of the catalyst, which can be increased by increasing the curing temperature.

As in the case of epoxy resins, the properties of polyester resins depend strongly on the cross-link density. The modulus, glass transition temperature, and thermal stability of cured polyester resins are improved by increasing the cross-link density, but the strain to failure and impact energy are reduced.

Polyester resins can be formulated with a variety of properties ranging from hard to brittle to soft and flexible. Their advantages are low viscosity, fast cure time, and low cost, but their properties are generally lower than those for epoxies. The principal disadvantage of polyesters over epoxies is their high volumetric shrinkage. Although this allows easier release of parts from the mold, the difference in shrinkage between the resin and the fibers results in uneven depressions on the molded surface. These depressions (sink marks) are undesirable for exterior surfaces requiring high gloss and good appearance (e.g., class A quality in automotive body components). One way of reducing these surface defects is to use low-shrinkage (low-profile) polyester resins containing a thermoplastic component (such as polystyrene or polymethyl methacrylate).

2.6.5.4 Vinyl ester. The starting material for a vinyl ester matrix is an unsaturated vinyl ester resin produced by the reaction of an unsaturated carboxylic acid, such as methacrylic or acrylic acid, and an epoxy resin.

Vinyl ester resins possess good characteristics of epoxy resins, such as excellent chemical resistance and tensile strength, and of unsaturated polyester resins, such as low viscosity and fast curing. However, the volumetric shrinkage of vinyl ester resins is in the range 5–10%, which is higher than that of the parent epoxy resins. They also exhibit only moderate adhesive strengths compared with epoxy resins.

2.6.5.5 Bismaleimides and other thermoset polyimides. BMIs, polymerized monomer reactants (PMRs), and acetylene-terminated polyimides (ACTPs) are examples of thermosetting PIs. Among these, BMIs are suitable for applications requiring a service temperature of 127–232°C. PMR and ACTP can be used at up to 288 and 316°C, respectively. PMR and ACTP also have exceptional thermooxidative stability and show only 20% weight loss over a period of 1000 h at 316°C in flowing air.

Thermosetting PIs are obtained by addition polymerization of liquid monomeric or oligomeric imides to form a cross-linked infusible structure. They are available either in solution form or in hot melt liquid form. Fibers can be coated with the liquid imides or their solutions before the cross-linking reaction. On curing, they not only offer high temperature resistance but also high chemical and solvent resistance. One useful method of reducing the brittleness of BMIs without affecting their heat resistance is to combine them with one or more tough thermoplastic PIs. The combination produces a semi-interpenetrating network polymer, which retains the easy processability of a thermoset and exhibits the good toughness of a thermoplastic. Although the reaction time is increased, this helps in broadening the processing window, which otherwise is very narrow for some of these PIs and causes problems in manufacturing large or complex composite parts.

BMIs are the most widely used thermosetting PIs in the advanced composite industry. PMR-15 is available as prepreg roving, as tape and broad goods, and as molding compounds.

In a recent study of the comparative performance of a range of high-temperature polymers, including various thermoplastic and thermoset PIs, researchers at United Technologies Research Center (UTRC)[48] concluded that in an overall assessment of long-term high-temperature performance, thermooxidative stability, and cost, PMR-15 offered the best compromise. DuPont's linear condensation thermoplastic PI, Avimid N, was ranked second, and a new PMR-based resin, PMR-II-30 from Dexter Composites, was ranked third. Third-generation PMR-based materials are being developed by Dexter that have been reported to offer a continuous temperature of up to 370°C.

Researchers in 3M's ceramic materials department have evaluated PMR-15 composites reinforced with Nextel Al_2O_3 ceramic fibers, and potential applications in radomes are being specifically targeted. National Starch and Chemical offers a range of Thermid addition-type PIs and is currently investigating modification of the polymer chain structure in order to achieve improvements in impact resistance.

DuPont's Avimid addition PI prepregs are being used in the prototype development of U.S. fighter aircraft. Avimid K-111, with a long-term service temperature of 350°C, is designed to tolerate the high temperatures developed during supersonic flight and is forecast to account for up to 50% of the structural weight of future fighter aircraft. A new family of high-temperature thermoplastic resins that are claimed to offer easy processing and mechanical properties superior to those of current BMIs and PIs are under development. Phenolic triazine (PT) resins overcome the processing difficulties encountered with PIs and offer the flame retardance of phenolic resins. It is reported that PT resin-based composites exhibit properties similar to those of PMR-15 PI composites at high temperature (330°C). The high combustion resistance and low smoke evolution suggest potential applications in aircraft interiors.

2.6.5.6 Cyanate ester. Cyanate ester resins offer an attractive combination of properties for aerospace and electronic applications, although the potential of these materials has yet to be fully realized. Key properties of cyanate ester resins are high heat resistance, excellent hot-wet performance, low moisture absorption, and low dielectric loss. As in the case of other major high-temperature thermoset resins, a major emphasis of current research activity is on alloys and blends of cyanate esters to achieve improved toughness. Enhanced toughness is being achieved by the addition of a second-phase material which may be an elastomer or a compatible engineering thermoplastic such as PPS, PES, PEI, or polyarylate.[48] Specific systems are being designed for such processes as filament winding, pultrusion, and resin transfer molding (RTM).

A Dow Chemical cyanate ester has been used in the production of a prototype missile body section. The composite part was produced by triaxial computer-controlled braiding to form the carbon fiber reinforcement preform and RTM. The selection of braiding, in preference to filament winding, was justified by the greater ease of obtaining a uniform thickness on a tapered form and the ability to introduce 0° fibers. Although the original specification required a BMI resin, the cyanate ester was used because it met the performance requirements and offered easier processing.

2.7 METAL AND CERAMIC MATRICES

In order for a metal matrix to effectively distribute the load borne by the composite, it must adhere well to the fibers or particulates (or other forms of reinforcement) and must have good shear strength. If these conditions are met, when the strong, stiff (but usually brittle) reinforcement fibers or particulates break, plastic flow occurs at the tip of any crack in the matrix and absorbs the energy, thus reducing the stress concentration. For a fiber-reinforced MMC, if the matrix-fiber interface is sufficiently strong and the matrix is ductile, any cracks occurring tend to be deflected parallel to the fibers rather than propagating across the fibers.[6] For whiskers or for short, discontinuous fibers, the amount of load that can be transferred from the matrix to the reinforcement is less than in the case of continuous fibers because the ends of the fibers cannot contribute to the support.[9] For particle-reinforced composites, strengthening results from the particles mechanically restraining matrix deformation.[10]

Wider usage of MMCs is beginning to occur as the result of an improved understanding of the factors controlling the constituent properties, as well as those of the overall composite, and an improved properties database.

2.7.1 MMC Advantages and Disadvantages

The primary advantage of advanced composites is their ability to provide higher mechanical properties and tailored physical properties at less weight than that of conventional materials. In the case of MMCs, these advantages are available for high-temperature applications (>316°C) where PMCs are inadequate. Furthermore, because metals are more electrically and thermally conductive, MMCs can be used in heat dissipation and electronic transmission applications.

The addition of a reinforcement to metal can be characterized as follows. (1) It can be ceramic in nature with high strength and high stiffness retention at high temperature, (2) it can have physical attributes such as high heat transfer or low thermal expansion characteristics, or (3) it can exhibit high wear resistance. Being very small in size, generally less than the diameter of a human hair (0.13 mm), these reinforcements do not suffer from the flaws and defects generally attributed to bulk materials of the same chemistry.

Although MMCs with widely different properties exist, some general advantages of these materials over monolithic (unreinforced) metals and PMCs can be cited. Compared with monolithic metals, MMCs have higher strength-to-density and stiffness-to-density ratios, better fatigue resistance and superior temperature properties, lower CTEs, and better wear resistance. The advantage of MMCs over PMCs are higher temperature capability, fire resistance, higher transverse stiffness and strength, no moisture absorption, higher electric and thermal conductivity, better radiation resistance, no outgassing in a service environment, and fabricability with conventional metalworking equipment.

The current disadvantages of MMCs are those associated with higher material and fabrication costs, lack of material databases, and limited service experience. In addition, because of the complexity of these multiphase materials and their less mature technology, the costs of design, analysis, and quality control are greater. However, because of their unique engineered properties, the use of MMC systems is projected to accelerate rapidly. As this happens, the cost and experience factors will improve.

2.7.2 Aluminum Matrices

Aluminum alloy is normally used as the metal matrix with B or borsic (SiC-coated B) filaments; 6061 aluminum is used more frequently than either 2024 or 1100 (pure aluminum). The 2024 alloy provides the highest strength, the 1100 alloy has superior Charpy impact resistance, and the 6061 alloy provides a good combination of strength, toughness, and corrosion resistance.

Aluminum alloys 2024 and 6061 have also been investigated for use with SiC fiber in yarn form, as have the 7000 series aluminum alloys and the 5000 series in applications for marine systems.

2.7.3 Magnesium and Lead Matrices

Magnesium alloys, AZ91C and AZ61A, have been used as matrices with graphite reinforcement, and typical longitudinal tensile strength for a unidirectional sample of this composite was 177 MPa with a modulus of 80–90 GPa. Comparable values for monolithic magnesium are 65 MPa and 44.8 GPa.

Magnesium matrices with B fibers have been found to be a model MMC system exhibiting an excellent interfacial bond and outstanding load redistribution characteristics. Tests have shown flexural strengths of 1117 MPa at 25% filament volume and 2937 MPa at 75%. A density of 2408 kg/m^3 has been attained, along with an ultimate tensile strength of 1303 MPa (compared with 965 MPa for steel) and a flexural modulus of 241.3 GPa.

To increase the strength of lead and its alloys and enhance lead's potential in the chemical, battery, building, and bearing industries, graphite fibers were added to improve composites for possible battery and bearing applications. Systems have been produced using PAN fiber and PAN graphite yarn. Lead containing 1.5–3 wt% magnesium (to aid wetting) reinforced with 39 vol% PAN fiber ($Pb/Gr/39_f$) has been made in wire form. Failure stress has averaged 758 MPa, and modulus of elasticity, 15.9 GPa.

Mechanical properties are very promising. Where pure lead has a strength of 14 MPa, and a typical lead-based bearing alloy, 72 MPa, the composites show values of 345–1103 MPa. The modulus of elasticity of the composites is also four to eight times higher than that of the bearing alloy. Another important property, fiber density, is low; thus, a 40 vol% loading has great impact on the total density of the composite. As a result, strength- and stiffness-to-weight ratios make these lead composites similar to low-alloy, medium-strength steels.

2.7.4 Titanium Matrix

Titanium has been used as a matrix with borsic fibers. Titanium matrix materials have included Ti-3A1-2.5V and Ti-13V-10Mo-5Zr-2.5Al.

Additionally, room-temperature strengths of up to 1172 MPa have been obtained with 40–5 vol% SiC-Ti-6Al-4V (0° filament orientation). The composite modulus was 262 GPa.

Finally, A-70Ti matrix composites have also been used and had a higher room-temperature modulus of elasticity (241.3 GPa) than either stainless steel (206.9 GPa) or titanium (124.1 GPa). This advantage persisted[60] up to and including test temperatures of 599°C.

Finally, the average room-temperature tensile strength of the SiC-A-70Ti composite was 655 MPa, which is nearly identical to that of the unreinforced A-70Ti, 669 MPa. At temperatures above 427°C the composite was superior in tensile strength to unreinforced unalloyed A-70Ti and had lower tensile strength than the Ti-6Al-4V.

2.7.5 Copper Matrix

Because the NASP vehicle makes extensive use of an actively cooled skin structure, there is particular interest in materials with high thermal conductivity. Heat exchangers or actively cooled skin panels must be designed to transfer large quantities of heat quickly and efficiently from one location to another. In addition, the high temperature differences that could exist across sections of heated skin structure could lead to unacceptably high thermal stresses. For these applications, the structural materials themselves must have adequate thermal conductivity or must be protected with an actively cooled, high-thermal-conductivity barrier layer.

Fiber-reinforced copper-matrix composites are potentially useful for these applications. Copper itself has good thermal conductivity, but it is heavy and its upper-use temperature is limited by its low mechanical properties. However, the addition of graphite fibers to a copper matrix to make a fiber-reinforced composite can reduce density, increase stiffness, raise the use temperature, and significantly improve the thermal conductivity of the composite compared to that of unreinforced copper.

Additional work on high-thermal-conductivity materials addresses discontinuously reinforced copper composites. These are made by melting and casting alloys of copper and elements such as niobium. Because niobium is essentially insoluble in copper in the solid state, it is possible to use appropriate mechanical processing methods to form a material where the copper matrix contains a fine strengthening distribution of niobium particles. The resultant composite retains the high thermal conductivity of the copper matrix but is strengthened by the dispersoid distribution.

2.7.6 Aluminide Matrices (Intermetallics)

FeAl and FeCrAlY matrices are being considered for advanced gas turbine engine component applications. Both matrices have been determined to be chemically compatible with Al_2O_3. Al_2O_3/FeAl had a low bond strength (55 ± 29 MPa), whereas the bond strength of Al_2O_3/FeCrAlY was measured to be significantly higher (191 ± 35 MPa). Because the oxidation resistance of FeAl and FeCrAlY are similar and the slightly lower density of FeAl doesn't outweigh the advantages of FeCrAlY in lower CTE and higher ductility, an assessment was made to focus future efforts on the Al_2O_3/FeCrAlY composite system for practical component applications and Al_2O_3/FeAl studied as a model system.[61]

Another intermetallic, NiAl, has excellent oxidation resistance as well as the low density and high melting temperature characteristic of intermetallics. For intermetallic composites to be completely successful, the fibers must provide both toughening and strengthening, be chemically compatible with the matrix, and have a similar CTE. Fibers tested included Al_2O_3, which created brittle or poorly bonded fibers, and Mo and W, which were ductile or strongly bonded fibers.[62]

2.7.7 CMC Matrix Groups

Matrix materials of interest are categorized in four main groups[63]: glass ceramics such as lithium aluminosilicate (LAS), oxides such as alumina and mullite, nitrides such as silicon nitride (Si_3N_4), and carbides such as silicon carbide (SiC).

Much of the processing work with CMCs has been strongly influenced by efforts with other composite systems. Many of the basic techniques have been adapted to accommodate differences such as the need for higher processing temperatures, and progress has been occurring at a rapid pace.

2.7.8 Silicon Nitride Matrix

For next-generation advanced heat engine applications, materials are required that are strong, tough, oxidation-resistant, and able to withstand high temperature and high heat fluxes. To meet this demand, high-temperature fiber-reinforced CMCs are being developed. One of these composite systems contains porous Si_3N_4 matrix (RBSN) and 30 vol% aligned SiC (SCS-6) fibers.

Unidirectional and two-dimensional SCS-6/RBSN laminates exhibit high specific stiffness and strength, high toughness, notch insensitivity, and thermal shock resistance to 1400°C.[64,65] The composite also shows excellent strength and graceful failure at temperatures to 1550°C after a 15-min exposure in air. Because RBSN composites can be fabricated with a minimum of glassy phase at the Si_3N_4 grain boundaries, these composites are expected to have better creep resistance than the fully dense Si_3N_4 materials. Furthermore, matrix porosity in the SCS-6/RBSN composites can be eliminted by high-temperature hot isotatic pressing or hot-pressing methods using sintering additives.

In summary, significant matrix strength enhancement is predicted by Bhatt[65] for RBSN composites reinforced by SiC fibers of much less than 40 μm in diameter.

2.7.9 Silicon Carbide Matrix

In the development of advanced gas turbine engines for supersonic aircraft, reaction-formed SiC is being investigated as a matrix for SiC fiber-reinforced composites.[66] Investigating SiC fiber-reinforced SiC matrix composites having strength, toughness, oxidation resistance, and high thermal conductivity for structural applications at temperatures greater than 1400°C is the aim of Behrendt and his associates.[66] The advantages of reaction-formed SiC matrices are

- Liquid polymer starting material for easy incorporation of fiber reinforcement
- A high-char-yield polymer as carbon source as opposed to carbon fillers which can be difficult to distribute uniformly around small-diameter reinforcements
- Good complex shape capability
- Little or no change in dimensions during siliconization
- Dense matrix with Si or Mo-Si alloy infiltration
- Reduced free Si with Mo-Si alloy infiltration
- Close thermal expansion match between SiC fiber and matrix
- High thermal conductivity.

SiC tubes with SiC reinforcement have also been successfully fabricated by CVD-chemical vapor infiltration (CVI) techniques.

2.7.10 Other Matrices

Ceramic matrices of ZrO_2, Al_2O_3, and mullite are under development with a variety of continuous and discontinuous reinforcement materials.

2.8 TOUGHENING MECHANISMS, INTERFACES, BLENDING, AND ADHESION

2.8.1 In Situ Matrix Failure

The weakest link in PMCs is the matrix. (For simplicity, the fiber-matrix interfacial bond strength is taken to be a matrix property.) For most good designs, failure of the matrix does not cause, at least under short-term loading conditions, structural failure. In fact, the reputation PMCs have for excellent fatigue resistance is due to their ability to tolerate a significant concentration of flaws without yielding catastrophically. Typical manifestations of matrix failure are delaminations between layers and microcracks within a layer. Delaminations usually result from out-of-plane loads such as those seen in lateral impact and at free edges. Microcracking is normally due to tensile failure transverse to the fiber direction. These cracks can be generated by residual thermal stresses arising from processing, as well as by in-service loads. Matrix failures in general are material damage that may grow in size or accumulate in number under long-term loading conditions. PMC structures for long-term use in commercial aircraft, marine structures, and automobiles must be designed for damage tolerance and aging resistance. The most mature of these industries, the commercial aircraft business, bases its designs on material tests for matrix-dominated properties (e.g., notched strength, interlaminar fracture toughness, and residual compression strength after lateral impact), (see Chapter 3).

2.8.2 Heat-Resistant Polymer Alloys

Multicomponent polymer systems are called polymer alloys in the same manner as the alloys of metals. The manufacture of polymer blends that can be obtained by mixing different polymers is a simple and useful means of modifying polymers and has been adopted widely in the past for the purpose of improving the impact resistance, and so on, of general-purpose polymer materials.

On the other hand, polymers whose principal chain consists of aromatic rings, such as PPS, PES, PEEK, aramid, and polyimide, possess properties that make them ripe for development as molecular composites. Initially the work was aimed at obtaining materials with satisfactory mechanical properties by dispersing such polymers as Kevlar and polyphenylene benzobisthiazole (PBT) in other matrix resins.

Molecular composites were originally introduced as the limiting form of fiber-reinforced resins, with emphasis on improvement of the mechanical properties, especially of the modulus of elasticity. Apart from this, from the conventional approach to polymer alloys, attempts have been made to improve the resin performance by alloying two kinds of heat-resistant polymers. Examples of this are PPS-PES, polyether ketone-PES, PEEK-PEI, PBI-polyimide, PBI-polyallylate systems. In the PEEK-PEI system blend, for example, an improvement in resistance to chemicals and thermal stability can be seen from the PEI side, an improvement in resistance to heat can be seen from the PEEK side, and the fracture toughness is higher than that of either component by itself.

Modifying heat-resistant polymers, such as PES, PEEK, and polyimide, by blending aramid with these materials has shown[67] that these systems exhibit unique blend behavior that was not known in the past. Based on these observations, new material design concepts have been proposed, for example, "micro composites which form fine reinforcement *in situ*" and "the possibility of designing compatible blends from self-associative polymers."[67] These results indicate that the outlook for the future is bright.

2.8.3 Fiber-Matrix Adhesion in PMCs

As composite materials have moved to the forefront of activity in the materials community in the last two decades, there has been an increasing interest in understanding the physical and chemical mechanisms responsible for fiber-matrix adhesion as well as the effect of fiber-matrix adhesion on composite properties. Early mechanical analyses showed that when composites are fabricated and the applied load is coincident with the fiber direction, a simple rule of mixtures expression can be used to predict composite properties from the properties of the constituents. There is almost always a component of the structural loads that is in a direction at an angle to the fiber axis. In these situations, fiber-matrix adhesion is important to ensure that a maximum stress level can be maintained across the interface and from fiber to matrix without disruption.

Most early work on the development of composite materials considered fiber-matrix adhesion a necessary condition to ensure good composite properties. The majority of effort was concentrated on increasing fiber-matrix adhesion through the use of surface treatments and coatings.[68] It is very desirable to have a testing method for measuring the adhesion between fiber and matrix that can provide a reproducible, reliable method for investigating and measuring fiber-matrix adhesion.

As a result, Drzal and Madhukar[69] drew the following conclusions.

- The embedded single-fiber fragmentation test has been shown to be both a predictor of changes in composite properties brought about by changes in fiber-matrix adhesion and a useful research tool that not only quantitatively measures fiber-matrix interfacial shear strength but also identifies failure modes.
- Comparison of results from single-fiber and composite tests indicates that two key parameters must be obtained from single-fiber tests in order to explain composite property data: level of adhesion and failure mode (interfacial or matrix).
- Fiber-matrix adhesion affects composite properties in different ways depending on the state of stress created at the fiber-matrix interphase.[69]

Studies similar to the one on the interfacial properties of a fiber-reinforced composite were conducted by Yue and Cheung.[70] They considered the existing theories for predicting stress distribution at the fiber-matrix interface and the applicability of these theories to real composite systems. In using the single-fiber pullout test, they determined the interfacial parameters and showed that the failure and pullout process, and hence the properties of a fiber-reinforced composite, can be characterized in terms of its interfacial parameters.

It is apparent that although the existing theories can predict the general shape of the interfacial stress distribution, more remains to be done before the stress distribution in practical composite systems (which often have interphase regions between the fiber and the matrix) can be understood, fiber-reinforced composites can be characterized properly, and the full potential of composites as advanced materials can be realized.[70]

2.8.4 Fiber-Matrix Adhesion in MMCs

Adherence of the matrix to the reinforcement is an important factor in determining composite properties.[12] The optimum adherence of the matrix to the reinforcement may involve some degree of chemical reaction at the interface. However, if the reaction becomes too extensive, the matrix-reinforcement bond may weaken because of the properties of any compounds formed, and it may also degrade the reinforcement. As examples, boron fiber-reinforced aluminum was observed to fail during fracture through debonding at the matrix-fiber interface. However, when this composite was exposed to a 500°C treatment in an argon atmosphere, the room-temperature composite strength was found to increase, with the failure then occurring in the fiber itself. Analytical studies of the interface showed that reaction products had formed during the high-temperature exposure. These products served to strengthen the matrix-fiber interface.[13] A more detailed study of the as-prepared B/Al composite found that the interface between the B and Al was covered by Al_2O_3 and that this oxide decreased the interfacial bonding. The 500°C heat treatment caused the oxide to be replaced by several B/Al compounds that provided much better interfacial strength.[14]

In general, to obtain an optimum matrix-reinforcement interface, it is necessary that the matrix at least wet the reinforcement. If the work of adhesion between the matrix and the reinforcement exceeds the surface tension of the matrix, wetting is assumed to occur. Experimentally, wetting is said to have occurred when a liquid spreads spontaneously on a solid surface.[16] Three approaches have been taken to improve wetting: pretreatment of the reinforcement, modification of the matrix through alloying, and coating the reinforcement.[17] Each of these approaches has met with some success. For example, Gr fibers must be cleaned chemically to remove contamination that interferes with wetting by metals.[16] Lithium additions apparently improve the wetting of aluminum to its reinforcement by weakening the $A1_2O_3$ film covering the aluminum.[16] Coatings on the reinforcement that enhance wetting apparently work also by disrupting the oxide on the molten metal matrix during fabrication of the composite.[17]

It is usual to modify the interface by applying a fine coating to the fibers. For example, SiC/SiC can be affected by creating a carbon-rich zone by precoating Nicalon fibers with a thin (<1-μm) layer of pyrolytic carbon via CVD. Other materials like BN have been used because they have the advantage of being more resistant to oxidation than carbon.

The reinforcement of titanium alloys with continuous-fiber SiC increases their strength at ambient and elevated temperatures.[71,72] In addition, SiC reduces the density of the titanium composite relative to the monolithic alloy. These improvements in properties render SiC-reinforced titanium alloys attractive for aerospace applications where even small reductions in weight or increases in elevated-temperature strength are significant.

Unfortunately, a potential deterrent to the use of SiC-reinforced titanium in structural applications is the formation of brittle reaction products at the fiber-matrix interface.[73,74]

Understanding the nature of the fiber-matrix reaction products can lead to their control or elimination.[75] Fiber-matrix reactions in two titanium alloy matrix composites, Ti-1100/SCS-6 SiC and β21S/SCS-6 SiC were studied.

Ti-1100 is a near-alpha titanium alloy designed for high-temperature applications (up to 593°C). β21S, on the other hand, is a metastable beta titanium alloy that was developed for improved oxidation resistance.

In the study by Rhodes and Spurling[75] there was an implication that beta titanium alloys like β21S react more rapidly with SCS-6-type SiC than do alpha titanium alloys like Ti-1100, a near-alpha alloy.

2.8.5 CMC Adhesion and Interfaces

In all CMCs the interface between the matrix and the fiber is very important, and the following situations must be avoided.[76]

- A chemical interaction occurring between fiber and matrix that results in fiber embrittlement and loss of strength
- Bonding between like materials such as SiC to SiC and Al_2O_3 to Al_2O_3

2.8.5.1 Interface/interphase. The nature of the fiber-matrix interface in a CMC is critical to the performance of the composite. The interface (or interphase where a finite thickness or another phase is present) must be understood and the properties controlled to achieve desired composite properties, particularly toughness.

The increase in toughness in CMCs then is a function of the matrix, the fiber, and the interphase. In MMC or PMCs, by contrast, load transfer is very important because of the large difference in stiffness between the matrix and the fibers or particulates. In ceramic matrices, however, this mechanism is probably of secondary importance.

Both the fiber and the matrix in a CMC have low strain to failure (typically <1%). With strong interfacial bonding, a crack propagating through the matrix to a fiber can continue unabated through the fiber. With weaker interface bonding, debonding can occur under stress and results in toughening by one or more mechanisms.

It is generally believed that fiber pullout and crack deflection are the primary toughening mechanisms in fiber-reinforced CMCs (Figure 2.5).

For optimization of fiber pullout, it is desirable to have a low interfacial shear strength (poor bond) between the fiber and the matrix and also a long critical length. Prestressing and microcracks may also contribute to higher toughness. Toughening mechanisms and stress states in brittle composites are two areas where research is required.

2.8.6 Toughening Mechanisms

2.8.6.1 Thermosets and thermoplastics. A technique has been developed[77] whereby laminated composites of carbon fibers and either PMR-15 or LARC RP-46 PIs (these PIs are thermosetting) can be toughened selectively by incorporating thin layers of Matrimid 5218 thermoplastic PI at the interfaces between the plies to form gradient semi-interpenetrating microstructures (Figure 2.6). The toughening of the composites is accompanied by acceptably small decreases in strength and stiffness. Thick laminates of the toughened composites can be fabricated easily at 3.45 MPa. These materials are expected to be useful, for example, in components of aircraft engines, where they could be exposed to high temperatures for short times.

The emerging morphological picture of these toughened composites is that of the gradient, single-phase, semi-interpenetrating form illustrated in Figure 2.6. Such a microstructure is likely to provide not only increased toughness and resistance to solvents but also better fatigue endurance and resistance to creep because of enhanced physical cross-linking arising from entanglement of molecules. By incorporating the toughening material at the most desired location and by providing the cross-linked semi-interpenetrating microstructure, one avoids having to toughen the bulk resin and thereby avoids the attendant increase in cost and difficulty in processing.[77]

In another new material development thermoplastics are being groomed as an alternative to thermosets for high-strength structural composites. In this work General Electric and Ford are collaborating in a program to use cyclic (ring structure) thermoplastics as the matrix resin in RTM and other liquid polymer processes. The idea is to take advan-

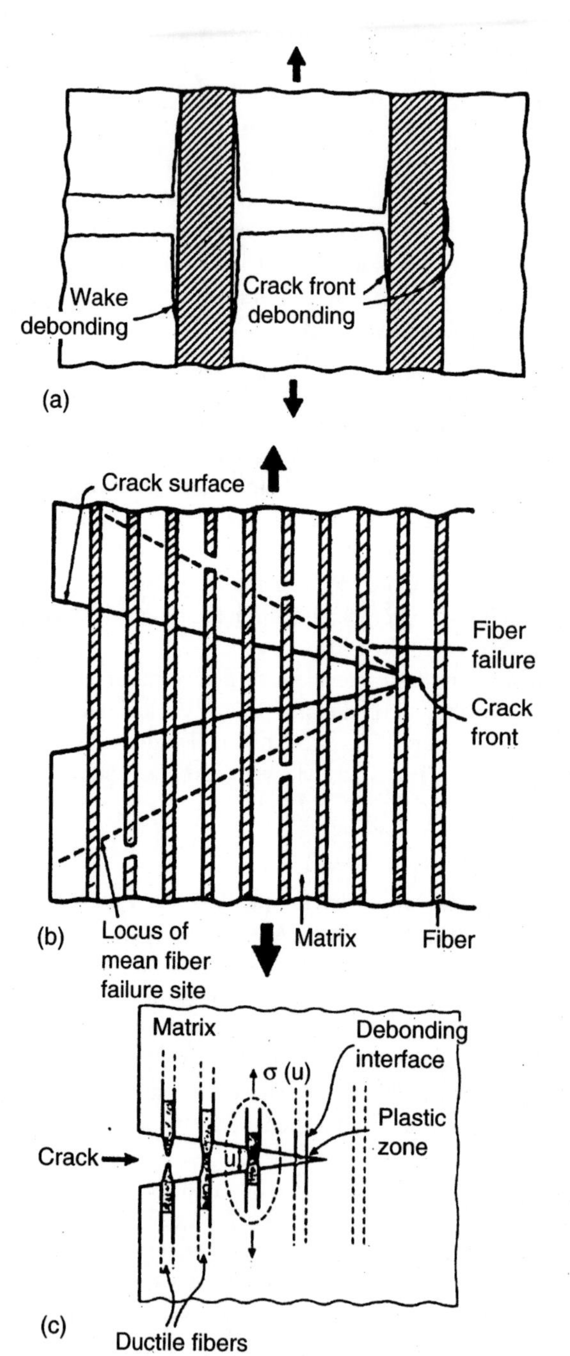

Figure 2.5. Schematic diagram showing composite toughening processes. (*a*) Crack front and wake bonding; (*b*) fiber failure and pullout; (*c*) crack propagation in a ductile phase-reinforced brittle matrix composite.[83]

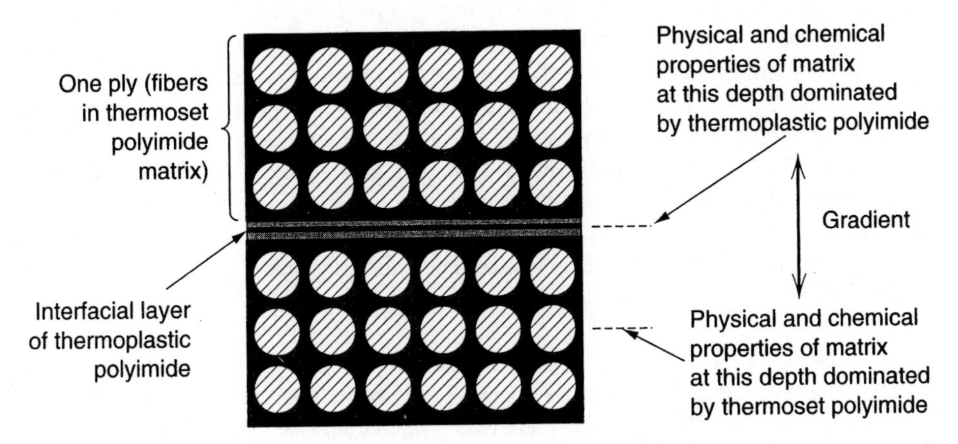

Figure 2.6. Thermoset and thermoplastic PIs form a single phase with a gradient between them at the interface between plies.[77]

tage of the relatively low viscosity of the cyclic molecule to penetrate the reinforcement and then break the ring to create a rigid chain-type molecule. The starting molecule is a cyclic ester, and the polymer is semicrystalline with a melting point above 204°C.

The heat-activated reaction takes about 10 s. The reaction is endothermic and so avoids nonuniform properties and potential degradation of the resin from exothermic heat, especially in thick sections, and the "cured" polymer, being thermoplastic, can be easily recycled.

Other matrix-toughening programs using thermoplastic PIs are discussed in reference 78, and reference 79 covers improvements in both thermoset (PMR family, 480°C) and thermoplastic (220°C) PIs that make them even better for high-temperature applications.

2.8.6.2 MMCs. Continuous-fiber composites usually have increased longitudinal strength and modulus compared with the unreinforced matrix. However, the toughness depends on the matrix and the fiber-matrix interface. Compared with the unreinforced state, composites with ductile metal matrices have lower toughness and composites with brittle matrices (aluminides and ceramics) have slightly higher toughness. In all composites, however, toughness values are much lower than the minimum values considered acceptable for critical structural components. This low toughness can be explained by the fracture mechanisms in the composite and the energy absorbed during fracture.[80]

In ductile MMCs cracks initiate at the fiber-matrix interface, often associated with a brittle reaction layer. In composites with brittle matrices (intermetallic or ceramic), the critical flaw size is usually smaller than the interfiber spacing, and cracks initiate in the matrix rather than at the fiber-matrix interface. Toughening of these composites involves increasing the energy absorbed in crack growth. In the toughening of continuous-fiber-reinforced composites the fiber-matrix interface strength plays a critical role. Ideally, for toughness in brittle matrix composites at room temperature, the fiber-matrix interface should be strong enough to ensure load transfer across the interface until matrix fracture and then be weak enough to allow debonding, thus preventing the composite from behaving as a monolithic material as shown in Figure 2.5a.[81] At elevated temperatures when brittle matrices become ductile, or for ductile MMCs at all temperatures, toughness is obtained when the fiber-matrix interface is strong enough to allow for high load transfer.

Current composite toughening mechanisms are based on microcrack branching in the matrix and on debonding at the fiber-matrix interface.[82] The interface strength, frictional load transfer, and fiber pullout stresses should not be too high. Fibers provide crack bridging behind the crack front either by pullout (Figure 2.5b) or by ductile extension (Figure 2.5c).[83,84] (Note that for other properties such as creep strength, a high interface strength is required.[85]) Small-diameter fibers and a reduction in grain size favor increased toughness in brittle composites.[86] On a coarser scale, ductile fibers, layers, or particles have been used to toughen ceramics, cermets, and intermetallic compounds, such as Ti aluminides.[87–92]

2.8.6.3 CMC. Three different forms of secondary phases are incorporated in CMCs: particulates, whiskers, and continuous fibers. The mechanisms by which these secondary phases affect reinforcement of ceramic composites is an area of intense interest and remains controversial, although a number of theories have been proposed.

Under suitable circumstances, particulates, whiskers and platelets can produce modest improvements in the fracture toughness of ceramic matrices through toughening mechanisms such as crack bridging, compressive matrix residual stresses, and transformation toughening (Figure 2.7). These mechanisms all increase the energy demand of cracks either before they grow or during growth, thus toughening the material.

The major toughening mechanism that has been exploited in CMCs is crack deflection by interaction of an advancing crack with the dispersed phase of hard refractory particulates or whiskers. When a crack approaches a hard secondary phase with a weakened interface, it is deflected from its original plane. This mechanism is associated with the stress field surrounding the dispersed secondary phase caused by thermal expansion and/or elastic modulus mismatch. As noted earlier, it is preferred that the CTE of the dispersed phase be higher than that of the matrix; this produces a residual radial tension and a tangential compression strain state in the matrix, which tends to divert the crack around the dispersed particle (Figure 2.7). In the case of randomly dispersed whiskers, because of differences in orientation of the local whiskers, the crack front twists; and this twist de-

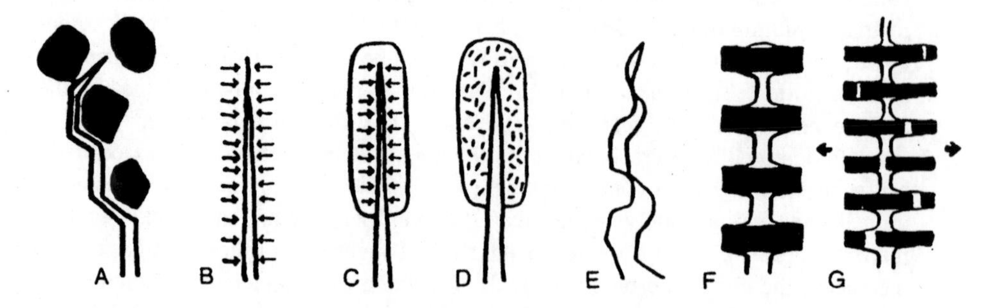

Figure 2.7. Schematics of toughening mechanisms in CMCs. (*a*) Crack deflection; (*b*) compressive matrix residual stress; (*c*) transformation toughening; (*d*) matrix microcracking; (*e*) frictional locking; (*f*) crack bridging; (*g*) fiber pullout.

flection is more important than the tilt deflection in determining the toughness enhancement. Thus, it has been suggested that tilting and twisting of the advancing crack front is an effective toughening mechanism. Theoretical analysis of this mechanism predicts that the shape of the dispersed particles is important in determining the actual level of fracture toughness increase. The major conclusions of the theoretical analysis can be summarized as follows.

- Rods (whiskers) are more effective toughening agents than disk-shaped particles, which are more effective than spherical particles.
- The degree of toughening reaches an asymptotic value of about 4 for high-aspect-ratio rods at a volume fraction of ≤ 0.15.

The absolute size of the rods does not affect the toughness calculations. However, the maximum benefit in strength is achieved with finer rods.

It is widely recognized that reinforcement with continuous fibers represents the most attractive option for the enhancement of fracture toughness, but the fabrication of such composites presents special difficulties. Theoretical analysis of composites in which the fibers are unidirectionally aligned is well developed. Crack growth in the matrix phase requires that energy be supplied from the elastic stored energy near the crack tip. In the presence of fibers, energy is absorbed as the crack grows, and the supply of energy to the crack tip growth region is limited. In addition, as a crack propagates through the matrix, the bonds between the matrix and the fiber are broken, and the subsequent separation of the matrix causes the fibers to be pulled out.

The total work of fracture is the sum of two terms: one representing fiber-matrix debonding and the other fiber pullout, which is the dominant term. Both terms are directly proportional to the volume fraction of fiber reinforcement; therefore, high volume fractions of fine-diameter fibers provide the highest degree of toughness benefit. The balance between strength and toughness is critically dependent on the magnitude of the interfacial shear strength between the fiber and the matrix; a high value causes suppression of matrix cracking and therefore favors high strength, whereas a low value is desirable for high toughness.

Fiber surface treatment is being investigated as a route for tailoring the interface such that fiber pullout is encouraged as the matrix begins to crack. The application of a 1-μm thickness coating of carbon on SiC whiskers has been shown to enhance the fracture toughness of whisker-reinforced SiC composites, with values of 10–12 MPa $m^{1/2}$ being reported. However, oxidation of the carbon results in a serious loss of strength at high temperature. Data from a study by researchers at Oak Ridge National Laboratory indicate a 40% reduction in strength at 1000°C compared with that at room temperature. In another study, sponsored by the U.S. Navy, surface treatment of whiskers by coating them with magnesium in a vapor deposition process and subsequent ignition in an oxygen atmosphere to produce a magnesium oxide layer has been described. The layer of magnesium protects the fibers from oxidation and provides enhanced fracture toughness.

In addition to crack deflection and fiber pullout (Figure 2.7), it is suggested that fibers, and to some extent whiskers, can prevent crack propagation by bridging a crack and thereby holding the two faces together. Obviously, the highest mechanical properties are obtained when the applied stress is parallel to the fiber alignment. The strength of the composite is at a minimun when the stress is perpendicular to the fiber direction. For most engineering applications, more complex stressing situations apply and multidirectional reinforced composites are required. Composites based on layered laminated structures, with different fiber directions in each layer, two-dimensional woven mats, and three-dimensional braided fibers, are being extensively evaluated. The theoretical analysis of these composites is not well developed, although the analytical techniques developed for plastic composites are being adopted. In plastic composites strain transfer between the matrix and the reinforcement is critical, and the matrix can be treated as a homogeneous continuum. This is not the case in ceramic composites. Not only is prediction of the performance of CMCs difficult, but it is also often difficult to produce composites in which the fibers are selectively placed to provide the desired stiffness and load-bearing properties. The actual performance of ceramic composite materials often deviates significantly from theoretical predictions, and this is commonly attributed to the introduction of flaws during fabrication and to changes in microstructure associated with the presence of the second-phase material.[93–98]

2.8.6.4 Modeling MMCs and CMCs. Numerous studies have been conducted and models prepared to examine the variety of toughening mechanisms, interface criteria, and fracture properties (fatigue) in order to understand the effects of reinforcements and matrices.

One study examined the toughening mechanism of continuous-fiber-reinforced CMCs using SiC (CVD) fiber-reinforced glass model materials.[99] The interfacial shear properties of the composite were evaluated by the single-fiber pullout method, and the toughening due to crack bridging of fiber was directly observed by a special lightening method.

The major mechanism of toughening of the composite was achieved by the stretching of bridging fibers between the matrix crack surfaces. It was found that at least two toughnesses should be defined and clearly distinguished by experiment to understand the toughness of fiber-reinforced ceramics: (1) Crack initiation toughness, K_{Ic}^{c}, and (2) crack propagation toughness, ΔK^{c}.

Crack initiation toughness was dependent on only the elastic modulus ratio between composite and matrix and on the critical stress intensity factor of the matrix. Whereas crack propagation toughness appears as *R*-curve behavior, in measuring full extension of *R*-curve behavior it is very important to understand the total toughening process of the composites.[100–104]

In another study by Masuda and Tanaka[105] fatigue fracture mechanisms were examined for three commercially fabricated aluminum matrix composites containing SiC_w and SiC_p under rotating bending fatigue conditions. They concluded that

1. The fatigue strength of SiC_w/A2024 and SiC_p/A357 composites increased in comparison to that of the matrix alloys, whereas for SiC_p/A356 composite at a volume fraction of 20% fatigue strength was equivalent to that of the matrix alloys.
2. Fatigue cracks initiated by slip deformation mechanism from the matrix for SiC_w/A2024 and SiC_p/A357 composites, whereas for SiC_p/A356 composite fatigue cracks initiated from many voids situated in the matrix.
3. For SiC_w/A2024 composites dimple patterns were predominant on the fatigue crack surfaces. It is suggested that the fatigue crack propagation rates would be retarded

by SiC whiskers for a 10%-SiC_w/A2024-T6 composite. On the other hand, for SiC_p/A356 and SiC_p/A357 composites dimple patterns and SiC particles were rarely observed on the fatigue fracture surfaces. It is suggested that fatigue crack propagation rates would hardly be influenced by SiC particles but that roughness-induced crack closure would occur near crack origin.[106–108]

REFERENCES

1. National Materials Advisory Board. 1992. High performance synthetic fibers for composites. NMAB-458.
2. Chou, T.-W., R. L. McCullough, and R. B. Pipes. 1986. Composites. *Sci. Am.* October:193–203.
3. Mallick, P. K. 1992. *Composite Materials Technology,* 2nd ed. Hanser, Munich.
4. Peters, S. T. 1992. Advanced composite materials and processes. In *Handbook of Plastics, Elastomers, and Composites,* 2nd ed., ed. C. Harper, 5.1–5.73. McGraw-Hill, New York.
5. *New Materials Society: Challenges and Opportunities,* vol. 2. New Materials Science and Technology, Bureau of Mines, 1990.
6. English, L. K. 1987. Fabricating the future with composite materials. Part I: The basics. *Mater. Eng.* 4(9):15–21.
7. England, J., and I. W. Hall. 1986. On the effect of the strength of the matrix in metal matrix composites. *Scr. Metall.* 20(5):697–700.
8. Shepard, L., and A. Klein. 1986. On the road with composites. *Adv. Mater. Process.* 130(6): 36–41.
9. Schoutens, J. E. 1982. Introduction to metal matrix composite materials, MMC no. 272. DOD Metal Matrix Composites Information Analysis Center, Santa Barbara, CA.
10. Weeton, J. W., D. M. Peters, and K. L. Thomas. 1987. *Engineers Guide to Composite Materials.* ASM, Metals Park, OH.
11. Schwartz, M. M. 1992. *Composite Materials Handbook,* 2nd ed. McGraw-Hill, New York.
12. Cohen, M. 1987. Progress and prospects in metallurgical research, In *Advancing Materials Research,* ed. P. A. Psaras, and H. D. Langford, 51–110. National Academy Press, Washington, DC.
13. Kyono, T., I. W. Hall, and M. Taya. 1986. The effect of isothermal exposure on the transverse properties of a continuous fibre metal-matrix composite. *J. Mater. Sci.* 21(12):4269–80.
14. Hall, I. W., T. Kyono, and A. Diwanji. 1987. On the fibre/matrix interface in boron/aluminum metal matrix composites. *J. Mater. Sci.* 22(5):1743–48.
15. Petrasek, D. W., D. L. McDanels, L. J. Westfall, et al. 1988. Fiber-reinforced superalloy composites provide an added performance edge. *Met. Prog.* 130(2):27–31.
16. Delannay, F., L. Froyen, and A. Deruyttere. 1987. The wetting of solids by molten metals and its relation to the preparation of metal-matrix composites. *J. Mater. Sci.* 22(1):1–16.
17. Mortensen, A., J. A. Cornie, and M. C. Flemings. 1988. Solidification processing of metal-matrix composites. *J. Met.* 40(2):12–17.
18. Partridge, P. G., and C. M. Ward-Close. 1993. Processing of advanced continuous fibre composites: Current practice and potential developments. *Int. Mater. Rev.* 38(1):1–24.
19. Kubel, E. J. *Adv. Mater. Process.* 132(4):47–54.

20. Metal matrix composites: Technology and industrial applications, 1990. Innovation 128, Paris.
21. Higgins, R. A. 1986. *Properties of Engineering Materials,* Hodder and Stoughton, London.
22. Majumder, B. S., A. H. Yegneswaran, and P. K. Rohatgi. 1984. *Mater. Sci. Eng.* 68:85.
23. Badia, F. A., and P. D. Merica. 1971. *AFS Trans.* 79:347.
24. Nieh, T. G., and D. J. Chellman. *Scr. Metall.* 18:925.
25. Das, S., T. K. Dan, S. V. Prasad, et al. *J. Mater. Sci. Lett.* 5:562.
26. Divecha, A. P., C. R. Crowe, and S. G. Fishman. 1978. In *Failure Modes in Composites IV,* ed. J. A. Cornie and F. W. Crossman, 406. AIME, Warrendale, PA.
27. Divecha, A. P., S. G. Fishman, and J. V. Foltz. 1979. The enigma of the eighties: Environment, economics, energy. *SAMPE J.* 2:1433.
28. Hasson, D. F., S. M. Hoover, and C. R. Crowe. 1985. *J. Mater. Sci.* 20:4147.
29. Aylor, D. M., and P. J. Moran. 1985. *J. Electrochem. Soc.* 6:1277.
30. El Baradie, M. A. 1990. Manufacturing aspects of metal matrix composites. *J. Mater. Process. Technol.* 24:261.
31. Girot, F. A., J. M. Quenisset, and R. Naslain. 1987. *Compos. Sci. Technol.* 30:155–184.
32. Wada, H. 1989. Silicon nitride and silicon carbide ceramic whiskers synthesis and phase stability. *Proc. Int. Symp. Adv. Struct. Mater., 27th Ann. Conf. Metall.,* ed. D. Wilkinson, pp. 149–56, Montreal, Canada. Pergamon Press, Oxford.
33. Milewski, J. V., F. D. Gac, J. J. Petrovic, et al. 1985. *J. Mater. Sci.* 20:1160.
34. Niwano, K. 1991. Properties and applications of SiC whiskers. In *Silicon Carbide Ceramics 2,* ed. S. Sōmiya and Y. Inomata, 99–116. Elsevier, London.
35. Akuyama, M., and M. Yamamoto. 1991. Silicon carbide whiskers (Toka-whiskers) and their application. In *Silicon Carbide Ceramics 2,* ed. S. Sōmiya and Y. Inomata, 117–138.
36. *High-Tech Mater. Alert.* 1988. Vol 5, p. 10, June 1988.
37. Lankford, J. 1989. Ceramics research at the Southwest Research Institute. *Ceram. Bull.* 68: 1418–23.
38. Birchall, J. D., J. A. A. Bradbury, and J. Dinwoodie. 1985. Alumina Fibers. In *Handbook of Composites,* ed. W. Watt and B. V. Perov, 115–55. North-Holland, Amsterdam.
39. Lock, J. 1990. *Prof. Eng.* April:21.
40. Kohara, S. 1990. *Mater. Manuf. Process.* 5:51.
41. Knowles, K. 1992. *Ceramic Matrix Composites,* 95–7. Sterling, London.
42. Millberg, L. S. 1987. The search for "ductile ceramics." *J. Met.* 39:10–13.
43. Prasad, S. V., and P. K. Rohatgi. 1987. Tribological properties of Al alloy particle composites. *J. Met.* 39:22–26.
44. Oh, S. Y., J. A. Cornie, and K. C. Russell. 1987. Particulate wetting and metal: Ceramic interface phenomena. Industrial Liaison Progress Report 6-44-87, Department of Material Science and Engineering, MIT, Cambridge, MA.
45. *High-Tech. Ceram. News.* 1988. Vol. 1, August 1988.
46. Kamat, S. V., J. P. Hirth, and R. Mehrabian. 1989. Mechanical properties of particulate-reinforced aluminum matrix composites. *Acta Metall.* 37:2395–402.
47. Kladnig, W. F. 1987. Fracture behavior of duplex Al_2O_3-ZrO_2 ceramics. *Mater. Chem. Phys.* 18:181–91.
48. Polymer matrix matrices: Technology and industrial applications, 1990. Innovations 128, Paris.
49. Nguyen, H. X., and H. Ishida. 1987. Poly(aryl-ether-ether-ketone) and its advanced composites: A review. *Polym. Compos.* 8:57.

50. Clemens, S., T. Hartness. 1988. Thermoplastics for advanced composites. *ASM/ESD Conf.* pp. 425–31, Dearborn, MI.
51. Siebert et al. 1973. *28th Soc. Plast. Inst. Reinforced Plast./Compos. Conf.,* February 1973.
52. Scola, D. A. 1992. United Techologies Research Corporation, unpublished work.
53. Almen, G. R., et al. 1988. *SAMPE J.* 33:979–89.
54. Browning, G. 1973. *28th Soc. Plast. Inst. Reinforced Plast./Compos. Conf.,* February 1973.
55. Pike, R. A., and R. Nowak. 1975. NASA Report ER-134763.
56. Nowak, R. 1975. Air Force Report AFML-TR-74-196.
57. Odigiri, N., T. Muraki, and K. Tobukuro. 1988. *SAMPE J.* 33:272–83.
58. Ciba-Geigy. 1985. High impact resin system. Product Data Sheet R-6376.
59. Fiberite/ICI. 1991. Data Sheet HY-E 1377-25.
60. Jech, R. W., and R. A. Signorelli. 1979. Evaluation of silicon carbide fiber/titanium composites. NASA Lewis Research Center Report TM-79232, Cleveland, OH. N79-31349.
61. Draper, S. L., D. J. Gaydosh, and A. Chulya. 1991. Tensile behavior of alumina-reinforced FeAl and FeCrAlY. *4th Ann. HITEMP Rev.,* CD-91-57303, pp. 42-1 to 42-14, October 1991, Cleveland, OH.
62. Bowman, R. R., R. D. Noebe, J. Doychak, et al. 1991. Effect of interfacial properties on the mechanical behavior of NiAl based composites. *4th Ann. HITEMP,* CD-91-57073, pp. 43-1 to 43-13, October 1991, Cleveland, OH.
63. Bray, D. J. 1989. Reinforced alumina composites. In *Composite Applications—The Future Is Now,* ed. J. T. Drozda. SME, Dearborn, MI.
64. Bhatt, R. T. 1988. The properties of silicon carbide fiber-reinforced silicon nitride composites. In *Whisker- and Fiber-Toughened Ceramics,* ed. R. A. Bradley, et al. 199–208. ASM International, NASA TM-101356.
65. Bhatt, R. T. 1991. Status and current directions for SiC/SiN composites, *4th Ann. HITEMP Rev.,* CD-91-57179, pp. 67-1 to 67-11, October 1991, Cleveland, OH.
66. Behrendt, D. R., and R. F. Dacek. 1991. Modeling of the effects of microporous carbon composition on free silicon in reaction-formed silicon carbide. *4th Ann. HITEMP Rev.,* CD-91-56189, pp. 74-1 to 74-5, October 1991, Cleveland, OH.
67. Imai, Y., and S. Nakata. 1993. Development of new heat-resistant polymer alloys. Journal Publications Research Service Report JST-93-040L, pp. 35–39.
68. Drzal, L. T., and P. J. Herrera-Franco. 1990. Composite fiber-matrix bond tests. In *Engineered Materials Handbook: Adhesives and Sealants,* vol. 3, 391–405. ASM International, Metals Park, OH.
69. Drzal, L. T., and M. Madhukar. 1993. Fibre-Matrix adhesion and Its relationship to composite mechanical properties. *J. Mater. Sci.* 28:569–610.
70. Yue, C. Y., W. L. Cheung. 1992. Interfacial properties of fibre-reinforced composites. *J. Mater. Sci. Lett.* 11(13):3843–55.
71. Metcalfe, A. G., ed. 1974. *Interfaces in Metal Matrix Composites,* vol. 1, 67–123. Academic Press, New York.
72. Smith, P. R., and F. H. Froes. 1984. Developments in titanium metal matrix composites. *J. Met.* 36(3):19–26.
73. Brewer, W. D., and J. Unnam. 1983. Metallurgical and tensile property analysis of several silicon carbide/titanium composite systems. In *Mechanical Behavior of Metal-Matrix Composites,* ed. J. E. Hack and M. F. Amateau, 39–50. Metallurgical Society, Warrendale, PA.

74. Smith, P. R., F. H. Froes, and J. T. Cammett. 1983. Correlation of fracture characteristics and mechanical properties for titanium-matrix composites. In *Mechanical Behavior of Metal-Matrix Composites,* ed. J. E. Hack and M. F. Amateau, 143–168. Metallurgical Society, Warrendale, PA.
75. Rhodes, C. G., and R. A. Spurling. 1991. Fiber/matrix interface reactions in SiC reinforced titanium alloys. In *Developments in Ceramic and Metal-Matrix Composites,* ed. K. Upadhya, 99–113. Minerals, Metals, and Materials Society, Warrendale, PA.
76. Bashford, D. 1990. A review of advanced metallic and ceramic materials suitable for high temperature use in space structures. *Acta Astronaut.* 22:137–44.
77. Toughened high-temperature thermoset composites. NASA Technical Brief, September 1993, pp. 94–98.
78. Monks, R. 1992. In advanced composites, RTM is focus of resin development. *Plast. Technol.* June:31–37.
79. Studt, T. 1992. Polyimides: Hot stuff for the '90s. *Res. Dev.* August:30–34.
80. Smith, P. R., C. G. Rhodes, and W. C. Revelos. 1989. In *Interfaces in Metal-Ceramics Composites,* ed. R. Y. Lin, et al, 35–58. Minerals, Metals and Materials Society, Warrendale, PA.
81. Evans, A. G. 1986. In *Ceramic Microstructures '86: Role of Interfaces,* ed. J. A. Pask, and J. A. Evans, 775–94, Plenum Press, New York.
82. Gupta, V., A. S. Argon, and J. A. Cornie. 1989. *J. Mater. Sci.* 24:2031–40.
83. Hayhurst, D. R., F. A. Lechie, and A. G. Evans. 1991. *Proc. Roy. Soc.* London A434:369–81.
84. Evans, A. G. 1988. *Mater. Sci. Eng.* A105:65–75.
85. Lu, T. C., J. Yang, and Z. Sou. 1991. *Acta Metall. Mater.* 39:1883–90.
86. Grisaffe, S. J. 1990. *Adv. Mater. Process.* 1:43–94.
87. Dève, H. E., M. J. Maloney. 1991. *Acta Metall. Mater.* 39:2275–84.
88. Lu, T. C., A. G. Evans, and R. J. Hecht. *Acta Metall. Mater.* 39:1853–62.
89. Xiao, L., Y. S. Kim, R. Abbaschian, et al. 1991. *Mater. Sci. Eng.* A144:277–85.
90. Dève, H. E., A. G. Evans, G. R. Odette, et al. 1990. *Acta Metall. Mater.* 38:1491–1502.
91. Bose, A., and J. Lankford. 1991. *Adv. Mater. Process.* 7:18–22.
92. Roe, K. T. V., G. R. Odette, and R. O. Ritchie. 1992. *Acta Metall. Mater.* 40:353–61.
93. Chiu, H.-P., S. M. Jeng, and J.-M Yang. 1993. Interface control and design for SiC fiber-reinforced titanium aluminide composites. *J. Mater. Res.* 8(8):2040–53.
94. Curtin, W. A. 1993. Ultimate strengths of fibre-reinforced ceramics and metals. *Composites* 24(2):98–102.
95. Becher, P. F., C. H. Hsueh, and P. Angelini. 1988. Toughening behavior in whisker-reinforced ceramic matrix composites. *J. Am. Ceram. Soc.* 71:1050–61.
96. Wei, G. C., and P. F. Becher. 1985. Development of SiC-whisker-reinforced ceramics. *Am. Ceram. Soc. Bull.* 64:298–304.
97. Buljian, S. T., et al. 1988. Dispersoid-toughened silicon nitride composites. Final Report, GTE, ORNL. Contract DE-ACO5-84OR21400, ORNL/Sub85-22011/1.
98. Corbin, N. D., G. A. Rossi, and P. M. Stephan. 1987. Making ceramics tougher. *Mach. Des.* July 23:84–89.
99. Kagawa, Y., and T. Kishi. 1990. Toughening mechanism of fiber-reinforced ceramic matrix composites. *Proc. Fr.-Jpn. Sem. Compos. Mater.: Process. Use, Database,* Tokyo University.
100. Phillips, D. C. 1974. *J. Mater. Sci.* 9(11):1847–54.
101. Prewo, K. M., and J. J. Brennan. 1980. *J. Mater. Sci.* 15(2):463–68.

102. Prewo, K. M., and J. J. Brennan. 1982. *J. Mater. Sci.* 17(4):1201–06.

103. Marshall, D. B., and A. G. Evans. 1985. *J. Am. Ceram. Soc.* 68(5):225–31.

104. Kagawa, Y., N. Kurosawa, T. Kishi, et al. 1989. *Ceram. Eng. Sci. Proc.* 10(9/10):1327–36.

105. Masuda, C., Y. Tanaka, and M. Fukazawa. 1990. Fatigue fracture mechanisms of SiC whiskers or SiC particulates reinforced aluminum matrix composites. *Proc. Fr.-Jpn. Sem. Compos. Mater.: Process. Use Database,* Tokyo University, Japan.

106. Nail, V., J. K. Tein, and R. C. Bates. 1985. *Int. Met. Rev.* 20:275.

107. Logsdon, W. A., and P. K. Liaw. 1986. *Eng. Fract. Mech.* 24:737.

108. Shang, J. K., W. Yu, and R. O. Ritchie. 1988. *Mat. Sci. Eng.* A102:181.

BIBLIOGRAPHY

ARSENAULT, R. J. Strengthening and deformation mechanisms of discontinuous metal matrix composites. In *Key Engineering Materials,* vol. 79–80, M. A. Taha and N. A. El-Mahallaway, 265–78. Trans Tech, Aldermannsdorf, Switzerland, 1993.

BENGISU, M., AND T. INAL. Whisker toughening of ceramic materials: A review. Los Alamos National Laboratory, 1993. LA-SUB-93-11/2.

COOKE, T. F. Inorganic fibers—A literature review. *J. Am. Ceram. Soc.* 74(12):2959–78, 1991.

COVEY, S. J. Study of fiber volume fraction effects in notched unidirectional SCS-6/Ti-15V-3CR-3Al-3SN Composite, NASA-CR-191165, 1993. NAS 1.26:191165, E-8116.

DAVIES, R., ED. *World Aerospace Technology '92.* Sterling, London, 1992.

FILATOVS, G. J. Graphite fiber textile preform/copper matrix composites, NASA-CR-193797, 1993. NAS 1.26: 193797.

KAUTE, D. A. W., H. R. SHERCLIFF, AND M. F. ASHBY. Delamination, fibre bridging and toughness of ceramic matrix composites. *Acta Metall. Mater.* 41(7):1959–70, 1993.

KNAUER, B. Construction and analysis of high performance composites. *Compos. Struct.* 24:181–91, 1993.

LOFTIN, T. A. Metal matrix composites materials for manufacturing. In *Composites Applications—The Future Is Now,* ed. T. J. Drozda, 59–70. SME, 1989.

MABUCHI, M., T. IMAL, K. KUBO, ET AL. New fabrication procedure for superplastic aluminum composites reinforced with Si_3N_4 whiskers or particulates. *Adv. Compos. Mater. Conf. Proc.,* pp. 275–82, Detroit, MI, September/October 1991.

MAI, Y. W. *Int. Workshop Adv. Inorg. Fibre Technol.,* Melbourne, Australia, August 1992. Sydney University, Centre for Advanced Materials Technology. ARO-30744.1-MS-CF.

MILEWSKI, J. V. Whiskers and short fiber technology polymer composites. *Polym. Compos.* 13(3): 223–36, 1992.

MITTNICK, M. A., AND J. MCELMAN. Continuous silicon carbide fiber reinforced metal matrix composites. In *Composites Applications—The Future Is Now,* ed. T. J. Drozda, 91–110. SME, Dearborn, MI, 1989.

OPSCHOOR, A. High-performance fibers in thermoplastic composites. *SME 10th Compos. Mfg.,* January 1991, Anaheim, CA. SME EM91-105.

PASCHAL, D. P. Application of metal matrix composites to military aircraft. In *Composites Applications—The Future Is Now,* ed. T. J. Drozda, 71–90. SME, Dearborn, MI, 1989.

PHILLIPS, L. N., ED. *Design with Advanced Composite Materials,* Design Council. Springer-Verlag, London, 1989.

Proc. Fr.-Jpn. Sem. Compos. Mater.: Process. Use Database, Tokyo University, 1990.

Proc. 6th Jpn.-U.S. Conf. Compos. Mater., June 1992, Orlando, FL. Technomic, Lancaster, PA.

USTUNDAG, E., R. SUBRAMANIAN, R. VAIA, ET AL. *In situ* formation of metal-ceramic microstructures, including metal-ceramic composites, using reduction reactions. *Acta Metall. Mater.* 41(7):2153–61, 1993.

WHITTAKER, D., ED. *Advanced Materials Technology International* 1992. Sterling, London, 1992.

3

Composite Modeling, Finite Element Analysis, and Design

3.0 INTRODUCTION

Composite performance is dependent on the relationships among the matrix, the reinforcing material, and the interface. For instance, prevailing models suggest that optimum performance is achieved with strong interfacial bonds in metals and polymer materials but that a weak interfacial bond is preferred in a ceramic. Both the shape of the reinforcing surface and the use of coatings on the reinforcing materials are two successful methods that have been used to optimize composite performance, especially coatings that improve the compatibility between matrix and reinforcement.

3.1 COMPOSITE MODELING

Composite modeling shows that platelets with higher aspect ratios should produce an even greater improvement in modulus. Currently, there is no universally accepted model showing how reinforcement materials interact with various matrices to achieve optimum performance. This is particularly true in ceramic materials, where extensive effects are being made to increase toughness. In terms of product size and shape, however, the following reinforcements have been tentatively identified as being optimum.

- For ceramics—discontinuous small fibers (whiskers) and thin, small-diameter platelets
- For metals—discontinuous small whiskers and thin platelets
- For polymer materials—large-diameter fibers and whiskers and large platelets.

The conventional approach for optimizing properties has been to tailor the matrix to the reinforcement. Composite materials incorporating reinforcements were made primarily by adapting C or SiC fibers or whiskers, which were available in fairly narrow size ranges, to function with as many matrix materials as possible. Now, researchers are starting to tailor discontinuous-reinforcement materials by controlling their composition, crystal structure, shape, and size.

By tailoring the reinforcement, it is possible to achieve optimum performance from the matrix. Some advantages of this approach are elimination of compatibility problems between reinforcement materials and the host matrix; optimization of mechanical properties such as strength, modulus of elasticity, and toughness; and ease of fabrication of composite materials.

The ability to control platelet or whisker size is critical to optimizing composite performance. As an example, platelets ranging from 20 to 70 μm in diameter should produce superior metal matrix moduli compared with much larger platelets in a similar matrix. Laboratory tests indicate that platelets with diameters greater than 50 μm are unsuitable as reinforcement for alumina. However, platelets ranging from 30 to 40 μm in diameter have produced matrix toughness values equal to those obtained with SiC whiskers. Even smaller platelets ranging from 20 to 30 μm in diameter should produce substantially better toughness values. Large platelets (>100 μm in diameter) should be an ideal reinforcement for polymer matrices.

Elastic moduli over 110 GPa are reported for platelet-reinforced aluminum matrix composites, which is a substantial increase over those for unreinforced materials. Preliminary tests of platelets in polymer matrices indicate that the platelet configuration is particularly effective. Results have proven that the material mixes well and bonds well in such matrices.

Not only have modeling techniques been developed for reinforcements and their matrices, but models have also been developed for and are now able to predict the high-temperature ultimate strength of a continuous-fiber metal matrix composite (CFMMC) material. The model developed by Barbero and Kelly[1] can predict how fiber-matrix interface properties, fiber diameter, and preexisting fiber failures will affect the high-temperature ultimate strength of a composite. The finite element model (FEM), developed in the form of a representative volume element (RVE), was used to calculate the time-dependent stress field surrounding a fiber break. Variables included in the calculation were process-related parameters such as fiber diameter, fiber-matrix interface strength, and interface roughness. The significant results of their work were that the FEM provided a time-dependent approximation of δ that cannot be determined analytically. The model results, coupled with a knowledge of the fiber break density that may exist in the composite before a load is applied, can be used to calculate values such as the time-dependent factor of safety and time to failure at a given applied load.[1–3]

While recent advances in composite modeling have made great contributions to understanding of the basic behavior of composite materials, most models consider only composites with reinforcing fibers of uniform diameter arranged in periodic architectures. The nonuniform microstructures of many types of composites, namely, vapor-deposited, cast, and ceramic matrix, do not resemble these models. Accounting for these discrepancies is believed to be important in order to predict accurately such mechanical properties as transverse strength, thermal residual stresses, fracture toughness, and fatigue crack initiation. Futhermore, the effects of fiber distribution and cross-sectional geometry on the deformation of composites have been studied[4,5] by FEM and were found to be significant. Thus, measuring the degree of nonuniformity is important, and the work in this area has been initiated and undertaken by Everett and Chu[6] who studied Nicalon SiC/ZrO_2 titanate composite materials.

Wider use of CMCs requires the development of advanced structural analysis technologies. The use of an interactive model to predict the time-independent reliability of a component subjected to multiaxial loads was investigated by Palko.[7] The deterministic three-parameter William-Warnke failure criterion served as the theoretical basis for the reliability model. The strength parameters defining the model were assumed to be random variables, thereby transforming the deterministic failure criterion into a probabilistic criterion. The ability of the model to account for multiaxial stress states with the same unified theory is an improvement over existing models. The new model was coupled with a public-domain finite element program through an integrated design program, which allowed a design engineer to predict the probability of failure of a component.

Modeling is used not only to assist in the design of composite structural components. For example, a recently initiated research project is intended to develop accelerated aging models that can predict the long-term strength and durability of composite materials planned for use in high-speed civil transport (HSCT).

3.2 FINITE ELEMENT METHOD AND FINITE ELEMENT ANALYSIS

3.2.1 Introduction

The finite element method is now a widely accepted technique for solving a wide variety of physical states in arbitrary structures. The method is a product of the digital computer age and thus is available on a wide range of hardware, from small 16-bit-word personal computers to extremely large, high-powered vector-processing machines. The visual display unit gives fast direct access to the hardware and has been exploited with respect to input and output data for the benefit of the user.

Finite element (FE) data are necessarily complex. Computer models of element meshes representing the often complicated shape of a structure can be very time-consuming to generate accurately, and so much software effort has been expended in creating designer- and user-friendly data generation and resultant viewing packages. The new viewer-designer-user equipment complements the older media of printers and digital plotters in this respect.

3.2.2 Merging FEM and Composites

FEM has become established in cases where critical molding areas of composite parts[8] cannot be estimated closely enough using empirical methods and where complex or critical loads are involved.

During mechanical analysis and structure optimization using FE computer programs, the generally inhomogeneous, anisotropic, and nonlinear material behavior of composites can be numerically approximated. Measurement of the necessary material properties for nonlinear analyses is extremely cost-intensive and time-consuming. Compared with the construction of prototypes, however, savings in time and costs can be achieved, in particular where changes in design and/or parameter studies are required.

In addition to a nonlinear FE program, as well as the preprocessors and postprocessors required for geometric generation and the display of results, such calculations are based on laws that define the material behavior of the composite with regard to the major parameters loading rate and temperature.[8]

Design analysis of a laminated composite structure almost invariably requires the use of computers to calculate stresses and strains in each ply and to investigate if the structure is "safe." For a simple structure, such as a plate or a beam, the design analysis can be performed on a microcomputer. A large variety of microcomputer software based on the lamination theory is available for this purpose.[9] If the structure and/or the loading are complex, it may be necessary to perform the design analysis on a large computer. Many user-oriented general-purpose FE packages, such as MSC-NASTRAN, NISA, and ANSYS, have the capability of combining the lamination theory with the FE codes. Many of these packages can calculate in-plane as well as interlaminar stresses, can incorporate more than one failure criterion, and contain a library of plate, shell, or solid elements with orthotropic material properties.

Although FEAs for both isotropic materials and laminated composite materials follow the same procedure, the problem of preparing input data and interpreting output data for composite structures is much more complex than in the case of metallic structures. Typical input information for an isotropic element includes its modulus, Poisson's ratio, and thickness. Its properties are assumed to be invariant in the thickness direction. An element for a composite structure may contain the entire stack of laminas. Consequently, in this case, the element specification must include the fiber orientation angle in each lamina, lamina thicknesses, and the location of each lamina with respect to the element midplane. Furthermore, the basic material property data in a fiber-reinforced composite material include four elastic constants: longitudinal modulus, transverse modulus, major Poisson's ratio, and shear modulus. Thus, the amount of input information even for a static load analysis of a composite structure is quite large compared with that for a similar analysis of a metallic structure.[9]

3.2.3 New Advances in FEM and Composites

Designers working with composite materials for structural applications use a mixed bag of hardware. It includes three primary categories: simple computer-aided design (CAD) and computer-integrated manufacturing (CIM) programs that are basically focused on geometry; computer-aided engineering (CAE) programs that are FE- or finite difference-based; and CAE programs that are not geometry-specific, such as laminate analysis and micromechanics.

A shipbuilder in Rhode Island may use Auto-CAD to draw a composite rudder, while an aerospace manufacturer in California uses a CADAM three-dimensional (3-D) wire frame database along with FE pre- and postprocessing to design C/Ep spacecraft equipment.[10]

In 1988 alone at least 57 composite design programs in each of the three categories mentioned were in existence, and since that time a number of new programs have been developed to serve the composites designer, including second-generation FEA packages.

First-generation composite analysis programs that ignored transverse shear deformation effects and hence did not provide accurate results for thick structures and laminated structures have been replaced by second-generation programs. By providing new element types, including those based on first-order shear deformation theory, and improving the basic shell element formulation, the newer FEA programs are better suited to the needs of the composites designer.

The newer FEA programs also offer more nonlinear capabilities because advances in computer hardware have reduced the cost of nonlinear analysis, and interactive graphics packages have made nonlinear codes more accessible. Such advanced nonlinear capabilities as contact, creep, and snap-through buckling analyses, as well as static normal mode dynamics, transient dynamic frequency response, shock spectrum, steady-state heat transfer, and buckling analyses, are available.

New enhancements provide composite capabilities: graphic displays of ply layers, fiber orientation, and section thicknesses for easy verification of the model.

FEA's forte is in estimating the structural performance of a new design before the prototype stage is reached. This capability allows the design to be fine-tuned for optimum performance in fewer design steps while reducing manufacturing and engineering costs. Reference 10 lists a series of laminate analysis, computer-integrated engineering, and group technology software programs that allow experienced composite designers and engineers to capture the total design intent and process knowledge and associativity required to support future designs.[10,11]

3.2.4 Other Uses of FEM

3.2.4.1 Consolidation of continuous-fiber MMCs. Goetz, Kerr, and Semiatin[12] investigated the consolidation of MMCs via hot isostatic pressing (HIP) of Ti-24Al-11Nb/SiC foil-fiber-foil lay-ups using FEM metal flow analysis. For this purpose, the deformation pattern for various fiber arrangements was determined using representative unit cells to describe extremes in behavior. For a given fiber architecture, the consolidation time was found to be heavily dependent on the ratio of HIP pressure to average flow stress and the friction conditions at the matrix-fiber interface. The specific influence of material properties such as the rate sensitivity of the flow stress appears to enter only as a second-order effect. The FEM solutions were used to construct HIP diagrams delineating temperature-time-pressure combinations for full composite consolidation, and laboratory trials on subscale foil-fiber coupons were used to validate the FEM predictions.[12]

3.2.4.2 Continuous-fiber-reinforced CMCs. Mital, Murthy, and Chamis[13] developed micromechanics theories for fiber-reinforced CMCs [SiC fiber- and reaction-bonded Si_3N_4 matrix (RBSN)] and corroborated experimental results with three-dimensional FEA analyses.[14,15] To analyze the behavior of these CMCs, a unique and novel substructuring technique was developed. This technique has four levels of substructuring—from laminate to ply, to subply, and then to fiber. The fiber is substructured into several slices, and the micromechanics equations are applied at the slice level. Although the basic philosophy can be applied to the analysis of any continuous-fiber-reinforced composite, emphasis was on the development of computer code to specifically analyze and simulate aspects unique to CMCs. The aspects of interest included varying degrees of interfacial bonding around the fiber circumference and accounting for the fiber breaks and local matrix cracking that may lead to rapid degradation of the interphase at higher temperatures as a result of oxidation. In addition, the multilevel substructuring technique can account for different fiber shapes and integrate the effects of all these aspects on composite properties and response and in turn provide greater detail in stress distribution.[13]

3.2.4.3 Numerical modeling ceramic-metal materials. Williamson and associates developed a FE numerical modeling approach that simulates residual stress development during gradient material production. Based on a simple cylindrical geometry, comparisons were made between a graded Al_2O_3/Ni material and a sharp Al_2O_3/Ni interface during cooling from fabrication temperatures to ambient conditions.[16]

The results indicated that significant reductions in residual stress are possible; however, the amount of stress reduction is strongly dependent on the specimen geometry and simulated boundary conditions. For the materials and temperature history assumed, plasticity effects are important in achieving residual stress reductions and should be included in analyses.[16]

3.2.4.4 Numerical modeling for TPCs. Beaussart, Pipes, and Okine have proposed a numerical technique for simulating the formation of fiber-reinforced thermoplastic sheets. The primary material considered was a composite system consisting of aligned, long, discontinuous fibers in a thermoplastic matrix. This composite was described at its forming temperature as an oriented assembly of discontinuous fibers suspended in a viscous matrix, and the constitutive equation for an equivalent viscous anisotropic medium was employed. Finally, a methodology for adapting existing FE programs to model various sheet-forming processes for advanced TPCs was developed.[17]

3.2.4.5 FEA tools for sporting equipment. FEA locates areas of high stress so engineers can eliminate them. FEA packages have successfully been used with software tool programs in the design of intricate broad seams of Kevlar and Mylar in sails used in an international America's Cup racing yacht. Designs have been developed for bobsleds and for bicyles used in the 1996 Summer Olympics which incorporate unibody frames of carbon-reinforced epoxy structures, as well as an oar blade for a rowing team.

3.2.4.6 Fastener and bolted composite joints. Numerous FEA programs have been formulated, tested, and implemented in determining methods to accurately predict, for example, multifastener thick-composite joint strength and the stress field distribution around the pin-loaded hole in both single-hole and multifastener joints.[18]

In another practical example typical applications of mechanically fastened joints in composite aircraft structures are the skin-to-spar and skin-to-rib connections in, for example, a wing structure, the wing-to-fuselage connection, and the attachment of fittings. Examples of such joints from the fighter aircraft JAS 39 Gripen are shown in Figure 3.1.

The conclusions drawn from this study[19] were

1. In the future, three-dimensional modeling will probably not be a commonly used tool in the design of mechanically fastened joints because of its complexity and the need for computer power.

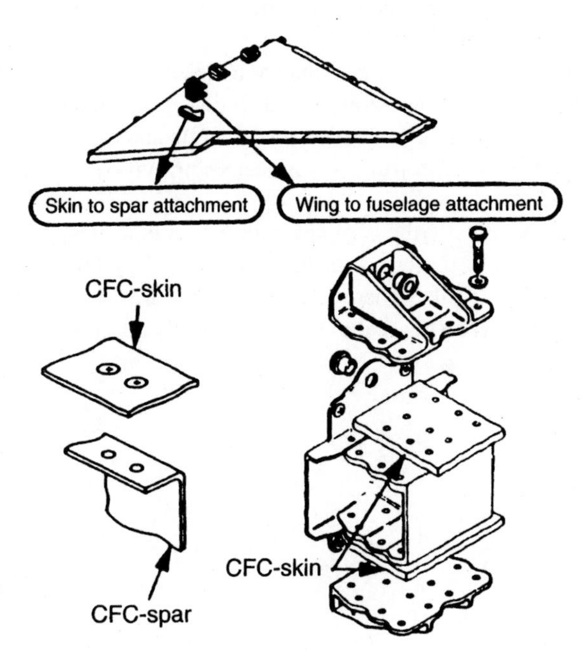

Figure 3.1. Examples of bolted composite joints in JAS 39 Gripen.[19] CFC, Carbon-fiber composite.

2. There will always be a need for fast, simple methods such as design diagrams and analytically based computer programs for the large volumes of joints in an aircraft.
3. Advanced three-dimensional FE models can be useful only in the design of critical primary joints.
4. For more advanced modeling and failure prediction one should generate correction factors for, for example, through-thickness effects, which can be used together with two-dimensional analyses in the design.

3.2.4.7 Discontinuous reinforced SiC-aluminum. DiGiovanni, Rosenberg, and Boyce[20] decided that they would use FEA to determine the locations and sizes of rivet assemblies so that all rivets carried approximately the same load in transferring the wing bending moment and transverse shear from a MMC wing core to the steel clevis. The MMC composite wing was fabricated from SiC_w/2124-T6/20% V/O and incorporated a 17-4 PH steel clevis mechanically fastened to the MMC wing with 10 hybrid steel bushing/titanium rivet fastener assemblies specifically designed for the composite wing.

Two-dimensional FE aerothermal and structural analyses were performed to determine the likely temperature distribution along chordwise stations and rivet loads. Analyses of spanwise stations included the steel clevis MMC as well as the monolithic MMC. Thermal trade-off studies were performed on a similar all-titanium wing. In all cases, the spanwise and chordwise temperatures exhibited gradients, whereas the through-wing-thickness temperature was isothermal at any instant in time. The heat transfer properties of the steel clevis resulted in higher temperatures at all clevis locations compared to the all-titanium wing. Because the through-thickness temperature distribution was isothermal, the MMC wing segment enclosed by the clevis experienced higher temperatures than the all-titanium wing for the same flight trajectory.

The design was adjusted within the limits of geometry to result in failure at the most outboard rivet location. This critical wing failure location was subsequently confirmed during high-temperature static testing. The distribution of rivets and sizes evolved from a two-dimensional FEA of the wing. The wing was modeled for purposes of structural analysis as shown in Figure 3.2.

It is not easy to describe mathematically the complex nature of composites, and designers use computer software systems to perform structural analysis of systems manufactured from laminated composite materials. There are more than 600 FE programs available to the structural engineer, but when the features are specifically addressed to ap-

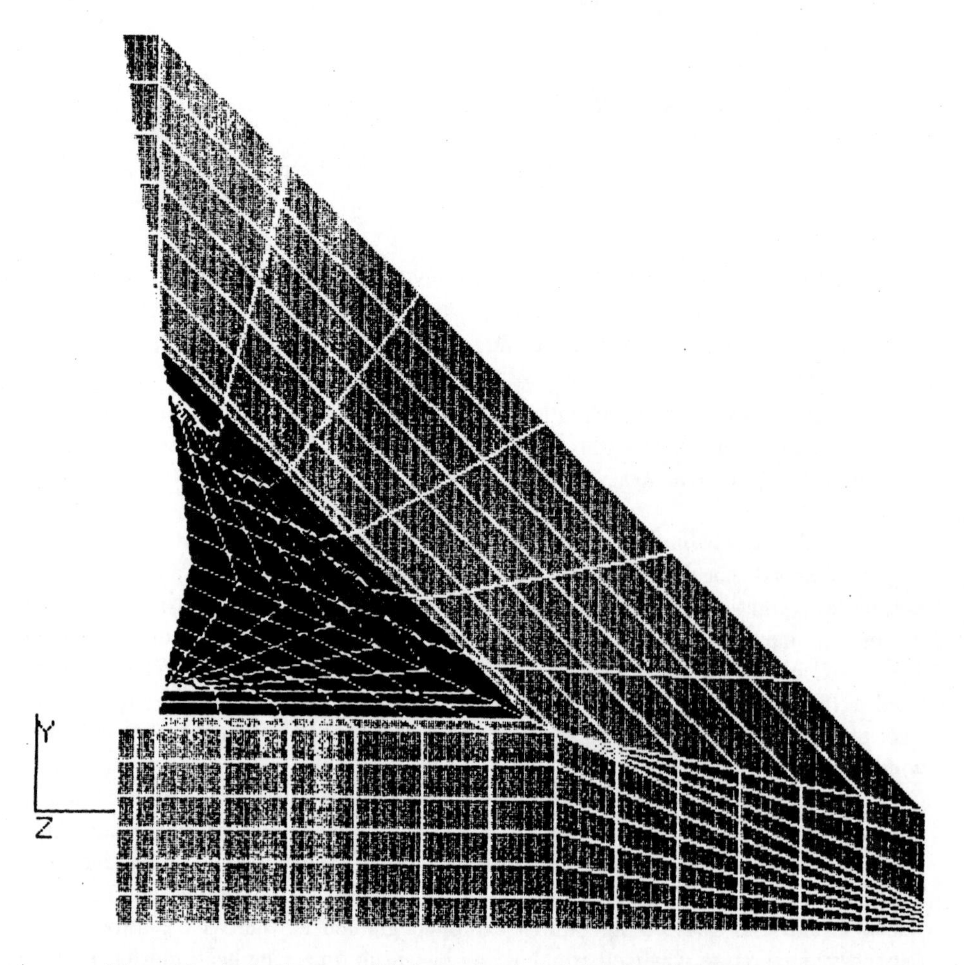

Figure 3.2. Finite element model of riveted steel clevis metal matrix wing.

plications of the structural analysis of composites, the number of suitable software systems decreases to about 10.

It is also possible to obtain software that has static, dynamic, and thermal capabilities and to be able to analyze hundreds of laminas with many different ply materials in the composite. Some software can evaluate the laminate stiffness and compliance matrices and calculate the laminate equivalent elastic and physical engineering constants.

3.2.5 Conclusions

The design of composite material structures and structural components is dominated by an increasing use of FEM. Because of the increased complexity of analyses coming from the use of composites of various materials, their architectures, and their failure modes, it is necessary to use FE computer code whenever possible. For instance, the U.S. Air Force Wright Aeronautical Laboratory has dozens of computer codes that have been developed by contractors and have been codified, and a large number are available for use.[21] In fact, Tsai[22,23] has selected and arranged a multitude of analytical formulas and empirical relations into a relatively simple format to enable designers and engineers, analysts, and others to design composite structures and components. The goal of this effort is to enable individuals with little prior knowledge of composites to have at hand the wherewithal and confidence to use composites or, at least, to examine their use as an alternative to metallic structures for various applications. Schoutens[24] has achieved the same goal for MMCs.

3.3 DESIGN METHODOLOGY

Good structural design is a compromise between design requirements and constraints. Criteria established by design and analysis determine the degree of emphasis to be placed on each factor considered. The designer now has at hand an additional class of materials that are very stiff and strong in a single direction. Moreover, these materials can be prepared so different fiber directions can be exploited, thus producing tailorable component properties. This requires that the load paths be well defined and reliability at least as great as that of the structure it replaces. Moreover, design factors must be considered, combined with experience and common sense, which can result in designs that demonstrate the potential of fiber-reinforced composite materials. To assist the designer in focusing on the principles of good design with composites and to promote design concepts based on the philosophy of "thinking composites," a set of design guidelines, adapted from June and Lager,[25] is presented in Table 3.1.

According to Graesser, Zabinsky, Tuttle, et al.,[26] it appears that composite materials offer an unprecedented flexibility that allows the engineer to *design the material.* However, many of the current design philosophies are based only on simple laminate orientations, such as quasi-isotropic or balanced and symmetric lay-ups. These restrictions adversely affect the design of a composite structure in the sense that they do not allow the design engineer to take full advantage of the composite.

As a result, Graesser, Zabinsky, Tuttle, et al. developed a design methodology for a laminated composite stiffened panel subjected to multiple in-plane loads and bending moments. Design variables included the skin and stiffener ply orientation angles and

TABLE 3.1. Recommendations for Good Composite Design Practice[25]

Fibers should be in the direction of the principal stresses.
Load must be transferred to the composite through shear.
Holes should not be cut in highly loaded regions of fiber-reinforced composites.
Isotropic unreinforced or reinforced metals[a] should be used in complex stress areas.
The highest payoffs are in areas where the load path is unidirectional and well defined.
Shear effects must be considered in compression stability analysis.
The matrix material, fiber, and fiber volume fraction should be chosen to optimize desired properties.
Allowance should be made for reasonable analysis capability.

[a] An example of an isotropic reinforced metal is SiC particulate- or whisker-reinforced Al.

stiffener geometry variables. Optimum designs were sought that minimized structural weight and satisfied mechanical performance requirements. Two types of mechanical performance requirements were placed on the panel, maximum strain and minimum strength. Minimum-weight designs were presented documenting that the choice of mechanical performance requirements caused changes in the optimum design. The effects of lay-up constraints that limited the ply angles to user-specified values, such as symmetric or quasi-isotropic laminates, were also investigated, and UWCODA, a computer program originally developed to optimize the lay-up of flat composite panels but expanded to include stiffened panel design solutions, was prepared and used.[26]

Rogers[27] recently discussed lessons learned and experiences with the V-22, the revolutionary tilt-rotor aircraft that has required extensive use of composites to decrease weight and increase cargo-carrying capacity. He claims that more reliable analytical techniques are needed to aid designers and manufacturing engineers. Advanced FEA systems, the laminate theory that takes into account out-of-plane deformations, and fracture mechanics methods for evaluating matrix cracking effects are some of these reliable analytical techniques. Manufacturing approaches also need to be improved both to lower cost and increase reliability. Designers, says Rogers, must "focus on integration of function and structure with an emphasis on tooled surfaces for next assembly mating." Closer tolerance fits can be ensured, for example, on a wing by molding the skin to the inside surface with the rib flange tooled to the same interface. Incorporating the complex interior features on the tool simplifies assembly and reduces labor intensity. The payoff in adapting these practices is higher usable strengths and lower costs. Composite designers have long been aware of some of the ideas discussed by Rogers; however, paying close attention to them in the future seems likely to become even more important.

One of the major differences in composite and metal design approaches lies in the material selection phase, and design with CMCs is no exception. In fact, CMC material selection is more involved because of the stringent conditions a CMC component is required to meet.

Another way in which a composite design differs from a metal design stems from design allowables. Unlike a metal designer, a composite designer does not have a "standard" for design allowables. In fact, there is no standard design methodology. Different organizations tend to follow different approaches for composite design allowables based on individual experiences. In order to compare different designs and improve on the quality of CMC applications, a common standard for design approaches and allowables is necessary.

As a result, no single design methodology in detailed form is applicable to all types of CMC materials; some general trends exist, and "classical" design techniques differ from CMC designs.[28]

There are many differences between a "classical" design and a CMC design at every phase of the design process, which includes (1) geometry and attachment, (2) load spectra and performance criteria, (3) environmental effects and metal properties, (4) inspection and repair, and (5) cost analysis.

REFERENCES

1. Barbero, E. J., and K. W. Kelly. 1993. Predicting high temperature ultimate strength of continuous fiber metal matrix composites. *J. Comp. Mater.* 27(12):1214–35.
2. Nimmer, R. P., R. J. Bankert, E. S. Russell, et al. 1989. Micromechanical modeling of fiber/matrix interface effects in SiC/Ti metal matrix composites. ASM Mater. Week, October 1989, Indianapolis, IN.
3. Rosen, B. W. 1964. *AIAA J.* 2(11):1985–91.
4. Brockenbrough, J. R., et al. 1991. Deformation of metal-matrix composites with continuous fibers: Geometrical effects of fiber distributions. *Acta Metall. Mater.* 39(5):735.
5. Brockenbrough, J. R., et al. 1992. A reinforced material model using actual microstructural geometry. *Scr. Metall.* 27:385.
6. Everett, R. K., and J. H. Chu. 1993. Modeling of non-uniform composite microstructures. *J. Comp. Mater.* 27(11):1128–44.
7. Palko, J. L. 1993. Interactive reliability model for whisker-toughened ceramics. Cleveland State University, OH. NAS 1.26:191048, NASA-CR-191048.
8. Menges, G., M. Michaeli, E. Baur, et al. 1988. Computer-aided plastic parts design for injection moulding. *RPI Workshop Compos. Mater. Struct.: Mater. Eng. Des.*, pp. 10.1–10.5, London. IOM, ARO 26588.1-EG-CF.
9. Mallick, P. K. 1993. *Fiber-Reinforced Composites,* 2nd ed. 471, App. A.10. Marcel Dekker, New York.
10. Leonard, L. 1992. *Adv. Compos.* March 4:44.
11. Paulsen, W. C. 1992. Unraveling the mysteries of FEA. *Mach. Des.* March 26:64–67.
12. Goetz, R. L., W. R. Kerr, and S. L. Semiatin. 1993. Modeling of the consolidation of continuous-fiber metal matrix composites via foil-fiber-foil techniques. *J. Mater. Eng. Performance* 2(3):333–40.
13. Mital, S. K., P. L. N. Murphy, and C. C. Chamis. 1993. Ceramic matrix composites properties/microstresses with complete and partial interphase bond. *38th SAMPE Symp. Exhibit.*, May 1993. Anaheim, CA. NASA TM-106138, E-7816.
14. Chamis, C. C., et al. 1990. METCAN verification status. NASA TM-103119.
15. Lee, H.-J., et al. 1991. METCAN updates for high temperature composite behavior: Simulation/verification. NASA TM-103682.
16. Williamson, R. L., J. T. Drake, and B. H. Rabin. 1991. Numerical modeling of interface residual stresses in graded ceramic-metal materials. In *Developments in Ceramic and Metal-Matrix Composites,* ed. K. Upadhya, 241–251. Minerals, Metals and Materials Society. Warrendale, PA.

17. Beaussart, A. J., R. B. Pipes, and R. K. Okine. 1992. Numerical modeling of sheet forming processes for thermoplastic composites. University of Delaware, Newark. CCM-92-12.

18. Cohen, D., M. W. Hyer, M. J. Shuart, et al. 1992. Failure criterion for thick multi-fastener graphite/epoxy composite joint. *Proc. 6th Jpn.-U.S. Conf. Compos. Mater.*, pp. 762–770, June 1992, Orlando, FL. Technomic, Lancaster, PA.

19. Ireman, T., T. Nyman, and K. Hellbom. 1993. On design methods for bolted joints in composite aircraft structures. *Comp. Struct.* 25(1–4):567–78.

20. DiGiovanni, P. R., E. M. Rosenberg, and D. A. Boyce. 1987. Design, fabrication, and test of a metal matrix composite missile wing. *Met. Matrix Carbon Ceram. Matrix Compos.* pp. 25–50, January 1986, Cocoa Beach, FL. NASA CP-2482.

21. Vinson, J. R. 1985. Recent advances in technology for composite materials in the United States. *Comp. Tech. Res.* 7(2):59.

22. Tsai, S. W. 1984. Designing with composites. *CMC Int. Symp.*, Newark, DE, September 1984.

23. Tsai, S. W., and H. T. Hahn. 1980. *Introduction to Composite Materials.* Technomic, Westport, CT.

24. Schoutens, J. E. 1989. Introduction to metal matrix composite materials, MMCIAC No. 272. In *Reference Book for Composite Technology,* ed. S. Lee, 175–269. MMCIAC Kaman Tempo, Santa Barbara, CA.

25. June, R. R., and J. R. Lager. 1973. Commercial aircraft: Applications of composite materials, ASTM STP 524. pp. 1–42.

26. Graesser, D. L., Z. B. Zabinsky, M. E. Tuttle, et al. 1993. Optimal design of a composite structure. *Comp. Struct.* 24(4):273–81.

27. Rogers, C. 1993. V-22 experience points the way to improved composite design practices. *Adv. Mater.* 15(2):3; New horizons in composite structure. *J. Comp. Technol. Res.* 14(4).

28. Phillips, L. N. 1992. *Design with Advanced Composite Materials,* Design Council. Springer-Verlag, London.

BIBLIOGRAPHY

Adhesive Bonding Handbook for Advanced Structural Materials. European Space Research and Technology Centre, Netherlands, ESA PSS 03-210, 1990; N91-32234, 1991.

ARNDT, S. M., AND J. W. COLTMAN. Design trade-offs for ceramic/composite armor materials. *22nd Int. SAMPE Tech. Conf. Proc.,* ed. R. P. Caruso, P. Adams, and L. D. Michelove, pp. 278–301, Boston, November 1990.

BAKUCKAS, J. G., AND W. S. JOHNSON. Modeling fatigue crack growth in cross ply titanium matrix composites, 1993. NASA TM-108988.

CHEN, D., AND S. CHENG. Analysis of composite materials: A micromechanical approach. *J. Reinforced Plast. Compos.* 12(12):1323–38, 1993.

ELZEY, D. M., AND H. N. G. WADLEY. Modeling the densification of metal matrix composite monotape. *Acta Metall. Mater.* 41(8):2297–2316.

EVANS, A. G., AND F. A. LECKIE. *Processing and Mechanical Properties of High Temperature/High Performance Composites.* Book I: *Constituent Properties of Composites;* Book II: *Constituent Properties and Macroscopic Performance: CMCs;* Book III: *Constituent Properties and Macroscopic Performance: MMCs;* Book IV: *Constitutive Laws and Design.* University of California, Santa Barbara, 1993.

GAO, Z., AND K. L. REIFSNIDER. Tensile failure of composites: Influence of interface and matrix yielding. *J. Comp. Technol. Res.* 14(4):201–10.

GOWAYED, Y. A., C. M. PASTORE, AND C. S. HOWARTH. An integrated approach to the geometrical and mechanical modeling of textile reinforced composites. *39th Int. SAMPE Symp. Exhibit.*, Anaheim, CA, April 1994.

GUNAWARDENA, S. R., S. JANSSON, AND F. A. LECKIE. Modeling of anisotropic behavior of weakly bonded fiber reinforced MMCs. *Acta Metall. Mater.* 41(11):3147–56, 1993.

GUO, Y., AND J. TANG. 1993. Finite element analysis of end notched flexure (ENF): Specimen for mode II strain energy release rate. *38th Int. SAMPE Symp. Exhibit.*, Anaheim, CA, May 1993.

HE, M. Y., AND A. G. EVANS. Finite element analysis of beam specimens used to measure the delamination resistance of composites. *J. Comp. Technol. Res.* 14(4):235–40.

HILLBERRY, B. M., AND W. S. JOHNSON. Prediction of matrix fatigue crack initiation in notched SCS-6/Ti-15-3 metal matrix composites. *J. Comp. Technol. Res.* 14(4):221–24, 1992.

HOUSE, M. B., AND R. B. BHAGAT. Nonlinear finite element analysis of metals and metal matrix composites: A local-global investigation. Applied Research Laboratory, 1992. TR-92-01.

KAUTE, D. A. W., H. R. SHERCLIFF, AND M. F. ASHBY. Delamination, fibre bridging and toughness of ceramic matrix composites. *Acta Metall. Mater.* 41(7):1959–70, 1993.

LU, G. Q. Modelling the densification of porous structures in CVI ceramic composites processing. *J. Mater. Process. Technol.* 37(1–4):487–98, 1993.

MALHOTRA, S. K. Report on design guidelines for mechanically fastened joints in composite laminates. Deutsche Forschungsanstalt fuer Luft- und Raumfahrt, Institut fuer Strukturm-mechanik, Cologne, 1992. DLR-1B-131-92/21.

Metals and Materials 1991. Institute of Metals, Cambridge University, 1991.

MOREL, M. R., D. A. SARAVANOS, AND C. C. CHAMIS. Tailored metal matrix laminates for high-temperature performance. *37th Int. SAMPE Symp. Exhibit.*, pp. 390–402, Anaheim, CA, March 1992.

National Center for Advanced Technologies. *National Advanced Composites Strategic Plan, 1990.* U.S. Department of Commerce, Washington, DC, 1990.

NAYFEH, A. H. Ultrasonic wave interaction with advanced complex materials for nondestructive evaluation applications, University of Cincinnati, 1992. AFOSR-TR-93-0394.

Selecting the software. *Adv. Compos. Eng.*, September 1987.

SORENSEN, N. J. A planar model study of creep in metal-matrix composites with misaligned short fibers. *Acta Metall. Mater.* 41(10):2973–83, 1993.

STELLBRINK, K. Preliminary design of composite joints. Deutsche Forschungs-und Versuchsanstalt fuer Luft- und Raumfahrt, Cologne, 1992. DLR-Mitt-92-05.

STELLBRINK, K. Preliminary design of composite joints. Deutsche Forschungsanstalt fuer Luft- und Raumfahrt. Gruppe Konstruktionssystematik, Stuttgart, 1992. DLR-MITT-92-05, ETN-92-92407.

TAYA, M., AND K. E. LULAY. Strengthening of a particulate metal matrix composite by quenching. *Acta Metall. Mater.* 39(1):73–87, 1991.

VEKINIS, G., M. F. ASHBY, H. SHERCLIFF, ET AL. The micromechanisms of fracture of alumina and a ceramic-based fiber composite—Modeling the failure processes. *Comp. Sci. Technol.* 48(1–4):325–30, 1993.

WITTNAUER, J., AND B. NORRIS. High temperature structural honeycomb materials for advanced aerospace designs. *J. Met.* March 1990: 36–41.

YAMADA, Y., M. TAYA, AND R. WATANABE. Strengthening of metal matrix composite by shape memory effect. *Mater. Trans. Jpn. Inst. Met. Trans.* 34(3):254–60, 1993.

4

PMC Properties

4.0 INTRODUCTION

The performance of PMCs is judged by their properties and behavior under tensile, compressive, shear, and other static or dynamic loading conditions in both normal and adverse test environments. This information is essential for selecting the proper material for a given application as well as for designing a structure with the selected material. There is a wealth of property data on thermoset and thermoplastic composites which can be found in a variety of published literature.[1–5] The database for thermoplastic composite materials is not as large; however, it is increasing as developments progress.

Material properties are usually determined by conducting mechanical and physical tests under controlled laboratory conditions. The orthotropic nature of fiber-reinforced composites has led to the development of test methods that are often different from those used for traditional isotropic materials. These unique test methods and their limitations are reviewed in relation to many of the properties considered in this chapter. The effects of environmental conditions, such as elevated temperature and humidity, on the physical and mechanical properties of composite laminates are also presented, as well as long-term behavior such as creep and stress rupture and damage tolerance.

4.0.1 Mechanical Property Spectrum

Polymer matrix composite materials occupy the peak in the mechanical property spectrum of bulk high-performance plastics. On a specific property basis, in which stiffness and strength are normalized by dividing by the specific gravity, continuous-fiber-reinforced plastics are exceeded only by ultraoriented polyethylene and rigid-rod polymers such as PBT; but these latter property levels have so far been achieved only in fibrous form.

4.0.2 Property Bounds

If we consider uniaxially aligned, continuous-fiber composites made from Kevlar (Kv), Gr, glass (Gl), and B fibers and a common polymer matrix such as epoxy or PEEK, the tensile modulus and strength properties in the fiber direction of the composites will have exactly the same properties as the fiber properties if a constant volume loading were used. Properties in the direction of such aligned continuous-fiber composites constitute the upper bound of composite material mechanical performance.

Short-fiber-reinforced polymers, on the other hand, were developed largely to fill the property gap between continuous-fiber laminates and unreinforced polymers used largely in non-load-bearing applications. In some respects the short-fiber systems couple advantages from each of these property-bounding engineering materials. If the fibers are sufficiently long, stiffness levels approaching those of continuous-fiber systems at the same fiber loading can be achieved, whereas the ability of the unreinforced polymer to be molded into complex shapes is at least partially retained in the short-fiber systems. Thus, short-fiber-reinforced polymers have found their way into lightly loaded secondary structures in which stiffness dominates the design but in which there must also be a notable increase in strength over the unreinforced polymer.

4.0.3 Toughness

The fracture toughness of polymeric composites is probably one of the least understood of all the mechanical responses. For most composites, including short-fiber systems, a sometimes espoused rule of thumb is that as the strength increases, the toughness decreases. Thus, it might be implied that as the degree of adhesion increases, the toughness decreases. While this is generally true for continuous-fiber-reinforced brittle matrices, it is not always the case for particulate-filled systems nor for short-fiber-reinforced thermoplastics.[6–8]

4.1 STATIC MECHANICAL PROPERTIES

Static mechanical properties, such as tensile, compressive, flexural, and shear properties, of a material are the basic design data in many if not most applications.

4.1.1 Tensile Properties

Tensile properties, such as tensile strength, tensile modulus, and Poisson's ratio of flat composite laminates, are determined by static tension tests in accordance with American Society for Testing and Materials (ASTM) D3039-76. The tensile specimen is straight-sided and has a constant cross section with beveled tabs adhesively bonded at its ends.

4.1.1.1 Unidirectional laminates. For unidirectional polymeric matrix laminates containing fibers parallel to the loading direction (i.e., $\theta = 0°$), the tensile stress-strain curve is linear up to the point of failure.[9] Specimens fail by tensile rupture of fibers, which is followed or accompanied by longitudinal splitting (debonding along the fiber-matrix interface) parallel to the fibers (Table 4.1).

TABLE 4.1. Typical Mechanical Properties of Unidirectional Continuous-Fiber Composites[9]

Property	Boron-Epoxy	AS Carbon-Epoxy	T-300 Epoxy	HMS Carbon-Epoxy	GY-70-Epoxy	Kevlar 49-Epoxy	E-Glass-Epoxy	S-Glass-Epoxy
Specific gravity	1.99	1.54	1.55	1.63	1.69	1.38	1.80	1.82
Tensile properties								
Strength (MPa)								
0°	1585	1447.5	1447.5	827	586	1379	1130	1214
90°	62.7	62.0	44.8	86.2	41.3	28.3	96.5	
Modulus (GPa)								
0°	207	127.5	138	207	276	76	39	43
90°	19	9.0	10	13.8	8.3	5.5	4.8	
Major Poisson's ratio	0.21	0.25	0.21	0.20	0.25	0.34	0.30	
Compressive properties								
Strength, 0° (MPa)	2481.5	1172	1447.5	620	517	276	620	758
Modulus, 0° (GPa)	221	110	138	171	262	76	32	41
Flexural properties								
Strength, 0° (MPa)		1551	1792	1034	930	621	1137	1172
Modulus, 0° (GPa)		117	138	193	262	76	36.5	41.4
In-plane shear properties								
Strength (MPa)	131	60	62	72	96.5	60	83	83
Modulus (GPa)	6.4	5.7	6.5	5.9	4.1	2.1	4.8	
Interlaminar shear strength, 0° (MPa)	110	96.5	96.5	72	52	48	69	72

Alumina fibers are generally not useful materials for reinforcing resins because their main features, such as heat resistance and stability in molten metals, are not required in the case of polymer matrix composites (PMCs). Rather, their drawbacks, such as high density and moderate tensile strength, limit their applications. Consequently, only a limited amount of work has been done on Al_2O_3 fiber-reinforced PMCs although some results are listed in Table 4.2.[10]

The data for γ-alumina-epoxy in Table 4.2 indicate that the PMCs also have unique features similar to those of the fibers. These can be summarized as follows.

1. They have high moduli of elasticity. The modulus is three times as high as that of glass-epoxy and is as high as that of carbon-epoxy.
2. They have high compressive strengths. Their compressive strength is four times larger than that of glass-epoxy.

Table 4.3 compares the tensile, compressive, and shear properties of Kevlar-epoxy unidirectional composite laminates with those containing glass and graphite fibers. It can be seen that Kevlar 49 provides a significant advantage in composite density and, with the exception of compressive strength, a good balance of laminate properties. The tensile strength/compressive strength ratio is 5 for Kevlar-epoxy composites, which is the same as for Kevlar fiber. This indicates the controlling effect of Kevlar fiber in a brittle epoxy matrix. However, it is interesting to note that there is no catastrophic failure of Kevlar 49-

TABLE 4.2. Properties of Unidirectional γ-Alumina-Epoxy[a]

Density g/cm^3		2.4
Tensile strength (GPa)		
0°[b]		1.35
90°		0.068
Tensile modulus (GPa)		
0°		137
90°		18
Flexural strength, 0° (GPa)		1.56
Flexural modulus, 0° (GPa)		122
Flexural fatigue strength, 0°, at 10^7 cycles (0°)		0.63
Compressive strength, 0° (GPa)		2.2
Modulus of rigidity (GPa)		7.25
Interlaminar shear strength (GPa)		0.13
Impact, 0°	(J cm^{-2})	20 (Izod)
Rockwell hardness		E85
Thermal expansion (K^{-1})		
0°		0.4×10^{-5}
90°		3.7×10^{-5}
Thermal conductivity (W/m · K)		
0°		2.32
90°		0.85
Volume electrical resistivity, 90° (Ω·m)		10^{13}
Arc resistance time (s)		130–180
Dielectric strength (kV/mm)		16–20
Dielectric constant		
10^6 Hz		4.9
10^{10} Hz		5.1
Loss tangent		
10^6 Hz		0.015
10^{10} Hz		0.016

[a] Fibers loading = 60 vol%; resin, DDM-type epoxy (Sumitomo's Sumiepoxy ELM-434); hardener, DDS.
[b] 0° and 90° indicate angles measured from the fiber direction.

epoxy composites at high bending as in the case of graphite and glass composites. The sustaining performance of Kevlar 49-epoxy composites at high compressive strains is consistent with the low strength loss of Kevlar fiber at high static and cyclic compressive strains.

The environmental stability of composites of Kevlar is generally good except in extremely hostile environments. Thus, the effect of thermal aging and moisture on the mechanical properties of Kevlar-epoxy composites has long been a continuing concern. Work reported in the literature recently suggests that moisture induces a plasticization effect and can also reduce mechanical properties by the formulation of internal microcavitation. With microcavitation, a composite can absorb more moisture than the theoretical solubility, which subsequently causes a crazing effect in the absence of external loading.[10]

The effect of other environments such as water, jet fuel, and lubricating oil is illustrated in Table 4.4. There are numerous other environmental effects on composite structures, and they are compiled in Table 4.5.[11]

TABLE 4.3. Kevlar-Epoxy Unidirectional Lamina Properties[10]

Property	Kevlar 49	E-Glass	Graphite
Density			
lb/in.^{2}	0.050	0.075	0.055
g/cm^3	1.38	2.08	1.52
Tensile strength, 0°[a]			
10^3 lb/in.^{2}	200	160	180
MPa	1380	1100	1240
Compressive strength, 0°			
10^3 lb/in.^{2}	40	85	160
MPa	276	586	1100
Tensile strength, 90°[b]			
10^3 lb/in.^{2}	4.0	5.0	6.0
MPa	27.6	34.5	41.4
Compressive strength, 90°			
10^3 lb/in.^{2}	20	20	20
MPa	138	138	138
In-plane shear strength			
10^3 lb/in.^{2}	6.4	9.0	9.0
MPa	44.1	62.0	62.0
Interlaminar shear strength			
10^3 lb/in.^{2}	7.10	12	14
MPa	48.69	83	97
Poisson's ratio	0.34	0.30	0.25
Tension and compression modulus, 0°			
10^6 lb/in.^{2}	11	5.7	19
MPa	75,800	39,300	131,000
Tension and compression modulus, 90°			
10^6 lb/in.^{2}	0.8	1.3	0.9
MPa	5500	8960	6200
In-plane shear modulus			
10^6 lb/in.^{2}	0.3	0.5	0.7
MPa	2070	3450	4830

[a] 0°, in the direction of the laminas.
[b] 90°, normal to the laminae direction.

A fiber percentage that is easily achievable and repeatable in a composite for several fibers is 60%. The properties of unidirectional fiber Gl/Ep laminates are listed in Table 4.6. These values are for individual laminas or for a unidirectional composite, and they represent the theoretical maxima (for that fiber volume) for longitudinal in-plane properties.[11]

Grimm[12] recently compiled a group of tables reflecting many properties of laminated composite materials because he feels that laminate properties are very relevant to the design and stress analysis of composite piece parts (Table 4.7).

At the Second International Workshop for Composite Materials and Structures for Rotorcraft, several engineers reported on the effects of environment on two helicopters that had been in service for over 5 years.[13] The first report covered the S-76 helicopter where residual strength of a stabilizer with 17 months' service was 220% of the design ultimate load, tail rotor spars retained 94% of baseline strength after 5 years, and residual

TABLE 4.4. Long-Term Environmental Stability of Composites of Kevlar and Fiberglass[10]

	Interlaminar Shear Strength (MPa)[a]		
Environmental Conditions	Kevlar 49 121°C (250° F) Epoxy[b]	Kevlar 49 177°C (350°F) Epoxy[c]	S-Glass 177°C (350°F) Epoxy[c]
Control	49.5	40.6	79.2
Boiling water			
3½ years	15.8	21.2	0
Salt water, 3½ years	32.2	40.6	75.8
Jet fuel (Texaco Abjet K-40), 3½ years	45.5	40.6	79.2
Lubricating oil (Skydrol), 3½ years	46.5	40.6	79.2

[a] ASTM D2344 short beam shear.
[b] Interior-grade general purpose.
[c] Exterior-grade high temperature.

flexure and short-beam shear strengths exceeded the accelerated tests after outdoor exposure. The excellent unidirectional properties of the AS-4 carbon fiber-PEKK laminate are shown in Table 4.8.

The second report described the Bell 206L study where composites performed well in service during 122,000 h of flight service, and in evaluating the vertical fin it was found that the required design strength was exceeded after service and 35% of the baggage doors exceeded the design strength after service.

There is a great deal of composite information available, part of which is summarized in Tables 4.9–4.11. The data are compiled from the literature and from suppliers' bulletins. Table 4.9 lists the effect of each type of carbon fiber.

Particularly noticeable is the high thermal conductivity of the graphite pitch-based fiber composite, which is 46 times that of the PAN-based material.

Table 4.10 presents the effect of fiber orientation on the properties. The large differences in tensile strength and modulus of unidirectional laminates tested at 0 and 90° should be noted.

Table 4.11 shows the effect of two types of polymer matrices, epoxy and PEEK, on the mechanical properties of a carbon fiber composite with two different fiber orientations: 0° (unidirectional) and 45° (three layers at 0, 45, and 90°, respectively). The fracture strain of the PEEK 45° composites is considerably larger than that of the others.

Other studies have shown that composite plies are up to twice as strong as high-strength steel and three to five times as strong as titanium and aluminum. In comparing specific strengths, composites are three to eight times stronger than the metals. In stiffness, the materials range from fiberglass composite at the low end to high-modulus carbon composite at the high end with the metals in between. Composites compare favorably to metals in specific modulus.

4.1.1.2 Cross-ply laminates. The tensile stress-strain curve for a cross-ply (0/90) laminate tested at $\theta = 0°$ is slightly nonlinear; however, it is commonly approximated as a bilinear curve.[4,14]

TABLE 4.5. Environmental Effects on Composite Structures[11]

Environment	Effects (Matrix or Fiber)	Comments
Moisture-related		
Rain	Matrix, softens and swells	Rain may cause water intrusion into composite and joints.
Humidity	Matrix, softens and swells	Effect can be aggravated by high (> 1%) void content.
Salt fog	Matrix, softens and swells	Corrosion of metal attached to graphite epoxy composite can be increased.
Deep submergence	Matrix	Interface may be affected; low void content is requirement for compressive inputs.
Rain and sand erosion	Both	Protect surfaces with paint or elastomer; composite response is significantly different from that of metals.
Galvanic corrosion	Neither	Graphite composites can cause corrosion of metals; composite is unaffected.
Radiation		
Solar radiation (earth)	Matrix	Unprotected aramid or polyethylene is affected.
Ultraviolet	Matrix	Unprotected aramid or polyethylene is affected.
Nonionizing space	Both	Composite can be destroyed; most metals are unaffected.
Solar (low earth orbit)	Both	Composites can be destroyed; most metals are unaffected.
Temperature		
High	Matrix	Fibers (except polyethylene) are generally more resistant than matrix.
Low	Matrix	
Extreme (ablation)	Both	
Thermal cycling (shock)	Matrix	
Miscellaneous environments		
Solvents, fuels	Matrix	Determine for each matrix, stressed and unstressed; effects may be aggravated by voids.
Vacuum	Matrix	
Fungus	Matrix	Each material must be evaluated; may affect organic fibers.
Fatigue	Both	Effects may be diminished by voids.

4.1.1.3 Multidirectional laminates. Tensile stress-strain curves for laminates with different fiber orientations in different laminas are in general nonlinear.[14,15]

4.1.1.4 Woven fabric laminates. The principal advantage of using woven fabric laminates is that they provide properties that are more balanced in the 0 and 90° directions than unidirectional laminates. Although multidirectional laminates can also be designed to produce balanced properties, the fabrication (lay-up) time for woven fabric laminates is less than that for multidirectional laminates. However, the tensile strength and modulus of a woven fabric laminate are, in general, lower than those of nonwoven laminates. The principal reason for lower tensile properties is the presence of fiber undulation in woven fabrics, as the fiber yarns in the fill direction cross over and under the fiber

TABLE 4.6. Properties of Unidirectional Glass-Epoxy Composites[11]

Property	E-Glass		S-Glass	
Elastic constants, GPa (10^6 lb/in.^{2})				
Longitudinal modulus E_L	45	(6.5)	55	(8.0)
Transverse modulus E_T	12	(1.8)	16	(2.3)
Shear modulus G_{LT}	5.5	(0.8)	7.6	(1.1)
Poisson's ratio ν_{LT} (dimensionless)	0.19		0.28	
Strength properties, MPa (10^3 lb/in.^{2})				
Longitudinal tension F_L^{tu}	1020	(150)	1620	(230)
Transverse tension F_T^{tu}	40	(7)	40	(70)
Longitudinal compression F_L^{cu}	620	(90)	690	(100)
Transverse compression F_T^{cu}	140	(20)	140	(20)
In-plane shear F_{LT}^{su}	60	(9)	60	(9)
Interlaminar shear F^{Lsu}	60	(9)	80	(12)
Ultimate strains, %				
Longitudinal tension ε_L^{tu}	2.3		2.9	
Transverse tension ε_T^{tu}	0.4		0.3	
Longitudinal compression ε_L^{cu}	1.4		1.3	
Transverse compression ε_T^{cu}	1.1		1.9	
In-plane shear			3.2	
Physical properties				
Specific gravity	2.1		2.0	
Density, lb/in.^{3}	0.075		0.72	
Longitudinal CTE, 10^{-6}m/m.·°C (10^{-6} in./in.·°F)	3.7	(6.6)	3.5	(6.3)
Transverse CTE 10^{-6} m/m.·°C (10^{-6} in./in.·°F)	30	(17)	32	(18)

[a] Fiber volume fraction $V_f = 0.60$.

yarns in the warp direction to create an interlocked structure. Under tensile loading, these crimped fibers tend to straighten out, which creates high stresses in the matrix. As a result, microcracks are formed in the matrix at relatively light loads.

The tensile properties of a woven fabric laminate can be controlled by varying the yarn characteristics (e.g., number of fiber ends, amount of twist in the yarn, and relative

TABLE 4.7. Design Properties of Unidirectional Composite Material (Laminates)

Property	E-Glass-Epoxy, 60 Vol% Fiber	Carbon HT-Epoxy, 60 Vol% Fiber	Aramid HT-Epoxy, 60 Vol% Fiber
Density (g/cm^3)	2.1	1.6	1.35
Tensile strength (N/mm^2)	900	1600	1300
Tensile modulus (kN/mm^2)	35	130	80
Flexural strength (N/mm^2)	900	1600	600
Flexural modulus (kN/mm^2)	35	120	50
Compression strength (N/mm^2)	600	1500	250
Interlaminar shear (N/mm^2)	28	90	40
Elongation at break (%/mm^2)	3.5	1.2	2.2
Coefficient of thermal expansion (10^{-6}/K)	8	0	−4

TABLE 4.8. Unidirectional Mechanical Properties of AS-4 Carbon Fiber-PEKK Laminates[13,a]

Property	LDF[b]	Continuous	ASTM Test
Tensile (MPa)			
Strength, 0°	1610	1676	D3039
Modulus, 0°	123.5	129.7	
Poisson ratio	0.35	0.33	
Strength, 90°	91	73.1	
Modulus, 90°	10.3	8	
Compressive (MPa)			
Strength, 0°	1262	1393	D695
Modulus, 0°	111	121.4	
Flexural (MPa)			
Strength, 0°	1655	1931	D790
Modulus, 0°	120	127.6	
Shear (MPa)			
In-plane strength	146	142	D3518
In-plane modulus	5.5	5.8	
Short-beam strength	110	117	D2344

[a] Fiber volume fraction = 65%, 23.9°C, dry.
[b] Average fiber length = 55.88 mm.

number of yarns in the fill and warp directions) and the fabric style. Tensile properties can also be controlled by changing the lamination pattern and stacking sequence.[14]

Chou and associates[16] evaluated BMI resin and three-dimensional fabrics. When woven fabrics are impregnated with a resin matrix and laminated to produce composite materials, it is possible to obtain high strength in the plane of the fabric, but such materials exhibit the drawbacks of low interlaminar fracture toughness and poor through-thickness properties.[17,18] Three-dimensional fabrics are textile fiber geometries having through-thickness reinforcement obtained by multiaxial arrangement of fibers in three orthogonal directions.[19–22] By using these reinforcements to fabricate materials, it is possible to obtain isotropic fiber arrangements that give the material a high strength in three

TABLE 4.9. Effect of the Type of Carbon Fiber on the Properties of Carbon-Epoxy Composites[a]

	Type of Fiber		
Property	T650/42[b]	P55[c]	P120[d]
Density (g/cm^3)	1.6	1.7	1.8
Tensile strength (MPa)	2585	896	1206
Modulus (GPa)	172	220	517
Compressive strength (MPa)	1723	510	268
Shear strength (MPa)	124	55	27
Thermal conductivity (W/m·K)	8.65	74	398

[a] The fibers are produced by Amoco Performance Products Inc. The composites have the same epoxy matrix with 62 vol% fiber. Fiber orientation is unidirectional (0°) and tested in the fiber direction.
[b] T650/42: a high-strength, intermediate-modulus, PAN-based fiber.
[c] P55: an intermediate-modulus, pitch-based fiber.
[d] P120: a high-modulus, pitch-based fiber.

TABLE 4.10. Effect of the Orientation of Carbon Fibers on the Properties of Carbon Polymer Composites[a]

	Unidirectional Laminate		Quasi-isotropic Laminate	
Testing Angle	0°	90°	0°	90°
Tensile strength (MPa)	793	20.0	379	241
Tensile modulus (GPa)	303	3.3	104	97
Tensile ultimate strain (%)	0.25	0.5	0.27	0.23
Compression strength (MPa)	400	158	172	200
Compression modulus (GPa)	255	6.7	76	88
Compressive ultimate strain (%)	—	—	0.55	0.86

[a] The carbon fiber is Amoco P75 high-modulus pitch fiber. Fiber content is 60 vol%. Polymer is epoxy PR500-2 from 3M. Isotropic laminate has 0°, 30°, 60°, 90°, 120°, 150° stacking sequence. Unidirectional laminate tested along the fibers (0°) and across the fibers (90°). Isotropic laminate tested in two directions perpendicular to each other.

orthogonal directions. As a result, it is necessary for advanced composites to be able to support multidirectional mechanical stresses.

Chou and associates designed a weaving process for three-dimensional fabrics and found that the tensile strength of 3-D composites increased with increasing weaving density, that is, a 3-D structure was higher in tensile strength than a five-dimensional (5-D) structure. Additionally they found that the short-beam shear strength of 3-D composites was higher than that of two-dimensional (2-D) composites and that the shear strength of materials loaded in the *X-Z* plane was higher than that of materials loaded in the *X-Y* plane by about 20%.

Finally, they showed that 5-D structures possessed greater toughness than 3-D structures, the impact properties of 3-D composites were better than those of 2-D composites, 5-D structures had more isotropic material properties and so had a higher rupture toughness, and 3-D composites retained about 50% of their mechanical strength, which reveals the high-temperature stability of BMI resins.

4.1.1.5 Interply hybrid laminates. Interply hybrid laminates are made of separate layers of low-elongation fibers (such as high-modulus carbon fibers) and high-elongation fibers (such as E-glass or Kevlar 49), both in a common matrix.

Other studies reported by Uhlmann[23] on thermoset and thermoplastic composites

TABLE 4.11. Mechanical Properties of Carbon-Fiber Composites with Epoxy and PEEK Polymer Matrices

Polymer Matrix	Fiber Orientation	Tensile Strength (MPa)	Tensile Modulus (GPa)	Fracture Strain (%)
Epoxy	0°	932	83	1.1
Epoxy	45°	126		1.3
PEEK	0°	740	51	1.1
PEEK	45°	194	14	4.3

show that textile technology (braiding, knitting, and weaving) (discussed further in Volume II) plays a key role in some structural composite manufacturing methods. Nearly final-shaped fabrics are made either from reinforcing fibers alone (to be injected with resin later) or from commingled fibers (both reinforcing and thermoplastic) in the near net shape mold. For composites made from woven fiber preforms, the damage tolerance throughout the part comes from fibers. At the joints in a laminated composite part, damage tolerance mostly comes from the resin.

The properties of various laminates employing different fibers and matrices are listed in Tables 4.12–4.14. For all matrices, translation of fiber modulus into composite modulus nearly follows a volume fraction rule of mixtures. However, fiber tensile strength is not as efficiently translated into composite tensile strength. The biggest difference between thermoset matrix (epoxy) and thermoplastic matrix composites is in fracture toughness, where thermoplastics are up to an order of magnitude tougher than simple thermosets. A study[23] also included some work on compression properties which are discussed in the next section.

BASF has been developing yarns that can be commingled such that the reinforcement and the matrix are interspersed at the individual filament level. Commingled yarns can be converted to various forms such as woven fabric, unidirectional tape, and braided sleeving. Illustrative properties of unidirectional laminates produced from commingled hybrid yarns based on AS-4 carbon fiber, S-2 glass, and Astroquartz with an aromatic polyketone matrix are presented in Table 4.15.

Researchers in the TRW Space and Technology Group[24] have reported the results of a comparison of graphite composites fabricated from thermoplastic PEEK/AS-4 graphite prepreg tapes with cowoven and commingled woven fabrics. The composites fabricated from traditional unidirectional prepreg tape exhibited the highest flexural strength and transverse properties, and this was attributed to the excellent fiber-matrix interfacial adhesion. Lower mechanical properties were obtained from the unidirectional cowoven and commingled fabrics, and this was attributed to fiber breakage during weaving, fiber misalignment during laminate consolidation, poor fiber matrix distribution and, in some cases, poor fiber-matrix adhesion. The commingled woven fabric composites had superior properties when compared with the cowoven fabric composite. Illustrative properties of these composites are presented in Table 4.16.

If the same amount of fiber were divided equally in the longitudinal and transverse

TABLE 4.12. Carbon Fiber-Epoxy Unidirectional Composite Properties[a]

Fiber	T300	T50	T650	T1000	P55	P100
Tensile (brittle resin)						
Strength	1862	1311	2413	3447	723	1138
Modulus	138	241	170	159	234	483
Tensile (ductile resin)						
Strength	2790	1414	3070	3795	890	1206
Modulus	138	241	170	234	483	
Compression						
Strength	1725	965	1650	1690	483	276
Modulus	124	234	151	199	505	

[a] All values are MPa.

TABLE 4.13. Unidirectional Composite Properties[a]

	XAS M-ATB (AF-8)		XAS M-ATS (AF-20)		Celion ST NARMCO 5245 C		Celion 6000 BMI 11795I	
	Strength (MPa)	Modulus (MPa)	Strength (MPa)	Modulus (MPa)	Strength (MPa)	Modulus (MPa)	Strength (MPa)	Modulus (MPa)
Tension								
−55 C	2280	140	1966	140	2410	152		
RT	2208	140	2068	138	2480	140	1414	117
93C					2447	140	1276[b]	124[b]
Flexure								
RT	1790	124	1897	124			1828	117
232 C	1103	110	758	117			1241[b]	117[b]
219 C wet	786	110	317	55				
Compression								
RT	1242	152	1286	131	1656	140	1380	124
93 C					1552	140		
232 C	724		490	124			793[b]	117[b]
219 wet	724		241	124				

[a] ATB and ATS are acetylene-terminated resins developed at Air Force Materials Laboratory (AFML); XAS and Celion are PAN-based carbon fibers developed by Courtaulds and BASF, respectively. NARMCO and 11795I are BMI matrix resins.
[b] 249°C.

directions (a bidirectional composite), then the two directions would have equal strength and stiffness. However, neither would be as high as in the unidirectional case. If the same amount of fiber were randomly laid (in-plane), then the resulting composite would have equal strength and stiffness in all directions (in-plane) but at lower values than for the bidirectional case (in the direction of the fibers).

TABLE 4.14. Unidirectional Fiber and Thermoplastic Composite Properties[a]

Resin	Fiber	Tensile Strength (MPa)	Tensile Modulus (MPa)	Compressive Strength (MPa)
PEEK (APC2)[b]	AS-4	2242	138	1069
APC (HTX)[b]	AS-4		138	1138
PEKK[b]	AS-4			1390
PPS[c]	AS-4	1656	138	655
Torlon-C[d]	C-6000	1390	140	1390
ULTEM 1000[e]	AS-4		138	
AVIMID K-III[f]	IM-6			
UDEL P 1700[g]	AS	1345	131	1035
J-2[h]	Kevlar 49	1310	76	276

[a]PEEK, polyether ether ketone; APC (HTX). PEKK, polyether ketone ketone; PPS, polyphenylene sulfide.
[b] Both PEEK, APC (HTX) and PEKK are different grades of aromatic polyketones.
[c] Polyphenylene sulfide, Ryton.
[d] Poly (amidlimide) (PAI).
[e] Poly (etherimide) (PEI).
[f] Polyimide (PI).
[g] Polysulfone (PS).
[h] Poly (arylamide).

TABLE 4.15. Typical Mechanical Properties for Unidirectional Laminates Fabricated from Hybrid Yarn[24]

Hybrid yarn composition				
Matrix	Victrex 150G, PEEK	Ultrapek, PEKEKK	Victrex 150G, PEEK	Victrex 150G, PEEK
Reinforcement type	AS4 3K	AS4 3K	S2 GLAS	Astroquartz
Reinforcement volume (%)	62	62	63	64
Laminate properties				
Flexural strength (MPa)	2380	2380	2030	2205
Flexural modulus (GPa)	138	134	58	46
Tensile strength, 90° (MPa)	82	84	50	78
Tensile modulus, 90° (GPa)	9.93	10.1	20.3	18.6
Elongation (%)	0.91	0.89	0.25	0.42

When only the fiber directionality is taken into account, the approximate relationships of unidirectional, bidirectional, and random (in-plane) modulus and strength are 1, ½, and ⅜, respectively. However, the amount of fiber that can be packed into a composite is a function of the degree of alignment. The typical fiber volume fraction for unidirectional composites is 65%. When the geometry is changed to arrange fibers in other directions, then the maximum fiber packing is reduced further. A typical fiber volume fraction for bidirectional reinforcement (woven fiber) is 50%, and a typical volume fraction for random in-plane reinforcement (chopped strand mat) is 20%.

Thus, the mechanical properties of a unidirectional laminate are very different from those of a random laminate even if the same fiber type and resin type are used in each. There is a factor of 1:0.375 to account for the directionality effect and a factor of 1:0.2 to account for the difference in the volume fraction of fibers. This gives an overall difference of 1:0.075, that is, a 13-fold difference in modulus.

Some of the properties of the composite may be predicted from known fiber and matrix properties using the simple rule-of-mixtures equation, most notably the tensile modulus.

Additionally, the rule of mixtures may be used to predict composite properties such as density and Poisson's ratio, as well as tensile modulus. However, it is a poor model of the tensile strength behavior of composite materials. Typical mechanical properties of a variety of composite materials are given in Table 4.17.[25]

TABLE 4.16. Comparison of Properties of Composite Laminates Fabricated from Unidirectional Prepreg Tapes and Fabrics[24]

Physical Properties of Laminates	Unidirectional Prepreg Tape	Comingled Fabric	Comingled Fabric	Cowoven Fabric
Number of plies	10	16	8	12
Specific gravity	1.56	1.55	1.53	1.53
Reinforcement content (vol%)	60	56.1	57.7	61.8
Void content (vol%)	1.9	2.1	3.2	4.4
Flexural strength (MPa)	1687	1514	1222	1150
Flexural modulus (GPa)	108	98	106	65
Transverse tensile strength (MPa)	91	64	25	
Moisture absorption (%)	0.15	0.22	0.17	2.0

TABLE 4.17. Typical Mechanical Properties of Composites[25]

	UD E-Glass-Epoxy	UD XAS Carbon-Epoxy	UD Kevlar 49-Epoxy	0.90 Woven E-Glass-Epoxy	±45 Woven E-Glass-Epoxy	0.90 Woven XAS Carbon-Epoxy	±45 Woven XAS Carbon-Epoxy	0.90 Woven Kevlar 49-Epoxy	CSM E-Glass-Polyester
Fiber volume, %	53	57	60	33	33	50	50	50	19
UTS, 0° (MPa)	1190	2040	1379	360	185	625	240	517	108
UTS, 90° (MPa)	73	90	30	360	185	625	240	517	108
UCS, 0° (MPa)	1001	1000	276	240	122	500	200	172	148
UCS, 90° (MPa)	159	148	138	205	122	500	200	172	148
USS (MPa)	67	49	60	98	137	130		110	85
E, 0° (GPa)	39	134	76	17	10	70	18	31	8
E, 90° (GPa)	15	11	5	17	10	70	18	31	8
G (GPa)	4	5	2	5	8	5	27	2	2.75
ILSS (MPa)	90	94	83	60	48	57	57	70	
Poisson's ratio		0.263	0.34	0.24	0.7				0.32
Density (g/cm^3)	1.92	1.57	1.38	1.92	1.92	1.53	1.53	1.33	1.45

4.1.2 Compressive Properties

Compressive properties of thin composite laminates are difficult to measure owing to sidewise buckling of specimens. Compression testing of composite materials is a process that is inconsistent from study to study and investigator to investigator. However, recommendations and guidelines can be deduced from a review of the published work in the area. Although no single test method can adequately satisfy the many objectives of compression studies, some investigations can benefit from the selection of tried and tested methods.

A number of test methods and specimen designs have been developed to overcome the buckling problem. Three of these methods include the Celanese test,[14] the Illinois Institute of Technology Research Institute (IITRI) test,[14] and the sandwich edgewise compression test.[14,26,27]

For testing unidirectional coupons, shear loading (Celanese, IITRI, sandwich beam) methods are by far the most popular and critically evaluated. The choice of one of these methods when specimen size and environmental conditions allow helps ensure the avoidance of problems that have been experienced in the past and results in data that can be compared to those from other studies. In summary, the three shear-loaded test methods just mentioned, the IITRI and Celanese methods are fundamentally the same test methods, with the IITRI being the more popular and an improved version of the Celanese. The sandwich beam method requires an expensive specimen and does not yield acceptable values of Poisson's ratio.

Compressive test data on fiber-reinforced composites are limited. From the available data on 0° laminates, the following observations can be made.

1. The compressive modulus of a 0° laminate is not equal to its tensile modulus as it is in ductile metals.
2. The longitudinal compressive strength of a 0° laminate depends on the fiber type, fiber volume fraction, matrix yield strength, fiber length/diameter ratio, and fiber straightness, as well as the fiber-matrix interfacial shear strength.[14,26]
3. Among the commercially used fibers, the compressive strength and modulus of Kevlar 49-reinforced composites are much lower than their tensile strength and modulus. Carbon and glass fiber-reinforced composites have slightly lower compressive strength and modulus than their respective tensile values, and boron fiber-reinforced composites exhibit virtually no difference between tensile and compressive properties.

A comparison of the mechanical properties of carbon-reinforced thermoplastic and thermoset composites indicates that the axial compressive strengths of all the carbon thermoplastic composites are consistently lower than those of carbon thermoset composites. For example, the axial compressive strength of carbon-reinforced thermoset composites (using the high-compressive-strength carbon fibers) is generally reported in the range 1380–1655 MPa, whereas for the same carbon fibers the carbon fiber-reinforced thermoplastic composites exhibit an axial compressive strength of 621–1380 MPa (621 MPa has been reported for PPS and 1380 MPa for PEEK; the fiber in both cases was AS-4). Other mechanical properties of the thermoplastic composites, such as flexural strength, flexural modulus, short-beam shear strength, fracture toughness, and longitudinal and transverse properties, are good. However, this is a very general statement considering the large number of thermoplastic matrices available with wide variations in properties. Lower properties, for example, a flexural strength of 120 MPa and a transverse tensile strength of 318 MPa for PPS/AS-4 composites appear to be the exception rather than the rule for the thermoplastic composites.

The low compressive strength of the carbon fiber-reinforced thermoplastic composites can be attributed to a number of factors, including fiber-matrix interfacial strength and matrix modulus. However, in this regard, processing temperature differences between thermosetting resins (generally lower than 177°C) and high-performance thermoplastics (343–454°C) may also be responsible. This is because of difference in the CTE of the fiber and the matrix, which causes residual compressive stresses in the fibers after lamination. These stresses are higher in the high-performance thermoplastic composites than in the thermoset composites because of the higher processing temperatures. The higher residual compressive stresses in the fiber may result in the lower observed compressive strength in the thermoplastic composites. Resin morphology changes may also be a contributing factor that has not been thoroughly explored.

Weeks and White[28] evaluated the mechanical and physical properties of 30 vol% carbon fiber-reinforced PEKK test parts. Room-temperature mechanical properties are shown in Figure 4.1, and of particular note are the excellent compressive properties, which are a result of uniform distribution of the individual carbon reinforcing fibers. The density was 1.44 g/cm^3.

The first test was performed according to ASTM D638M where tensile strength was 272 MPa, modulus was 24.3 GPa, and Poisson's ratio was 0.34. The compressive test was according to ASTM D695 where strength was 327 MPa and compressive modulus was 22.2 GPa. The final tests were shear, employing ASTM D4255, and results were 118 MPa for strength and 14.1 GPa for shear modulus.

Camponeschi studied the effects of the mechanical response of composites greater than 6.4 mm in thickness for the U.S. Navy which is interested in these types of thick-section composite structures, especially in their compression strength.[29]

In tension, Kevlar composites exhibit one of the highest specific strengths available in structural materials; however, compressive strengths are only one-fifth of tensile strength, and off-axis strengths are 100 times lower than in the filament direction. Near

Density = 1.44 g/cm^3

Property	Test	Value
Tensile	ASTM D638M	
Strength		272 MPa
Modulus		24.3 GPa
Poisson's ratio		0.34
Compressive	ASTM D695	
Strength		327 MPa
Modulus		22.2 GPa
Shear	ASTM D4255	
Strength		118 MPa
Modulus		14.1 GPa

Figure 4.1. Room temperature test properties of carbon-PEKK.[28]

zero in-plane thermal expansion coefficients can be obtained with Kevlar-epoxy composites because of the negative axial expansion of the filaments. The tendency of the filament rigid-rod morphology to fibrillate and, in so doing, to absorb large quantities of energy results in high-impact-resistant composites.

4.1.3 Flexural Properties

Flexural properties, such as flexural strength and modulus, are determined by ASTM test method D790-81. In this test, a composite beam specimen of rectangular cross section is loaded in either a three-point bending mode or a four-point bending mode.[10]

The flexural strength (three-point) of three LAS ceramic matrix–Nicalon fiber composite systems is shown in Figure 4.2 as a function of temperature when tested in an inert environment. As in the case of glass matrix composites, the eventual loss of strength in these composites at elevated temperature is due to softening of the matrix and is thus highly dependent on the percentage of matrix still left in the glassy state. Testing in an inert argon environment was found to be required because composite fracture morphology and strength were strongly related to the test environment.[30] The data in Figure 4.3 show this dependence of strength on the test atmosphere.

Other composite properties have been determined for the various LAS matrix–Nicalon fiber systems including fracture toughness, Charpy and ballistic impact, elevated-temperature creep, thermal and mechanical fatigue, thermal expansion, and thermal conductivity. Some of these properties have been reported in reference 31. The results of these measurements have shown that these composite systems exhibit excellent potential for structural use up to temperatures in excess of 1100°C. For example, the flexural strength of a unidirectional LAS–Nicalon fiber composite was measured after thermally shocking the material from elevated temperature into a water quench and comparing the

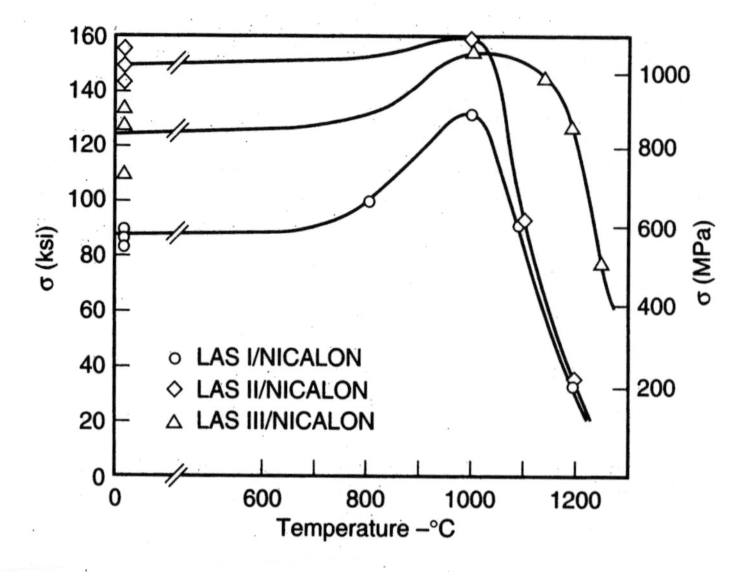

Figure 4.2. Flexural strength (three-point) versus temperature in argon for unidirectional LAS matrix–Nicalon fiber composites.

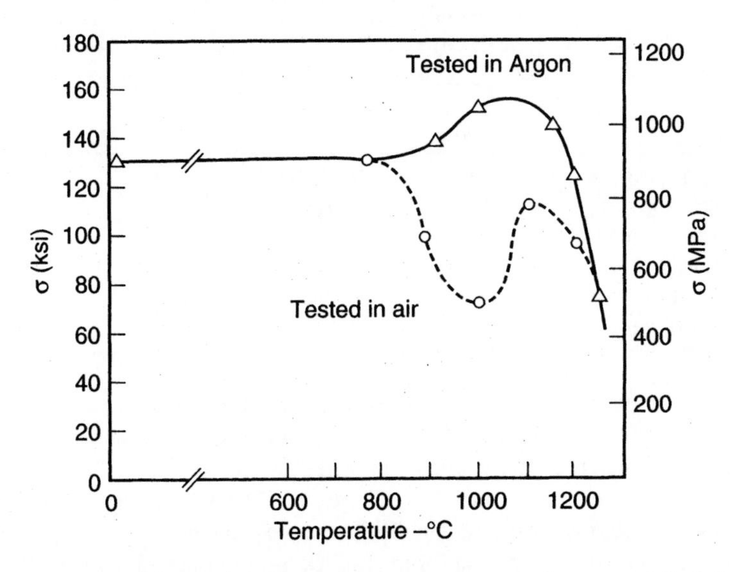

Figure 4.3. Flexural strength (three-point) versus temperature in argon and in air for unidirectional LAS-III matrix–Nicalon fiber composites.

results to strengths obtained for hot-pressed Si_3N_4 and monolithic LAS quenched under identical conditions. The results of these tests are presented in Figure 4.4 and show that the thermal shock properties of this type of composite are excellent.

Scola[32] reviewed the synthesis and characterization of several PI materials intended for application up to 371°C. He found that over the last 7 years several approaches to the development of PIs for 317°C applications have emerged. Most of these approaches involve modification of an existing high-temperature PI, called Avimid-N, which is considered the most thermooxidatively stable PI material for application in composites up to 371°C, and Table 4.18 reflects some flexural and interlaminar shear strength test results.

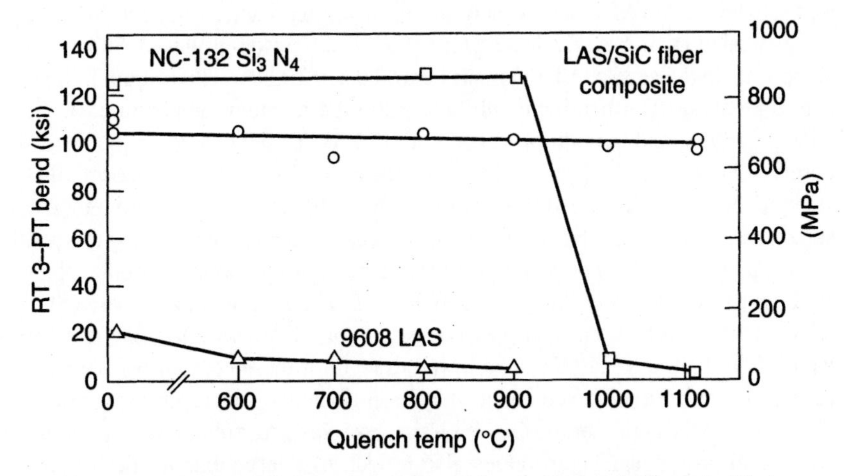

Figure 4.4. Residual RT flexural strengths of water-quenched LAS matrix–Nicalon fiber composite. Corning 9608 monolithic LAS and NC-132 hot-pressed Si_3N_4.

TABLE 4.18. Composite Mechanical Properties at 371°C

Composite System	Flexural Strength after Aging at 371°C in Air (MPa)		Interlaminar Shear Strength after Aging at 371°C in Air (MPa)		
	100 h	200 h	100 h	200 h	300 h
Celion 6K/LaRC-RP46	710		32.4		
Celion 6K/PMR-15	310	689	21.4	42	
Celion 6K/PMR-II-30		841		47	
T4OR/PMR-II-50	350	517	28	33	
T4OR/VCAP-50		400			28
Quartz/AFR-700B	393		51.7		

The flexural unilateral properties were investigated in the use of SiC yarn, Al_2O_3, and Nextel fibers with either a SiC or a Si_3N_4 matrix, which can result in a more oxidation-resistant material than a C/C composite. Composites of SiC yarn with a silicon matrix have also been made, and the SiC yarn–CVD SiC and Nextel 440 cloth–CVD SiC systems have also been investigated.[33] The bend strength of the Nextel cloth–SiC composite was about 0.14 GPa, and for the SiC yarn–SiC composite it was about 0.4 GPa.

Stinson et al.[34] produced composites with fiber contents of 35–60 vol%, with 45% being selected as the best for infiltration. The flexural strength of unidirectional composites ranged from 0.21 to 0.41 GPa, of cloth-reinforced composites from 0.18 to 0.49 GPa, and of chopped fiber (only 25%) or whisker-reinforced composites from 0.05 to 0.18 GPa. Although these values were much lower than those for dense, sintered α-SiC, the ultimate tensile strength of these composites was found at a strain of 1% compared with 0.1% for monolithic SiC. Fracture toughness tests on the SiC composite gave results of 3.5 MPa $m^{1/2}$ which was the same as that for monolithic SiC.

Fitzer et al.[35] also studied the CVI SiC composites and obtained a flexural strength of 0.35 GPa for a polycarbosilane-derived SiC fiber, 2-D reinforced material. They measured much higher strengths, 0.9 GPa for a unidirectional SiC monofilament-reinforced SiC and 0.70 GPa for a high-modulus carbon fiber-reinforced SiC composite. Good fracture toughness (9.5 $MN/m^{3/2}$) was found for a carbon reaction-bonded SiC test part.

Lamicq et al.[36] studied the properties of a Cerasep-produced SiC fabric–CVD SiC composite. In the tensile test, the matrix cracked at 0.10 GPa with a strain failure of 0.23 GPa before breaking. The bend strength was 0.30 GPa at room temperature, increased to 0.40 GPa at 1300°C, and then dropped off at up to 2200°C. No change in the bend strength was observed after an 1100°C exposure to air for up to 500 h. The composite also retained its strength after being cycled 100 times from 700 to 1000°C. The fracture toughness was measured to be as high as 25 MPa $m^{1/2}$, but care must be taken in interpreting this result.

Caputo et al.[37] conducted investigations necessary to obtain reproducible results and to scale up to the SiC/CVD SiC formation process. The average flexural strength of 0.25–0.35 GPa was more consistent than previous results of 0.07–0.48 GPa. The strength at 1000°C varied from 0.08 to 0.28 GPa. The lower strength of these composites at elevated temperatures compared to the Lamicq results may be attributed to the carbon layer being oxidized.

For a polymeric composite, the transverse flexural strength and modulus are matrix-dominated properties. Their values also reflect, to a large extent, the interfacial bonding

TABLE 4.19. Transverse Flexural Properties of APC-2 and Other Thermoplastic Composite Materials[a]

	Flexural Strength		Flexural Modulus	
Material	ksi[b]	MPa	msi[b]	(GPa)
APC-2/AS-4, melt-impregnated (ICI)	19.7 (1.5)	136	1.17 (0.15)	8.07
P-1700/AS-4, solvent-impregnated (U.S. Polymeric)	4.33 (0.37)	29.9	0.623 (0.044)	4.30
P-1700/T-300, solvent-impregnated (AMOCO)	9.49 (1.1)	65.4	0.604 (0.068)	4.16
PPS/AS-4, melt-impregnated (PHILLIPS)	3.13 (0.66)	21.6	0.606 (0.097)	4.18
PAS-2/AS-4, melt-impregnated (PHILLIPS)	6.23 (1.2)	43.0	0.772 (0.048)	5.32
PEI/AS-4, hybrid weave (BASF)	11.1 (0.73)	76.5	0.755 (0.13)	5.21
PEEK/AS-4, hybrid weave (BASF)	18.2 (1.7)	125	0.981 (0.13)	6.76
PEEK/G30-500, powder prepreg (BASF)	14.1 (1.4)	97.2	0.910 (0.13)	6.27

[a] ASTM D790 method 1, nominal $L/D = 25$.
[b] Standard deviation is shown in parentheses.

strength between the fibers and the polymer matrix. In Table 4.19, data for nine composite laminates are summarized. Flexural tests are usually conducted according to ASTM D790 (Method 1, L/D-25). The results rank APC-2/AS4 the best among the materials evaluated.

4.1.4 In-Plane Shear Properties

A variety of test methods[13–15] have been used for measuring in-plane shear properties, such as shear modulus G_{12} and the ultimate shear strength τ_{12U} of unidirectional fiber-reinforced composites.

4.1.5 Interlaminar Shear Strength

Interlaminar shear strength (ILSS) refers to the shear strength parallel to the plane of lamination and is measured in a short-beam shear test. Because of its simplicity, the short-beam shear test is widely accepted for material screening and quality control purposes.[15,16]

Despite the limitations of the short-beam shear test, interlaminar shear failure is recognized as one of the critical failure modes in fiber-reinforced composite laminates. Interlaminar shear strength depends primarily on matrix properties and fiber-matrix interfacial shear strengths rather than on fiber properties. The interlaminar shear strength is improved by increasing the matrix tensile strength as well as the matrix volume fraction. Because of better adhesion with glass fibers, epoxies in general produce higher ILSS values than vinyl ester and polyester resins in glass fiber-reinforced composites. The interlaminar shear strength decreases, often linearly, with increasing void content. Fabrication defects, such as internal microcracks and dry strands, also reduce the interlaminar shear strength.

Morgan and Allred[38] evaluated the various mechanical properties of polyaramid composites, and some of the findings indicated that polyaramid composites have low off-axis strengths. The surface treatments and coupling agents developed for glass and graphite reinforcements increase interfacial strengths by two to three times those of untreated composites.[39,40] Consequently, interface-sensitive properties are weaker in polyaramid-containing systems than in their glass or graphite counterparts. Transverse tensile strengths of polyaramid-epoxy are only 40–50% as strong as those of glass-epoxy or graphite-epoxy. In interlaminar shear the polyaramid-epoxy is only 30–40% as strong as the glass-epoxy and the graphite-epoxy.[40] Thus, the interphase rather than the cohesive strength of the filaments or matrix limits the off-axis strength of Kevlar-epoxy.

4.2 FATIGUE PROPERTIES

The fatigue properties of a material represent its response to cyclic loading, which is a common occurrence in many practical applications. It is well recognized that the strength of a material is significantly reduced under cyclic loads. Metallic materials, which are ductile in nature under normal operating conditions, are known to fail in a brittle manner when subjected to repeated cyclic stresses (or strains). The cycle to failure depends on a number of variables, such as stress level, stress state, mode of cycling, process history, material composition, and environmental conditions.

Fatigue behavior of a material is usually characterized by an *S-N* diagram showing the relationship between the stress amplitude or maximum stress and number of cycles to failure on a semilogarithmic scale. This diagram is obtained by testing a number of specimens at various stress levels under sinusoidal loading conditions. For many fiber-reinforced composites, a fatigue limit may not be observed; however, the slope of the *S-N* plot is markedly reduced at low stress levels. In these situations, it is common practice to specify the fatigue strength of the material at very high cycles, say, 10^6 or 10^7 cycles.

Fatigue damage in short glass fiber-reinforced thermoplastics is characterized by the interaction of several damage mechanisms. Matrix deformation, matrix cracking, fiber debonding, and fiber fracture may all contribute to the failure of these materials. *S-N* curves show the relationship between these mechanical properties and the fatigue lives of the materials.

4.2.1 Fatigue Resistance

Fatigue resistance is another important material property to consider when designing a product. Fatigue describes the phenomenon of reduced strength over a time of load application (static fatigue) or the number of times a load is applied (cyclic fatigue). Figure 4.5 lists the cyclic strength of various metals and composites. The failure loads shown are given as stress density, which is the test stress divided by the material density. For the materials tested, composites have one to six times the specific fatigue resistance of metals.

Other studies have shown that composite plies are up to twice as strong as high-strength steel and three to five times as strong as titanium and aluminum. When comparing specific strengths, composites are three to eight times stronger than the metals. In stiffness the materials range from fiberglass composite at the low end to high-modulus

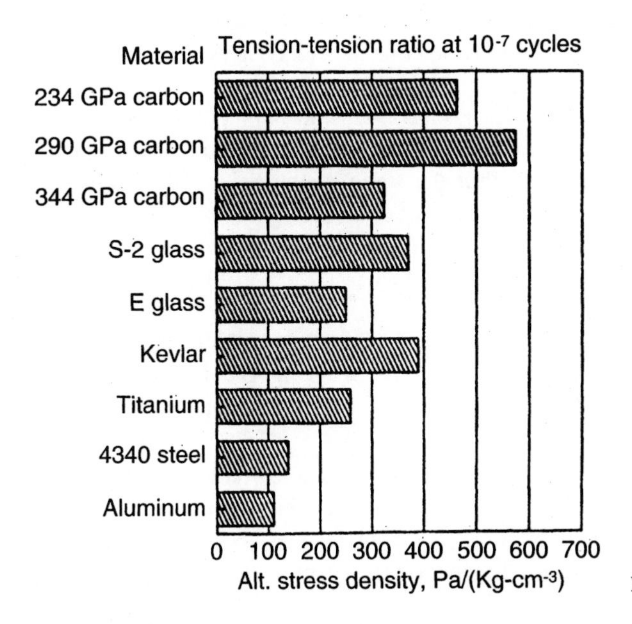

Figure 4.5. Fatigue of composites.

carbon composite at the high end with metals in between. Composites compare favorably to metals in specific modulus.

In metal parts, flaws introduced during the fabrication processes can propagate under cyclic loading and eventually result in failure of the structure. A composite material is less sensitive to flaws because the large number of fibers create redundant load paths through the composite layer.

4.2.2 Fatigue Test Methods

The majority of fatigue tests on fiber-reinforced composite materials have been performed with uniaxial tension-tension cycling. Tension-compression and compression-compression cycling are not commonly used because failure by compressive buckling may occur in thin laminates. Completely reversed tension-compression cycling is achieved by flexural fatigue tests.

The tension-tension fatigue cycling test procedure is described in ASTM D3479-76. It utilizes a straight-sided specimen with the same dimensions and end tabs as in static tension tests. At high cyclic frequencies, PMCs may generate appreciable heat as a result of internal damping, which in turn increases the test part temperature. Because a frequency-induced temperature rise can affect the fatigue performance adversely, low cyclic frequencies (less than 10 Hz) are preferred.

Hoppel[41] tested fatigue parts in tension-tension ($R = 0.1$) to evaluate the ways the interface controls the change in the mechanical properties of the composite. Tests also included a series of materials that were examined under both "wet" (immersed in water at 25°C) and "dry" (50% relative humidity at 25°C) conditions. The fatigue performance (life curves and cyclic stress-strain curves) was directly related to the fiber-matrix interface. Because water weakens the interface, especially in composites without a silane coupling agent, composites tested in the aqueous environment exhibited shorter fatigue lives than parts tested in the dry environment. Interfacial coatings that included a silane couplant were found to reduce the effects of water on the fatigue lives of the composites.

4.2.2.1 Tension-tension fatigue. Tension-tension fatigue tests on unidirectional 0° ultrahigh-modulus carbon-fiber-reinforced thermosetting polymers produce *S-N* curves that are almost horizontal and fall within the static scatter band. The fatigue effect is slightly greater for relatively lower-modulus carbon fibers.

The fatigue behavior for AS4/PEEK and IM6/PEEK laminates was examined by Curtis et al.,[42] and they found the following.

1. Tensile fatigue results for carbon fiber-PEEK at 23 and –55°C showed that the fatigue behavior at low temperatures need be of no more concern than the performance at 23°C. This is particularly encouraging in contemplating the performance of carbon fiber-PEEK at high altitudes.
2. Compressive fatigue results show that unidirectional carbon fiber-PEEK at 23 and 1200°C are satisfactory. These data provide a foundation for a designer exploring the feasibility of high-temperature applications for these composites where a buckling failure criterion might be a critical design aspect.

In addition, Curtis et al. have started to try to understand fatigue behavior by identifying the fundamental failure mechanisms that occur during the fracture of various laminates. It is apparent that this understanding is incomplete. By examining the influence of different material parameters and test parameters, the start of some useful observations can be made.

Unidirectional 0° boron and Kevlar 49 fiber composites also exhibit exceptionally good fatigue strength in tension-tension loading (Figure 4.6). Other laminates, such as

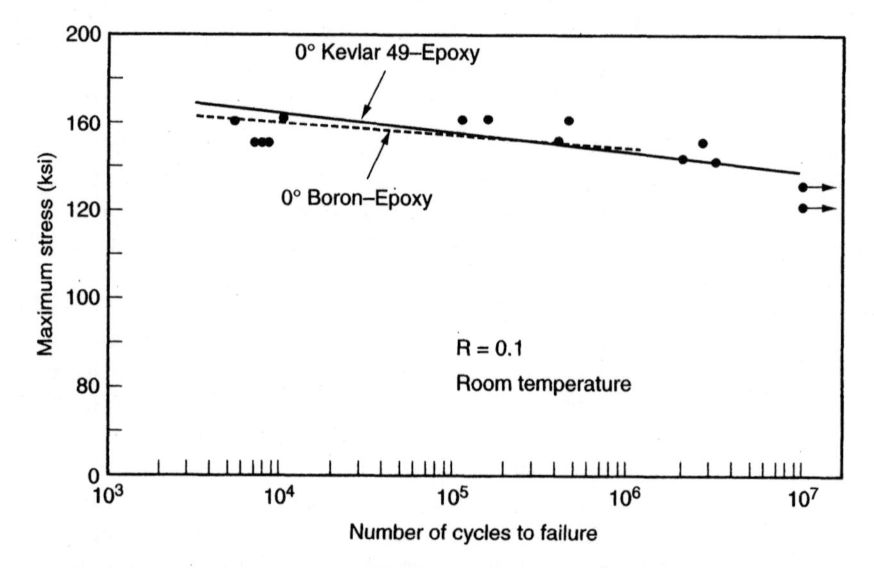

Figure 4.6. Tension-tension *S-N* diagrams for 0° boron and Kevlar 49 fiber-epoxy composites.

$[0/\pm45/90]_s$ carbon, $[0/90]_s$ carbon, $[0/\pm30]_s$ carbon, and $[0/\pm45]_s$ boron, show very similar *S-N* diagrams, although the actual fatigue effect depends on the properties of the fibers aligned with the loading axis, the stacking sequence, and mode of cycling.[42]

The fatigue performances of both E- and S-glass fiber-reinforced composites are inferior to those of carbon, boron, and Kevlar 49 fiber-reinforced composites. Both types of fibers produce steep *S-N* plots for unidirectional 0° composites. An improvement in their fatigue performance can be achieved by hybridizing them with other high-modulus fibers such as T-300 carbon.[43–45]

4.2.2.2 Flexural fatigue. The flexural fatigue performance of fiber-reinforced composite materials is in general less satisfactory than their tension-tension fatigue performance. In flexural *S-N* curves the slope is greater than that of the tension-tension *S-N* curve for high-modulus carbon fibers. The lower fatigue strength in flexure is attributed to the weakness of composites on the compression side.[46]

4.2.2.3 Interlaminar shear fatigue. Fatigue characteristics of fiber-reinforced composite materials in the interlaminar shear τ_{xz} mode have been studied by Pipes[47] and several other investigators.[48,49] For a unidirectional 0° carbon fiber-reinforced epoxy, the interlaminar shear fatigue strength at 10^6 cycles was reduced to less than 55% of its static ILSS even though its tension-tension fatigue strength was nearly 80% of its static tensile strength. The interlaminar shear fatigue performance of a unidirectional 0° boron-epoxy system was similar to that of a unidirectional carbon-epoxy system. However, a reverse trend was observed for a unidirectional 0° S-glass-reinforced epoxy. For this material, the interlaminar shear fatigue strength at 10^6 cycles was approximately 60% of its static ILSS, but the tension-tension fatigue strength at 10^6 cycles was less than 40% of its static tensile strength. Unlike the static interlaminar strengths, the fiber volume fraction[48] and fiber surface treatment[49] did not exhibit any significant influence on the high cycle interlaminar fatigue strength.

4.2.2.4 Torsional fatigue. The torsional fatigue strength of ±45° specimens of carbon fiber-reinforced epoxy thin tubes is approximately 3.7–3.8 times higher than that of 0° specimens after an equivalent number of cycles. The 0° specimens failed by a few longitudinal cracks (cracks parallel to fibers), and the ±45° specimens failed by cracking along the ±45° lines and extensive delamination. Although the 0° specimens exhibited a lower torsional fatigue strength than the ±45° specimens, they retained a much higher postfatigue static torsional strength.

Torsional fatigue data for a number of unidirectional 0° fiber-reinforced composites are compared in Figure 4.7. The data in this figure were obtained by shear-strain cycling of solid rod specimens.[50] Fatigue testing under pure shear conditions clearly has a severe effect on unidirectional composites, all failing at approximately 10^3 cycles at approximately half the static shear strain to failure. Short-beam interlaminar shear fatigue experiments do not exhibit such rapid deterioration.

4.2.2.5 Compressive fatigue. Compression-compression fatigue *S-N* diagrams for various E-glass fiber-reinforced polyester and epoxy composites are shown in reference 51. Similar trends are also observed for T-300 carbon fiber-reinforced epoxy systems.[51]

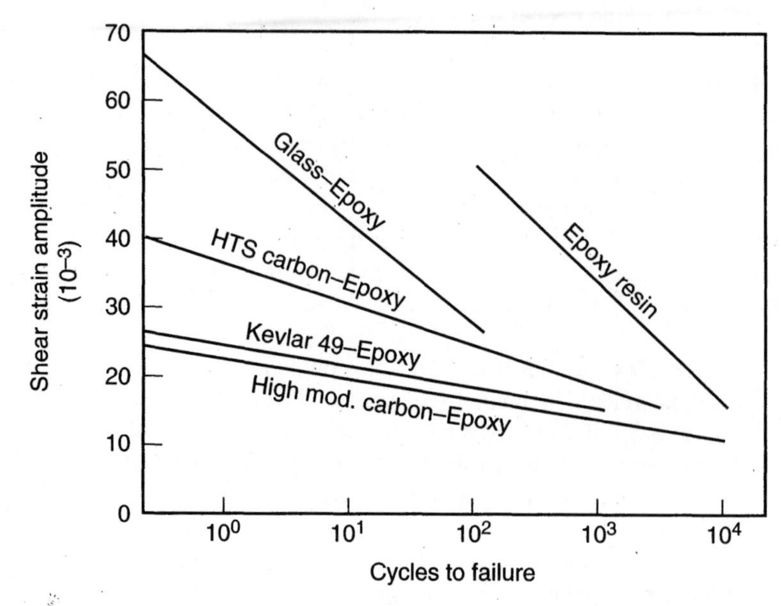

Figure 4.7. Torsional shear strain cycle diagrams for various 0° fiber-reinforced composites.[49,50]

4.3 IMPACT PROPERTIES

The impact properties of a material represent its capacity to absorb and dissipate energies under impact or shock loading. In practice, the impact condition may range from the accidental dropping of a hand tool to a high-speed collision, and the response of a structure may range from localized damage to total disintegration. If a material is strain-rate-sensitive, its static mechanical properties cannot be used in designing against impact failure.

A variety of standard impact test methods are available for unreinforced plastics (ASTM D256-78). Some of these tests have also been adopted for fiber-reinforced composite materials. However, as in the case of unreinforced plastics, the impact tests do not yield basic material property data that can be used for design purposes. They are useful in comparing the failure modes and energy absorption capabilities of two different materials under identical impact conditions. They can also be used in the areas of quality control and materials development.

4.3.1 Material Parameters

The primary factor influencing the longitudinal impact energy of a unidirectional 0° composite is the fiber type. E-glass fiber composites have the highest impact energy because of the relatively high strain to failure of E-glass fibers. Carbon and boron fiber composites have low strains to failure which lead to low-impact energies for these composites. Increasing the fiber volume fraction also leads to higher impact energy.

The next important factor influencing the impact energy is the fiber-matrix interfacial shear strength. Several investigators[52–54] have reported that impact energy is reduced when fibers are surface-treated for improved adhesion with the matrix. At high levels of adhesion, the failure mode is brittle and relatively little energy is absorbed. At very low levels of adhesion, multiple delamination may occur without significant fiber failure. Although the energy absorption is high, failure may take place catastrophically. At intermediate levels of adhesion, progressive delamination occurs, which in turn produces high-impact energy absorption.

The matrix can influence the impact damage mechanism because delamination, debonding, and fiber pullout energies depend on fiber-matrix interfacial shear strength. Because epoxies have better adhesion with E-glass fibers than polyesters, E-glass-epoxy composites exhibit higher impact energies than E-glass-polyester composites when the failure mode is a combination of fiber failure and delamination.

In unidirectional composites, the greatest impact energy is exhibited when the fibers are oriented in the direction of the maximum stress, that is, at 0° fiber orientation.

The most efficient way of improving impact energy of a low strain-to-failure fiber composite is to hybridize it with high strain-to-failure fiber laminas. Mallick and Broutman[55] have shown that a hybrid sandwich composite containing GY-70 fiber laminas in the outer skin and E-glass fiber laminas in the core has an impact energy 35 times higher than that of the GY-70 fiber composite. This is achieved without much sacrifice in either flexural strength or flexural modulus. The improvement in impact energy is due to delamination of the GY-70/E-glass interface, which occurs after the GY-70 skin on the tension side has failed. Even after the GY-70 skin sheds owing to complete delamination, the E-glass laminas continue to withstand higher stresses, preventing brittle failure of the whole structure. By varying the lamination configuration as well as the fiber combination, a variety of impact properties can be obtained. Furthermore, the impact energy of a hybrid composite can be varied by controlling the ratio of various fiber volume fractions.

4.4 OTHER PROPERTIES

4.4.1 Pin-Bearing Strength

Pin-bearing strength is an important design parameter for bolted joints and has been studied by a number of investigators.[56–61]

For 0° laminates, failure in pin-bearing tests occurs by longitudinal splitting because such laminates have poor resistance to in-plane transverse stresses at the loaded hole. The bearing stress at failure for 0° laminates is also quite low. Inclusion of 90° layers [100], ±45° layers, or ±60° layers [101] at or near the surface improves the bearing strength significantly. However, $[\pm 45]_s$, $[\pm 60]_s$, and $[90/\pm 45]_s$ laminates have lower bearing strengths than $[0/\pm 45°]_s$ and $[0/\pm 60°]_s$ laminates. A number of other observations on the pin-bearing strength of composite laminates are as follows.

1. Stacking sequence has a significant influence on the pin-bearing strength of composite laminates. Quinn and Matthews[59] have shown that a $[90/\pm 45/0]_s$ lay-up is nearly 30% stronger in pin-bearing tests than a $[0/90/\pm 45]_s$ lay-up.
2. Collings[58] has shown that a $[0/\pm 45]_s$ laminate attains its maximum pin-bearing strength when the ratio of 0 to 45° layers is 60:40.
3. Fiber type is an important material parameter in developing high pin-bearing strength in various laminates. Kretsis and Matthews[56] have shown that for the same specimen geometry, the bearing strength of a $[0/\pm 45]_s$ carbon fiber-reinforced epoxy laminate is nearly 20% higher than that of a $[0/\pm 45]_s$ E-glass fiber-reinforced epoxy.

4. The pin-bearing strength of a composite laminate can be increased significantly by adhesive-bonding a metal insert (preferably an aluminum insert) at the hole boundary.[60]
5. Lateral clamping pressure distributed around the hole by washers can significantly increase the pin-bearing strength of a laminate.[61] The increase is attributed to the lateral restraint provided by the washers, as well as by frictional resistance against slip. The lateral restraint contains the shear cracks developed at the hole boundary within the washer perimeter and allows the delamination to spread over a wider area before final failure occurs. The increase in pin-bearing strength levels off at high clamping pressure. If the clamping pressure is too high, causing the washers to dig into the laminate, the pin-bearing strength may decrease.

4.4.2 Damping Properties

The damping properties of a material represent its capacity to reduce the transmission of vibration caused by mechanical disturbances to a structure. The measure of damping of a material is its damping factor η. Increasing the value of η is desirable in reducing the resonance amplitude of vibration in a structure. Fiber-reinforced composites, in general, have a higher damping factor than metals. However, its value depends on a number of factors, including fiber and resin types, fiber orientation angle, and stacking sequence.

4.4.3 Coefficient of Thermal Expansion

The CTE represents the change in unit length of a material due to unit temperature rise or drop. Its value is used in calculating dimensional changes as well as thermal stresses caused by temperature variation.

The CTEs of unreinforced polymers are higher than those of metals. The addition of fibers to a polymeric matrix generally lowers its CTE. Depending on the fiber type, orientation, and fiber volume fraction, the CTE of fiber-reinforced polymers can vary over a wide range of values. In unidirectional 0° laminates, the longitudinal CTE α_{11} reflects the fiber characteristics. Thus, both carbon and Kevlar 49 fibers produce a negative CTE, and glass and boron fibers produce a positive CTE in the longitudinal direction. As in the case of elastic properties, the CTEs for unidirectional 0° laminates differ in the longitudinal and transverse directions.[14] Compared with carbon fiber-reinforced epoxies, Kevlar 49 fiber-reinforced epoxies exhibit a greater anisotropy in their CTE because of greater anisotropy in the Kevlar 49 fibers.[62]

In quasi-isotropic laminates, as well as in randomly oriented discontinuous fiber laminates, the CTEs are equal in all directions in the plane of the laminate. Furthermore, with proper fiber type and lamination configuration, the CTE in the plane of the laminate can be made close to zero. An example is shown in Figure 4.8, in which the proportions of fibers in 0, 90, and ±45° layers were controlled to obtain a variety of CTEs in the $[0/\pm45]_s$, and $[90/\pm45]_s$ laminates.[63]

4.4.4 Thermal Conductivity

The thermal conductivity of a material represents its capacity to conduct heat. Polymers in general have low thermal conductivities, which makes them useful as insulating materials. However, in some circumstances, they may also act as a heat sink by not being able to dissipate heat efficiently. As a result, there may be a temperature rise within the material.

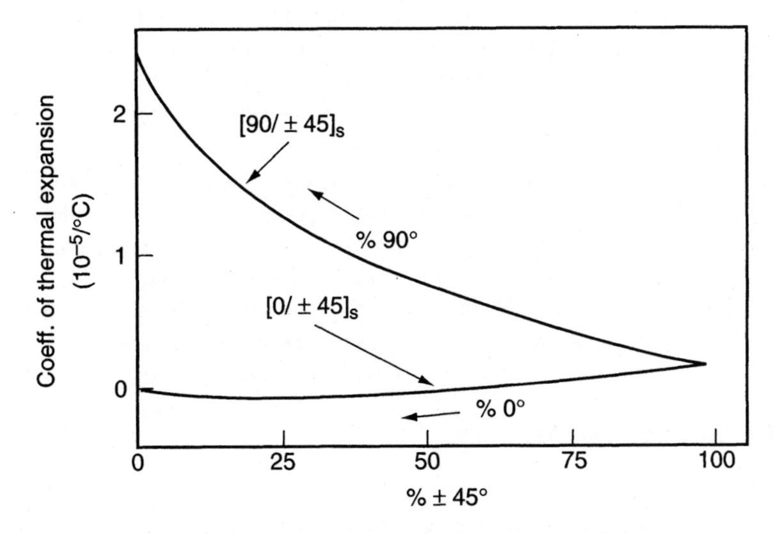

Figure 4.8. Coefficients of thermal expansion of $[0/\pm45]_s$ and $[90/+45]_s$ laminates.[63]

The thermal conductivity of a fiber-reinforced polymer depends on the fiber type, orientation, fiber volume fraction, and lamination configuration.[14] With the exception of carbon fibers, fiber-reinforced polymers in general have low thermal conductivities. Carbon fiber-reinforced polymers possess relatively high thermal conductivities as a result of the highly conductive nature of carbon fibers. For unidirectional 0° composites, the longitudinal thermal conductivity is controlled by the fibers and the transverse thermal conductivity is controlled by the matrix. This is reflected in widely different values of thermal conductivities in these two directions.

The electric conductivities of fiber-reinforced polymers are similar in nature to their thermal counterparts. For example, E-glass fiber-reinforced polymers are poor electric conductors and tend to accumulate static electricity. For protection against static charge buildup and the resulting electromagnetic interference (EMI) or radiofrequency interference (RFI), small quantities of conductive fibers, such as carbon fibers, aluminum flakes, or aluminum-coated glass fibers, are added to glass fiber composites.

4.5 THE INFLUENCE OF ENVIRONMENTAL EFFECTS

The influence of environmental factors, such as elevated temperatures, high humidity, corrosive fluids, and ultraviolet (UV) rays, on the performance of PMCs is of concern in many applications. These environmental conditions may cause degradation in the mechanical and physical properties of a fiber-reinforced polymer because of one or more of the following reasons.

1. Physical and/or chemical degradation of the polymeric matrix, for example, reduction in modulus owing to increasing temperature, volumetric expansion resulting from moisture absorption, and scission or alteration of polymer molecules owing to chemical attack or ultraviolet rays. However, it is important to note that different groups of polymers or even different molecular configurations within the same group of polymers may respond differently to the same environment.

2. Loss of adhesion or debonding at the fiber-matrix interface, which may be followed by diffusion of water or other fluids into this area. In turn, this may cause a reduction in fiber strength as a result of stress corrosion. Many experimental studies have shown that compatible coupling agents are capable of either slowing down or preventing the debonding process even under severe environmental conditions such as exposure to boiling water.
3. Reduction in fiber strength and modulus. For a short-term or intermittent temperature rise up to 150–294°C, reduction in the properties of most commercial fibers is insignificant. However, depending on the fiber type, other environmental conditions may cause deterioration in fiber properties. For example, moisture is known to accelerate static fatigue in glass fibers. Kevlar 49 fibers are capable of absorbing moisture from the environment, which reduces their tensile strength and modulus. The tensile strength of Kevlar 49 fibers is also reduced with direct exposure to ultraviolet rays.

4.5.1 Elevated Temperatures

When a polymer specimen is tension-tested at elevated temperatures, its modulus and strength decrease with increasing temperature because of thermal softening. In a PMC, the matrix-dominated properties are more affected by increasing temperature than the fiber-dominated properties. For example, the longitudinal strength and modulus of a unidirectional 0° laminate remain virtually unaltered with increasing temperature, but its transverse and off-axis properties are significantly reduced as the temperature approaches the T_g of the polymer. For a randomly oriented discontinuous-fiber composite, strength and modulus are reduced in all directions.[14]

High temperatures are very problematic. As temperature is increased, all resins soften and reach a stage where the polymer passes from a glasslike state to a rubbery state. This is the aforementioned glass transition temperature T_g. Beyond T_g the properties of the polymer change dramatically. Hence this is the limit of the usable temperature range for all normal applications. However, the effect the fibers have is to improve the T_g of the composite above that of the resin matrix.

Thermal aging due to long-term exposure to elevated temperatures without load can cause deterioration of the properties of a PMC. Kerr and Haskins[64] reported the effects of 100–50,000 h of thermal aging on the tensile strength of AS carbon fiber-epoxy and HTS carbon fiber-polyimide unidirectional and cross-ply laminates. For the AS carbon-epoxy systems, thermal aging at 121°C produced no degradation for the first 10,000 h. Matrix degradation began between 10,000 and 25,000 h and was severe after 50,000 h. After 5000 h the matrix was severely embrittled. Longitudinal tensile strength was considerably reduced for aging times of 5000 h or longer. The HTS carbon-polyimide systems were aged at higher temperatures but showed less degradation than the AS carbon-epoxy systems.

Concern for the reduction in mechanical properties of thermoplastic composites at elevated temperatures is greater than for thermoset composites because the properties of thermoplastic polymers are reduced significantly at or slightly above their glass transition temperatures. As in thermoset composites, the effect of increasing temperature is more pronounced for matrix-dominated properties than for fiber-dominated properties.

As a general rule both the strength and the stiffness of composites are unimpaired by the effect of low temperature, and under some circumstances they may be enhanced.

However, low temperatures tend to make polymers less flexible, and therefore there may be a tendency toward damage by fatigue.

4.5.2 Moisture

When exposed to humid air or water environments, many PMCs take up moisture by instantaneous surface absorption followed by diffusion through the matrix. Analysis of moisture absorption data for epoxy and polyester matrix composites shows that the moisture concentration through a diffusion process increases initially with time and approaches an equilibrium (saturation) level after several days of exposure to a humid environment.[14] The rate at which the composite laminate attains the equilibrium moisture concentration is determined by its thickness as well as by the ambient temperature. On drying, the moisture concentration is continually reduced until the composite laminate returns to the original as-dry state.

All polymers are susceptible to the effects of moisture, which in general result in a reduction in mechanical properties and glass transition temperature. The loss of properties is a function of the degree of moisture pickup. Resin systems can be selected that have excellent resistance to the effects of moisture.[25] However, the glass transition temperature can be reduced to 75% of its dry value by a moisture content of 4%. Flexural strength can be reduced to 50% of the dry value by a 1.5% moisture content.

The designer must therefore take the effects of moisture into account at the design stage. If the composite is to be subjected to a wet environment, then "wet properties" must be used in the design process.

A 2-year specimen test program was undertaken to generate data and design allowables for the severe environmental conditions of a material that consisted of Celion 3000 (C3000) PAN precursor carbon tows woven into an eight-harness satin-weave fabric and impregnated with PMR-15 polyimide resin. Allowables were to include data at 204 and 316°C after moisture conditioning (wet). Specimens tested as fabricated were considered "dry," although their actual moisture content was 0.1–0.4% by weight. Wet specimens typically had a moisture content of 1.21–1.33% by weight. Ply properties were evaluated and then used to predict laminate behavior under load. The properties of the laminate were then verified by tests.

Researchers found that absorbed moisture caused a significant reduction in elevated-temperature resin-dominated properties. This reduction in mechanical properties was significant even at moderate levels of moisture, such as 40–50% of saturation at the time of specimen failure. Elevated temperature exposure, even without prior moisture conditioning, also significantly reduced mechanical properties, especially at temperatures greater than 204°C.

They also found that hot-wet conditions can be obtained by graphite-PMR-15 laminates during service; therefore, the effects of these conditions must be known and considered in the design. Hot-wet specimens should be heated quickly to retain sufficient moisture at the time of failure.

Finally, when designing graphite-PMR-15 structures, several variables must be determined and considered including the following: temperature, moisture absorption, and load time history relationships. Design allowables should then be developed using appropriate moisture levels and test temperatures that simulate the service environment.

4.5.2.1 Performance changes due to moisture and temperature. The effects of temperature and moisture content on the tensile strength and modulus of carbon and boron fiber-reinforced epoxy laminates have been investigated widely. From the summary of available data[65,66] the following conclusions can be made.

For 0 and [0°/±45/90]s quasi-isotropic laminates, changes in temperature up to 107°C have negligible effects on tensile strength and modulus values regardless of the moisture concentration in the material. Although the effect on modulus is negligible up to 177°C, there may be up to a 20% decrease in tensile strength as the temperature increases from 107 to 177°C.

For 0 and $[0°/\pm45/90]_s$ laminates, the tensile strength and modulus are not affected by moisture absorption below a 1% moisture concentration. Although the modulus is not affected by an even higher moisture concentration, the tensile strength may decrease by as much as 20% for moisture concentrations above 1%.

For 90° laminates, increasing temperature and moisture concentration reduce both the tensile strength and the modulus by significant amounts. Depending on the temperature and moisture concentration, the reduction may range as high as 60–90% of the room temperature properties under dry conditions.

The interlaminar shear strength of composite laminates is also reduced by increasing the moisture absorption. For example, short-beam shear tests of a unidirectional carbon fiber–epoxy show nearly a 10% reduction in ILSS at a moisture concentration of 1.2 wt%.[67,68]

Jones et al.[69] reported the effect of moisture absorption on the tension-tension fatigue and flexural fatigue properties of a $[0/90]_s$ cross-ply epoxy matrix composite reinforced with E-glass, HTS carbon, and Kevlar 49 fibers. Conditioning treatments included exposure to humid air (65% relative humidity) and immersion in boiling water. The fatigue resistance of carbon fiber-epoxy was found to be unaffected by the conditioning treatment. Exposure to humid air also did not affect the fatigue response of E-glass fiber-epoxy composites; however, immersion in boiling water reduced the fatigue strength by significant amounts, principally as a result of the damage incurred by the glass fibers from the boiling water. On the other hand, the fatigue response of Kevlar 49-epoxy composites was improved owing to moisture absorption, although at high cycles there appears to be a rapid deterioration as indicated by the sharp downward curvature of the *S-N* curve.[69]

4.6 LONG-TERM PROPERTIES

4.6.1 Creep

Creep is defined as the increase in strain with time at a constant stress level. In polymers, creep occurs because of a combination of elastic deformation and viscous flow commonly known as viscoelastic deformation. The resulting creep strain increases nonlinearly with time. When the stress is released after a period of time, the elastic deformation is immediately recovered. The deformation caused by the viscous flow recovers slowly to an asymptotic value called the recovery strain.

Creep strain in polymers and polymeric composites depends on the stress level and temperature. Many unreinforced polymers can exhibit large creep strains at room temperature and at low stress levels. At elevated temperatures or high stress levels, the creep phenomenon becomes even more critical. In general, highly cross-linked thermosetting polymers exhibit lower creep strains than thermoplastic polymers. With the exception of Kevlar 49 fibers, commercial reinforcing fibers, such as glass, carbon, and boron, do not creep.[70]

As would be expected, unidirectional reinforcement is the most creep-resistant construction. This is followed by bidirectional woven construction which has the disadvantage that the fibers tend to straighten out, thus increasing creep. The construction least resistant to creep is chopped strand mat (random in-plane), which has relatively short-length fibers, thus making the composite creep performance more dependent on the creep resistance of the matrix.

The creep performance of carbon fiber-reinforced composite materials in the direction of the fibers is remarkably good; it compares very favorably with "low-relaxation" steel and is significantly better than that of "standard" steel.[71]

When subjected to a tensile load, carbon fiber composites are considered to have a major advantage in their ability to resist long-term creep. Here the totally elastic performance of the fibers dominates the behavior of the composite. However, when under compression or off-axis, the matrix properties become more significant. Polymers are viscoelastic materials, and they deflect continuously with time under load.

In spite of their high inherent tensile strength, and even in unidirectional configurations, aramid fiber composites have creep rates generally very much higher than for glass or carbon composites.

4.6.2 Stress Rupture

Stress rupture is defined as the failure of a material under sustained constant load. It is usually tested by applying a constant tensile stress to a specimen until it fractures completely. The time at which the fracture occurs is termed the lifetime or stress rupture time. The objective of this test is to determine a range of applied stresses and lifetimes within which the material can be considered "safe" for long-term static load applications.

Glass, Kevlar 49, and boron fibers and their composites exhibit failure by stress rupture. Carbon fibers, on the other hand, are relatively less prone to stress rupture failure. Chiao and his coworkers[72,73] have gathered the most extensive stress rupture data on epoxy-impregnated S-glass and Kevlar 49 fiber strands. For both materials the lifetime at a given stress level varied over a wide range. However, the rate of degradation under sustained tensile load was lower in Kevlar 49 strands than in S-glass strands.[14]

As an example of the maximum likelihood estimates of lifetimes for Kevlar 49 strands is when one considers the first percentile lifetime estimate of Kevlar 49 strands at 1380 MPa, which is also 39.5% of the fiber's UTS. Furthermore, a maximum likelihood estimate for the lifetime is 126,000 h = 14.4 years.[14]

4.6.3 Damping Characteristics

The damping characteristic of a material, that is, its ability to reduce induced vibrations rapidly, is an important aspect of the material selection process in certain areas. Fishing rods, for instance, benefit from good damping properties because vibrations die out more rapidly, reducing frictional resistance of the line against the guides.

Polymer composites, particularly carbon fiber-reinforced composites and aramid-reinforced composites, have a very good damping performance which can be used to reduce vibration resonance.

4.6.4 Fire Resistance

The factors by which fire performance is assessed include

- Surface spread of flame
- Fire penetration
- Ease of ignition
- Fuel contribution
- Oxygen index (i.e., the minimum oxygen content that supports combustion).

Each factor varies in significance depending on the particular circumstances. In short, there is no single simple test by which composites can be assessed and compared in terms of their fire resistance.

The fire performance of composites covers a wide spectrum, ranging from highly flammable to nonburning. The degree of flammability of a composite is governed by the following factors.

- Matrix type
- Quantity and type of fire retardant additives included
- Quantity and type of fillers included
- Reinforcement type—volume fraction and construction.

The dominant factor is the polymer matrix. For the polymers commonly used in composites the approximate order of flammability without fire retardant additives is as follows.

Polyester	Burns readily
Vinyl ester	↓
Epoxy	
Phenolic	Excellent fire resistance

Polyesters, in spite of their relatively poor intrinsic fire resistance, can have very low flammability after the incorporation of suitable fire-retardant additives and a reasonably high glass fiber content.

4.6.5 Galvanic Corrosion

Glass fiber and aramid fiber composites do not suffer from galvanic corrosion. However, carbon fiber is a conductor of electricity, and relative to those various metals its composites are highly inert. Therefore, if carbon composites are used in contact with metals, galvanic corrosion can take place. The severity of the corrosion depends on the distance between the composite and the metal, the extent of the polarization, and the efficiency of the electrolyte. If the conditions are favorable, then the metal, being more anodic, will corrode away. This would be disastrous if the two were bonded together as a structural joint. The situation is overcome simply by ensuring, by the application of a suitable coating, that the carbon composite and the metal do not come into contact.

4.6.6 Chemical Corrosion

Composites are inherently corrosion-resistant and can show substantial cost benefits when used in environments that are extremely aggressive. Hence glass-reinforced plastics have been used extensively in chemical plant structures for many years. The polymers are selected on the basis of their ability to withstand the aggression of the chemical environment, be it strong acid, strong alkali, or perhaps simply salt water.

4.6.7 Weathering

The mechanical performance of composites is degraded by the effects of temperature, moisture, sunlight, wind, dust, and acid rain. These effects come under the general heading of weathering. There are several mechanisms attributed to weathering, the most notable being leaching from the resin of chemical constituents; sunlight attack on the resin, causing embrittlement and erosion as a result of its degradation; and the effect of wind- and airborne particles. Thus, the environment is allowed access to the fibers. However, for a carbon-epoxy composite in flying service, "generally there is no indication of increasing degradation with length of time in service, the maximum deterioration occurring in the first year."[69]

4.6.8 Erosion

A significant aspect of weathering is the erosive effect of rain in rapidly moving air. This is the case, for example, in high-speed flight and with helicopter blades, where radomes, antenna covers, fairings, the leading edge of the blades, and so on, are susceptible to erosion from rain, snow, and ice. The rate of erosion is a function of speed, angle of impact, mass, and frequency. For example, an unpainted edge cap on an aircraft radome that had been exposed for 19 years had a strength retention of 68% according to Lubin[74] and Blackford.[75] The most common method of countering these effects is the use of rain-erosion-resistant finishes such as polyurethane. These can be so beneficial that rain erosion can be reduced from "badly degraded" to "no drop in tensile or flexural strength or modulus."

4.6.9 Lightning

Carbon fiber-reinforced composites are more susceptible to damage from lightning than is aluminum. Hence, the use of carbon fiber-reinforced plastics (CFRPs) in aircraft has prompted investigations into methods of protection from lightning. It is well known that aircraft are at risk and have been lost as a result of damage from lightning strikes. The losses have been due both to structural damage to the aircraft and to electrical effects on systems and elements that are critical.

A CFRP is a poor conductor of electricity. It therefore has neither the ability to dissipate electricity, as does aluminum, nor the insulating ability of a graphite-reinforced plastic (GRP).

The most common method of protecting CFRPs against "arc root" damage from lightning is by the use of surface metallization. This is achieved either by flame spray or by molding in a layer of aluminum mesh or tissue. The CFRP is insulated from the layer by a thin layer of GRP. An excellent review is given in Payne et al.[76]

4.6.10 Stress Corrosion

The stress corrosion of glass fiber composites is a mechanism of failure where a laminate under stress in an acid or basic environmnent can fail catastrophically at very low stresses compared to the fracture strength in air.

The phenomenon is dominated by the resistance of the glass fiber to the aggressive environment. The matrix has several roles: it must act as a chemical barrier, but more importantly, to avoid stress corrosion, it must be tough.

It is thought that the mechanism requires that the environment gain access to the glass fiber either by diffusion or via surface cracks. Work by Caddock, Evans, and Hull[77] has shown that polyester resin is impermeable to hydrochloric acid and thus discounts the permeability mechanism for these particular laminates. This implies that the propagation of cracks between the fiber and the laminate surface is generally the route of environmental ingress. Having gained access, the environment may or may not degrade the fiber very rapidly.

Hence the fiber is not able to perform one of its primary duties, to act as a crack-stopping device. Matrix cracks propagate, allowing the acid access to the next fiber. This fiber is attacked by the acid, fails, and so on.

There appear to be three types of specimen failure.[77] At high loads failure is similar to dry failure under tension. At lower loads the failure surface is stepped, but at very low loads the fracture surfaces are very smooth.

A further complication is that some aggressive agents do not attack the fiber but severely degrade the polymer. Tests by Steard and Jones[78] with E-glass-epoxy showed rapid microcracking of the laminate, but the fibers alone were found to be virtually crack-free. It is therefore dangerous to generalize in this field, and it is necessary to refer to relevant data or to carry out accelerated stress corrosion trials.

4.6.11 Blistering

It has been observed that gel-coated polyester laminates reinforced with glass fiber may form unsightly blisters on their surfaces after lengthy contact with water.

The mechanism by which blisters occur is thought to be basically osmotic. The laminating resin and gel-coat resin are permeable to water molecules, which pass into the resin, albeit very slowly. Any voids in the laminate pick up moisture. If the resin system is hydrolytically unstable (i.e., able to decompose by reaction with water), then salts may be present in the water to allow the gel coat to act as a semipermeable membrane and thus create an osmotic cell. Sufficient pressure builds up eventually and creates a blister.

Considerable experimentation has shown that in the prevention of blistering, there are two major considerations: raw material selection and production techniques. The use of water-resistant resins and preferably "powder-bound" rather than "emulsion-bound" chopped strand mat is the most important aspect of material selection. Production techniques must ensure thoroughly impregnated and wetted glass fiber and the correct resin/glass ratio per Birley et al. and Crump.[79,80]

4.7 FRACTURE BEHAVIOR AND DAMAGE TOLERANCE

The fracture behavior of materials is concerned with the initiation and growth of critical cracks that may cause premature failure of a structure. In fiber-reinforced composite materials, such cracks may originate at manufacturing defects, such as microvoids, matrix microcracks, and ply overlaps, or at localized damages caused by in-service loadings, such as subsurface delaminations due to low-energy impacts and hole edge delaminations due to static or fatigue loads. The resistance to the growth of cracks that originate at localized damage sites is frequently referred to as the damage tolerance of the material.

4.7.1 Improved Damage Tolerance

The damage tolerance of laminated composites is improved if the initiation and/or growth of delamination can be either prevented or delayed. Efforts to control delamination have focused on both improving interlaminar fracture toughness and reducing interlaminar stresses by means of laminate tailoring. Material and structural parameters that influence the damage tolerance are matrix toughness, fiber-matrix interfacial strength, fiber orientation, stacking sequence, laminate thickness, and support conditions.

4.7.1.1 Matrix toughness. The fracture toughness of epoxy resins commonly used in the aerospace industry is 100 J/m^2 or less. Laminates using these resins have an interlaminar (mode I delamination) fracture toughness in the range 100–200 J/m^2. Increasing the fracture toughness of epoxy resins has been shown to increase the interlaminar fracture toughness of the composite. However, the relative increase in the interlaminar fracture toughness of the laminate is not as high as that of the resin itself.

The fracture toughness of an epoxy resin can be increased by adding elastomers (e.g., CTBN), reducing cross-link density, increasing resin chain flexibility between cross-links, or a combination of all three. Addition of rigid thermoplastic resins also improves fracture toughness. Another alternative is to use a thermoplastic matrix, such as PEEK, PPS, PAI, and so on, which has a fracture toughness value in the range 1000 J/m^2, 10-fold higher than that of conventional epoxy resins.[81]

4.7.1.2 Interleaving. A second approach to enhancing interlaminar fracture toughness is to add a thin layer of tough, ductile polymer or adhesive between consecutive plies in the laminate. Although the resin-rich interleaves increase interlaminar fracture toughness, fiber-dominated properties, such as tensile strength and modulus, may decrease as the result of a reduction in overall fiber volume fraction.

4.7.1.3 Stacking sequence. High interlaminar normal and shear stresses at the free edges of a laminate are created by mismatches between Poisson's ratio and coefficients of mutual influence in adjacent layers. Changing the ply stacking sequence may change the interlaminar normal stress from tensile to compressive so that opening mode delamination can be suppressed. However, the growth of delamination may require the presence of an interrupted load path.

4.7.1.4 Interply hybridization. This technique can be used to reduce the mismatch between Poisson's ratio and coefficients of mutual influence in consecutive layers and thus reduce the possibility of interply edge delamination. For example, replacing the 90° carbon fiber-epoxy plies in $[\pm45/0_2/90_2]_s$ AS-4 carbon fiber-epoxy laminates with E-glass fiber-epoxy plies increases the stress level for the onset of edge delamination due to interlaminar normal stress from 324 MPa to 655 MPa. The ultimate tensile strength is not affected, because it is controlled mainly by the 0° plies which are carbon fiber–epoxy for both laminates.[82]

4.7.1.5 Through-thickness reinforcement. Resistance to interlaminar delamination can be improved by means of through-thickness reinforcement which can be in the form of stitches, metallic wires and pins, or three-dimensional fabric structures.

4.7.1.6 Ply termination. High interlaminar stresses created by mismatching plies in a narrow region near the free edge are reduced if they are terminated away from the free edge. Chan and Ochoa[83] found that to be the case and the resultant ultimate tensile strength of a laminate with 90° ply termination. In the baseline laminate, they found that free-edge delamination between the central 90° layers was observed at 49% of the ultimate load.

4.7.1.7 Edge modification. Chu and Sun[84] introduced a series of narrow and shallow notches along the laminate edges and observed a significant increase (25% or higher) in tensile failure load for laminates prone to interlaminar shear failure. Delamination was either eliminated or delayed in laminates prone to opening mode delamination, but there was no improvement in tensile failure load. The presence of notches disrupts the load path near the free edges and reduces the interlaminar stresses. However, they also introduce high in-plane stress concentration. Thus, suppression of delamination by edge notching may require proper selection of notch size and spacing.

4.8 TRIBOLOGICAL PROPERTIES

The anisotropic physical properties and unique specific strength and modulus of advanced composites have made them an important class of material for structural as well as tribological applications. Systematic studies have been carried out to generate data for the usual mechanical properties, and Vishwanath et al.[85] have evaluated tribological properties.

Researchers have shown the dependence of friction and wear of fiber-reinforced polymeric composites on the material and form of the reinforcing fiber, its orientation, the type of matrix, the fiber volume fraction, the counterface material, the sliding conditions, and so on. Addition of glass, carbon, and Kevlar fibers to polymers is known to improve their mechanical and tribological properties. Vishwanath and associates[85] investigated the effect of reinforcement materials on the wear and friction characteristics of fabric-reinforced polymeric composites. Plain-weave fabric reinforcement of three different fibers, namely, E-glass, high-strength carbon, and Kevlar 49 with polyvinyl butadiene (PVB) modified phenolic resin matrix, were used to fabricate the composites.

They found that

The rate of increase in the specific wear rate of the glass-phenolic composite was higher than that of the carbon-phenolic or Kevlar-phenolic composites.

At all sliding velocities, the coefficients of friction obtained in the Kevlar-phenolic composite were higher than those in the glass-phenolic and carbon-phenolic composites.

The critical sliding velocity, where the mechanism of wear changes from sliding to impactlike wear, is low (i.e., less than 2.93 m/s^1) in all three composites.

Changes in the roughness of the counterface surface depend on the type of reinforcing material and the resulting wear debris and influence the wear of the composites.

The resulting wear particles in the glass-phenolic composite are the most abrasive and cause increased surface roughness of the counterface and consequently an increase in the specific wear rate.

The low wear rates in the Kevlar-phenolic composites are due to the nonabrasive nature of the worn debris and the subsequent smoothing of the counterface.

It should be noted that wear and tribological characteristics are system properties rather than material properties. Operating wear mechanisms depend on the contact type (rolling, sliding, etc.), the operating conditions, the environment, the material characteristics of the test material, and the mating material. In the case of MMCs, the hard ceramic constituents such as SiC are more wear-resistant than the metal matrix, and the soft matrix keeps the reinforcement in place and transfers the applied load to the reinforcement.

Of the various types of tests for wear resistance, lubricated rolling and sliding wear are two common ones. For rolling contact tests, a roller-on-cylinder is used; the roller is made of the composite test material, and the cylinder is made of cast iron. The sliding test, called the block-on-ring test, involves a stationary test block (composite material) and a rotating steel ring. The wear test is done at a specific applied normal load and rotational speed while a jet of lubricant is sprayed on the steel ring. The wear on the test composite material is measured by measuring weight loss as a function of sliding distance.

As in other areas, one would expect continuous fiber-reinforced composites to show anisotropy in their wear behavior as well.[85,86] The wear rate is high when the fibers are parallel to the wear surface but perpendicular to the sliding direction. Ceramic particle (Al_2O_3, SiC, Gr, etc.)-reinforced aluminum alloys show excellent abrasive resistance with a variety of mating surfaces.[87–89]

Analyses of the wear tracks and subsurface microstructure of composite materials showed that adhesive wear occurred through plastic deformation under dry conditions, and abrasive wear through plowing and polishing under lubricated conditions.

REFERENCES

1. Reinhart, T. 1987. *Engineered Materials Handbook,* vol. 1: *Composites.* ASM International, Metals Park, OH.
2. *Materials Selector: Materials Engineering.* 1993.
3. Weeton, J. W., ed. 1986. *Engineer's Guide to Composite Materials.* ASM International, Metals Park, OH.
4. Schwartz, M. M. 1992. *Composite Materials Handbook,* 2nd ed. McGraw-Hill, New York.
5. Zweben, C., W. S. Smith, and M. M. Wardle. 1979. Test methods for fiber tensile strength, composite flexural modulus, and properties of fabric–reinforced laminates. *5th Conf. Compos. Mater.: Test. Des.,* STP 674:228. ASTM, Philadelphia.
6. Bramuzzo, M., A. Savadori, and D. Bacci. 1985. *Polym. Compos.* 6(1).
7. Wambach, A., K. Trachte, and A. T. DiBenedetto. 1968. *J. Compos. Mater.* 2:266.
8. Kardos, J. L. 1990. Mechanical properties of polymeric composite materials, In *High Performance Polymers,* ed. E. Baer and A. Moet, 199–237. Hanser, Munich.
9. Chamis, C. C. 1977. In *Hybrid and Metal Matrix Composites.* AIAA, New York.
10. Bunsell, A. R. 1988. *Fibre Reinforcements for Composite Materials,* Composite Materials Series 2, 313–315, 436–445. Elsevier, Amsterdam.
11. Harper, C. A. 1991. *Handbook of Plastics, Elastomers, and Composites,* 2nd ed. McGraw-Hill, New York.
12. Grimm, R. 1994. Private communication.
13. Army Reserve Office. 1980. Compendium of abstracts and viewgraphs. *2nd Int. Workshop Compos. Mater. Struct. Rotorcraft,* September 1989. ARO 26588-1 EGCF, AHS. AD-A217 189.
14. Mallick, P. K. 1992. *Fiber-Reinforced Composites,* 2nd ed. Marcel Dekker, New York.
15. Buckley, J., ed. 1991. *5th Conf. Adv. Eng. Fibers Text. Struct. Compos.: Fiber Tex. 1991,* October 1991, Raleigh, NC. NASA CP 3176.
16. Chou, S., H.-C. Chen, and C.-C. Wu. 1992. BMI resin composites reinforced with 3D carbon-fibre fabrics. *Compos. Sci. Technol.* 43:117–28.
17. Ko, F. K. 1982. Three-dimensional fabrics for composites: An introduction to the magnaweave structure. *Int. Conf. Compos.,* Tokyo, October 1982.
18. Li, W., and A. El-Shiekh. 1988. The effect of processes and processing parameter on 3-D braided preforms for composites. *SAMPE Q.* 19(4):22.
19. Bruno, P. S., D. O. Keith, and A. A. Vicario. 1986. Automatically woven three-dimensional composite structures. *SAMPE Q.* 17(4):10.
20. Ko, F. K. 1986. Tensile strength and modulus of a three-dimensional braid composite. *Compos. Mater.: Test. Des.,* STP 893. ASTM, Philadelphia.
21. Chung, W., B. Jang, and R. Wilcox. 1988. A study on multidirectional composites, 1630. ANTEC, Montgomery, AL.
22. Aboudi, J. 1984. Minimechanics of tri-orthogonally fiber-reinforced composites: Overall elastic and thermal properties. *Fiber Sci. Technol.* 21:277.
23. Uhlmann, D. R. 1992. Ultrastructure processing of advanced materials. *Proc. 4th Int. Conf. Ultrastruct. Process. Ceram. Glasses Compos.,* Tucson, AZ. Wiley-Interscience, New York. AFOSR-TR-92-0977, AFSOR-89-0236.
24. Polymer matrix composites: Technology and industrial applications, technical trends, 1990. Interim Reports on Advanced Technology. Innovation 128, Paris.

25. Phillips, L. N. 1989. *Design with Advanced Composite Materials,* Design Council. Springer-Verlag, London.
26. Wilson, D. W. 1990. Evaluation of V-notched beam shear test through an interlaboratory study. *J. Compos. Technol. Res.* 12:131.
27. Madan, R. C. 1991. Influence of low-velocity impact on composite structures. In *Composite Materials: Fatigue and Fracture,* vol. 3, ed. T. K. O'Brien, 457–75, STP 1110. ASTM, Philadelphia.
28. Weeks, G. P., and A. H. White. Advances in thermoplastic compression molding material and technology. *Proc. 8th Ann. ASM/ESD Adv. Compos. Conf.,* p. 81, November 1992, Chicago.
29. Camponeschi, E. T., Jr. 1991. Compression testing of thick-section composite materials. In *Composite Materials: Fatigue and Fracture,* vol. 3, ed. T. K. O'Brien, 439–56, STP 1110. ASTM, Philadelphia.
30. Prewo, K. M., and J. J. Brennan. 1990. Fiber reinforced glasses and glass ceramics for high performance applications. In *Reference Book for Composites Technology,* vol. 1, ed. S. M. Lee, 102–115. Technomic, Lancaster, PA.
31. Brennan, J. J., and K. M. Prewo. 1982. Silicon carbide fibre reinforced glass-ceramic matrix composites exhibiting high strength and toughness. *J. Mater. Sci.* 17:2371.
32. Scola, D. 1990. Synthesis and characterization of polyimides for high temperature applications. *2nd Eur. Tech. Symp. Polyimides Other High-Temp. Polym.,* ed. M. J. M. Abadie and B. Sillion, 265–278. Elsevier, Amsterdam.
33. Galasso, F. S. 1989. *Advanced Fibers and Composites.* Gordon and Breach, New York.
34. Stinton, D. D., A. J. Caputo, and R. A. Lowden. 1986. *Am. Ceram. Soc. Bull.* 65:345.
35. Fitzer, E., and R. Gadow. 1986. *Am. Ceram. Soc. Bull.* 65:326.
36. Lamicq, J., G. A. Bernhart, M. M. Daunchier, et al. 1986. *Am. Ceram. Soc. Bull.* 65:336.
37. Caputo, A. J., D. P. Stinton, R. A. Lowden, et al. 1987. *Am. Ceram. Soc. Bull.* 66:368.
38. Morgan, R. J., and R. E. Allred. 1990. Aramid fiber reinforcements. In *Reference Book for Composites Technology,* vol. 1, ed. S. M. Lee, 152–166. Technomic, Lancaster, PA.
39. Pluddemann, E. P. 1974. Mechanism of adhesion through silane coupling agents. In *Interfaces in Polymer Matrix Composites,* 173–216. Academic Press, New York.
40. Riggs, D. M., R. J. Shuford, and R. W. Lewis. 1982. Graphite fibers and composites. In *Handbook of Composites,* ed. G. Lubin, 196–271. Van Nostrand Reinhold, New York.
41. Hoppel, C. P. The effects of tension-tension fatigue on the mechanical behavior of short fiber-reinforced thermoplastics. *6th Ann. Compos. Conf.,* pp. 509–18, Orlando, FL, October 1991. Technomic, Lancaster, PA, 1991.
42. Curtis, D. C., M. Davies, D. R. Moore, et al. 1991. Fatigue behavior of continuous carbon fiber-reinforced PEEK. In *Composite Materials: Fatigue and Fracture,* vol. 3, ed. T. K. O'Brien, 581–95, STP 1110. ASTM, Philadelphia.
43. Agarwal, B. D., and J. W. Dally. 1975. Prediction of low-cycle fatigue behavior of GFRP: An experimental approach. *J. Mater. Sci.* 10:193.
44. Miner, L. H., R. A. Wolfe, and C. H. Zweben. 1975. Fatigue, creep, and impact resistance of Kevlar 49 reinforced composites. In *Composite Reliability,* STP 580. ASTM, Philadelphia.
45. Ramani, S. V., and D. P. Williams. 1976. Axial fatigue of $[0/\pm30]_{6s}$ graphite/epoxy. In *Failure Mode in Composites III,* 115. AIME, New York.
46. Hahn, H. T., and R. Y. Kim. 1976. Fatigue behavior of composite laminates. *J. Compos. Mater.* 10:156.
47. Pipes, R. B. 1974. Interlaminar shear fatigue characteristics of fiber reinforced composite materials. *ASTM 3rd Conf. Compos. Mater.: Test. Des.* STP 546:419.

48. Dharan, C. K. H. 1978. Interlaminar shear fatigue of pultruded graphite fibre-polyester composites. *J. Mater. Sci.* 13:1243.

49. Owen, M. J., and S. Morris. 1972. Some interlaminar-shear fatigue properties of carbon-fibre reinforced plastics. *Plast. Polym.* 4:209.

50. Wilson, D. W. 1980. Characterization of the interlaminar shear fatigue properties of SMC-R50. Technical Report 80-05, Center for Composite Materials, University of Delaware, Newark, DE.

51. Fujczak, B. R. 1974. Torsional fatigue behavior of graphite-epoxy cylinders. U.S. Army Armament Command Report WVT-TR-74006.

52. Hancox, N. L. 1971. Izod impact testing of carbon-fibre-reinforced plastics. *Composites* 3:41.

53. Bader, M. G., J. E. Bailey, and I. Bell. 1973. The effect of fibre-matrix interface strength on the impact and fracture properties of carbon-fibre-reinforced epoxy resin composites. *J. Phys. D: Appl. Phys.* 6:572.

54. Yeung, P., and L. J. Broutman. 1978. The effect of glass-resin interface strength on the impact strength of fiber reinforced plastics. *Polym. Eng. Sci.* 18:62.

55. Mallick, P. K., and L. J. Broutman. 1977. Static and impact properties of laminated hybrid composites. *J. Test. Eval.* 5:190.

56. Kretsis, G., and F. L. Matthews. 1985. The strength of bolted joints in glass fibre/epoxy laminates. *Composites* 16:92.

57. Mallick, P. K., and R. E. Little. 1985. Pin bearing strength of fiber reinforced composite laminates. *Proc. Conf. Adv. Compos. ASM.*

58. Collings, T. A. 1977. The strength of bolted joints in multidirectional CFRP laminates. *Composites* 8:43.

59. Quinn, W. J., and F. L. Matthews. 1977. The effect of stacking sequence on the pin-bearing strength in glass fibre reinforced plastic. *J. Compos. Mater.* 11:139.

60. Nilsson, S. 1989. Increasing strength of graphite/epoxy bolted joints by introducing an adhesively bonded metallic insert. *J. Compos. Mater.* 23:641.

61. Stockdale, J. H., and F. L. Matthews. 1976. The effect of clamping pressure on bolt bearing loads in glass fibre-reinforced plastics. *Composites* 7:34.

62. Strife, J. R., and K. M. Prewo. 1979. The thermal expansion behavior of unidirectional and bidirectional Kevlar/epoxy composites. *J. Compos. Mater.* 13:264.

63. Parker, S. F. H., M. Chandra, B. Yates, et al. 1981. The influence of distribution between fibre orientations upon the thermal expansion characteristics of carbon fibre-reinforced plastics. *Composites* 12:281.

64. Kerr, J. R., and J. F. Haskins. 1982. Effects of 50,000 h of thermal aging on graphite/epoxy and graphite/polyimide composites. *AIAA J.* 22:96.

65. Shen, C. H., and G. S. Springer. 1977. Effects of moisture and temperature on the tensile strength of composite materials. *J. Compos. Mater.* 11:2.

66. Shen, C. H., and G. S. Springer. 1977. Environmental effects on the elastic moduli of composite materials. *J. Compos. Mater.* 11:250.

67. Gillat, O., and L. J. Broutman. 1978. Effect of an external stress on moisture diffusion and degradation in a graphite-reinforced epoxy laminate. In *Advanced Composite Materials—Environmental Effects,* STP 658:61. ASTM, Philadelphia.

68. Joshi, O. K. 1983. The effect of moisture on the shear properties of carbon fibre composites. *Composites* 14:196.

69. Jones, C. J., R. F. Dickson, T. Adam, et al. 1983. Environmental fatigue of reinforced plastics. *Composites* 14:288.

70. Sturgeon, J. B. 1978. Creep of fibre reinforced thermosetting resins. In *Creep of Engineering Materials,* ed. C. D. Pomeroy. Mechanical Engineering, London.

71. Holmes, M., and D. J. Just. 1983. *GRP in Structural Engineering*. Elsevier, Barking, U.K.
72. Glaser, R. E., R. L. Moore, and T. T. Chiao. 1983. Life estimation of an S-glass/epoxy composite under sustained tension loading. *Compos. Technol. Rev.* 5:21.
73. Glaser, R. E., R. L Moore, and T. T. Chiao. 1984. Life estimation of aramid/epoxy composites under sustained tension. *Compos. Technol. Rev.* 6:26.
74. Lubin, G., ed. 1982. *Handbook of Composites.* Van Nostrand Reinhold, New York.
75. Blackford, R. W. 1985. Improved durability airline paint schemes. *Proc. 6th Int. Conf. SAMPE,* Scheveningen, Netherlands. Elsevier, Amsterdam.
76. Payne, K. G., B. J. Burrows, and C. R. Jones. 1987. Practical aspects of applying lightning protection to aircraft and space vehicles. *Proc. 8th Int. Conf. SAMPE,* La Baule, France.
77. Caddock, B. D., K. E. Evans, and D. Hull. 1986. The role of diffusion in the micromechanisms of stress corrosion cracking of E glass/polyester composites. *Proc. I Mech. Eng. 2nd Int. Conf. Fibre Reinforced Compos.*
78. Steard, P. A., and F. R. Jones. 1986. The effects of the environment on stress corrosion of single E-glass filaments and the nucleation of damage in GRP. *Proc. I Mech. Eng. 2nd Int. Conf. Fibre Reinforced Compos.*
79. Birley, A. L., J. V. Dawkins, and H. E. Strauss. 1984. Blistering in glass reinforced polyester laminates. *Proc. Brit. Plast. Forum Cong.*
80. Crump, S. 1986. A study of blister formation in gel coated laminates. *Proc. 41st Conf. Reinforced Plast./Compos. Inst.* Society of Plastics Institute.
81. Jordan, W. M., W. L. Bradley, and R. J. Moulton. 1989. Relating resin mechanical properties to composite delamination fracture toughness. *J. Compos. Mater.* 23:923.
82. Sun, C. T. 1989. Intelligent tailoring of composite laminates. *Carbon* 27:679.
83. Chan, W. S., and O. O. Ochoa. 1989. Suppression of edge delamination by terminating a critical ply near edges in composite laminates. Key Engineering Materials Series, vol. 37, 285. Trans Tech. Aedermannsdorf, Switzerland.
84. Sun, C. T., and G. D. Chu. 1991. Reducing free edge effect on laminate strength by edge modification. *J. Compos. Mater.* 25:142.
85. Vishwanath, B., A. P. Verma, and C. V. S. Kameswara Rao. 1993. Effect of reinforcement on friction and wear of fabric reinforced polymer composites. *Wear* 167:93–99.
86. Eliezer, Z., V. D. Khanna, and M. F. Amateau. 1979. *Wear* 53:387–89.
87. Greenfield, I. G., and R. R. Vignaud. 1985. *Advances in Composites Proceedings,* 213–21. ASM, Metals Park, OH.
88. Pan, Y. M., M. E. Fine, and H. S. Cheng. 1992. *Tribol. Trans.* 35:482–90.
89. Hosking, F. M., S. V. Portillo, R. Wunderlin, et al. 1982. *J. Mater. Sci.* 17:477–98.
90. Rohatgi, P. K., P. J. Blau, C. S. Yust, eds. 1990. *Tribology of Composite Materials.* ASM International, Materials Park, OH.

BIBLIOGRAPHY

ARONSSON, C. G. Strength of carbon/epoxy laminates with countersunk hole. *Compos. Struct.* 24(4):283–89, 1993.

BERCHTOLD, G., AND J. KLENNER. Integrated design and manufacturing of composite structures for aircraft using an advanced tape laying technology, MBB-LME251-S. MBB, Munich, 1992.

BERTIN, Y. A., L. POUSSIN, AND A. HARDY. Crystal texture: An original composite's test. *15th Int. SAMPE Eur. Conf.,* Toulouse, France, June 1994.

BROCKE, P., H. SCHURMANS, AND J. VERHOEST. *Inorganic Fibers and Composite Materials: A Survey of Recent Developments.* Pergamon Press, Oxford, 1984.

BUCK, M. E., AND M. L. DORF. Boron fiber reinforced composite materials. *25th Int. SAMPE Tech. Conf.,* Philadelphia, October 1993.

BULL, S. J. Advanced coatings and surface treatments for polymers and polymer composites. *14th SAMPE Eur. Conf.,* Birmingham, U.K., October 1993.

CARLSONN, J. O. 1986. Silicon carbide fibers. In *Encyclopedia of Materials Science and Engineering,* ed. M. B. Bever. Pergamon Press, Oxford, 1986.

CHEN, P., A. EL SHEIKH, AND S. DE TERESA. Effect of lay-up on the mechanical properties of laminates. *38th Int. SAMPE Symp.,* Anaheim, CA, May 1993.

CHEN, P. W., AND D. D. L. CHUNG. Concrete reinforced with up to 0.2 vol% of short carbon fibres. *Composites* 24(1):33–52, 1993.

DEATON, J. W., AND S. M. KULLERD. Mechanical characterization and damage tolerance behavior of 3-D multilayer and conventional 2-D triaxial braided/RTM materials. *Proc. 4th NASA/DOD Adv. Compos. Technol. Conf.,* ed. J. G. Davis, J. E. Gardner, M. B. Dow, vol. 1, part 2, 555–77, 1993. NASA-CP-3229.

DING, Y. Q., W. WENGER, AND R. MCILHAGGER. Structural characterization of mechanical properties of 3D woven GRP composites. *14th SAMPE Eur. Conf.,* Birmingham, U.K., October 1993.

DONNELLAN, T. M., P. A. MEHRKAM, A. YEN, ET AL. Structure-property relationships in high temperature resins and composites. *Proc. Amer. Soc. Compos. 6th Tech. Conf.,* pp. 3–12, Albany, NY, October 1991.

EASTERLING, K. *Tomorrow's Materials.* Institute of Metals, London, 1988.

EICHBERGER, W. Temperature resistant glass fibre-epoxy composites. *Compos. Struct.* 24(3): 205–12, 1993.

ELDRIDGE, J. I. New testing tool for composite interfaces. NASA Technical Brief, pp. 49–50, January 1994.

FAROUK, A., AND N. A. LANGRANA. Fracture properties of PMR-15/graphite fiber composites. *Proc. Am. Soc. Compos. 6th Tech. Conf.,* pp. 525–37, Albany, NY, October 1991.

FLECK, N. A., K. J. KANG, AND M. F. ASHBY. The cyclic properties of engineering materials: Overview 112. *Acta Metall. Mater.* 42(2):365–81, 1994.

GER, G. S., J. C. CHEN, W. Y. CHEN, ET AL. Dynamic tensile properties of carbon/Kevlar hybrid composites. *38th Int. SAMPE Symp.,* Anaheim, CA, May 1993.

GOTO, A., M. MATSUDA, H. HAMADA, ET AL. Vibration damping and mechanical properties of continuous fiber reinforced various thermoplastic composites. *38th Int. SAMPE Symp.,* Anaheim, CA, May 1993.

GRIMES, G. C. *ASTM 10th Conf. Compos. Mater. Test. Des.,* STP 1120. ASTM, Philadelphia, 1992.

GRULKE, E. A. *Polymer Process Engineering.* Prentice-Hall, Englewood Cliffs, NJ, 1994.

JACOBS, O., AND K. FRIEDRICH. Fretting wear performance of glass-, carbon-, and aramid-fibre/epoxy and PEEK composites. *Wear* 135:207–16, 1990.

JONES, D. E. Flexural test methods and results for Dow thermoplastic sandwich panels. *38th Int. SAMPE Symp.,* Anaheim, CA, May 1993.

KELLEHER, P. G. Reinforced thermoplastics, composition, processing, and applications. *Rapra Rev. Rep.* 6(6), 1993.

KERN, K., P. C. STANCIL, W. L. HARRIS, ET AL. Simulated space environmental effects on a polyetherimide and its carbon fiber-reinforced composites. *SAMPE J.* 29(3):29–44, 1993.

KINMAN, E. B. Hot/wet testing of Celion 3000/PM-15 coupon specimens. In *Composite Materials: Testing and Design,* vol. 10, 131–41, STP 1120. ASTM, Philadelphia, 1992.

KLETT, M. W., AND E. A. MARTILLE. Mechanical property relationships of pultruded fiber glass reinforced plastic (FRP) unidirectional composites. *Proc. Am. Soc. Compos. 6th Tech. Conf.,* pp. 269–80, Albany, NY, October 1991.

LEE, S., M. MUNRO, AND T. DICKSON. Evaluation of test methods for in-plane shear modulus of composite materials for aerospace applications. *38th Int. SAMPE Symp.,* Anaheim, CA, May 1993.

MA, C. C. M., H. C. KUO, S. H. WU, ET AL. Creep properties of carbon fiber reinforced polyetheretherketone (PEEK) laminated composites (III). *38th Int. SAMPE Symp.,* Anaheim, CA, May 1993.

MA, C. C. M., H. C. KUO, M. J. CHANG, ET AL. Fatigue Behavior of Carbon fiber reinforced polyetheretherketone (PEEK) laminated composites (IV). *38th Int. SAMPE Symp.,* Anaheim, CA, May 1993.

MALARIK, D. C., AND R. D. VANNUCCI. High molecular weight first generation PMR polyimides for 343°C applications. *SAMPE Q.* 23(4):3–8, 1992.

MALLICK, P. K. *Fiber-Reinforced Composites: Materials, Manufacturing, and Design,* Mechanical Engineering Series, vol. 62. Marcel Dekker, New York, 1988.

MESSIER, D. R. High temperature chemistry of fibers and composites, Army Materials Technology Laboratory, Watertown, MA, 1992. MTL-TR-92-62.

MILEWSKI, J. V. Whiskers. In *Encyclopedia of Materials Science and Engineering,* ed. M. B. Bever. Pergamon Press, Oxford, 1986.

MOORE, G., K. POSTON, AND A. MATTOUSCH. Impact properties and related applications of glass fibre metal laminates. *15th Int. SAMPE Eur. Conf.,* Toulouse, France, June 1994.

MORLEY, J. G. *High-Performance Fiber Composites.* Academic Press, New York, 1987.

ODAGIRI, N., H. KISHI, AND T. NAKAE. T800H/3900-2 toughened epoxy prepreg system: Toughening concept and mechanism. *Proc. Am. Soc. Compos. 6th Tech. Conf.,* pp. 43–52, Albany, NY, October 1991.

PILIPOVSKIY, Y. L., T. V. GRUDINA, ET AL. Composite materials in machine building, FASTC-ID(RS)T-0379-92, AD-B168012, 92-27282. Tekhnika, Kiev, 1990.

Proc. 6th Fr.-Jpn. Sem. Compos. Mater.: Process. Use Databases, Tokyo, 1990.

SHEU, C., R. H. DAUSKARDT, AND I. DE JONGHE. Toughness and fatigue of encapsulation processed silicon carbide-polymer matrix particulate composites. *J. Mater. Sci.* 28(8):2196–2206, 1993.

SMITH, M. D. Surface modification of high-strength reinforcing fibers by plasma treatment, FR 7/91. Allied-Signal, 1991. KCP 613-4369, DE-AC04-76DP00613.

Staff. Sea duty for composites. *Adv. Mater. Process.* 142(2):16–20, 1992.

STRONG, A. B. *High Performance and Engineering Thermoplastic Composites.* Technomic, Lancaster, PA, 1993.

WOLFF, E. G. Moisture effects on polymer matrix composites. *SAMPE J.* 29(3):11–19, 1993.

5

MMC Properties

5.0 INTRODUCTION

MMCs are a materials class that consists of metal alloys reinforced with a variety of materials including fibers, whiskers, particulate, platelets, and metal wires. These composites are being widely developed because their unique physical and mechanical characteristics are extremely attractive for many structural and nonstructural applications. Typical of the types of applications being considered are those dependent on the control of physical properties such as CTE, those dependent on the increased stiffness that can be achieved, those dependent on enhanced elevated-temperature properties, and those requiring improved wear resistance. Often the main driving force is the replacement of existing monolithic materials at a savings in component weight. Many projects are currently under way to apply the unique physical and mechanical property capability of MMCs to actual component production.

5.1 PROPERTY PREDICTION

Property predictions of MMCs can be obtained from mathematical models that require as input a knowledge of the properties and geometry of the constituents. For metals reinforced by straight, parallel, continuous fibers, three properties that are frequently of interest are elastic modulus, CTE, and thermal conductivity in the fiber direction. Reasonable values can be obtained from rule-of-mixture expressions for Young's modulus:[1]

$$E_c = E_f v_f + E_m v_m$$

CTE:[2]

$$\alpha_c = \alpha_f v_f E_f + \alpha_m v_m \frac{E_m}{E_f v_f + E_m v_m}$$

and thermal conductivity:[3]

$$k = k_f v_f + k_m v_m$$

where v is volume fraction and E, α, and k are the modulus, CTE, and thermal conductivity in the fiber direction, respectively. The subscripts c, f, and m refer to composite, fiber, and matrix, respectively.

5.2 MECHANICAL PROPERTIES

The mechanical strength of metal-based systems is controlled by the ease with which bulk coordinated dislocation motion is achieved. The particle reinforcement in MMCs changes this motion through its influence on microstructure. This occurs principally through three mechanisms: grain size control, quenched-in dislocation density, and dispersoid strengthening. The inability of the composite production route chosen to incorporate certain volume fractions or particle sizes in turn restricts the strengthening achieved.

Grain size control arises from the ability of a second-phase particle to both nucleate recrystallization and pin grain growth. In this case, based on the simplest assumption of one grain per particle, the grain size is given by

$$\text{Grain size} - \text{particle size}\left(\frac{1-F_v}{F_v}\right)^{1/3}$$

where F_v is volume fraction of reinforcement.

Using a Hall–Petch-type relationship to relate this to yield strength predicts that fine particle sizes and high volume fractions lead to the greatest strength. As an example, a 3-μm reinforced composite should be 30 MPa stronger than a 20-μm composite at 20 vol% ceramic.

Quenched-in dislocations are generated because of the widely differing thermal expansivities of the matrix and the reinforcement. Assuming that the linear expansion difference divided by the Burgers vector gives the number of dislocations generated per particle, then,

$$\text{Dislocation density} - f\left(\frac{\text{volume fraction}}{\text{Particle size}}\right)$$

Once again, as yield strength increases with dislocation content, the strongest composites are those with high volume fractions of fine particles. Estimates to quantify this effect from simple geometries predict that a 3-μm composite is about 25 MPa stronger than a 20-μm composite of 20 vol% ceramic.

The level of direct dispersion strengthening can be estimated from Orowan looping of dislocations between adjacent ceramic particles. In this case it is quickly seen that for all practical purposes the reinforcement is too coarse and widely dispersed to provide more than 1–2 MPa strengthening.

The tensile strengths of commercially successful aluminum matrix particulate composites need to be high to take full advantage of their increased stiffness in engineering structures. A typical requirement may be a yield strength of 450 MPa. The strengthening mechanism just discussed is likely to produce as much as 100 MPa of this figure. It is therefore clear that the greater majority has to arise from the inherent matrix strength. As in conventional systems this is achieved by alloying to produce precipitates, dispersoids, and matrix strengthening. Because of its inherent rapid solidification, the powder metallurgy route is clearly the most suitable for achieving these demanding goals.

The minority strengthening produced by the particle reinforcement is, however, effective in complete optimization. The "reserve" of up to 100 MPa extra strength should enable further matrix chemistry development and hence increase the toughness and ductility achievable. In addition, these increases are further emphasized by the increased resistance of fine reinforcements to brittle cracking over their coarser equivalents.

The mechanical properties of MMCs are inevitably a compromise between the properties of the matrix and of the reinforcement phases. For melt-produced composites such compromises are compounded because the phase mixtures are thermodynamically unstable and interfacial reactions occur rapidly to degrade the interfaces and diminish their load-bearing capabilities. Any binder phase in the reinforcement preform can complicate matters further. Much attention is now being focused on interfacial monitoring and control in MMC systems. Interfacial stability can be improved either by creating a stable microenvironment around each fiber using barrier coatings (e.g., surface-treated SiC fibers) or by altering the macroenvironment of the reinforcement phase by changing the chemistry of the matrix phase (e.g., adding at least 7% Si to the aluminum phase for SiC particulate composites).

Research on the mechanical properties of composite materials has to date been restricted by the limited availability of a small variety of reinforcement materials and by the limited amounts of MMC material that have been available to researchers for testing. This is not to say that little research has been performed but that work has been selective and specific rather than comprehensive and progressive. Except for the early work on model systems, most recent mechanical property investigations have been carried out using MMC material produced using some type of squeeze-casting process or an alternative, less efficient, lower-pressure process.

Finally, it is clear that the composition and properties of the matrix phase affect the properties of the composite both directly, by normal strengthening mechanisms, and indirectly, by chemical interactions at the reinforcement-matrix interface.

5.2.1 Fiber Uniformity

Fiber uniformity is important in ensuring the overall mechanical properties of a composite. The conditions for achieving uniform fiber distribution in solid-state consolidated composites, according to Guo and Derby,[4] are shown by the uniformity which is influenced by initial fiber spacing, fiber packing, and foil thickness before consolidation and matrix flow during consolidation.

5.3 MMC PROPERTY LIMITATIONS

The limits of the properties and performance of MMCs are described by the maximum strength that can be achieved through the strengthening mechanisms related to the matrix and the inclusions, and their interaction. The limits are also governed by the relaxation mechanisms that become operative as a result of mechanical and thermal effects. These are caused both by external and internal loadings, for example, external thermal treatments and internal stresses.

In MMCs strength is related to both the temperature and the plastic deformation of the composite. Calculation of the maximum strength is essentially an elastic analysis in the sense that the stresses are calculated from the elastic strains. The strength caused by plastic deformation, that is, via work-hardening, is treated through an equivalent elastic strain related to the plastic applied strain. This analysis is related to a dislocation configuration, which in this context is the optimum configuration in that it leads to maximum stresses in the composite.

The reduction in the maximum strength level of the composite is caused by several mechanisms that all tend to reduce the internal stresses to numerically lower levels. The relaxations can both affect the normally elastic strains of the inclusions and lead to rearrangements of the optimum dislocation configuration into a lower energy structure. The operative mechanisms of relaxation can be long-range atom movements, dislocation movements both locally and globally, and internal fracture of the composite.[5]

5.4 MMC TAILORING

A driving force for the existence of composite materials is their ability to be designed to provide a needed type of material behavior, a feature often referred to as tailorability. Classically, stiffness and strength have received the most attention as tailorable properties. Composites have also been tailored to provide enhanced elevated-temperature performance compared to that of their unreinforced metal counterparts, such as a wider range of operating temperatures or a lower creep rate. In recent years, however, interest has arisen in tailoring such attributes as CTE, thermal conductivity, dimensional stability, friction characteristics, and, through the incorporation of ductile phases, fracture toughness. MMCs may be made highly directional or virtually isotropic.[6]

Tailoring is primarily achieved by selecting the reinforcement—type, shape, size, and concentration—and the matrix metal. The categories of materials utilized as reinforcements include various forms of carbon, oxides, carbides, borides, and metal alloys. Their aspect (length-to-diameter) ratio may be very large (continuous filaments), moderate (whiskers, short fibers), or small (particles). Most conventional structural metals—that is, alloys of Mg, Al, Ti, Cu, Ni, or Fe—have been successfully employed as matrices, and research has turned to fabricating MMCs with ordered alloys (intermetallic compounds) as matrices for high-temperature applications.

5.5 MMC BEHAVIOR

Owing to the natural complexity of composites, together with their nonlinear behavior and environmental effects, even a reasonable prediction of the behavior of MMCs is a major concern of the composites community.

A series of issues of research interest in MMCs have been addressed and include

1. Fabrication and processing
2. Test methods
3. Material properties up to the highest use temperature
4. In situ properties
5. Fatigue
6. Creep
7. Coupled fatigue and creep
8. Long-term behavior
9. Fracture toughness
10. Environmental effects.

Progress has been made on all of these fronts, both theoretically and experimentally. Computational simulation for materials and structural assessment of MMCs has become a necessity for rapid progress in the application of MMCs.

5.5.1 Fatigue Behavior

The use of MMCs reinforced with continuous fibers is projected for high-temperature, stiffness-critical parts that will be subjected to cyclic loads. However, fatigue of a MMC can be quite complex. The matrix, because of its relatively high strength and stiffness compared to those of the fiber, plays a very active role compared to a polymer matrix.

Depending on the relative fatigue behavior of the fiber and the matrix and on the properties of the interfaces between the fibers and the matrix, the modes of failure of MMCs can be grouped into four categories: (1) matrix-dominated, (2) fiber-dominated, (3) self-similar growth of damage, and (4) fiber-matrix interfacial.

Matrix-dominated damage occurs when the matrix material has a lower fatigue endurance strain range than the fiber. The result is the development of matrix cracks that can cause significant losses in stiffness in laminates with off-axis plies.

Fiber-dominated damage occurs when the fiber has a lower fatigue endurance strain range than the matrix material. In this case, numerous fiber breaks may occur within the laminate, yet the stiffness may be relatively unaffected. This type of damage results in sudden laminate failure.

Self-similar growth of cracks can occur if the fiber and matrix materials have similar fatigue endurance strain ranges. The material can experience growth of cracks much like the growth of a crack in a homogeneous material.

Fiber-matrix interfaces fail if they are weaker in the transverse direction than the fiber and the matrix are. The higher the strength of the matrix material, the greater the chance of interfacial failures in the off-axis plies.

As new continuous-fiber-reinforced MMCs are developed, projections of their fatigue behavior can be made by understanding the relative strengths of the fiber, matrix, and fiber-matrix interface.

Second, considerable progress has been made in processing technology for developing microstructures that confer required properties. In most cases, however, this has been achieved empirically with little scientific input or understanding of the salient micromechanisms.

With respect to mechanical property characterization of MMCs, a database of *S-N* and fatigue-crack propagation test results is growing for both discontinuously and continuously reinforced materials. Unfortunately, comparatively little information has been published on crack initiation properties. In the areas of damage- and failure-mode characterization, several salient micromechanisms have been identified and modeled, such as crack trapping and uncracked-ligament bridging in particulate-reinforced alloys[7] and crack bridging in continuous-fiber-reinforced alloys.[8] However, for many potential applications, much more work is needed to develop an understanding of the performance of these materials at elevated temperatures and in service environments.[9]

5.6 MMC CONTINUOUS-FILAMENT CHARACTERISTICS

All continuous-filament composites, whether boron-aluminum, graphite-aluminum, or any other combination, have certain characteristics in common. The most important is the variation of properties as stresses are applied away from the direction of the fibers.[10]

Loads applied in the direction of the fibers see a high-strength–high-stiffness material. The composite takes on the properties of the reinforcement, usually superior to those of the matrix. As the loads rotate off-axis, the properties fall quickly once the direction of the fibers and the direction of the load differ. The composite then takes on the properties of the matrix. Transverse strengths are typically 10% of longitudinal strengths. Continuous-filament composites are superb in applications where the loads encountered are strongly oriented in one direction and very high strength-to-weight ratios can be achieved (Figure 5.1).

A second characteristic of continuous-filament composites is the necessity for forming structures of flat sheets, flat sheets stacked one atop another, or bent or formed flat sheets. Thin-shell structures, such as tubes, can also be produced. The limiting factor is the movement available to the long fibers after they are locked within the matrix. After-bonding processing of continuous-filament composites is restricted to those that don't require much internal movement: bending parallel to fiber direction, gentle rolling perpendicular to fiber direction, and joining.

It must also be remembered that any machining that cuts the fibers, for example, milling and drilling, can reduce the strength of the structure significantly if load-bearing fibers are broken. Not much of the load is distributed from fiber to fiber through the matrix, and this must be taken into consideration when designing structures to be made of continuous-filament MMCs.

Some properties of the composites depend on the characteristics of the reinforcement. For example, graphite in a metal matrix is generally a benign material and presents no problem to steel-cutting tools like drills or shear blades. Boron, on the other hand, is extremely hard and rapidly dulls ordinary high-speed steel cutting edges.

Differences between the thermal properties of the fiber materials can also be used to vary the properties of the composite. Graphite fibers, for example, typically have negative CTEs; they contract when heated. By combining the negative expansion of the

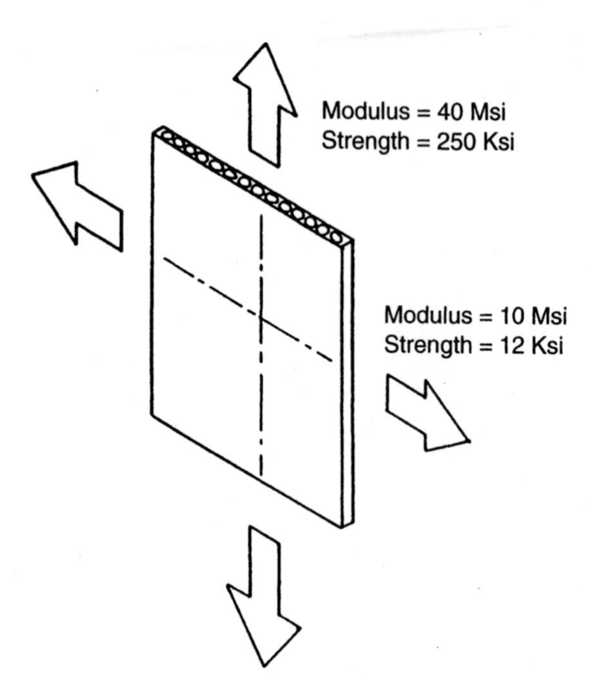

Figure 5.1. Properties of continuous-filament MMCs—anisotropic.

graphite with the positive expansion of aluminum, composite structures can be engineered that undergo no expansion as the temperature changes. Structures with this kind of extreme thermal stability, coupled with the inherent stiffness of the composite, can be very useful where dimensional stability is critical.

5.7 MMC DISCONTINUOUS REINFORCEMENT

Discontinuous-reinforcement MMCs are all characterized by the presence of relatively small particles of reinforcing material spread uniformly throughout the metallic matrix. The reinforcement can be in the form of a particulate where the dimensions of the particle are roughly equal in all directions, in the form of a whisker where one dimension is significantly longer than the other two, or in the form of a platelet where one dimension is significantly shorter than the other two. The reinforcement can range from several hundred microns to less than a micron in size.

Reinforcements are usually some form of ceramic powder. SiC, Al_2O_3 and B_4C are common reinforcements for discontinuous MMCs. Methods of making discontinuous reinforcements as well as composites are discussed in Volume II.

The particles are normally randomly oriented in the matrix, with no direction receiving preferential reinforcement. The properties of the composite are therefore isotropic. It should be noted that for reinforcements such as whiskers and platelets, which are not the same size in all dimensions, some operations such as extrusion and rolling may cause some particle realignment. This can introduce some anisotropy in the properties of the composite (Figure 5.2).

5.7.1 Discontinuous-MMC Characteristics

The primary similarity between particulate-reinforced composites is the isotropic nature of their properties. The high stiffness and strength of the ceramic particulate augments the stiffness of the matrix. The orientation of the particles is random, and so the reinforcement aids the matrix no matter in what direction the forces are applied. Certain processes, however, can cause preferential alignment. For example, rolling can cause flat platelets to align themselves in parallel layers, like leaves in a book, making the stiffness higher in the two long dimensions of the sheet. This requires a high degree of material flow during the forming and doesn't occur in more moderate treatments.

Another similarity is the dependence of the composite's properties on the amount of reinforcement present. The modulus of elasticity of aluminum with no reinforcement is about 70 GPa. When enough SiC particulate is added to equal 15% of the overall volume, the modulus increases to 98 GPa. The modulus increases monotonically as the reinforcement loading increases until, at 40 vol% it is 147.6 GPa, more than twice the stiffness of the unreinforced material.

As the percentage of ceramic increases, the composite displays more and more properties of the ceramic. One of these properties is improved stiffness, and another is decreased ductility. As the volume of ceramic increases, the amount of matrix between the particles decreases. The ability of the matrix to deform is the basis of the composite's ductility. The ductility of the unreinforced matrix in the case of aluminum can be over 10% as measured by the strain to failure of test samples. As the percentage of reinforcement increases to 40 vol%, the strain to failure decreases to less than 2%. For comparison, a typical graphite-aluminum continuous-filament composite containing 40 vol% graphite has less than 1% strain to failure.

As a result of the reduced ductility of heavily loaded composites, manufacturing processes requiring the large-scale movement of materials must be carried out at high

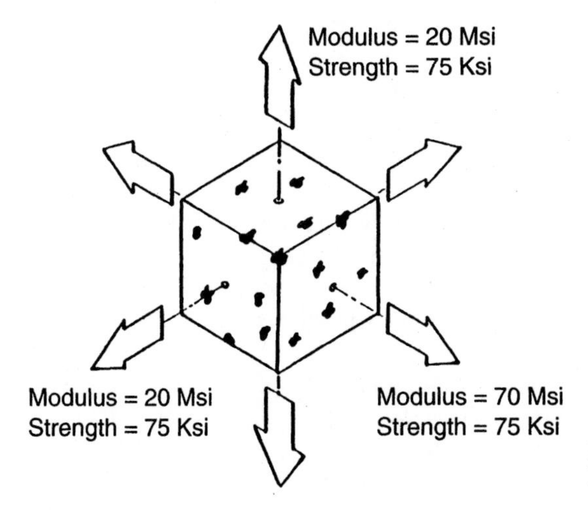

Figure 5.2. Properties of discontinuous MMCs—isotropic.

temperatures and the material moved slowly, to allow the limited matrix present to conform. Processes must be designed to keep the material subjected to compressive loads during its movement to avoid cracking.

5.7.2 Choosing MMCs

MMC is the material to choose when strength, stiffness, and low weight are required. This can apply to structures that must be moved frequently or to structures that must be moved quickly, subjected to high acceleration levels. High stiffness increases the resonant frequency of the structure and allows it to be moved by applying lower forces. Reducing the weight of vehicles, whether airborne, landborne, or seaborne, increases the available payload or increases the fuel efficiency.

MMC is the material to choose when abrasion resistance is required. Parts made of a ceramic-reinforced composite are able to operate in abrasive environments, being protected from wear by their semiceramic nature.

MMC is the material to choose when extreme dimensional stability under load and thermal cycling is required. Structural elements can be engineered of continuous-filament composite that not only concentrates the benefits of the reinforcement in the direction where it is most needed but also combines the thermal properties of graphite fibers and aluminum to achieve zero thermal expansivity.

MMC is the material to choose when the strength of a composite is required with either the thermal or electric conductivity of a metal. The electric conductivity of the matrix makes it suitable for such electrically active structures as waveguides and feedhorns in microwave installations. The thermal conductivity of the matrix makes it suitable for such thermally active structures as heat sinks and heat pipes.

MMC is the material to choose when the strength of a composite is required with a higher damage tolerance than can be expected of epoxy-type matrices. The ductility of the metallic matrix, even at high-reinforcement loading levels, allows the composite to accept punctures and impacts without shattering. The material maintains good fracture toughness, inhibiting crack growth in stressed elements.

The consideration of MMCs can give the designer important options while balancing the requirements of the job at hand with the properties of the materials available to meet these requirements.

These MMC materials range from aluminum and its alloys through titanium and its alloys, and several are highlighted here.

5.8 ALUMINUM CONTINUOUS MMCs

Continuous SiC fibers (SiC_c) are now commercially available; these fibers are candidate replacements for boron fibers because they have similar properties and offer a potential cost advantage. One such SiC fiber is SCS, which can be manufactured using any of several surface chemistries to enhance bonding with a particular matrix such as aluminum or titanium.[11–14]

SiC/Al MMCs exhibit increased strength and stiffness as compared with unreinforced aluminum, and with no weight penalty. Selected properties of SCS-2/Al are given in Table 5.1. In contrast to the base metal, the composite retains its room-temperature tensile strength at temperatures up to 260°C (Figure 5.3).

5.8.1 Silicon Carbide

The development of MMCs has resulted in great interest in 6061 aluminum alloy reinforced with high-modulus and high-strength fibers such as SiC. The continuous unidirectional (UD) SiC fiber-reinforced 6061 aluminum alloy can be manufactured by diffusion bonding, and in practice no subsequent heat treatment is normally carried out.

The SiC/Al composite is usually diffusion-bonded at a temperature of about 550°C and then air-cooled to room temperature. The bonding takes about 30 min. This process is similar to a solution treatment. However, whereas after a normal solution heat treatment the material is quenched at room temperature, in the case of diffusion bonding, the composite is usually air-cooled to room temperature.[10]

An important requirement in modeling the mechanical behavior of a composite by either a continuum-mechanics approach or with finite element micromechanical modeling is a knowledge of the in situ constitutive properties of the individual phases, that is, the fibers and the matrix. In the case of fibers such as SiC, the elastic properties are not appreciably altered by the fabrication process. However, the actual in situ response of the aluminum matrix presents a different problem. Even though the elastic properties are thought not to differ significantly from the original bulk properties after the fabrication process, the yield stress and subsequent strain-hardening may change.

The mechanical behavior of the matrix material during the manufacture of the composite is a complex process because of the differences in the thermal expansion coefficients of the fibers and the matrix. The thermal stresses built up can be so high as to load the matrix to its yield point, which leaves considerable residual stress and strain in the matrix. The magnitudes of the stresses are dependent on the fiber and matrix elastic properties, their thermal expansion coefficients, and the matrix plasticity behavior, which are all functions of temperature. Metallurgical factors, such as precipitation and recovery in the aluminum, are also important. In addition, matrix creep may occur at elevated temperatures.

Sheng Li and his associates[10] concluded that air-cooling results in a small reduction in the yield stress the solution-treated material and a more significant reduction in the aged condition.

The main effect of constraint during cooling is to cause plastic deformation of the

TABLE 5.1. Room-Temperature Properties of Unidirectional Continuous-Fiber Aluminum-Matrix Composites

Property	B/6061 Al	SCS-2/6061 Al	P100 Gr/6061 Al	FP/Al-2Li[a]
Fiber content (vol%)	48	47	43.5	55
Longitudinal modulus, GPa (10^6 psi)	214 (31)	204 (29.6)	301 (43.6)	207 (30)
Transverse modulus, GPa (10^6 psi)	—	118 (17.1)	48 (7.0)	144 (20.9)
Longitudinal strength, MPa (ksi)	1520 (220)	1462 (212)	543 (79)	552 (80)
Transverse strength, MPa (ksi)	—	86 (12.5)	13 (2)	172 (25)

[a] FP is the proprietary designation for an alpha alumina (α-Al_2O_3) fiber developed by E. I. Du Pont de Nemours & Company, Inc.

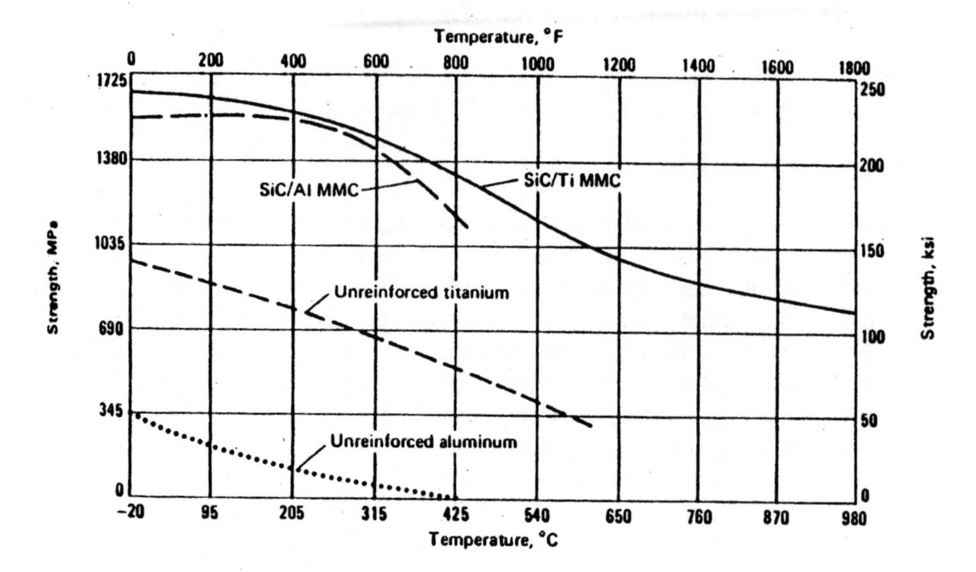

Figure 5.3. Effect of temperature on tensile strength for two continuous-fiber MMCs and two unreinforced metals.[14]

aluminum alloy and hence strain-hardening. This produces a significant increase in the yield strength and also a much sharper yield point.

Japan initiated the Jisedai Project[15,16] several years ago which has involved the development of SiC/Al MMCs by CVD. SiC_{CVD} fiber heat-resistant Al-based SCS-2/Al-4Ti and SCS-2/Al-8Cr-1Fe have displayed the same superior high-temperature strengths as Ti MMCs strengthened by SiC fibers (SCS-6) using the same base as the previously mentioned materials.[14] Their specific strengths are astonishingly high, showing values of more than 55 km at 427°C (Figure 5.4).

Also developed in this project were heat-resistant Al-based MMCs for precursor-type Si-C-O-based fibers (Nicalon) and Si-Ti-C-O-based fibers (Tyranno). Both have Al-Al3Ni eutectic alloy as the matrix (Nicalon-Al-5.7 Ni,[17] Tyranno-Al-8Ni).[18] Their strengths are inferior to that of SiC_{CVD}-based MMCs up until about 427°C. They displayed normal temperature strengths but, on the other hand, exhibited superior heat resistance. These MMCs are UD materials, and their 90° strengths are characteristically higher than those of other Al-based MMCs.[19]

In a recent development Nippon Carbon[20] produced preform wires for MMCs based on Nicalon fibers coated with a thin layer of aluminum. The SiC fibers were pretreated to remove the sizing prior to surface treatment. The rovings were then spread by a vibration technique and infiltrated with molten aluminum; after cooling, the continuous wires, up to 500 m in length, were passed through a die to produce a smooth, uniform shape. The wires were reported to exhibit a tensile strength of 1500 MPa and to have diameters in the range 0.3–0.5 mm. It is anticipated that these preform wires will be used in monotape fabrication for laminates and for filament winding of complex-shaped components.

According to the company, the preform wires retained more than 95% of room-temperature strength after exposure to a temperature of 350°C for 1000 h, and hot-pressed or rolled composites are reported to have no voids or defects.

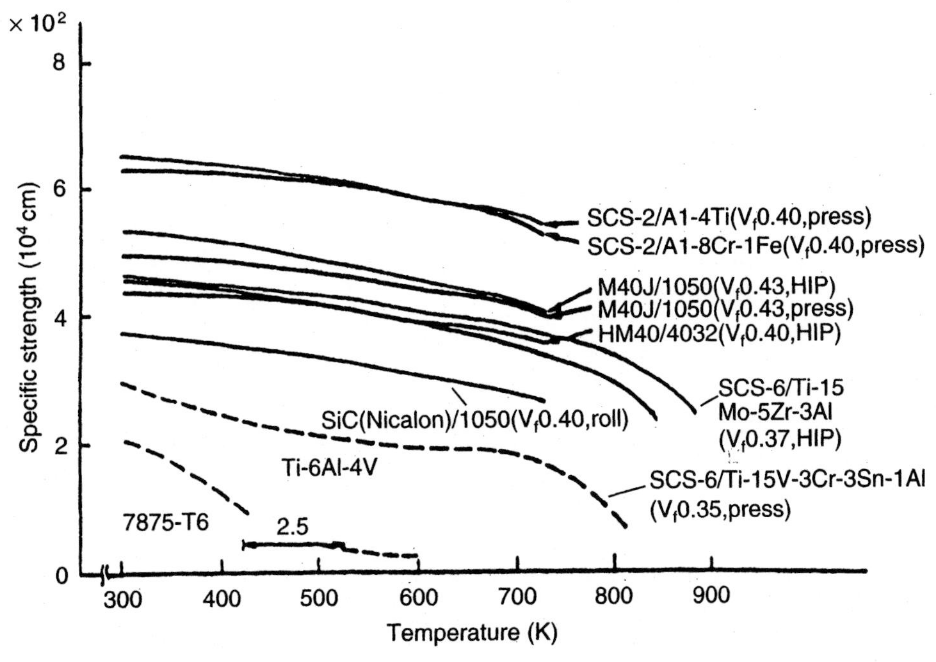

Figure 5.4. Specific strength versus temperature of MMC developed under the Jisadei project.[16] $T_C = T_K - 273.15$.

Illustrative properties of MMCs, as quoted by the fiber manufacturers, are presented in Table 5.2. The high strength parallel to the fiber direction is a notable feature: the anisotropy of strength values is marked, but the transverse strength is nevertheless quite high. The effect of volume loading of Nicalon fibers on strength at room temperature and at elevated temperatures is illustrated in Figure 5.5. The strength of the compos-

TABLE 5.2. Properties of Nicalon and SCS-6 Silicon Carbide Continuous-Fiber-Reinforced Composites[20]

Fiber type	Nicalon	Nicalon	SCS-6	SCS-6
Volume fraction	35%	35%	48%	Ti-6A1-4V
Matrix type	A17075	A16061	A16061	35%
Lay-up (directional)	Uni-	Bi-	Uni-	Uni-
Density (g/cm^3)	2.6	2.6	2.84	3.86
Elastic modulus, 0° (GPa)	100–110		220	210
Tensile strength (MPa)				
0°	800–900	400–500	1750	1750
90°	70–80		105	410
Flexural strength (GPa)	1.0–1.1	0.70		
Coefficient of expansion (×10^{-6}/°C)				
0°	12			
90°	25			

Source: Nippon Carbon, Textron.

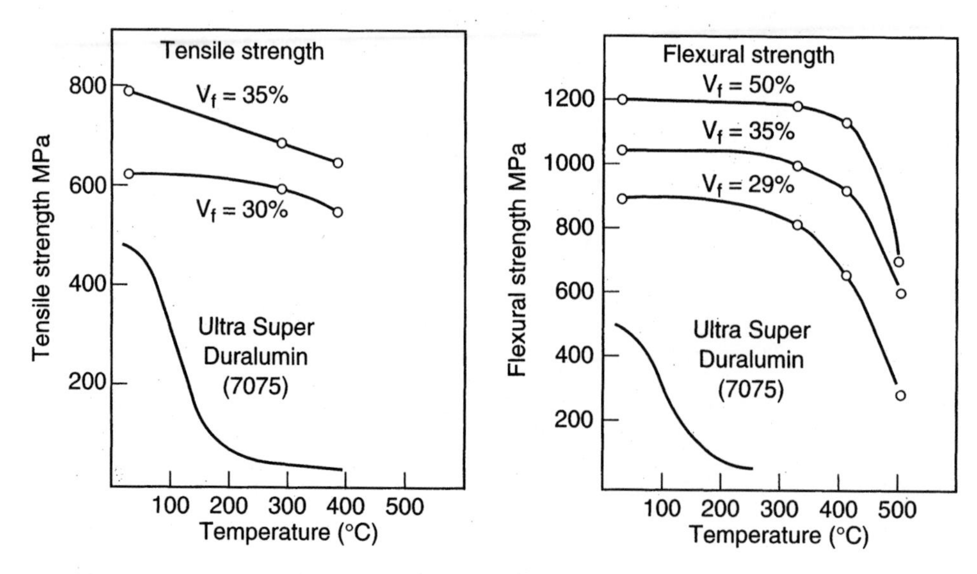

Figure 5.5. Strength properties of unidirectional Nicalon fiber-reinforced aluminium matrix composites.[20]
Source: Nippon Carbon, Nicalon technical data.

ite materials is shown to be maintained above that of the room-temperature strength of the monolithic aluminum matrix at temperatures up to 450°C.

Ube Industries in Japan has described in the patent literature[21] an aluminum alloy that is as compatible with ceramic fibers (especially the company's own Tyranno fibers) as pure aluminum. Although pure aluminum does not react with ceramic fibers to form brittle intermetallic compounds at the fiber-matrix interface, its intrinsic strength is low. As the transverse strength of UD composites is determined by the properties of the matrix, the transverse strength of pure aluminum matrix composites is low. According to the patent, the addition of 8 wt% nickel increases the transverse strength of a typical UD Tyranno fiber-reinforced aluminum matrix composite by a maximum of 80%. In the longitudinal direction the strength is at a maximum with 8–9 wt% nickel, with the value being raised by about 10% compared with that of a pure aluminum matrix composite.

5.8.2 Aluminum Oxide

Rapid solidification rate (RSR) AlFeCe and AlFeMo alloy matrices reinforced with Al_2O_3 (Saffil) fibers and SiC whiskers were investigated in a study by Pollock et al.[22] They examined the use of reinforcements under two conditions: (1) as produced and (2) with a thin cobalt coating to enhance bonding between reinforcement and matrix. The latter condition permits lower fabrication temperatures for the composites, thereby avoiding a decrease in shear strength.

Reinforcement of RSR AlFeMo and AlFeCe matrices is an effective strengthening mechanism at elevated temperatures. In the majority of tensile tests (Table 5.3), the composite samples had higher strengths than the unreinforced matrix despite a reduction in the aspect ratio of the fibers during processing to the degree that they were essentially particles.

TABLE 5.3. Tensile Strengths of Composites for the Conditions ksi (MPa)

	AlFeMo				AlFeCe			
Test Temperature (°C)	Unreinforced	SIC_w	Al_2O_3	Co/Al_2O_3	Unreinforced	SIC_w	Al_2O_3	Co/Al_2O_3
20	50 (353)	57 (401)	51 (355)	57 (403)	51 (357)	62 (434)	44 (308)	48 (336)
200	[a]	[a]	[a]	[a]	38 (266)	50 (350)	33 (231)	37 (259)
300	[a]	[a]	[a]	30 (210)	31 (217)	42 (294)	24 (168)	29 (203)
350	[a]	[a]	[a]	[a]	27 (189)	32 (224)	19 (133)	23 (161)
400	12 (86)	24 (171)	10 (69)	23 (162)	21 (147)	23 (161)	11 (77)	14 (98)
450	8 (58)	17.6 (123)	7.4 (52)	13.4 (94)	14 (98)	17 (119)	7 (49)	12 (84)
500	5 (37)	[a]	4.6 (32)	[a]	[a]	[a]	[a]	[a]

[a] Not tested because of insufficient material.

The cobalt coating appears to be beneficial in increasing the strength of the composite. Tensile strengths for composites with cobalt-coated Al_2O_3 were greater at all temperatures than for composites with uncoated reinforcements. The mechanism of this improvement is not completely clear, but is probably a result of improved bonding and/or inhibition of a matrix attack on the fiber.

The surface layer of cobalt on the reinforcing fibers remains intact at the interface through the fabrication stage of the composite. During thermal exposure, however, the cobalt diffuses away from the interface into the matrix. This probably contributes to the observed growth of cobalt particles with elevated-temperature exposure and subsequent reduced composite strength (Figure 5.6).

Schulte and associates[23] investigated the Al_2O_3 fiber (fiber FP from DuPont)-reinforced Al/2.5 Li composite. They found that fatigue testing of the MMCs showed that with fiber reinforcement a pronounced improvement in the fatigue behavior could be achieved. They also found that the alumina grains close to the fiber surface were completely surrounded by a thin layer of lithium aluminate ($LiAlO_2$) and lithium spinel ($LiAl_5O_8$). A strong bonding can be expected at this interface because the matrix not only adheres to the fibers but also interconnects the first layers of alumina grains.[13]

Researchers Musson and Yue[24] compared and contrasted the mechanical properties obtainable from two Saffil-based MMCs, the first with a high-strength heat-treatable 7010 aluminum alloy matrix and the second with a low-strength non-heat-treatable alloy Al-5Mg where the composition is in approximate weight percent. They found that the incorporation of Saffil into aluminum alloys reduced the elongation at failure; this was seen most severely in the 7010 composites where the elongation was 0.2% compared with 10% for the matrix alloy. The corresponding values for the Al-5Mg composite and alloy were 1.5% and 13.8%, respectively. This premature failure of the 7010-Saffil material was due to its higher yield stress.

The 7010-Saffil was shown to have a higher fatigue crack propagation rate than the

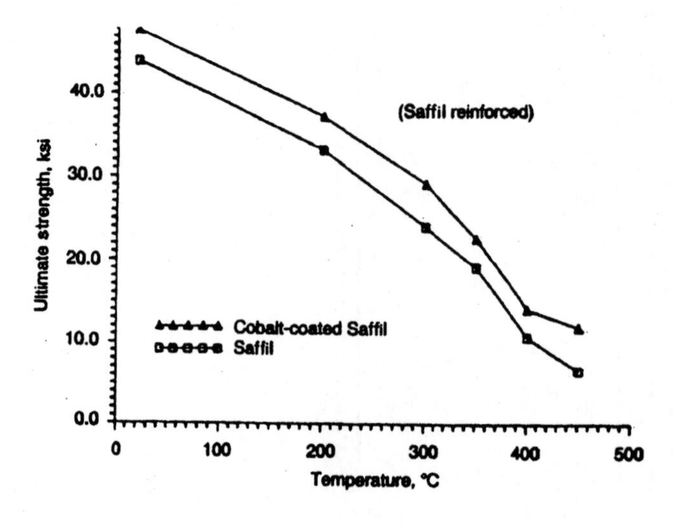

Figure 5.6. Comparison of tensile data for cobalt-coated and uncoated Al_2O_3.[22]

7010 matrix material. The (Al-5Mg)-Saffil was found to have a fatigue-crack propagation rate comparable to that of Al-5Mg matrix material.

The higher fatigue-crack propagation resistance observed in the (Al-5Mg)-Saffil material compared with the 7010-Saffil material was found to be due to a rougher fatigue surface and also to the large shear cliffs observed in the former material.

The composite materials had lower fracture toughnesses (20 MPa $m^{1/2}$ compared with 36 MPa $m^{1/2}$ for the alloy 7010 and 29 MPa $m^{1/2}$ compared with more than 46 MPa $m1^{1/2}$ for Al-5Mg), resulting in the very short observed critical crack lengths of the composite compared with those of the matrix material (1.5 mm compared with more than 15 mm for the alloy 7010).

Rasmussen, Hansen, and Hansen[25] examined fiber-reinforced aluminum castings with AlSi9Cu3 as matrix material; Saffil-based preforms containing 20 vol% fibers were used as reinforcement.

With AlSi9Cu3 as matrix material, the test specimens were submitted to tensile testing at 200, 300, 400, and 450°C. The reinforcement offered a considerable rise in the tensile strength for temperatures above 300°C. At 400°C a tensile strength of 120 MPa was achieved compared with 46 MPa for the AlSi9Cu3 alloy; that is, it was almost double. At 200°C the tensile strength of the composite was lower than that of the matrix material (198 MPa versus 229 MPa).

Flinn and his colleagues[26] fabricated metal-reinforced ceramic composites by infiltrating metal (Al-4 wt% Mg alloy) into ceramic preforms containing continuous open channels. The fracture resistance behavior of the composites was characterized using both *R*-curve measurements and work-of-rupture tests. The latter tests indicated that the steady-state composite toughness was almost 10 MPa $m^{1/2}$, a value that was about three times that for the Al_2O_3 ceramic matrix.

In a series of wear property tests Kainer and Kaczmar[27] prepared 6061 aluminum composite materials with 10, 15, 20, and 30 vol% δ-alumina Saffil fibers by a powder metallurgical route. They observed that composite materials strengthened by ceramic δ-alumina fibers were characterized by significantly lower CTEs than sintered 6061 material.

This coefficient depends mainly on the very low coefficient α of δ-alumina ceramic fibers. The lower CTEs of composite materials are the result of bearing stresses on the fiber-matrix interface, lowering their total expansion. Theoretically, composite materials show very poor bonding at the interface between reinforcement and matrix, and thus the influence of reinforcement content on the CTE would be much smaller. On the other hand, comparisons of CTE $\alpha_{20\text{–}400°C}$ and $\alpha_{20\text{–}300°C}$ for composite materials containing 15, 20, and 30 vol% strengthening δ-alumina fibers can be explained by the significant improvement in bonding at the interface as a result of diffusional transport of matter to the interface and relaxation of stresses during heating at relatively high temperatures up to 400°C.

Illustrative mechanical properties of an A380 aluminum alloy reinforced with Saffil RF fiber are shown in Figure 5.7. The fibers were incorporated in an approximately random planar orientation. It is seen that the room-temperature tensile strength is not improved but that the retention of strength at temperatures above 200°C is markedly improved. At a fiber loading of 24 vol% the tensile modulus is increased by about 50% compared with that of the unreinforced matrix alloy.

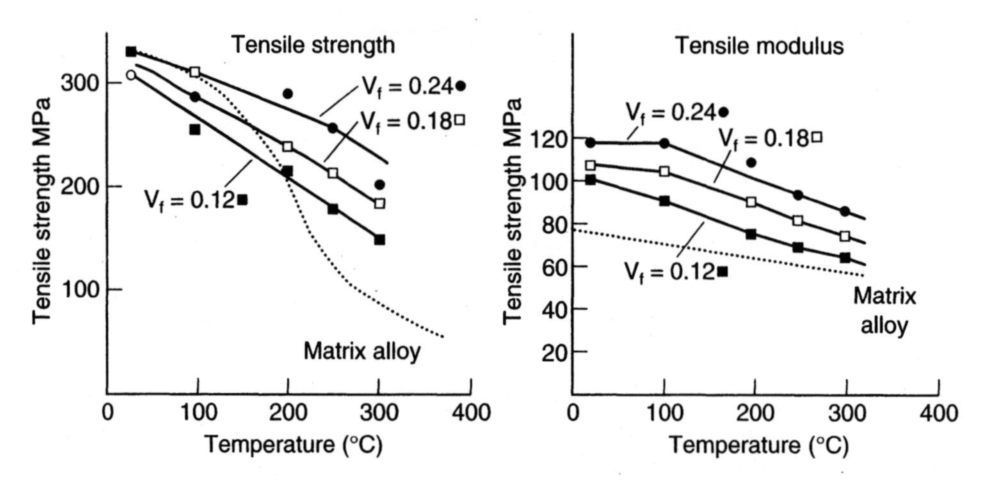

Figure 5.7. Reinforcement of an A380 aluminium alloy with Saffil RF alumina fiber.[20]

The effect of alumina continuous-fiber (Fiber FP) reinforcement of aluminum is also illustrated in Figure 5.8. The continuous fibers provide the capability of attaining higher volume loadings: with short discontinuous fibers, volume loadings are limited to a maximum of about 30%, whereas with continuous fibers loadings of up to 60–65% can be achieved. A more than threefold increase in strength in the direction of fiber orientation is achieved at a loading of 50%. The transverse strength is governed by the strength of the matrix alloy, whereas the modulus is increased in both the longitudinal and transverse directions, with the highest increase being in the longitudinal direction as anticipated by a rule of mixtures.

The carbon-aluminum interface is of the greatest importance, and the reaction taking place at the carbon-aluminum interface at temperatures above 500°C to form aluminum carbide, Al_4C_3, has long been considered critical in affecting the strength of C/Al composites.[28–32]

5.8.3 Special Properties of SiC and Al_2O_3 Fibers

In summary,[33] the most mature SiC/Al consolidation technique, hot-molding, has also produced the largest mechanical property database developed using this material. The design database for hot-molded SCS-2/6061Al aluminum includes static tension and compression properties, in-plane and interlaminar shear strengths, tension-tension fatigue strengths (*S-N* curves), flexure strengths, notched tension data, and fracture toughness data. Most of the data have been developed over a temperature range of –55 to 75°C with static tension test results up to 480°C. A high-performance, continuous SiC fiber in 6061 aluminum yields a very high-strength (1378 MPa) and high-modulus (207 GPa) anisotropic composite material having a density just slightly greater (2.85 g/cm^3) than that baseline aluminum. As in organic matrix composites, cross- or angle-plying produces a range of properties useful to the designer.

The property data that have been developed for investment-cast SCS-aluminum to date have been limited to static tension and compression. Fiber volume fractions are

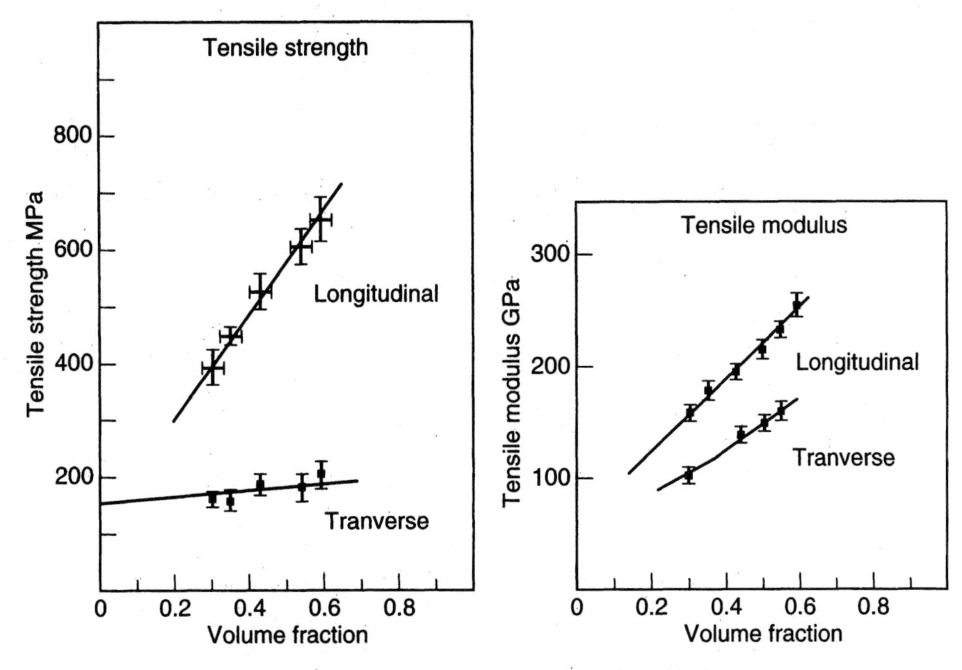

Figure 5.8. Unidirectional reinforcement of aluminium alloy with alumina continuous fiber FP.[20]

lower (40% maximum) than those of the hot-molded laminates (47% typical) because of volumetric constraints in dry-loading the shell molds; however, good rule-of-mixture (ROM) tensile strengths and excellent compression strengths (twice the tensile strength) have been achieved.

Corrosion resistance, another important consideration for MMCs was tested at the David W. Taylor Naval Ship R&D Center.[34] Testing on SCS-2/6061 under marine atmosphere, ocean splash and spray, alternate tidal immersion, and filtered seawater immersion conditions for periods of 60–365 days took place. The SCS-aluminum material performed well in all the tests, exhibiting no more than pitting damage comparable to that in the baseline 6061 aluminum alloy.

Al/Al_2O_3 composites containing 0.48–0.68% Mg in the matrix alloys have been fabricated by pressure infiltration casting. The addition of Mg was found to improve the infiltration and to decrease the porosity of the composites, whereas the tensile strength of the composites increased with increased Mg content and elongation remained almost constant (10%).[35]

5.8.4 Graphite

Carbon and aluminum in combination are difficult materials to process into a composite. A deleterious reaction between carbon and aluminum, poor wetting of the carbon by molten aluminum, and oxidation of the carbon are significant technical barriers to the production of these composites.[28]

Precision aerospace structures with strict tolerances on dimensional stability need stiff, lightweight materials that exhibit low thermal distortion. Graphite-aluminum MMCs have the potential to meet these requirements. UD P100 Gr/6061Al pultruded tube[12] exhibits an elastic modulus in the fiber direction significantly greater than that of steel, and it has a density approximately one-third that of steel (Table 5.1). Reference 29 and Table 5.4 contain additional data for P100 Gr/Al.

The advent of pitch-based graphite fibers with three times the thermal conductivity of copper[30] suggests that a high-conductivity, low-CTE version of Gr/Al can be developed for electronic heat sinks and space thermal radiators.

One of the most important properties being considered for Gr/Al is the ability to withstand alternate heating and cooling in space applications. It seems obvious that MMCs, which are mixtures of two materials with greatly different CTEs, might show some effect from the thermal cycling. Therefore, MMCs were exposed between –196 and 121°C, thermally cycled for 2000 and 5000 cycles, and then tested for mechanical properties by Sherman.[36]

Some of his results showed that

- P100/6061 in the as-fabricated and in the T6 condition has a 10% lower modulus at –193°C than at room temperature.
- This reduction in modulus is due to the aluminum being at a stress greater than its yield strength so that its contribution to the modulus is significantly reduced.
- The change in modulus exhihited by P100/6061 during tensile testing (proportional limit) is due to the aluminum being stressed beyond its yield strength.
- P100/6061 has a higher proportional limit than the as-fabricated material because heat-treating 6061 significantly increases the yield strength.

TABLE 5.4. Typical Mechanical Properties of MMCs

Material	Tensile strength (MPa)[a]	Tensile modulus (GPa)[a]
6061-T6 aluminum alloy	306	70
T-300 carbon-6061 Al alloy (V_f = 35–40%)	1034–1276 (L)	110–138 (L)
Boron-6061 Al alloy (V_f = 60%)	1490 (L)	214 (L)
	138 (T)	138 (T)
Particulate SiC-6061-T6 Al alloy (V_f = 20%)	552	119.3
GY-70 carbon-201 Al alloy (V_f = 37.5%)	793 (L)	207 (L)
Al_2O_3-Al alloy (V_f = 60%)	690 (L)	262 (L)
	172–207 (T)	152 (T)
Ti-6Al-4V titanium alloy	890	120
SiC-Ti alloy (V_f = 35–40%)	820 (L)	225 (L)
	380 (T)	
SCS-6-Ti alloy (V_f = 35–40)	1455 (L)	240 (L)
	340 (T)	

[a] L, longitudinal; T, transverse.

- P100/6061 has a significantly higher proportional limit at 121°C than at room temperature because the aluminum is under a severe compression stress during the initial portion of the test.
- Thermal cycling P100/6061 between −193°C and 121°C for 2000 cycles degrades the material as a result of twisting or warping of the material. This distress is significant for lengths of 20.52 cm or more.

Finally, Cheng, Akiyama, Kitahara, et al.,[37] examined the behavior of high-modulus carbon fiber-reinforced Al/Si composites after thermal exposure at 500°C and found, first, that the longitudinal tensile strength of the composites at first increased and then decreased as the thermal exposure time was further prolonged. After exposure for 216 h, the longitudinal tensile strength reached the maximum value of 899 MN/m^{-2} and the apparent and real transfer efficiencies of fiber strength were 87 and 109%, respectively.

Second, at a temperature of 500°C, the transverse tensile strength of the composites increased monotonically as the thermal exposure time was increased. This behavior also indicates that the interfacial bonding strength of the composites increased with increasing exposure time.

Third, single-fiber tensile tests showed that at a temperature of 500°C, long-term thermal exposure reduced the strength of the carbon fibers only slightly. The strength of the fibers extracted from the composite exposed for 300 h was 88% of that of the fibers in the as-cast composites. The microstructural observations gave some evidence that chemical interactions occurred at the interfaces of the composites after prolonged thermal exposure.

Finally, the fiber pullout on the tensile fracture morphologies of the composites had an important effect on the longitudinal properties of the composites. Moderate fiber pullout corresponded to higher longitudinal tensile strength of the composite. Too long and too little fiber pullout resulted in deterioration of the longitudinal strength of the composites.

5.8.5 TiNi

Researchers[38,39] motivated by the idea of enhancing the tensile properties of a $TiNi_f$/1100 aluminum matrix TiAl composite proposed the design concept shown in Figure 5.9. In the figure[38] an as-fabricated composite is (1) heated to shape-memorizing temperature, (2) cooled to the martensitic phase, (3) subjected to tensile prestrain ε, and (4) heated to and above the austenite finish temperature (Af) where the TiNi fiber shrinks to the length shape-memorized at step (1). The final step (4) induces the compressive stress in the matrix along the fiber axis while the fiber is in tension. Hence, when the composite at step (4) is tested to obtain the tensile stress-strain curve, one should observe the improvement in tensile yield stress over that of the unreinforced metal by two mechanisms: back stress strengthening[39] and compressive stress induced by the shape memory effect.

The authors[38] found that mechanical tensile properties such as stiffness and yield strength were improved by the strengthening mechanisms: back stress in the aluminum matrix induced by stiffness of the TiNi fibers and the compressive stress in the matrix caused by shape memory shrinkage of TiNi fibers. The damping capacity of the compos-

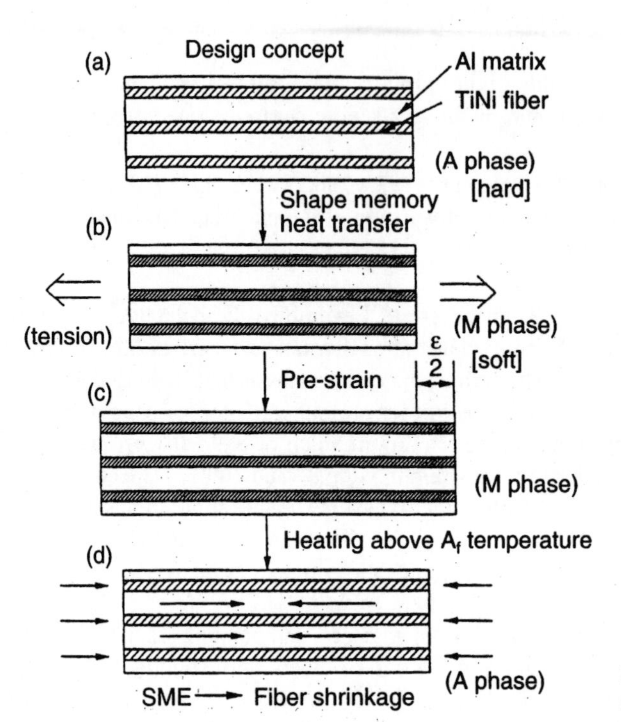

Figure 5.9. Design concept of the shape memory fiber MMC.[38]

ite was also increased. These results suggest that a composite with prestrain can be applicable and is suitable for machinery, especially engine components where the material becomes stronger at higher temperatures owing to the shape memory effect.

5.9 COPPER CONTINUOUS MMCs

5.9.1 Tungsten

Continuous tungsten-reinforced copper composites were studied in the 1950s as models for stress-strain behavior, stress rupture and creep phenomena, and impact strength and conductivity in MMCs.[40] On the basis of their high strength at temperatures up to 925°C, W/Cu MMCs are now being considered for use as liner material for the combustion chamber walls of advanced rocket engines.[41]

5.9.2 Graphite

Pitch-based graphite fibers have been developed that have room-temperature axial thermal conductivity properties better than those of copper.[30] The addition of these fibers to copper reduces density, increases stiffness, raises the service temperature, and provides a mechanism for tailoring the CTE. Table 5.5 compares the thermal properties of aluminum

TABLE 5.5. Thermal Properties of Unreinforced and Reinforced Aluminum and Copper

Material	Reinforcement content (vol%)	Density (g/cm^3)	Axial thermal conductvity (W/m • °C)	Axial coefficient of thermal expansion (10^{-6}/°C)
Al	0	2.71	221	23.6
Cu	0	8.94	391	17.6
SiC_p/Al	40	2.91	128	12.6
P120 Gr/Al	60	2.41	419	–0.32
P120 Gr/Cu	60	4.90	522	–0.07

and copper MMCs with those of unreinforced aluminum and copper. Gr/Cu MMCs have the potential to be used for thermal management of electronic components,[42] satellite radiator panels,[43] and advanced airplane structures.[44]

5.10 NICKEL CONTINUOUS MMCs

5.10.1 Tantalum Carbide

An advanced composite designated NiTaC14B was developed in the late 1970s and early 1980s. This composite was made into a hollow turbine blade and successfully engine-tested in 1983. Testing included 180 accelerated mission test cycles in GE's GE23 engine. The total running time for these high-pressure blades was 184 h and included 2290 full thermal cycles. Maximum temperatures in excess of 1150°C were reached. After 15 years of R&D, no other composite has exceeded this performance.[45]

The composite was an in situ directionally solidified eutectic—a nickel-based superalloy matrix reinforced by about 3 vol% of tantalum carbide fibers—and the succesful engine test confirmed the technical capability of an advanced eutectic composite for the high-pressure blade application.

5.11 MAGNESIUM CONTINUOUS MMCs

5.11.1 Graphite

The graphite continuous fiber-reinforced composite has been used in space structures.[46] Testing of a UD-reinforced Gr/Mg MMC in the fiber direction recorded modulus values in agreement with Equation 1 and a tensile strength of 572 MPa. A P100 Gr/AZ91C Mg UD laminate was shown to have a lower CTE and smaller residual strain than those of a P100 Gr/6061 Al MMC after both composites had undergone thermal cycling between −155 and 120°C.[47]

A follow-up study by Dries and Tompkins[48] has determined that graphite fiber-reinforced aluminum and magnesium composites have the potential to meet the property requirements for space structural components in dimensionally critical applications (high specific stiffness with a near zero CTE). Graphite fiber-reinforced 6061 Al and AZ91C Mg composite systems appear to be candidates that meet these criteria.

The thermal and mechanical properties of these materials must remain stable during exposure to the space environment for periods extending up to 20 years.[49] The temperature range over which these composites must be dimensionally stable can be as wide as 121 to −157°C (worst case), depending on the thermal control systems used.

The research reported that the thermal expansion behavior of P100 pitch-graphite fiber-reinforced aluminum composites made with high-strength aluminum alloys 201, 2024, and 7075 and high-strength Mg alloys QH21A and ZK60A can be thermally processed to eliminate thermal strain hysteresis. UD magnesium composites were tested before and after they had been thermally cycled 500 times between 121 and −157°C; the UD Gr/Al composites were tested after 1500 thermal cycles.

For the P100 Gr/Al composites, the thermal strain hysteresis common to the post-consolidation thermal processing was shown to eliminate the thermal strain hysteresis in high-strength Al alloy matrix composites in both the UD laminates and in the low-angle-ply configurations necessary for a zero CTE. In the UD laminates, once the hysteresis was eliminated, the thermal expansion behavior was unchanged after 1500 thermal cycles.

For the P100 Gr/Mg composites, the thermal strain hysteresis common to the as-consolidated material was unaffected by postconsolidation processing of composites with the Mg alloys AZ91C, QH21A, and ZK60A. This was attributed to the insensitivity of the elastic strength of the matrix alloy to precipitation strengthening. Thermal cycling up to 500 cycles had no significant effect on the magnitude of hysteresis. Currently available high-strength Mg alloys lack the necessary elastic limit strengths to eliminate hysteresis. New Mg alloys are needed with high elastic strengths to obtain dimensionally stable Gr/Mg compositions.

Other work at Cordec, Lorton, Virginia, involved fibers of E120 high-modulus graphite from Tonen or Mitsubishi which were embedded in a polymer matrix and coated with magnesium using a plasma vapor deposition (PVD) process. The sheet had a density of 1.5–1.8 g/cm^3 and could operate at temperatures up to 427°C. For even higher operating temperatures, 15-ply panels of P100/AZ31B Gr/Mg were clad with Ti-6Al-4V foil. The result was panels 0.45 mm thick, with tensile yield strength of 616 MPa and Young's modulus of 399 GPa.

5.12 TITANIUM CONTINUOUS MMCs

5.12.1 Silicon Carbide

In comparison with aluminum, titanium retains its strength at higher temperatures; it has increasingly been used as a replacement for aluminum in aircraft and missile structures as the operating speeds of these items have increased from subsonic to supersonic. Silicon carbide is now the accepted reinforcement for titanium; the SCS-6 fiber is an example of one commercially available type.

Properties of a representative UD SiC/Ti composite are significantly stronger than those of unreinforced titanium (Figure 5.3). The main type of reinforcement fiber discussed earlier is carbon-coated as this improves the fiber-matrix compatibility. Ti-based MMCs are gradually being shifted from Ti-6Al-4V matrix alloy to Ti-15V-3Al-3Cr-3Sn (Ti-15-3) or Ti-l0V-2Fe-3Al (10-2-3) β alloys.

Investigators have also evaluated the traditional Ti-6Al-4V alloy found susceptible to degradation in tensile strength after high-temperature thermal exposure, a feature that is attributed to reactions at the fiber-matrix interface. Improved performance is reported to have been obtained with a Ti-6Al-4V-2Ni (Ti-6-4-2) alloy as the matrix material. Composites have been consolidated at 843°C without obvious signs of fiber degradation. An SiC continuous-filament-reinforced Ti-6-4-2 composite, containing 10.5 vol% SiC, exhibited a room-temperature tensile strength of 1136 MPa. This value compares favorably with a conventional state-of-the-art Ti-6Al-4V composite containing 35 vol% SiC, which exhibits a strength of 1455 MPa.

Researchers Jeng and Yang[50] evaluated and investigated the flexural creep behavior and damage mechanisms of an unnotched SCS-6 fiber-reinforced Ti-6 at % Al-2 at % Mo-4 at % Zr-2 at % Sn (Ti-6242) matrix composite at temperatures between 550 and 700°C and stresses ranging from 700 to 1300 MPa. Their results revealed by microstructural observation that multiple fiber breakage, microcracking along the reaction zone-matrix interface and matrix cracking, extending from the broken fiber ends, were the major damage mechanisms during quasi-steady-state creep. The stress dependence of the quasi-steady-state creep rate indicates that the stress exponent increased with temperature.

Another series of tests at the University of California at Los Angeles[51–53] were conducted that examined the fatigue crack growth behavior of a UD SCS-6 fiber-reinforced Ti 6242 composite as well as several fatigue crack growth modes in UD fiber-reinforced titanium composites (Ti-15-3, Ti-6-4, Ti-24Al-l1Nb, and Ti-25Al-l0Nb-3Cr-lMo) under axial tension-tension loading. They found that the critical parameters controlling the fatigue crack growth were applied stress levels, specimen geometry, testing conditions, fiber strength, and interfacial bond strength. Their results also showed that the accuracy of analytical predictions relies explicitly on the interfacial properties. However, the variations in interfacial properties and fiber strength subject to extensive cycling are difficult to determine.

In another study at NASA Lewis Research Center, Cleveland, Ohio, Gayda and Gabb[54] reported on an evaluation of the transverse fatigue behavior of a UD SiC/Ti-15-3 composite [35 vol% (v/o)] SiC, $(90°)_8$ at 426°C, along the fiber direction $[(0°)_8]$ and of an unreinforced Ti-15-3 alloy for purposes of comparison. The $(90°)_8$ composite fatigue life was much shorter than that of the $(0°)_8$ composite. Furthermore, the $(90°)_8$ fatigue life was also found to be far lower than that of the unreinforced Ti-15-3 alloy. A simple, one-dimensional model for $(90°)_8$ fatigue behavior indicated that the short life of the composite in this orientation resulted, in large part, from weak fiber-matrix bond strength.

5.13 ALUMINUM DISCONTINUOUS MMCs

Discontinuous silicon carbide/aluminum (SiC_d/Al) is a designation that encompasses materials with SiC particles, whiskers, nodules, flakes, platelets, or short fibers in an aluminum matrix. Several companies are currently involved in the development of powder metallurgy and/or powder metal (P/M) SiC_d/Al, using either particles or whiskers as the reinforcement phase.[55] A casting technology exists for this type of MMC, and melt-produced ingots can be procured in whatever form is needed—extrusion billets, ingots, or rolling blanks—for further processing.[56] Arsenault and Wu[57] compared P/M and melt-produced discontinuous SiC/Al composites to determine if a correlation exists between strength and processing type. They found that if the size, volume fraction, distribution of reinforcement, and bonding with the matrix are the same, then the strengths of the P/M and melt-produced MMCs are the same.

Various researchers have concluded that

1. Particle reinforcement can lead directly to an increase in composite strength by up to 1000 MPa.
2. The majority of particulate composite strength arises from the matrix chemistry.
3. The particle-induced strengthening can lead to a greater choice of matrix composition and hence increased toughness at a particular strength level.

5.13.1 Silicon Carbide

5.13.1.1 Whiskers. Whiskers in discontinuously reinforced MMCs can be oriented in processing to provide directional properties. McDanels[58] evaluated the effects of reinforcement type, matrix alloy, reinforcement content, and orientation on the tensile behavior of SiC_d/Al composites made by P/M techniques. He concluded that these composites offered a 50–100% increase in elastic modulus as compared with unreinforced aluminum. He also found that these materials had a stiffness approximately equivalent to that of titanium but with one-third less density. Tensile and yield strengths of SiC_d/Al composites are up to 60% greater than those of the unreinforced matrix alloy. Selected properties of SiC_d/Al MMCs are given in Tables 5.5–5.7. Studies on the elevated-temperature mechanical properties of SiC_d/Al with either 20% whisker or 25% particulate reinforcement indicate that SiC_d/Al can be used effectively for long-time exposures to temperatures of at least 200°C and for short exposures at 260°C.[59]

Tydings[60] reported on work performed by DWA and ACMC where they showed that SiC in both whisker and particulate form is the most commonly used reinforcement and when combined with conventional 2000, 6000, and 7000 series aluminum alloys in additions from 10 to 25 v/o (the range for most structural composites), stiffness can be increased up to 60% whereas densities increase only a few percent. Thus, this MMC with enhanced specific properties and the ability to be readily formed and machined without a compromise in or loss of properties would benefit from fabrication techniques similar to those applied to continuous lay-up materials.

TABLE 5.6. Room-Temperature Properties of a Unidirectional SiC_c/Ti MMC

Property	SCS-6/Ti6Al-4V
Fiber content (vol%)	37
Longitudinal modulus (GPa)	221
Transverse modulus (GPa)	165
Longitudinal strength (MPa)	1447
Transverse strength (MPa)	413

TABLE 5.7. Properties of Discontinuous SiC/Alum Composites

Property	SiC_p/Al-4Cu-1.5Mg[a]	SiC_w/Al-4Cu-1.5Mg[b]
Reinforcement content (vol%)	20	15
Longitudinal modulus (GPa)	110	108
Transverse modulus (GPa)	105	90
Longitudinal tensile strength (MPa)	648	683
Transverse tensile strength (MPa)	641	545
Longitudinal strain to failure (%)	5	4.3
Transverse strain to failure (%)	5	7.4

[a] 12.7-mm plate.
[b] 1.8- to 3.2-mm sheet.

Walker[61] reported on the treatment and preparation of SiC whiskers in the manufacture of P/M discontinuously reinforced aluminum (DRA) composites. He showed that before SiC_w was intermixed with spherical aluminum powder particles attention was given to their treatment and preparation. To show the effect of whisker processing on the properties of DRA, test data were analyzed from Al-4%Cu-1.2%Mg-15%SiC_w-T6 (SXA 24E/15_w-T6) sheet made via simple and complex methods of preparing untreated and treated whiskers. The results indicated that substantial enhancement of ductility could occur, with little or no loss of strength and stiffness, if the proper method of treating and preparing whiskers was selected in manufacturing the composite.

He also showed that the greatest difference in tensile properties due to whisker processing resulted when complex methods of preparing treated whisker were used instead of simple methods of preparing untreated whisker. Ductility was increased 71%, whereas strength and stiffness were reduced only 4%. Whisker processing has a pronounced effect on ductility but only a modest effect on strength. Moreover, the effect that whisker processing has on the ductility of SXA 24E/15_w-T6 is of no less significance than the effects of other process and compositional variables like (1) the billet process conditions, (2) the copper and magnesium content of the matrix alloy, (3) the volume fraction of reinforcement, (4) the hot-rolling parameters, and (5) the temper.

House, Meinert, and Bhagat[62] recently concluded a series of studies on the high-temperature performance of aluminum composites, including their aging characteristics, strength at room and elevated temperatures, and creep. They evaluated 6061/SiC/20_w, 6061/Al_2O_3/15_p, and 6061-T6Al and found that the hardening rates of 6061/SiC/20_w and 6061/Al_2O_3/15_p were significantly higher than that of wrought 6061 Al. The whisker composite retained its tensile strength over a broad range of aging times (up to 500 h). Under identical aging conditions, the tensile strength of the whisker composite was about 65% higher than that of the particulate composite and wrought aluminum. Likewise, the stiffness of the whisker composite was approximately 60% higher than that of the other two materials (which have similar stiffnesses). At 350°C, the whisker composite had about 55% higher strength than the particulate composite and the wrought aluminum, in comparison with a 111% advantage at room temperature. The whisker composite showed better resistance creep than the particulate composite and wrought aluminum; the improvement was almost two orders of magnitude in terms of the steady-state creep rate.

Masuda and Tanaka of the National Research Institute of Japan and Fukazawa of the Fuji Institite of Tokai Carbon[63] studied the fatigue fracture mechanisms of three commercially fabricated materials; A2024/SiC_w, A357/SiC_p and A356/SiC_p.

The SiC_w/A2024 composite was fabricated by P/M and extruded, whereas the SiC_p/A356 and SiC_p/A357 composites were made by casting methods. The volume fractions V_f of SiC_w and SiC_p were 10 and 20%, respectively. AC4CH aluminum alloy, which designated by Japan Industrial Standard (JIS) as a matrix alloy and has a chemical composition very similar to that of A356 aluminum alloy, was used to obtain the reference data for SiC_p/A356 composite (see Table 5.8 for the tensile properties of the three composites under heat-treated conditions as fabricated or as cast.[58]) Fatigue strengths were higher for SiC_w/A2024 and SiC_p/A357 composite than those for matrix alloys, whereas for SiC_p/A356 composite with a V_f of 20% the fatigue strength was equivalent to that of the matrix alloy.

The tensile strengths of the three composites were higher than those of the matrix alloys, whereas the effects of the V_f of SiC whiskers or SiC particles were slightly significant in a V_f range of 10 to 20%, while the ductility for composites was less than that for matrix alloys. The elastic modulus increased with an increase in the V_f.

González-Doncel and Sherby conducted a study on all the investigations on the creep behavior of SiC/Al composites in the 230–525°C temperature range, as well as whisker- and particulate-reinforced SiC/Al composites from 10 vol% up through 30 vol%. Their conclusions[64] and observations for these composites were as follows.

- Flow stress of these MMC materials depends on the temperature of testing and the morphology of the reinforcing phase. The strength of the whisker materials was found to be higher (by a factor of 2) than that of the particulate materials at any given temperature.

TABLE 5.8. Tensile Properties of Composites

Material	Thermal Treatment	0.2% Yield Stress (MPa)	Tensile Strength (MPa)	Elongation (%)	Modulus (GPa)
SiC_w/A2024	Fabric	221	407	2.6	89
10%	T6	408	630	4.3	92
20%	T6	515	720	1.6	121
SiC_p/A356	Cast	103	193	9.2	80
10%	T6	353	387	2.2	76
SiC_p/A356	Cast	125	190	4.5	92
20%	T6	370	394	0.8	96
SiC_p/A357	Cast	149	165	7.6	83
10%	T6	345	372	1.6	79
SiC_p/A357	Cast	124	196	5.1	100
20%	T6	378	402	0.95	98
A2024	T6	390	480	10.4	75
A356	T6	206	284	10.0	
A357	T6	275	343	10.0	
AC4CH	T6	215	275	6.9	

- The calculated threshold stress (σ_o/E) decreases with increasing temperature and is higher for whisker-reinforced composites than for particulate composites. The temperature above which the threshold stress disappears is very similar for both composites (between 460 and 470°C). For the whisker composites, the σ_o/E value is essentially independent of the direction of testing.

Because of the ease of fabrication many researchers have investigated the properties of 6061Al/SiC_w. Among them are Ma, Lui, and Yao[65] from the People's Republic of China who observed dynamically by using scanning electron microscopy (SEM) the processes of tensile fracture in SiC_w/6061 Al composite. They found that

1. The off-axis angle has a great effect on the tensile strength of the composite. For SiC_w/6061 Al composite the strength decreased as the off-axis angle increased.
2. The low-ductility fracture in SiC_w/6061 Al composite may be caused by initiation, propagation, and joining of a great number of microcracks in the matrix.
3. There are two patterns of microcrack initiation, namely, void initiation around the whisker ends for smaller off-axis angles and debonding at the whisker-matrix interface for larger off-axis angles.
4. There are four patterns of microcrack propagation: (a) bypassing whiskers, (b) debonding at the interface, (c) pulling out of whiskers, and (d) joining of microcracks.

In another series of tests at other Chinese institutes, Cao, Wang, and Yao[66] found that the SiC_w/Al composite exhibits fairly good wear resistance, especially for higher-sliding velocities and/or higher loads in testing. The results are considered to be due to the high hardness of the SiC whiskers, the rotation of the SiC whiskers, and the constraint of the SiC whiskers in the aluminum matrix.

Bhagat and House[67] evaluated SiC_w/6061 Al composite with a $V_f = 0.20$ for its high-temperature performance. The composite showed an accelerated aging response in comparison with wrought 6061 Al alloy. The peak hardness for this composite was 148 kg_f/mm^2 after aging for 18 h at 150°C and 138 kg_f/mm^2 after aging for 1 h at 200°C following solutionization for 15 min at 530°C. The maximum tensile strength was retained after aging for 18–55 h at 150°C, an improvement of 40% over that of wrought aluminum. The tensile strength of the tested composite decreased from 578 MPa at room temperature to 229 MPa at 350°C. In terms of the use temperature, the composite provided an advantage of approximately 200°C over that of wrought aluminum. Young's modulus for the composite remained almost constant (104 MPa) from 150°C to 350°C and was 58% higher than that of the wrought aluminum at 150°C.

The composite exhibited steady-state creep deformation at applied stresses below 100 MPa at temperatures of 232–356°C. An improvement of almost two orders of magnitude in the steady-state strain rate of the composite was found over the wrought aluminum alloy under identical conditions of applied stress and temperature. The composite was highly stress-sensitive under creep conditions as evidenced by its high stress exponent value of 16. SiC_w/Al composites offer substantial advantages at elevated temperatures compared with unreinforced aluminum alloys in terms of increased strength and stiffness, use temperature, and resistance to creep deformation.

As a result of the composite whisker work with 6061 Al, the 2XXX and the 7XXX families of aluminum alloys that characteristically possess higher strength properties than 6061 Al have been examined by many researchers. Some recent findings are as follows.

Prasad and McConnell[68] performed a study to examine the tribological compatibility of two commercial composites, namely, a 2014 aluminum alloy (Al-4.5 wt% Cu) reinforced with 20 V_f SiC_w and an Al-Si alloy (Al-7.0 wt% Si) dispersed with 20 V_f of SiC particles with graphite as a solid lubricant. Significant reductions in wear resulted from both types of reinforcement.

Second, pronounced differences were observed in the adhesion of graphite films burnished on various surfaces. Films on Al-Cu alloys withstood much higher loads than those on Al-Si alloys. The highest failure load was recorded for the film on whisker-reinforced Al-Cu alloys, and it was found that protrusions of eutectic silicon on the etched surfaces of Al-Si alloys anchored the burnished films.

Finally, SiC_p in Al/Si/SiC particle composites caused abrasion damage to the steel counterface in testing, whereas the much smaller SiC_w in Al-Cu alloys acted like a polishing medium.

Schueller and Wawner[69] carried out various tests on AA2124 (Al-4Cu-1.4Mg-0.2Fe)/SiC/15_w composites to develop a better understanding of the failure mechanisms involved at high temperatures. It was discovered that the loss in strength of aluminum composites with increasing temperature was caused mainly by a decrease in matrix shear strength. On evaluating the literature (Figure 5.10), the researchers found that considerable data relating tensile strength to temperature were similar.[70–74] Their findings showed that the elevated-temperature tensile strength behavior of AA2124/SiC/15_w composites was similar to that of other discontinuous-fiber-reinforced aluminum alloys reported in the literature. They observed that at temperatures above 300°C the whiskers contributed negligibly to the composite strength because of the extremely low shear strength of the matrix. They also found that the precipitates offered no contribution to the strength above 350°C as the precipitates become dissolved in the material.

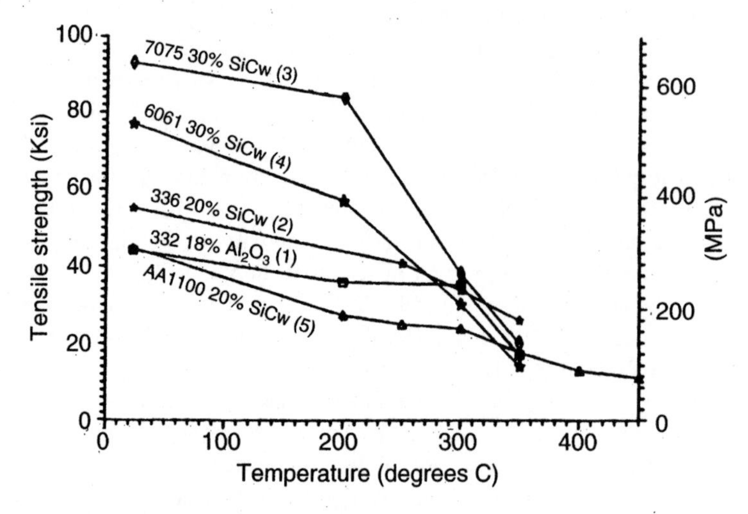

Figure 5.10. Tensile strengths of various aluminum matrix composites at different temperatures.[64]

Thus, the strength values for the various composites begin to converge at temperatures above 300°C, and at 400°C they are all identical. At these high temperatures, the matrix is the controlling feature. A thorough understanding of the mechanism of failure at temperatures above 0.6 T_m^1 is necessary in the development of aluminum-based composites with high-temperature capabilities.

Other researchers[75] examined the cyclic load fatigue damage in 2124/SiC_w by ultrasonic wave techniques. ACMC has developed an aluminum alloy 2009 (known formerly as 24E) reinforced with 15% SiC_w which exhibits ultimate tensile strength (UTS) of 638 MPa, tensile modulus of 105 GPa, and compression yield strength of 427 MPa. This alloy has been shown to have a superior balance of strength, ductility, and toughness compared to conventional 2XXX/SiC composites.

Prior to the development of alloy 2009, high-strength DRA composites exhibited low ductility and toughness which limited their usage. Alloy 2009 is an Al-3.6Cu-1.3Mg alloy that is virtually free of insoluble elements such as manganese and iron. When 2009 was blended with SiC_w, the resulting composite provided substantial toughness (>43.92 $MN/m^{3/2}$), the strength and density of 7XXX series aluminum (TYS > 483 MPa and $\rho = 2.75$ g/cm^3), and the stiffness of titanium (E > 105 GPa).

2009/SiC composite is suited for structural components of advanced aircraft, spacecraft, and optic systems. Use of this composite can improve system performance by reducing the weight of structures by up to 30%.

Wang and Rack[76] investigated the abrasive wear behavior of SiC particulate- and whisker-reinforced aluminum matrix composites under two-body abrasion conditions utilizing a pin-on-disk apparatus. Both SiC_p- and SiC_w-reinforced composites exhibited greater wear resistance than the unreinforced 7091 alloy.

They concluded that

1. The abrasive wear resistance of aluminum alloys was improved by both SiC_p and SiC_w reinforcement, the improvement being greatest against finer abrasive particles.
2. SiC_w reinforcement was found to be more effective than SiC_p reinforcement in enhancing wear resistance over the entire range of abrasive sizes.
3. For fine-particle wear, composites reinforced with SiC_w and oriented with the reinforcement normal to the contact face were more resistant to abrasive wear than were those with whiskers oriented within the contact surface: the effect was reversed when coarser abrasives were used.

5.14 MAGNESIUM DISCONTINUOUS MMCs

5.14.1 Silicon Carbide

5.14.1.1 Whiskers. Inert-gas atomized powders of magnesium alloy ZK60A have been blended with 20 V_f SiC_w or 30 V_f B_4C_p, vacuum-pressed, and extruded. UTS for the SiC/ZK60A was 579 MPa, elongation was 1.58%, and elastic modulus was 98 GPa, whereas UTS of the B_4C/ZK60A was 510 MPa, elastic modulus was 95.2 GPa, and elongation was 0.93% (Table 5.9). The magnesium composites compare favorably with

TABLE 5.9. Properties of Selected Aerospace Composites

Material	Density (g/cm^3)	UTS (ksi)	TYS (ksi)	Elastic Modulus (msi)	Comments
Ti-6Al-4V	4.4	149	128	17.3	Most widely used Ti alloy
Mg ZK60A (T6)	1.8	45	28	6.5	T6 condition: solution heat-treated and artificially aged
Al6061 (T6)	2.7	45	40	10	
Be (annealed)	1.8	40–100	30–60	42	
AlLi 8090	2.54	70.6	60.4	11.6	Elongation 6–7%
Beta-21S (Ti-3Al-15Mo-2.6Nb-0.2Si)	4.94	124	122	10.7–12.3	Three times more resistant to hydrogen absorption than other Ti alloys
10% TiC/Ti (Alloy Tech)	4.53				Hardness 50-52 HRC; 10 to 30 μm particles
40% BeO/Be (Brush)	2.3	42		42	CTE is 7.5 ppm/°C; thermal condition 380–430 Btu/ft·h°F
17% SiC/8090 (BP composites)	2.66	78.6	65.5	15.0	In the T6 condition; elongation 3–4%
Mg-clad C/Mg (FT 700/ZK60A) (Cordec)			150–160	57	CTE can be tailored from zero
SiC/Mg (20 vol% SiC/ZK60A) (ACMC)	2.04	84	66	14	CTE is 6.5 ppm/°F; elongation is 1.58%; compressive extruded strength 74 ksi
SiC/Mg (25.4 vol% SiC/AZ91) (Dow)	2.06	35	33.1	9.3	Low-cost extrusion

the aluminum composites currently used for structural sleeves, brackets, and spacers on Navy satellites. The aluminum composite has a tensile strength of 448 MPa, compared with 579 MPa for the magnesium composite. In addition, the magnesium has a density of 2.04 g/cm^3, compared with 2.96 g/cm^3 for aluminum.

5.15 ALUMINUM DISCONTINUOUS MMCs

5.15.1 Silicon Carbide

5.15.1.1 Particulates. Moderate-strength DRAs recently developed are aluminum alloys 6090 and 6090H/SiC/25_p. The composites have achieved UTS of 484 MPa, tensile modulus of 1183 GPa, and compression yield strength of 405.3 MPa.

De Bondt, Froyen, and Deruyttere[77] evaluated the homogeneity parameters in the production of 6061 Al/SiC_p composites and resultant tensile and stiffness properties.

They found that homogeneity could be increased by decreasing the fiber volume fraction, by changing the squeeze-casting procedure, and by introducing particles.

The average strength and stiffness of samples without particles were 260 MPa and 144 GPa, respectively. The stiffness was in agreement with the rule of mixtures; the low strength was attributed to the high volume fraction of fibers (70 V_f), which implies that fiber contacts act as failure initiation points.

The average strength and stiffness of samples with particles were 450 MPa and 150 GPa, respectively. Again the stiffness was in agreement with the rule of mixtures. The V_f was on average 33%.[77]

Lim et al.[78] evaluated the fabrication and mechanical properties of aluminum matrix composite materials and found that mechanical properties, such as Young's modulus and UTS, were improved up to 80% by the addition of reinforcements. As expected, the strength of MMCs decreased with increased testing. Their evaluation included 6061Al, SiC Tokawhisker, and Saffil RF Grade Al_2O_3.

It was found that tensile strengths at 150 and 300°C were maintained at 85% and 70% of room-temperature strengths in Al/SiC composites, and 80% and 40% in Al/Al_2O_3 composites, respectively. Still, these composites have reasonable values for high-temperature applications. It was suggested, based on an SEM study and experimental data, that strength reduction at elevated temperature was mainly caused by overaging and softening of the matrix alloy.

Fractographic studies show a ductile failure mode of discontinuous MMCs at the microstructural level. As the temperature increases, the failure mode becomes progressively more ductile.

Nardone, Strife, and Prewo[79,80] have developed a fabrication approach that subsequently improves the damage tolerance of materials that would normally exhibit failure at room temperature with minimal absorption of energy. The approach, termed microstructural toughening (MT), consists of controlling the composite microstructure such that continuous ductile toughening regions are incorporated throughout the composite microstructure. They examined the tensile and impact behavior of both ductile and brittle matrix composites. The ductile matrix composites consisted of SiC_p-reinforced 6061 Al alloy with commercially pure titanium toughening regions. The brittle matrix composites consisted of B_4C_p-reinforced NiAl with type 304 stainless steel toughening regions.

With the MT approach as much as an order-of-magnitude increase in notched Charpy impact energy absorption capability was demonstrated relative to conventional Al alloy 6061/SiC_p composites. The longitudinal tensile properties of the composites were shown to be independent of the scale of the microstructure or the CP titanium V_f and could be estimated using a simple rule-of-mixtures approach.

The brittle matrix composites showed an increase in density resulting from type 304 stainless steel being compensated for by the low density of B_4C_p so that the overall composite density was within 5% of the value for monolithic NiAl. The notched Charpy impact energy absorption of the NiAl-B_4C-type 304 stainless steel composites was in the range 15–90 J/cm^2, compared with a value of 0.8 J/cm^2 for unnotched NiAl. Significant elongation (15–35%) was measured during tensile testing of the brittle matrix composite as a result of constrained yielding of the type 304 stainless steel which prevented composite failure after the NiAl regions had cracked.

Bhat, Surappa, and Sudhaker Nayak[81] studied the corrosion behavior of 6061

Al/SiC_p composites (in as-cast and extruded form) in seawater and acid media and the effects of the temperature of both the media and the concentration of the acid medium. They found, first, that 6061 Al/SiC_p composites exhibited more active open-circuit potentials compared to the base alloy. Second, 6061 Al/SiC_p composites showed localized attack both in the as-cast and extruded states. However, in the as-cast condition the extent of corrosion damage was greater.

Third, 6061 Al/SiC_p composites in the as-cast condition suffered localized corrosion in acid medium, and the attack increased with acid concentration and temperature of the media.

Finally, 6061 Al/SiC_p composites in the extruded state showed relatively less attack compared to those in the as-cast condition, because of fewer defects in the matrix and fewer agglomerates of SiC_p. The increase in acid concentration and temperature increased the attack.

Other corrosion work was conducted at the University of Texas at Austin by Sun, Koo, and Wheat.[82] Here the corrosion behavior of 6061 $Al/SiC/15\text{-}40_p$ MMCs was studied in chloride solution by means of electrochemical techniques, scanning electron microscopy, Auger electron spectroscopy (AES), energy-dispersive spectroscopy (EDS), and x-ray diffraction. It was observed that the corrosion potentials did not vary greatly or show definite trends in relation to the amounts of SiC_p reinforcement. However, the degree of corrosion increased with increasing SiC_p content, and the presence or absence of oxygen as well as the concentration of the NaCl solution affected corrosion potentials.

Wu, Goretta, and Routbort[83] studied the solid-particle erosion of 2014 Al, 2014 $Al/SiC/20_p$, and 2014 $Al/Al2O3/20_p$ at room temperature. The alloys were tested under the as-cast, annealed, as-quenched, and T6 conditions, whereas experiments with Al_2O_3 and SiC abrasives were conducted in vacuum over ranges of abrasive size, velocity, and angle of impact.

Erosion data and SEM observations revealed that weight loss was more severe for Al/SiC and Al/Al_2O_3 composites than for the unreinforced 2014 Al alloy. Lack of ductility was the primary cause of the reduced erosion resistance of the composites. Erosion rates were generally greatest for the T6 heat-treated materials. T6 heat treatments increased hardness but decreased ductility. The heat treatments that preserved ductility, annealing, and quenching from 502°C had little influence on erosion rates. Abrasive shape affected material removal in that flat abrasives induced less wastage.

Bonollo et al.[84] evaluated the effect of quenching on the mechanical properties of P/M-produced 1100 Al/SiC_p MMCs with various reinforcement V_f (10–25%). Their results showed that the quenching treatment produced a general increase in the UTS of Al/SiC_p composites. Such an increase, they determined, depended on several factors, such as the process followed, the average size of the SiC particles and their dispersion in the matrix, and the oxygen content of the composites. The basic phenomenon acting to achieve such strengthening was the increase in dislocation density determined, during the quenching treatment, by the mismatch between the CTE of aluminum and SiC. Finally, the UTS increase was fitted by the parameters describing the Orowan bowing mechanism, which governs the particle dislocation interaction at room temperature, and good results were achieved.

Sarkar, Brewer, and Lisagor[85] conducted a series of experiments on 1100 $Al/SiC/20_p$ to improve the mechanical properties through increased microstructure uniformity. The composites were fabricated with powders of various particle sizes. Microstructural homogeneity was significantly affected by the size of the particles in the starting matrix powder. Coarse powders resulted in large free areas in the composites, whereas fine powders produced composites with more uniform distributions and improved mechanical properties. Tensile strength and modulus, proportional limit, and ductility were all greatest for the finest particle composite. Ductility was affected most by variations in the powder; –600-mesh powder resulted in composites with a strain to failure about three times that of the composite made with –100 mesh powder.

Srivatsan[86] wished to understand the low-cycle fatigue and cyclic fracture behavior of the aluminum alloy 2124 with varying volume fractions of SiC_p. The Al/SiC_p composites were cyclically deformed over a range of strain amplitudes giving lives of less than 10^4 cycles to failure. The specimens were cycled using tension-compression loading and under total strain control. The 2124 + SiC_p, composites, in the as-extruded condition, displayed cyclic hardening at all strain amplitudes and for the different volume fractions of reinforcement in the ductile metal matrix. The material rapidly softened prior to failure. The cyclic hardening behavior increased with an increase in carbide particle content, and a degradation in the low-cycle fatigue resistance occurred with an increase in V_f of SiC_p reinforcement in the 2124 matrix (Figure 5.11).

Healy, Wright, Halliday, et al.[87] investigated the fatigue crack growth characteristics of 8090 reinforced by different V_f of SiC_p.

The mechanical properties of the materials tested are shown in Table 5.10. In all cases fracture in specimens manufactured from the 8090/SiC/14_p occurred within the elastic region of the stress-strain curve as stress values substantially below the expected yield stresses predicted on the basis of hardness measurements.

Short-crack growth data obtained for material containing 3% and 14% by volume of SiC were conducted at $R = 0.1$ with peak stresses ranging from 224 to 344 MPa.

Short-crack growth in both materials was found to be matrix-dominated with little or no evidence of SiC_p fracture or decohesion of the particle-matrix interface. The role of the particulate in the 8090/SiC/14_p composite was to promote local crack deflection, and in regions of high reinforcement density to retard crack growth rates.

As a result, the mechanical properties obtained for the 8090/SiC/14_p composite were attributed to the presence of a high degree of reinforcement in conjunction with the processing route, which resulted in low ductility and premature failure.

Alcoa has developed a series of P/M composites. Gas-atomized aluminum powder is combined with SiC_p in the proportions required for the desired properties. Alcoa's Innometal 2080/SiC composite, with matrix composition Al-3.8Cu-1.8Mg-0.2Zr, exhibited highly refined microstructure and damage tolerance. The alloy is heat-treatable, allowing a range of strengths depending on both the reinforcement level and thermomechanical treatment. With 15 V_f SiC, modulus is 101.5 GPa and density is 0.028 g/cm^3. In the T4 condition (solution heat-treated and naturally aged), UTS is 483 MPa, TYS is 365 MPa, and elongation is 7.5%. In the T8 condition (solution heat-treated, cold-worked, and artificially aged), UTS is 545 MPa, TYS is 517 MPa, and elongation is 4%.[88]

Several other aluminum-SiC-reinforced composites are shown in Tables 5.11 and 5.12.

Bloyce and Summers[89] squeeze-cast A357/SiC/15_p and 20_p aluminum alloy matrix composites. They applied a series of heat treatments to the 20 V_f SiC_p composite, result-

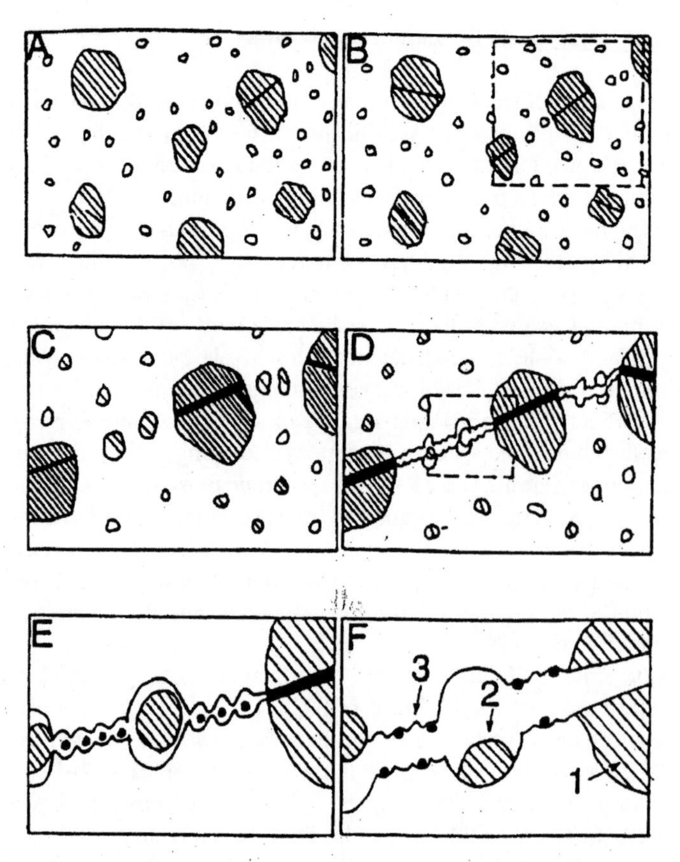

Figure 5.11. Fracture sequence in the 2124 + SiC composite: (*a*) Initial state of the composite. (*b*) At low strain, coarse SiC particles crack. (*c*) At high strain, fine SiC particles decohere and form voids. (*d*) Finer particles in the 2124 matrix decohere and form voids. (*e*) Voids coalesce. (*f*) Primary fractographic features are formed: (1) fractured coarse SiC particle, (2) dimples around fine SiC particles, (3) fine dimples in matrix.[86]

ing in the general trend of a higher UTS, giving a less tough and less ductile material. UTS values of up to 416 MPa were obtained with Young's modulus of approximately 100 GPa. Elongations of up to 5% were obtained but at the expense of UTS and proof strength.

The *S-N* curves demonstrated that A357/SiC/20_p -T6 has a fatigue strength comparable to that of the monolithic A357-T6 and, more importantly, the presence of SiC_p does not appear to induce premature ignition of fatigue cracks.

TABLE 5.10. Tensile Properties, Transverse Direction[87]

Material	0.2% YS (MPa)	UTS (MPa)	Elongation (%)	*E* (GPa)	VHN[a]
8090 3% SiC	409	487	≈6	74	129
8090 14% SiC	N/A	278	N/A	100	141

[a] VHN, Vickers hardness number.

TABLE 5.11. Tensile Properties of Discontinuously Reinforced Aluminum Alloy Sheets and Extrusions

Material (Matrix/Reinforcement/Volume Fraction, and Type Alloy Temper)	Orientation[a]	UTS (MPa)	YS (MPa)	Elongation (%)	Modulus (GPa)
$2009/SiC/15_w$-T8, sheet	L	634	483	6.4	106
	T	552	400	8.4	98
$2009/SiC/20_p$-T8, sheet	L	593	462	5.2	109
	T	572	421	5.3	109
$6013/SiC/15_p$-T6, extrusion	L	517	434	6.3	101
$6013/SiC/20_p$-T6, extrusion	L	538	448	5.6	110
$6013/SiC/25_p$-T6, extrusion	L	565	469	4.3	121
$6013/SiC/15_w$-T6, extrusion	L	655	469	3.2	119
$6090/SiC/25_p$-T6, extrusion	A	483	393	5.5	117
$6090/SiC/40_p$-T6, extrusion	A	538	427	2.0	138
$7475/SiC/25_p$-T6, extrusion	A	655	593	2.5	117

Source: A. L. Geiger and J. A. Walker, The processing and properties of discontinuously reinforced aluminum composites, *J. Metals.* 43:8 (1991). W. C. Harringan, Jr. Scaling up particulate-reinforced aluminum composites for commercial production, *J. Metals.* 43:32 (1991).
[a]L, longitudinal; T, transverse; A, average.

The fatigue-crack growth rate of a certain stress intensity factor was higher in the composite than in the matrix alloy. Metallographic examination of the fatigued surface of the composite indicated that the crack does not seek to propagate along or through the SiC_p. In monolithic Al-Si cast alloys[90] a coarse structure allows the crack to grow further in a crystallographic plane. With a more refined structure, the crack encounters a silicon particle more frequently and is constrained to a straighter path. This effect is further exaggerated in the A357Al/SiC_p MMC, where both silicon and silicon carbide can effectively straighten the crack path. These materials (A357, A357 + SiC_p) differ from some wrought materials (6000 series, 6XXX + SiC_p), where the addition of SiC_p reduces the crack growth rate for a ΔK at the intermediate values of crack growth rate (10^{-6}–10^{-4} mm/cycle).[91] However, the results for the squeeze-cast A357/SiC/20_p -T6 lie within the data reviewed in reference 91.

Wang and Zhang[92] carried out mechanical tests in their study of the deformation behavior of A356/SiC/15_p matrix composites produced through direct casting using the molten-metal mixing method. They found that the A356/SiC/15_p composite had a higher elastic constant and yield stress than the matrix alloy A356Al. In addition to the effect of high particle strength, the strengthening mechanisms in this type of material were found to be related directly to particle distribution. The microscopic nonuniformity of the particle distribution, created by dendrite structure formation and usually in the form of clustering, was considered the reason for internal stresses and also stress triaxiality, which was responsible not only for the special hardening behavior but also for the early appearance of particle cracking, particle interface debonding, and void formation in the matrix. These behaviors led to a low overall ductility.

Selvaduray, Hickman, Quinn, et al.[93] examined the mechanical properties of $6061/Al_2O_3/10_p$ and 20_p-T6 and $A356/Al_2O_3/20_p$-T6 and measured shear properties of the

TABLE 5.12. Mechanical Properties of Some Cast Particulate-Reinforced MMCs

	Reinforcement				Property			
Matrix	Type	Shape	Amount	Process	UTS (MPa)	YS (MPa)	Elongation (%)	E (GPa)
Al-12Si-Ni-Cu			0	Squeeze-casting	297	210	NA	71.9
	Al_2O_3 (Saffil)	Short fibers	20 vol%		312	283	NA	95.2
	SiC (Tokai)	Whiskers	20 vol%		384	298	NA	111
Al-3Mg	Graphite	Particle 75–125 μm	1.5 wt%	Gravity-casting	72	48	4.8	NA
			4.5 wt%		49	40	4.5	NA
		Particle 180–250 μm	1.5 wt%		85	64	5.3	NA
Al-3Mg	Zircon[a]	Particle 125–180 μm	10 wt%	Gravity-casting	78	68	5.5	NA
			30 wt%		68	62	2.6	NA
		Particle 180–250 μm	10 wt%		86	70	9.3	NA
Al-7Si-Mg-Fe			0	Gravity-casting	121	73	2.9	NA
	SiC	Particle 50 μm	12 wt%		81	65	0.4	NA
			0	Squeeze-casting	190	96	5.8	NA
	SiC	Particle 50 μm	12 wt%		163	61	2.2	NA
Al-7Si-Mg-Fe			0	Gravity casting + aging-T6	302	148	16	77
	SiC (Alcan)	Particle 5–10 μm	15 vol%		280	183	1	98
Al-7Si-Mg-Be			0	Squeeze casting	214	106	10.3	77
	SiC (Alcan)	Particle 5–10 μm	15 vol%		195	127	1.5	NA
		Particle 5–10 μm	20 vol%		216	148	1.3	NA
			0	Squeeze casting + aging-T6	359	302	9.6	70
		Particle 5–10 μm	15 vol%		328	345	0.6	91
		Particle 5–10 μm	20 vol%		408	377	1.4	105
Al-4.4Cu-Si-Mn-Mg	SiC (Alcan)	Particle 5–10 μm	15 vol%	Gravity casting + extrusion + aging-T6	450	342	1.6	107

[a] 65.9% ZrO_2, 32.2% SiO_2 0.3% TiO_2, 0.07% Fe_2O_3, 1.53% volatiles.

6061-T6 MMC parallel and perpendicular to the extrusion direction. The microstructure showed banding and reinforcement agglomeration along the grain boundaries for the 6061-T6/Al_2O_3 MMC and the presence of defects in the A356-T6 MMC.

The strength and abrasion resistance of both these alloys improved with the addition of reinforcement particles, whereas the ductility and fracture toughness decreased. In the case of the 6061-T6 alloy, doubling the reinforcement from 10% to 20% did not always produce a corresponding change in mechanical properties. The extent of change was found to depend on the microstructural mechanism controlling the particular property. Abrasion resistance and shear strength (in the longitudinal direction) showed a distinct improvement, whereas ultimate strength showed a marginal improvement at best.

Hochreiter and associates[94] examined the fatigue lifetime behavior for extruded 6061/SiC/15_p and 6061/Al_2O_3/10_p composites. The peak particle sizes were about 4.5 and 6.0 μm, within measured *S-N* curves the fatigue lifetime at given stress amplitudes of SiC_p/AA6061 was superior to that of Al_2O_{3p}/AA6061 in the low-cycle fatigue region as well as in the high-cycle fatigue region.

They further concluded from their study that the interaction of particles and the propagating fatigue crack was determined by the residual stress field in the region near the particles and the strength of the interfacial bonding and the particles themselves. These quantities ultimately depend on the chemistry, size, and shape of the particles. In order to reduce the fatigue crack propagation rate, crack closure mechanisms should be encouraged.

Although the fatigue lifetime of aluminum alloys is decreased in most cases by particle reinforcement, an improvement in fatigue properties can be expected if the compatibility and the size of the reinforcement is well selected. Furthermore, the homogeneity of the particle distribution, which depends largely on the production route, determines the efficiency of the reinforcement.

5.15.2 Al_2O_3

5.15.2.1 Particulates. Discontinuous Al_2O_3/Al MMCs are made using short fibers, particles, or compacted single-fiber preforms as reinforcements.

A recent development in MMC fabrication technology is the proprietary Lanxide PRIMEX process, which involves pressureless metal infiltration into a ceramic preform. This process has been used to produce an Al_2O_3/Al composite by infiltrating a bed of Al_2O_{3p} with a molten alloy that has been exposed to an oxidizing atmosphere. The matrix material of the resultant composite is composed of a mixture of the oxidation reaction product and unreacted Al alloy.[95] The Lanxide process and the properties of composites produced by this method can be tailored to fit specific applications. This subject is further discussed in Volume II under composite processing.

Table 5.13[96] reflects some of the composite material properties that can be tailored through the selection of matrix system; filler material; filler geometry, size, and volume percent; nature and amount of dopants; and process conditions. As an example, Table 5.14 shows the effects of process temperature on the room-temperature three-point bend strength and fracture toughness of an Al_2O_3/Al matrix system with a 500-mesh SiC abrasive-grade particulate filler.[97,98]

The elevated-temperature fatigue crack growth rates of alloy 2519 [Al-(5.3–6.4)Cu-

TABLE 5.13. Properties of Selected Lanxide Composite Products[96]

Reinforcement and Matrix System	Tensile Strength (MPa)	Bend Strength (MPa)	Young's Modulus (GPa)	Fracture Toughness (MPa/m^2)	Density (g/cm^3)	CTE ($10^{-6}/K$)	Compressive Strength (MPa)	Thermal conductivity (W/m • K)	Rockwell Hardness (R_A)
NX-5201 silicon carbide-reinforced aluminum (SiC/Al)	210		200	10	2.95	8.5		160	
NX-5101 alumina-reinforced aluminum (Al_2O_3/Al)	400		150	17	3.3	11.5	760		
NX-1201 silicon carbide-reinforced aluminum oxide (SiC/Al_2O_3)	150		313	6.3	3.32	5.4	1193	147	80
NX-3401 zirconium diboride platelet-reinforced zirconium carbide (ZrB_2/ZrC)		800–900	300	16–18	6.2	7		50	
NX-1010 alumina-reinforced aluminum titanate (Al_2O_3/$AlTiO_3$)		20	14		2.6	1.9	330	1.0	

TABLE 5.14. Effects of Process Conditions on the Mechanical Properties of an Alumina-SiC Particulate System

Growth Temperature (°C)	Strength (MPa)	Fracture Toughness (MPa $m^{1/2}$)
900	350	7.8
1000	390	5.4
1150	525	4.7

0.3Fe-0.25Si-(0.1–0.5)Mn] and 2519/Al_2O_3/15_p were reported by Wang et al.[99] Compared with room-temperature performance, no appreciable change in fatigue crack growth kinetics was observed for 2519 at 150°C, but a significant reduction in fatigue-crack growth threshold stress occurred in the composites. The threshold stress reduction in the composites was due to cyclic strain-induced overaging, which greatly reduced the level of crack closure. At 300°C, both materials showed much higher fatigue-crack growth rates.

Other Al_2O_3/Al matrix composites, 6061/Al_2O_3/10_p and 20_p-T6, were studied for their dynamic response by subjecting them to a single high-amplitude stress pulse of extremely short duration (approximately 500 ns).

Perng, Huang, and Doong[100] examined three MMCs consisting of 6061/Al_2O_3/10_p, 15_p, and 20_p-T6. From their results, they concluded that (1) the Al_2O_3-reinforced composites were more sensitive to strain rate than the unreinforced material, (2) the strain rate sensitivity of the UTS of these composites increased with increasing temperature, and (3) the addition of Al_2O_{3p} reduced the ductility of 6061 Al-T6 irrespective of the strain rate and temperature.

5.15.3 Graphite

5.15.3.1 Particulates. Pillai, Pai, Kelukutty, et al.[101] evaluated the properties of an Al alloy-Gr-reinforced composite pressure die-cast versus gravity die-cast. The Al-Si-Mg (LM25)/Gr/15_p was initially made by the rheocasting technique. The composite ingots were used as a master alloy and diluted further with LM25 alloy to obtain a 5 wt% dispersion of graphite in the matrix.

The strength properties of the pressure die-cast plates were superior to those of the gravity die-cast samples. The UTS evaluated in this composite was superior (132–136 MPa) to that of gravity die-cast (85–95 MPa) composites, and the plane-strain fracture toughness of the composites was in the range 8–10 MPa $m^{1/2}$.

Zhang, Perez, Gungor, et al.[102] evaluated the effects of four different types of Gr_p on the resultant damping behavior of 6061 Al alloy MMCs. The MMCs were processed by a spray atomization and codeposition technique, and the damping characterization was conducted on a dynamic mechanical thermal analyzer. The damping capacity, in terms of loss tangent (tan ϕ), and storage modulus were measured at frequencies of 0.1, 1, 10, and 50 Hz over the temperature range 30–250°C. The experimental damping capacity results for the 6061 Al/Gr MMCs were compared with those for as-received 6061-T6 Al alloy and as-deposited 6061 Al alloy. It was shown that the damping capacity of 6061 Al alloy could be significantly improved by the addition of Gr_p.

Crystalline graphite is superior to amorphous graphite in reinforcing the damping capacity of MMCs. For natural crystalline flake Gr_p, the larger the Gr_p size, the higher the damping capacity of the corresponding 6061 MMCs. The relatively high damping capacity of the 6061 Al/Gr MMCs was also attributed to the behavior of the particulate-matrix interface, the presence of the micrometer-sized pores, and the fine-grained microstructure.

5.15.4 Titanium Carbide (TiC)

5.15.4.1 Particulates. The low- and high-cycle fatigue behavior and cyclic response of naturally aged and artificially aged 2219/TiC/15_p and unreinforced 2219 Al were investigated utilizing plastic strain-controlled and stress-controlled testing.

The higher ductility of the unreinforced material made it more resistant to fatigue failure at high strains, and thus, at a given plastic strain, it had a longer fatigue life. It should be noted that the tensile ductilities of the 2219/TiC/15_p were significantly higher than those previously reported for 2XXX series composites. During stress-controlled testing at stresses below 220 MPa, the presence of TiC particles led to an improvement in fatigue life. Above 220 MPa, no influence of TiC reinforcement on fatigue life could be detected. In both the composite and unreinforced materials, the low- and high-cycle fatigue lives were found to be virtually independent of matrix microstructure.

5.16 COPPER DISCONTINUOUS MMCs

5.16.1 Al_2O_3

5.16.1.1 Particulates. Sun, Orth, and Wheat[103,104] reported on the corrosion behavior of Al_2O_3 dispersion-strengthened (DS) copper in a 3.5% NaCl solution and compared it to the behavior of pure Cu using electrochemical techniques, SEM, x-ray diffraction, and inductively coupled argon plasma atomic emission spectroscopy (ICP-AES). Two categories of composites were studied: the previously mentioned commercially available Al_2O_3/Cu or DS Cu, developmental Cu/Gr/1_p, 2_p, 5_p, 15_p, 25_p, and 40_p, and +30° angle-plied continuous P100/Cu/50_f MMCs.

While the anodic polarization curves for the monolithic and composite materials were quite similar, the corrosion behavior varied. The commercially available material was found to possess corrosion resistance comparable to that of pure Cu. This was due largely to the stability of the protective film in the presence of finely distributed Al_2O_3 particles. The developmental composites reinforced with graphite showed that graphite content and the presence of dissolved oxygen affected corrosion severity, and all the developmental composites exhibited uniform corrosion and some localized galvanic corrosion at the reinforcement-Cu interface during polarization.

Some factors that may influence this behavior are the much larger particles than in the case of DS Cu, residual stresses that may result from the high-energy high rate (HEHR) consolidation processing, and the fact that there is only a mechanical bond between the Cu and the graphite.

5.17 MAGNESIUM DISCONTINUOUS MMCs

5.17.1 Particulates

5.17.1.1 SiC. Krishnadev, Angers, Krishnadas Nair, et al.[105] fabricated Mg and Mg-alloy composites with SiC_p as the reinforcement using different techniques with a view of comparing the effects of different processing routes on the final microstructure and the mechanical properties with special emphasis on evaluating the relative importance of the particle-matrix interface, the reinforcement, and the metal matrix on crack initiation and propagation. The processing routes that were used were P/M and ingot metallurgy (IM). In addition, for the P/M route, two approaches were used. In the first method, mechanical alloying (MA) via a low-energy ball mill was employed to form composite particles that were subsequently hot-pressed and hot-extruded. In the second method, simple dry mixing of the constituents was followed by hot-pressing and hot extrusion. For the IM route, a fluxless melting technology was used.

Results of tensile properties are listed in Table 5.15. It is apparent that MA leads to better yield strength, UTS, and ductility than the dry mixing process. This can be related to the greater microstructural refinement and homogenization brought about by the low-energy ball milling employed.

Mechanical properties of the composites produced via the IM route are shown in Table 5.16. For the sake of comparison, properties of Mg-Al alloys from the literature are also shown. Compared to the unreinforced materials, the composites have higher modulus and yield strength; however, tensile strength remains unchanged, and there is a drastic reduction in the ductility.

By comparing the properties of $Mg/SiC/10_p$ composites obtained via the MA route with the properties obtained via the IM route (Tables 5.15 and 5.16), the improvements brought about by MA are quite evident. This is because of the improved microstructural refinement and improved particle-matrix adhesion brought about by MA.

Norway's Magnesium Technical Center[106] evaluated ceramic fiber preforms of var-

TABLE 5.15. Tensile Properties of Composites Made by Mechanical Alloying and Dry-Powder Mixing

Alloy	Yield Strength (MPa)	Tensile Strength (MPa)	Elongation (%)	Young's Modulus (GPa)
Wrought Mg	80	160	6	NA
Unreinforced Mg	215	275	8.5	39
Mechanically alloyed				
10 vol% SiC	270	315	3.3	50
20 vol% SiC	304	350	1.0	66
30 vol% SiC	SF	303	0.2	74
Dry mixed				
10 vol% SiC	222	280	1.2	57
20 vol% SiC	232	250	0.3	59
30 vol% SiC	SF	217	0.1	77

NA, not available; SF, sample failed prematurely due to porosity.

TABLE 5.16. Tensile Properties of Composites Fabricated via the Ingot-Metallurgy Route

Alloy	Yield Strength (MPa)	Tensile Strength (MPa)	Elongation (%)	Young's Modulus (GPa)
Mg/10 vol% SiC	120	160	2	45
Mg-Al (extruded)	190	290	18	44
Mg-Al/10 vol% SiC	250	290	5	55
Mg-Al/15 vol% SiC	225	280	2.5	60

ious compositions (ZrO_2, SiC, Al_2O_3, and Al_2O_3-SiO_2) for infiltrating with molten Mg during squeeze-casting. The optimum fiber composition was an alumina-based (97 wt% Al_2O_3 + 3 wt% SiO_2) discontinuous form at 20 V_f concentration. Tensile strength improved significantly, over 25%, compared with that of the unreinforced material.

Dow Chemical[106] has successfully manufactured 70-kg billets of Mg alloy composites reinforced with SiC_p. Two compositions have been examined based on their excellent properties: AZ31B/SiC/20_p-600 grit and AZ31B + 6% Zn/SiC/20_p-1000 grit composites. The latter is significantly stronger than any other commercially available unreinforced Mg currently on the market. Room-temperature properties have been improved as follows: elastic modulus by 69%, tensile yield strength by 42%, UTS by 53%, and compressive yield strength by 34% (see Table 5.17).

Schröder and Kainer[107] found that liquid infiltration of ceramic fiber preforms was one of the most successful methods of producing Mg MMCs with discontinuous reinforcements. They used two hybrid composites: Mg alloy MSR (2.5% Ag, 2.0% rare earth, 0.6% Zr) + 10 V_f Saffil Al_2O_{3f} + 15 V_f SiC and MSR + 5 V_f Al_2O_3 + 20 V_f SiC_p.

The particle fiber reinforcements led to an increase in hardness. Compared with the unreinforced matrix alloy, an improvement of 70–85% was observed. The reinforcement addition was found to lead to a significant increase in Young's modulus from 45 GPa for the unreinforced Mg to 77 GPa for the hybrid composite. The highest value was obtained for the composite with 10 V_f fibers and 15 V_f particles.

TABLE 5.17. Tensile Properties of Magnesium-Based Materials

Material	Temperature (°C)	Elastic Modulus (GPa)	Tensile Yield Strength (MPa)	Ultimate Tensile Strength (MPa)	Elongation (%)
AZ31B + 600-grit SiC[a]	25	79 (0.7)	251 (0.7)	330 (4.8)	5.7 (0.7)
	150	56 (4.8)	154 (0.0)	215 (0.7)	10.4 (1.4)
AZ31B + 1000-grit SiC[a]	25	79 (4.8)	270 (4.8)	341 (10.8)	4.0 (0.8)
	150	68 (4.8)	167 (0.7)	215 (0.7)	9.2 (1.1)
AZ31B typical[b]	25	45	165	250	12.0
	150		105	170	39.0
ZK60A-T5 typical[b]	25	45	270	340	12.0
	150	36	140	175	53.0

Source: Dow Chemical.

[a] Three tensile bars were tested at each temperature for the composite materials. Mean properties are reported. Standard deviations are given in parentheses.

[b] Room-temperature properties represent typical properties for large tubing. High-temperature properties are not necessarily typical of large tubing.

Magnesium hybrid composites were characterized by a significant increase in mechanical properties compared with the unreinforced magnesium.

Additionally 0.2% proof stress and UTS at room temperature for the composite material exhibited the highest strength, 330 MPa compared with 240 MPa for the unreinforced matrix. At elevated temperature the highest strength (190 MPa) was obtained for the composites. Discontinuous reinforcements also led to an increase in bending and compression strengths, and the highest bending strength was obtained for the composite material: 560 MPa. This represents an increase of 190 MPa compared with that for the unreinforced matrix (370 MPa).

A process has been developed for making MMCs that blends liquid magnesium alloy with ceramic particles, such as SiC and Al_2O_3. The method is essentially the same as those that have been developed for aluminum composites in that blending is accomplished via a high-shear process. Major differences in the new process result from the increased general reactivity of magnesium and the difference in surface chemistry between the Al-SiC and Mg-SiC systems.

The particulate-reinforced magnesium MMCs produced by this new technology are lightweight and demonstrate a significant increase in modulus and tensile strength at both ambient and elevated temperatures over those of the unreinforced material, leading to a significant improvement in the operating envelope. In addition, the CTE of these Mg MMCs can be reduced by up to one-third by varying the SiC content, enabling the CTE to be matched to that of other materials.

The process has recently been scaled up to a prototype stage capable of producing extrusions, foundry ingots, and sand castings.

To study the effects of alloy type on composite properties, ZM21 (2 Zn,1 Mn), AZ61 (6 Al,1 Zn, 0.3 Mn), AZ80 (8.5 Al, 0.5 Zn, 0.3 Mn), and ZC71 (6.5 Zn, 0.7 Mn, 1.2 Cu) wrought Mg alloys reinforced with V_f 9-μm SiC grit were cast in water-cooled steel molds.

Test results indicated that ZC71 had the best tensile properties (406 MPa), and the modulus was found to be 63 GPa. This is a 43% increase over the values for the unreinforced alloy and, if specific modulus is considered, the value for the composite is 32 MN m/kg, which is higher than that for either unreinforced Mg (24 MN m/kg) or unreinforced Al (26 MN m/kg).

Rotating bend fatigue testing of the composite showed a fatigue runout at 50 million cycles at 125 MPa, the same value obtained for the unreinforced alloy.

Addition of SiC_p significantly increased the ambient-temperature UTS of ZC71, and the operating-temperature envelope was raised substantially. However, at temperatures approaching 200°C, where the matrix alloy had relatively little strength, the composite was also weakened.

Hihara and Kondepudi[108,109] have spent considerable time examining the corrosion of Mg MMCs. They found that SiC monofilament-ZE41 Mg MMC corroded in a 0.5 $NaNO_3$ solution at 30°C at higher rates in oxygenated than in deaerated solutions. This is uncharacteristic of the corrosion behavior of Mg and its alloys which is normally unaffected by the presence of dissolved oxygen. The SiC_{mf}, which were inert electrodes on which proton and oxygen reduction could occur, induced galvanic corrosion in the MMC. The alloying elements in ZE41A Mg should not have significant effects on galvanic corrosion rates because anodic polarization curves of ZE41A and pure Mg were very similar. The cathodic current densities (CDs) of the SiC_{mf} were generally much greater than that of hot-pressed SiC and closer to that of P-100 graphite fiber. The carbon core and carbon-rich surface of the SiC_{mf} apparently caused the monofilaments to behave more like graphite than SiC. Because hot-pressed SiC sustained significantly lower cathodic CDs than the SiC_{mf}, the fabrication of Mg MMCs with SiC reinforcement is electrochemically similar to that with hot-pressed SiC and could significantly reduce galvanic corrosion rates in these materials.

5.17.1.2 Al_2O_3. Llorca, Bloyce, and Yue[110] conducted a study to evaluate fatigue behavior by squeeze-casting AZ91 Mg alloy with ceramic reinforcements based on Al_2O_3 (Saffil) fiber two-dimensional preforms with 0.05, 0.16, and 0.25 V_f Al_2O_3 reinforcements.

Results showed that

1. Both the yield strength and elastic modulus of Mg AZ91 alloy were greatly enhanced by Al_2O_3 fiber reinforcement. The measured increases in Young's modulus for the composite material agreed well with that predicted by the rule of mixtures.
2. Low tensile properties were obtained when the composite material was solutionized at 420°C. However, cast material exhibited potentially useful tensile properties with high stiffness and reasonable strength.
3. The fatigue strength of the composite material, tested at room temperature, increased with increasing V_f of the fiber, and an improvement of 85% in fatigue strength was obtained for the 0.25 V_f composite material over the monolithic alloy. The improvement was believed to be the result of the higher fatigue crack initiation resistance of the composite material.
4. The critical crack depth of the composite material was similar for the three composites (0.05, 0.16, and 0.25 V_f). This suggested that the fracture toughness of the AZ91/Al_2O_3 composite was likely to be unaffected by a fiber V_f between 0.05 and 0.25.
5. The fatigue crack growth rate of the composite material, on the contrary, was inferior to that of the matrix material in the high-ΔK region (above 5 MPa $m^{1/2}$).

Towle and Friend[111] determined the mechanical properties of short-fiber-reinforced Mg and Mg RZ5 alloy (nominally 4.2% Zn, 0.35% Zr, 1.3% rare earth metals, balance Mg) MMCs reinforced with Saffil fibers of approximate composition 95% δAl_2O_3 and 5% SiO_2, and the fibers had nominal dimensions of 3 μm diameter and 150 μm length. Tension and compression were the main properties evaluated.

The data show that certain property values are highly dependent on the test method employed for their determination. In particular, values of UTS are consistently greater in compression than in tension, and MMCs are considerably more "deformable" in compression. However, irrespective of the type of test method employed, the MMC materials exhibit higher Young's moduli, proof stresses, and UTS than the corresponding unreinforced materials, and much lower ductilities.

In general, the averaged data for the MMCs in tension and compression exhibited the following properties compared with the data for unreinforced metals.

1. Young's moduli were ~30% higher for both MMCs.
2. 0.1% proof stresses were ~100% higher for both MMCs.
3. UTSs were ~100% higher for the Mg MMC.
4. UTSs were ~30% higher for the RZ5 MMC.

However, the MMCs exhibited very low total deformation. In compression the deformations were ~25–30% of the values for the unreinforced metals, and the differences in tension were even greater. The different improvements in UTS exhibited by the two types of MMCs were due to a much greater reinforcement effect in the relatively weaker Mg matrix.

5.17.1.3 Graphite. Carbon fiber-reinforced MMCs have been considered attractive candidates for engineering structural materials because of their high specific stiffness, high specific strength, and elevated-temperature properties. Among the fiber-reinforced metals, a Mg-base matrix offers some advantages over an Al base matrix because of the low matrix density.

Concerning the mechnical properties of carbon fiber-reinforced Mg matrix composites, Diwanji and Hall[112] recently reported that carbon fiber-incorporated Mg-based composites fabricated under optimum condition retained the original strength and that the composite thus showed good mechanical properties.

By squeeze-casting under optimum conditions, Kagawa and Nakata[113] fabricated a high-strength-type carbon fiber-reinforced commercially pure Mg (99.5 wt%) matrix composite to investigate its mechanical properties.

They developed plots of tensile strength and Young's modulus versus fiber V_f, with transverse tensile strength results shown as a function of test temperature. The transverse strength decreased gradually with increasing test temperature below 300°C. At 300°C the transverse strength of the composite decreased rapidly, and the strength was nearly the same as the tensile strength of the unreinforced matrix. Additionally, flexure strength was examined, and the flexure strength of the composite gradually decreased with the test temperature.

Kagawa and Nakata[113] claim that carbon fiber-reinforced commercially pure Mg matrices have excellent mechanical properties when the composite is fabricated under optimum conditions; however, more-detailed experimental work is required to understand clearly the mechanical behavior of carbon fiber-reinforced Mg matrix composites.

5.17.1.4 Boron carbide (B_4C). Some recent work has been performed with B_4C-reinforced Mg composites fabricated with inert gas-atomized powders, and they had improved mechanical properties compared to those of similar composites made with ground powders. Tensile properties were greatly influenced by the purity of the ZK60A Mg powders, whereas compressive properties had not changed. This ZK60A/$B_4C/30_p$ composite was compared to a ZK60A/SiC/20_p composite, and results for tensile modulus were slightly higher for the SiC reinforced composite, 95.8 GPa, compared to that for the B_4C-reinforced composite, 93.8 GPa.

In compression, the ZK60A/$B_4C/30_p$ had higher yield strength but lower elastic modulus than the ZK60A/SiC/20_p composite. The tensile properties are shown in Table 5.18.

5.17.1.5 Diamond. As higher-power electronic components are developed, an increasing need will arise for high-thermal-conductivity packaging materials. Many electronics materials are either too low in thermal conductivity (150 W/m • K) or not isotropic. A novel material, ZK60A/diamond/25_p, diamond particulate-reinforced Mg, is under development for use in electronic packaging as reported by Stevenson, Whatley, and Clark.[114] This fully isotropic material combines high thermal conductivity, low CTE, low density, and good mechanical properties. Preliminary work has been performed to

TABLE 5.18. Tensile Properties of Mg Composites with SiC and B_4C

Composite	Process	Yield Strength (MPa)	Ultimate Tensile Strength (MPa)	Elongation (%)	Modulus (GPa)
20% SiC/ZK60A	Extruded	455	579	1.58	95.8
30% B_4C/ZK60A	Extruded	455	510	0.93	93.8

optimize diamond particle size, coatings, particulate loading, and processing parameters. Materials fabricated to date have demonstrated thermal conductivity above 196 W/m • K, CTE values as low as 7.4×10^{-6} m/m • K, and densities of only 2.38 g/cm^3.

The high-V_f diamond (35–50%) resulted in significantly higher CTE and lower thermal conductivity than predicted. Density measurements indicated that the level of porosity in composites with reinforcement fractions below 35% was low. As the reinforcement fraction was increased to 50 vol% the porosity also increased. These results indicated that there is a bonding problem to be solved that is the cause of the lower thermal properties of the composites with 50% diamond.

5.18 TITANIUM DISCONTINUOUS MMCs

5.18.1 Particulates

5.18.1.1 Graphite. Warrier, Blue, and Yin[115] evaluated carbon fiber-reinforced titanium matrix composites (TMCs) for applications in the aerospace industries, primarily for their light weight and high strength and modulus at elevated temperatures.

The flexural strength and modulus of the composites (Table 5.19) were consider-

TABLE 5.19. Tensile Properties of Titanium Alloys and Composites[115]

Material	Tensile Strength, σ (MPa)	Modulus, E (GPa)	Density, ρ (Mg/m^3)	Specific Strength, σ/ρ (MPa m^3/kg^1)	Specific Modulus, E/ρ (MPa m^3/kg^1)
C fibers	1730	379	2.00	0.87	189[a]
Ti 80 wt%	512	47	5.38	0.10	9
Ti 85 wt%	1600	83	5.00	0.32	17
Ti 80 wt%/C (30 vol%)	660	123	4.29	0.15	29
Ti 85 wt%/C (40 vol%)	1100	140	3.80	0.29	37
Ti-6Al-4V	890	110	4.44	0.20	25[b]
Ti-6-4/SiC (35 vol%)	820	225	3.75	0.22	60[b]
Ti/SCS-6	1680	81			[c]

[a] Hansen.[116]
[b] Smith et al.[117]
[c] Blue and Lin.[118]

ably higher than those of the constituent matrix alloys. The strength and specific strength of Ti 80 wt%/C composites were higher than those of the alloy. Although the strengths of Ti 85 wt%/C composites were lower than those of the alloy, the specific strengths were almost equal. The lower strengths of Ti 85 wt%/C composites compared with that of the alloy may be due to either initiation of cracks at the reaction sites because of unfavorable morphology of the reaction products or to residual stresses developed between the fiber and the matrix during fabrication. As shown in Table 5.19, this kind of strength reduction was also observed with Ti-6Al-4V/SiC composites.[117,119] In addition, the Ti 85 wt%/C composites showed a higher strength (by about 35%) than Ti-6Al-4V/SiC composites,[117,119] a strength almost equal to that of Ti-14Al-21Nb/SiC composites,[120] and a higher modulus (by about 50%) than that of Ti matrix SCS-6 0-90° composites from the NASA Lewis Center.[118]

In summary, controlled-reaction titanium alloy-matrix composites produced by a rapid infiltration process showed small, controllable reaction zones and room-temperature tensile values comparable to those of currently available titanium-matrix composites.

5.18.1.2 Titanium carbide (TiC). Creep behavior of the TiC particulate-reinforced Ti alloy composite was investigated by Zhu, Lu, Wang, et al.[121] at temperatures from 500 to 650°C and stresses from 230 to 430 MPa. Creep strain rates of the composite were lower than those of the matrix alloy by one order of magnitude. The lower creep rates of the composite can be attributed to a Young's modulus effect.

However, this TiC_p-reinforced Ti alloy has been developed recently because TiC is completely compatible with titanium and its alloys, and high-quality TiC_p are readily available.[122] Yield strength, tensile strength, and Young's modulus for the TiC_p-reinforced P/M Ti-6Al-4V composite were superior to those of the matrix alloy.[122] However, the fracture toughness of the composite was four times lower than that of the matrix alloy. Moreover, the fatigue crack propagation behavior of the composite was comparable to that of the matrix alloy at intermediate and near-threshold stress intensities below 10 MPa $m^{1/2}$.[122]

5.19 BERYLLIUM DISCONTINUOUS MMCs

5.19.1 Particulates

5.19.1.1 Beryllium oxide (BeO). Three beryllium MMCs have been designed for avionics packages and constraining cores on high-density SMT circuit boards. The composites were composed of single-crystal BeO in a matrix of structural Be-grade S200F. Alloy E20 had 20 V_f BeO, E40 had 40 V_f, and E60 had 60 V_f. Compared to those of conventional materials, the density of the composites was 10–20% less than that of Al, about one-fifth that of Kovar, and one-fourth that of Cu-Invar-Cu.

The new materials provided improved thermal conductivity, a higher elastic modulus, and a more favorable thermal expansion match to the Al_2O_3 substrate. This allowed for the constraint of the printed wiring board to which the composite was joined. Some of the properties of these Be matrix composites are shown in Table 5.20.

TABLE 5.20. Selected Properties of Be Composites[a]

Property	E20	E40	E60
Thermal condition (Btu/h·ft°F)	116	121	124
Elastic modulus (Msi)	40	44	46
CTE (ppm/F)	4.58–5.13	3.79–4.30	3.13–3.58
Density (g/cm^3)	2.045	2.277	2.513
Specific heat (cal/g°C)	0.379	0.336	0.301

[a] Preliminary specifications.

5.20 NICKEL ALUMINIDE MMCs

5.20.1 Particulates

5.20.1.1 TiC. The most compatible reinforcements for a Ni_3Al matrix are NbC, TiC, HfN, TiN, and Y_2O_3; however, through XD addition, the TiC composite appears to enhance the cast properties at room temperature and high temperature and can probably be used for certain cast applications. The most recent work[123] on Ni_3Al-based composites using NbC, HfN, and TiN particulates showed no significant strengthening. Thus, the only hope for beneficial reinforcement appears to be TiC through XD addition.

Recent data[124] confirm that TiC additions to Ni_3Al via P/M increase its strength at both low and high temperatures.

5.21 SPECIAL COATINGS

The development of titanium-based composites has been hindered largely because of interfacial diffusion. When SiC is in contact with titanium at elevated temperatures, metal silicide (Ti_5Si_3) and carbide (TiC) are formed. The aim of a study[125] was to reduce the fiber-matrix interaction at high temperatures by interposing compounds between the metal and the SiC reinforcement.

Procedures have been developed for the formation of thin layers on SiC fibers (Sigma) by sputter deposition. Sputtering offers good control over density and stress state, thickness, and composition. Past work has shown that sputtered Y_2O_3 deposited on SiC, SiC-Si, or metal can reduce the degree of ceramic-metal reaction. Because $TiSi_2$ and Si are also present in the Ti-Si diagram, they could act as diffusion barriers. Coatings of $TiSi_2$, Y_2O_3 and ZrO_2 have been produced by sputter deposition.

Therefore, the study by Bilba, Manaud, Le Petitcorps, et al.[125] on fiber-matrix interactions in Ti matrix composites reinforced by SiC filaments shows that interposition of a $TiSi_2$ interphase is more efficient in protecting the filaments than is coating them with an oxide such as Y_2O_3 or ZrO_2.

Xiao, Kim, Abbaschian, et al.[126] described a framework for the processing of Nb-reinforced $MoSi_2$ composites. As a part of the study, composites containing coated and uncoated Nb reinforcements were produced. The Nb reinforcements were coated with inert diffusion barriers. A chemical compatibility study of $MoSi_2$ with Al_2O_3 as a potential coating material showed that they were chemically compatible in the absence of SiO_2 impurity particles. The presence of SiO_2 was found to cause extensive reaction between the second phase (SiO_2) and $MoSi_2$. $Al_2O_3 \cdot ZrO_2$ and mullite were also found to be chemically compatible with $MoSi_2$.

When Al_2O_3 and ZrO_2 coatings were applied on Nb prior to incorporation in MoSi2, they were effective in eliminating Nb and Mo diffusion and retarding the interaction between Nb and $MoSi_2$.

Chemical compatibility between the coatings and the matrix was studied, and the effect of interface modification by coating on the fracture toughness of the composites was investigated. The results indicated that the coatings had a significant effect on the debonding at the reinforcement-matrix interface, which in turn could affect the damage tolerance of the composite.

Composites containing uncoated and Al_2O_3-coated Nb foils showed improved damage tolerance, with a K_{IC} of about 14–15 MPa $m^{1/2}$. On the contrary, the composite reinforced by ZrO_2-coated Nb foils showed a lower damage tolerance (8.6 MPa $m^{1/2}$) with a brittle cleavage fracture of the foils.

REFERENCES

1. Hashin, Z., and B. W. Rosen. 1964. The elastic moduli of fiber-reinforced materials. *J. Appl. Mech.* June:223.
2. Bowles, D. E., and S. S. Tompkins. 1989. Prediction of coefficients of thermal expansion for unidirectional composites. *J. Compos. Mater.* 23:370.
3. Springer, G. S., and S. W. Tsai. 1967. Thermal conductivities of unidirectional materials. *J. Compos. Mater.* 1:166.
4. Guo, Z. X., and B. Derby. 1993. Fibre uniformity and cavitation during the consolidation of metal-matrix composite via fibre-mat and matrix-foil diffusion bonding. *Acta Metall. Mater.* 41(11):3257–66.
5. Lilholt, H. 1991. Aspects of deformation of metal matrix composites. *Mater. Sci. Eng.* A135:161–71.
6. Ritchie, R. O., W. Yu, and R. J. Bucci. 1989. Fatigue crack propagation in ARALL laminates: Measurement of the effect of crack-tip shielding from crack bridging. *Eng. Fract. Mech.* 32:361–77.
7. Everett, R. K., and R. J. Arsenault, eds. 1991. *Metal Matrix Composites: Mechanisms and Properties.* Academic Press, San Diego, CA.
8. Cox. B. N., and D. B. Marshall. 1991. Crack bridging in the fatigue of fibrous composites. *Fatigue Fract. Eng. Mater. Struct.* 14:847–61.
9. Dauskardt, R. H., and R. O. Ritchie. 1993. Fatigue of advanced materials. Part 1 *Adv. Mater. Process.* 144(1):26–31.
10. Li, D. S., and M. R. Wisnom. 1991. The mechanical properties of the matrix in continuous-fibre 6061 aluminum-alloy metal-matrix composites. *Compos. Sci. Technol.* 42(4):413–27.
11. Textron. 1989. Silicon carbide composite materials. Data Sheet, Textron Specialty Materials.
12. Francini, R. B. 1988. Characterization of thin-wall graphite-metal pultruded tubing. In *Testing and Technology of Metal Matrix Composites,* ed. P. R. DiGiovanni and N. R. Adsit, 396, STP 964. ASTM.

13. Folgar, F. 1988. Fiber FP/metal matrix composite connecting rods: Design, fabrication and performance. *Ceram. Eng. Sci. Proc.* 9(7/8):561.
14. Hughes, D. 1988. Textron unit makes reinforced titanium, aluminum parts. *Aviat. Week Space Technol.* November 28:65–67.
15. NASA. 1991. Report on Space Plane Development.
16. Joint Publications Research Service. JST-92-007L, July 14, 1992, pp. 53–55.
17. Imai, G. 1989. *Sixth Symp. Jisedai Proj.* p. 327.
18. Waku, Y., et al. 1989. *Proc. 34th Int. SAMPE Symp.* p. 2278.
19. Yamamura, T., et al. 1987. *Proc. 8th SAMPE Eur. Chap. Meet.*, p. 19.
20. Metal Matrix Composites: Technology and Industrial Applications. Tech Trends: Interim Reports on Advanced Technologies. Innovation 128, Paris, 1990.
21. Japan Patent EP 0335 692.
22. Pollock, W. D., T. D. Bayha, F. E. Wawner, et al. 1989. High temperature discontinuously reinforced, P/M aluminum composites. *Ind. Heat.* October:16–19.
23. Schulte, K., A. Bockheiser, F. Girot, et al. 1990. Characterization of aluminium oxide (FP) fibre reinforced Al-2.5 Li-composites. *Fourth Eur. Conf. Compos. Mater.* pp. 293–300, September 1990, Stuttgart.
24. Musson, N. J., and T. M. Yuc. 1991. The effect of matrix composition on the mechanical properties of squeeze-cast aluminium alloy-Saffil metal matrix composites. *Mater. Sci. Eng.* A135:237–42.
25. Rasmussen, N. W., P. N. Hansen, and S. F. Hansen. 1991. High pressure die casting of fibre-reinforced aluminium by preform infiltration. *Mater. Sci. Eng.* A135:41–43.
26. Flinn, B. D., F. W. Zok, F. F. Lange, et al. 1991. Processing and properties of Al_2O_3-Al composites. *Mater. Sci. Eng.* A144:153–57.
27. Kainer, K. U., and J. W. Kaczmar. Development and wear properties of 6061 Al base P/M composite materials strengthened with alumina fibres. In *Advances in Powder Metallurgy—1991,* vol. 6. *Proc. P/M Conf. Exhibit.,* comp. L. F. Pease, III, and R. J. Sansoucy, pp. 211–25, June 1991, Chicago. Metal Powder Industries Federation, Princeton, NJ.
28. Islam, M. U., and W. Wallace. 1988. Carbon fibre reinforced aluminium matrix composites: A critical review. *Adv. Mater. Manuf. Process,* 3(1):1.
29. Rubin, L. 1989. Data base development for P100 graphite aluminum metal matrix composites. Aerospace Corporation Report TOR-0089 (4661-02)-1.
30. Sheppard, L. M. 1988. Challenges facing the carbon industry, *Ceram. Bull.* 67(12):1897.
31. Hull, D. 1981. *An Introduction to Composite Materials.* Cambridge University Press, Cambridge.
32. Metcalfe, A. G., and M. J. Klein. 1974. In *Composite Materials,* vol. 1, ed. L. J. Broutman and R. H. Krock, 125. Addison-Wesley, Reading, MA.
33. Mittnick, M. 1990. Continuous SiC fiber-reinforced metals. *Fourth Eur. Conf. Compos. Mater.,* pp. 325–38, September 1990, Stuttgart.
34. Kumnick, A. J., R. J. Suplinskas, W. F. Grant, et al. 1983. Filament modification to provide extended high temperature consolidation and fabrication capability and to explore alternative consolidation techniques. Naval Surface Weapons Center, Silver Spring, MD. Contract N00019-82-C-0282.
35. Liu, T. M., and C. G. Chao. 1993. Effect of magnesium on mechanical properties of alumina-fiber-reinforced aluminum matrix composites formed by pressure infiltration casting. *Mater. Sci. Eng.* A169:79–84.
36. Sherman, R. G. 1990. Behavior of metal matrix composite materials at cryogenic tempera-

tures. Nevada Engineering and Technology Corporation, Final Report, August 1988–May 1990. NAVSWC-TR-90-272.

37. Cheng, H. M., S. Akiyama, A. Kitahara, et al. 1992. Behaviour of carbon fibre reinforced Al-Si composites after thermal exposure. *Mater. Sci. Technol.* 8(3):275–81.

38. Furuya, Y., A. Sasaki, and M. Taya. 1993. Enhanced mechanical properties of TiNi shape memory fiber/Al matrix composite. *Mater. Trans. Jpn. Inst. Met.* 34(3):224–27.

39. Arsenault, R. J., and M. Taya. 1987. *Acta Metall.* 35:651.

40. McDanels, D. L. 1989. Tungsten fiber reinforced copper matrix composites: A review. NASA TP 2924.

41. Westfall, L. J., and D. W. Petrasek. 1988. Fabrication and preliminary evaluation of tungsten fiber reinforced copper composite combustion chamber liners. NASA TM 1000845.

42. Foster, D. A. 1989. Electronic thermal management using copper coated graphite fibers. *SAMPE Q.* October:58.

43. McDanels, D. L., and J. O. Diaz. 1989. Exploratory feasibility studies of graphite fiber reinforced copper matrix composites for space power radiator panels. NASA TM 102328.

44. Ronald, T. M. F. 1989. Advanced materials to fly high in NASP. *Adv. Mater. Process. Met. Prog.* May:29.

45. Woodford, D. A. 1990. Critical property evaluation of high-temperature composites: A case study in materials design. *J. Met.* 42(11):50–55.

46. McLean, B. J., and M. S. Misra. 1982. Thermal-mechanical behavior of graphite/magnesium composites. In *Mechanical Behavior of Metal-Matrix Composites,* ed. J. E. Hack and M. F. Amateau, 195. Metallurgical Society of AIME., New York.

47. Tompkins, S. S., and G. A. Dries. 1988. Thermal expansion measurements of metal matrix composites. In *Testing Technology of Metal Matrix Composites,* ed. P. R. DiGiovanni and N. R. Adsit, 248, STP 964. ASTM, Philadelphia.

48. Dries, G. A., and S. S. Tompkins. 1988. Development of stable composites of graphite-reinforced aluminum and magnesium. *Proc. Joint NASA/DOD Conf. Met. Matrix Carbon Ceram. Matrix Compos.,* pp. 97–112, Cocoa Beach, FL. NASA CP 3018.

49. Tenney, D. R., G. F. Sykes, and D. E. Bowles. 1983. Space environmental effects on materials. *AGARD Meet. Environ. Effects Mater. Space Appl.,* pp. 6-1 to 6-24, March 1983. AGARD CP 327.

50. Jeng, S. M., and J.-M. Yang. 1993. Creep behavior and damage mechanisms of SiC-fiber-reinforced titanium matrix composite. *Mater. Sci. Eng.* A171:65–75.

51. Jeng, S. M., T.-H. Bruce Nguyen, O. Dana, et al. 1993. Fatigue cracking of fiber-reinforced titanium matrix composites. *J. Comp. Technol. Res.* 15(3):217–24.

52. Ghosn, L. J., J. Telesman, and P. Kantzos. 1990. Fatigue crack growth in unidirectional metal matrix composite. NASA TM 103102, E 5426.

53. Kantzos, P., J. Telesman, and L. Ghosn. 1989. Fatigue crack growth in unidirectional SCS-6/Ti-15-3 composite. NASA TM 103095, E5413.

54. Gayda, J., and T. P. Gabb. 1992. Isothermal fatigue behavior of a $(90°)_8$ SiC/Ti-15-3 composite at 426°C. *Int. J. Fatigue* 14(1):14–20.

55. Erich, D. L. 1987. Metal-matrix composites: Problems, applications, and potential in the P/M industry. *Int. J. Powder Metall.* 23(1):45.

56. Schuster, D. M., M. Skibo, and F. Yep. 1987. SiC particle reinforced aluminum by casting. *J. Met.* November:60.

57. Arsenault, R. J., and S. B. Wu. 1988. A comparison of P/M vs. melted SiC/Al composites. *Scr. Metall.* 22:767.

58. McDanels, D. L. 1985. Analysis of stress-strain fracture and ductility behavior of aluminum matrix composites containing discontinuous silicon carbide reinforcement. *Metall. Trans. A* 16A:1105.

59. Boland, P. L., P. R. DiGiovanni, and L. Franceschi. 1988. Short-term high-temperature properties of reinforced metal matrix composites. In *Testing Technology of Metal Matrix Composites,* ed. P. R. DiGiovanni and N. R. Adsit, 346 ASTM, Philadelphia.

60. Tydings, J. 1990. Metal matrix composites session overview. *Proc. Conf. Exhibit. P/M Aerosp. Def. Technol.,* ed. F. H. Froes, pp. 189–191. Metal Powder Industries Federation, Princeton, NJ.

61. Walker, J. A. 1991. The treatment and preparation of silicon carbide whiskers for the manufacture of powder metallurgical discontinuously reinforced aluminum composites. *Proc. Conf. Exhibit. P/M Aerosp. Def. Technol.,* ed. F. H. Froes, pp. 207–11. Metal Powder Industries Federation, Princeton, NJ.

62. House, M. B., K. C. Meinert, and R. B. Bhagat. 1991. The aging response and creep of DRA composites. *J. Met.* August:24–28.

63. Tanaka, Y. 1994. Private communication.

64. González-Doncel, G., and O. D. Sherby. 1993. High temperature creep behavior of metal matrix aluminum-SiC composites. *Acta Metall. Mater.* 41(10):2797–805.

65. Ma, Z. Y., J. Liu, and C. K. Yao. 1991. Fracture mechanism in a SiC_w-6061 Al composite. *J. Mater. Sci.* 26(7):1971–76.

66. Cao, L., Y. Wang, and C. K. Yao. 1990. The wear properties of an SiC-whisker-reinforced aluminium composite. *Wear* 140:273–77.

67. Bhagat, R. B., and M. B. House. 1991. Elevated-temperature mechanical properties of silicon-carbide-whisker-reinforced aluminum matrix composites. *Mater. Sci. Eng.* A144:319–26.

68. Prasad, S. V., and B. D. McConnell. 1991. Tribology of aluminum metal-matrix composites: lubrication by graphite. *Wear* 149:241–53.

69. Schueller. R. D., and F. E. Wawner. 1991. An analysis of high-temperature behavior of AA2124/SiC whisker composites. *Compos. Sci. Technol.* 40(2):213–23.

70. Dinwoodie, J., E. Moore, E. Langman, et al. 1985. The properties and applications of short staple alumina fibre reinforced aluminum alloys. *Proc. 5th Int. Conf. Compos. Mater.,* ed. W. Harrigan, J. Strife, and A. Dhingra, pp. 671–85. Metallurgical Society, Warrendale, PA.

71. Ackerman, L., J. Charbonnier, G. Desplanches, et al. 1985. Properties of reinforced aluminum foundry alloys. *Proc. 5th Int. Conf. Compos. Mater.,* ed. W. Harrigan, J. Strife, and A. Dhingra, pp. 687–98. Metallurgical Society, Warrendale, PA.

72. Sakamoto, A., H. Hasegawa, and Y. Minoda. 1985. Mechanical properties of SiC whisker reinforced aluminum composites. *Proc. 5th Int. Conf. Compos. Mater.,* ed. W. Harrigan, J. Strife, and A. Dhingra, pp. 699–705. Metallurgical Society, Warrendale, PA.

73. Sakamoto, A., H. Hasegawa, and Y. Minoda. 1985. Mechanical properties of SiC whisker reinforced aluminum composites. *Proc. 5th Int. Conf. Compos. Mater.,* ed. W. Harrigan, J. Strife, and A. Dhingra, pp. 705–7. Metallurgical Society, Warrendale, PA.

74. Pollock, W., and F. Wawner. 1988. Microstructure and strength properties of high temperature Al composites. *12th Conf. Compos. Mater. Struct.,* pp. 64–75, January 1988, Cocoa Beach, FL.

75. Achenbach, J. D., M. E. Fine, I. Komsky, et al. 1992. Ultrasonic wave technique to assess cyclic-load fatigue damage in silicon-carbide whisker reinforced 2124 aluminum alloy composites. In *Cyclic Deformation, Fracture, and Nondestructive Evaluation of Advanced Materials,* ed. M. R. Mitchell and O. Buck, 241–50. STP 1157. ASTM, Philadelphia.

76. Wang, A., and H. J. Rack. 1991. Abrasive wear of silicon carbide particulate- and whisker-reinforced 7091 aluminum matrix composite. *Wear* 146(2):337–48.

77. DeBondt, S., L. Froyen, and A. Deruyttere. 1991. Squeeze casting of hybrid Al-SiC fibre-particle composites. *Mater. Sci. Eng.* A135:29–32.

78. Lim, T., Y. H. Kim, C. S. Lee, et al. 1992. Fabrication and mechanical properties of aluminum matrix composite materials. *J. Comp. Mater.* 26(7):1062–86.

79. Nardone, V. C., J. R. Strife, and K. M. Prewo. 1991. Processing of particulate reinforced metals and intermetallics for improved damage tolerance. *Mater. Sci. Eng.* A144:267–75.

80. Nardone, V. C., J. R. Strife, and K. M. Prewo. 1991. Microstructurally toughened particulate-reinforced aluminum matrix composite. *Metall. Trans. A* 22(1):171–82.

81. Bhat, M. S. N., M. K. Surappa, and H. V. Sudhaker Nayak. 1991. Corrosion behavior of silicon carbide particle reinforced 6061/Al alloy composites. *J. Mater. Sci.* 26(18):4991–96.

82. Sun, H., E. Y. Koo, and H. G. Wheat. 1991. Corrosion behavior of SiC_p/6061 Al metal matrix composites. *Corrosion* 47(10):741–53.

83. Wu, W., K. C. Goretta, and J. L. Routbort. 1992. Erosion of 2014 Al reinforced with SiC or Al_2O_3 particles. *Mater. Sci. Engr.* A151:85–95.

84. Bonollo, F., R. Guerriero, E. Sentimenti, et al. 1991. The effect of quenching on the mechanical properties of powder metallurgically produced Al-SiC (particles) metal matrix composites. *Mater. Sci. Engr.* A144:303–9.

85. Sarkar, B., W. D. Brewer, and W. B. Lisagor. 1988. Improvements in mechanical properties of SiC whisker reinforced aluminum composites through increased microstructural uniformity. *Conf. Met. Matrix Carbon Ceram. Matrix Compos.,* pp. 65–71, January 1988, Cocoa Beach, FL. NASA CP-3018.

86. Srivatsan, T. 1992. The low-cycle fatigue behavior of an aluminum-alloy-ceramic-particle composite. *Int. J. Fatigue* 14(3):173–82.

87. Healy, J. C., M. D. Wright, M. D. Halliday, et al. 1988. The fatigue characteristics of 8090 reinforced with silicon carbide particulate. *Conf. Met. Matrix Carbon Ceram. Matrix Compos.* pp. 1365–71, January 1988, Cocoa Beach, FL. NAAS CP-3018.

88. Hunt, M., and M. Horgan. 1992. Metal and ceramic composites come down to earth. *Mater. Eng.* July:12–13.

89. Bloyce, A., and J. C. Summers. 1991. Static and dynamic properties of squeeze-cast A357-SiC particulate Duralcan metal matrix composite. *Mater. Sci. Eng.* A135:231–36.

90. Pitcher, P., and P. J. E. Forsyth. 1982. The influence of microstructure on the fatigue properties of an aluminium casting alloy. Royal Aircraft Establishment Technical Report 82107.

91. Roebuck, B., and J. D. Lord. 1990. Plane strain fracture toughness test procedure for particulate metal matrix composites. National Physical Laboratory Report DMM(A).

92. Wang, Z., and R. J. Zhang. 1991. Mechanical behavior of cast particulate SiC/Al (356) metal matrix composites. *Metall. Trans. A* 22(7):1585–93.

93. Selvaduray, G., R. Hickman, D. Quinn, et al. 1990. Relationship between microstructure and physical properties of Al_2O_3 and SiC reinforced aluminum alloys. *Proc. Int. Conf. Interfaces Met.-Ceram. Compos., TMS Ann. Meet.,* ed. R. Y. Lin, R. J. Arsenault, G. P. Martins, and S. G. Fishman, pp 271–89, February 1990, Anaheim, CA.

94. Hochreiter, E., M. Panzenböck, and F. Jeglitsch. 1993. Fatigue properties of particle-reinforced metal-matrix composites. *Int. J. Fatigue* 15(6):493–99.

95. Newkirk, M. S., A. W. Urquhart, H. R. Zwicker, et al. 1986. Formation of Lanxide™ ceramic composite materials. *J Mater. Res.* 1(1):81.

96. Johnson, T. L. 1993. Lanxide Corporation opens new materials frontiers. Metal Matrix Composites Information Analysis Center, 9(3):1–9.

97. Urquhart, A. W. 1991. Novel reinforced ceramics and metals: A review of Lanxide's composite technologies. *Innovative Inorg. Compos. Symp.* October 1990, Detroit, MI. *Mater. Sci. Eng.* A144:75–82.

98. *Adv. Mater. Process.* 141(1):25–27.

99. Wang, G., and M. Dahms. 1993. Synthesizing gamma-TiAl alloys by reactive powder processing. *J. Met.* 45(5):52–56.

100. Perng, C.-C, J.-R. Hwang, and J.-L. Doong. 1993. High strain rate tensile properties of an (Al_2O_3 particles)—(Al alloy 6061-T6) metal matrix composite. *Mater. Sci. Eng.* A171:213–21.

101. Pillai, U. T. S., B. C. Pai, V. S. Kelukutty, et al. 1993. Pressure die cast graphite dispersed Al-Si-Mg alloy matrix composites. *Mater. Sci. Eng.* A169:93–98.

102. Zhang, J., R. J. Perez, M. N. Gungor, et al. 1992. Damping characterization of graphite particulate reinforced aluminum composites. *Proc. Develop. Ceram. Met.-Matrix Compos., TMS Ann. Meet.,* ed. K. Upadhya, pp. 203–17, March 1992, San Diego, CA.

103. Sun, H., and H. G. Wheat. 1993. Corrosion study of Al_2O_3 dispersion strengthened Cu metal matrix composites in NaCl solutions. *J. Mater Sci.* 28(20):5435–42.

104. Sun, H., J. E. Orth, and H. G. Wheat. 1993. Corrosion behavior of copper-based metal-matrix composites. *J. Met.* 45(9):36–41.

105. Krishnadev, M. R., R. Angers, C. G. Krishnadas Nair, et al. 1993. The structure and properties of magnesium-matrix composites. *J. Met.* 45(8):52–4.

106. Staff. *Ceram. Bull.* 69(6):1010–20.

107. Schröder, J., and K. U. Kainer. 1991. Magnesium-base hybrid composites prepared by liquid infiltration. *Mater. Sci. Eng.* A135:33–6.

108. Hihara, L. H., and P. K. Kondepudi. 1992. Corrosion of magnesium-matrix composites. *Proc. 6th Jpn.-U.S. Conf. Compos. Mater.,* pp. 511–16, June 1992, Orlando, FL. Technomic, Lancaster, PA.

109. Hihara, L. H., and P. K. Kondepudi. 1993. The galvanic corrosion of SiC monofilament ZE41 Mg metal-matrix composite in 0.5 *M* $NaNO_3$. *Corrosion Sci.* 34(11):1761–72.

110. Llorca, N., A. Bloyce, and T. M. Yue. 1991. Fatigue behavior of short alumina fibre reinforced AZ91 magnesium alloy metal matrix composite. *Mater. Sci. Eng.* A135:247–52.

111. Towle, D. J., and C. M. Friend. 1993. Comparison of compressive and tensile properties of magnesium based metal matrix composites. *Mater. Sci. Technol.* 9(1):35–41.

112. Diwanji, A. P., and I. W. Hall. 1987. *Proc. 6th ICCM 2nd ECCM,* ed. F. L. Matthewis, N. C. R. Buskell, J. M. Hodgkinson, et al. Elsevier, London.

113. Kagawa, Y., and E. Nakata. 1992. Some mechanical properties of carbon fibre-reinforced magnesium-matrix composite fabricated by squeeze casting. *J. Mater. Sci. Lett.* 11(3): 176–8.

114. Stevenson, R. D., W. J. Whatley, and J. B. Clark. 1993. Diamond particulate reinforced magnesium alloys for Navy applications. Final Report, May 1991–March 1993. U.S. Naval Surface Warfare Center, Silver Spring, MD. NSWCDD-TR-93-148.

115. Warrier, S. G., C. A. Blue, and R. Y. Lin. 1993. Infiltration of titanium alloy-matrix composites. *J. Mater. Sci. Lett.* 12(11):865–8.

116. Hansen, N. L. 1987. Carbon fibers. In *Engineered Materials Handbook,* vol. 1: *Composites,* ed. T. J. Reinhart, 112. ASM International, Metals Park, OH.

117. Smith, P. R., F. H. Froes, and J. T. Cammett. 1977. *Fracture Modes of Composites,* vol. VI,

ed. J. A. Cornie and F. W. Crossman, 143. Metallurgy Society of the American Institute of Mining, Metallurgy, and Petroleum Engineers, Warrendale, PA.

118. Blue, C. A., and R. Y. Lin. 1991. Unpublished research.
119. Arsenault, R. P. 1991. Metal matrix composites: Mechanisms and properties, ed. R. K. Everett and R. J. Arsenault, 133. Academic Press, San Diego, CA.
120. Brindley, P. K., S. L. Draper, M. V. Nathal, et al. 1990. Fundamental relationships between microstructure and mechanical properties of metal matrix composites. *TMS Conf. MMCs,* ed. P. K. Liaw and M. N. Gungor, 387. TMS, Warrendale, PA.
121. Zhu, S. J., Y. X. Lu, Z. G. Wang, et al. 1992. Creep behavior of TiC-particulate-reinforced Ti alloy composite. *Mater. Lett.* 13(4–5):199–203.
122. Shang, J. K., and R. O. Ritchie. 1990. *Scr. Metall. Mater.* 24:1691.
123. McKamey, C. G., and C. A. Carmichael. 1991. Microstructure and mechanical properties of Ni_3Al-based alloys reinforced with particulates. In *High Temperature Ordered Intermetallic Alloys IV,* ed. L. A. Johnson, D. P. Pope, and J. O. Stiegler, *Mater. Res. Soc. Proc.,* vol. 213, pp. 263–72. *TMS,* Boston.
124. Fuchs, G. E. 1991. The chemical compatibility and tensile behavior of an Ni_3Al-based composite. *J. Mater. Res.* 5(8):1649.
125. Bilba, K., J. P. Manaud, Y. Petitcorps, et al. 1991. Investigation of diffusion barrier coatings on SiC monofilaments for use in titanium based composites. *Mater. Sci. Eng.* A135:141–44.
126. Xiao, L., Y. S. Kim, R. Abbaschian, et al. 1991. Processing and mechanical properties of niobium-reinforced $MoSi_2$ composites. *Mater. Sci. Eng.* A144:277–85.

BIBLIOGRAPHY

Adv. Compos. Mater. 1992 Sem. Sci. Technol., Osaka, Japan. Journal Publications Research Source JST-93-007, March 1993.

Alahelisten, A., F. Bergman, M. Olsson, et al. On the wear of aluminum and magnesium metal matrix composites. *Wear* 165(2):221–26, 1993.

Alonso, A., A. Pamies, J. Narciso, et al. Evaluation of the wettability of liquid aluminum with ceramic particulates (SiC, TiC, Al_2O_3) by means of pressure infiltration. *Metall. Trans.* 24A:1423–32, 1993.

Ananth, C. R., N. Chandra, K. Murali, et al. Effect of residual stress on interfacial debonding in metal-matrix composites—A computational study. *ASM Int. Ann. Meet.,* Pittsburgh, PA, October 1993.

Annigeri, R., T. S. Srivatsan, and W. H. Hunt, Jr. Influence of reinforcement content on low cycle fatigue behavior of X2080 aluminum alloy-ceramic particle composite. *ASM Int. Ann. Meet.,* Pittsburgh, PA, October 1993.

Appendini, P., C. Badini, F. Marino, et al. 6061 aluminium alloy-SiC particulate composite: A comparison between aging behaviour in T4 and T6 treatments. *Mater. Sci. Engr.* A135:275–9, 1991.

Arsenault, R. Strengthening and deformation mechanisms of discontinuous metal matrix composites. In *Advances in Metal Matrix Composites,* ed. M. A. Taha and N. A. El-Mahallawy, 265–78. Trans Tech, Aedermannsdorf, Switzerlad, 1993.

Ashbaugh, N. E., M. Khobaib, G. A. Hartman, et al. Mechanical properties for advanced engine materials. Research Institute, Dayton University, Dayton, OH, 1992. UDR-TR-91-149.

BAGGERLY, R. G., AND T. ARCHIBOLD. Characterization of the tensile, fatigue and fracture properties of nodular iron composite materials. *ASM Int. Ann. Meet.*, Pittsburgh, PA, October 1993.

BAKUCKAS, J. G., AND W. S. JOHNSON. Application of fiber bridging models to fatigue-crack growth in unidirectional titanium matrix composites. *J. Compos. Technol. Res.* 15(3):234–41.

BALIS, C. D., D. R. CURRAN, AND S. S. WANG. Elevated temperature transverse creep of FP/Al-2Li metal-matrix composite. *Proc. 4th Jpn.-U.S. Conf. Compos. Mater.*, pp. 148–78, Washington DC, June 1988. Technomic, Lancaster, PA.

BAO, G., AND K. T. RAMESH. The deformation and fracture of a tungsten-based metal matrix composite. *ASM Int. Ann. Meet.*, Pittsburgh, PA, October 1993.

BARBERO, E., AND K. KELLY. Predicting longitudinal creep and strength of a continuous fiber metal matrix composite. *38th Int. SAMPE Symp. Exhibit.*, Anaheim, CA, May 1993.

BARDAL, A., AND R. HOEIER. Interfaces in cast Al-SiC composites: Effects of alloying elements and oxide layers. Seiskapet for Industriell og Teknisk Forskning, Trondheim, Norway, 1991.

BARNEY, C., C. J. BEEVERS, AND P. BOWEN. Fatigue crack propagation in SiC continuous fibre-reinforced Ti-6Al-4V alloy metal-matrix composites. *Composites* 24(3):229–34.

BAXTER, W. J. The strength of metal matrix composites reinforced with randomly oriented discontinuous fibers. *Metal. Trans. A* 23(11):3045–53, 1992.

BAYOMI, M. A., AND M. SUERY. Structural and mechanical properties of SiC particle reinforced Al-alloy composites. Institute National Polytechnique de Grenoble, 1993.

BHAGAT, P. B., A. H. CLAUER, P. KUMAR, ET AL. *Metal and Ceramic Matrix Composites.* TMS, Warrendale, PA, 1991.

BIGELOW, C. A., AND W. S. JOHNSON. Effect of fiber-matrix debonding on notched strength of titanium metal matrix composites. NASA Langley Research Center, Hampton, VA, 1991.

BOND, C. D., F. J. CAMPBELL, AND D. P. SMITH. Lightning strike tests of composite connectors. NRL, Washington DC, June 1992. NRL/MR/4654-92-6986.

BUCKLEY, J. D. *Proc. 15th NASA DOD Conf. Met. Matrix Carbon Ceram. Matrix Compos.*, Cocoa Beach, FL, January 1991. NASA CP 3133.

CARVALHINHOS, H., M. H. CARVALHO, AND T. MARCELO. Comparative properties of Al 6061/SiC and Al 6061/Al_2O_3 composites produced by a sintering and forging route. In *Advances in P/M and Particulate Materials—1992,* vol. 9. 157–69. Metal Powder Industries Federation, Princeton, NJ.

CASTELLI, M. G., J. R. ELLIS, AND P. A. BARTOLOTTA. Thermomechanical testing techniques for high temperature composites: TMF behavior of SiC(SCS-6)/Ti-15-3. *ASTM 10th Symp. Compos. Mater.: Test. Des.*, San Francisco, CA, April 1990. NASA TM 103171, E-5543.

CLYNE, T. W., AND F. H. GORDON. Thermal cycling creep of metal matrix composites. *ASM Int. Ann. Meet.*, Pittsburgh, PA, October 1993.

COOK, J., AND E. A. FEAST. Control of fibre matrix interactions in SiC/Ti MMC. Commission of the European Communities, Directorate-General, Science, Research, and Development, Final Report EUR 13614 EN, 1991.

COTTERILL, P. J., AND P. BOWEN. Fatigue crack growth in a fibre-reinforced titanium MMC at ambient and elevated temperatures. *Composites* 24(3):214–21, 1993.

DAUSKARDT, R. H., R. O. RITCHIE, AND B. N. COX. Fatigue of advanced materials. *Adv. Mater. Process.*, 144(1):26–31, 1993.

DAVIDSON, D. L. Fatigue and fracture toughness of aluminum alloys reinforced with SiC and alumina particles. *Composites* 24(3):248–55, 1993.

DIGIOVANNI, P. R., E. M. ROSENBERG, AND D. A. BOYCE. Design, fabrication, and test of a metal matrix composite missile wing. *Raytheon Proc. Met. Matrix Carbon Ceram. Matrix Compos.* ed. J. D. Buckley, pp. 25–49, October 1987. NASA CP 2482.

DOEL, T. J. A., M. H. LORETTO, AND P. BOWEN. Mechanical properties of aluminum-based particulate metal-matrix composites. *Composites* 24(3):270–75.

DONG, H., AND A. W. THOMPSON. Creep of a fiber-reinforced titanium alloy. *ASM Int. Ann. Meet.*, Pittsburgh, PA, October 1993.

DRIVER, D. Metal matrix composites and powder processing for aero engine applications. Rolls-Royce Report PNR 90617, 1989.

DURANKO, B., AND L. JOESTEN. The application of advanced magnesium alloys to aerospace system components. *49th Ann. World Conf. Int. Magnesium Assoc.*, Chicago, IL, May 1992.

ENGELSTAD, S. P., AND J. N. REDDY. Probabilistic methods for the analysis of metal-matrix composites. *Mater. Sci. Technol.* 9(10):91–107, 1993.

EVANS, A. G., ET AL. High performance brittle matrices and brittle matrix composites, vols. 1 and 2. University of California, Santa Barbara, 1989. FR 9/86-9/89, N00014-86-K-0178.

FISHMAN, S. G. Reaction processing: route for controlling interfacial behavior. *Proc. 6th Jpn.-U.S. Conf. Compos. Mater.*, pp 163–72, Orlando, FL, June 1992.

FOO, K. S., AND W. M. BANKS. The effect of volume fraction and the size of the particulate reinforcement on the mechanical properties of a SiC particulate reinforced 6061 aluminum alloy composite. *38th Int. SAMPE Exhibit.*, Anaheim, CA, May 1993.

FRAZIER, W. E., AND G. J. LONDON. In situ liquid-crystal-polymer fiber reinforced aluminum matrix composites. Final Report, October 1989–October 1990. NRDC, Warminster, PA, 1991.

FRILER, J. B., A. S. ARGON, AND J. A. CORNIE. Strength and toughness of carbon fiber reinforced aluminum matrix composites, *Mater. Sci. Eng.* A162:143–52, 1993.

FUKUNAGA, H., AND K. GODA. Strength reliability and compatibility of B_4C/B fiber for aluminum matrix composite. *Proc. 4th Jpn.-U.S. Conf. Compos. Mater.*, pp. 653–60, Washington DC, June 1988. Technomic, Lancaster, PA.

GABRIELE, M. C. Lanxide ventures set to make matrix composite parts. *Metalwork. News,* p. 26. February 15, 1988.

GOTO, S., AND M. MCLEAN. Role of interfaces in creep of fibre-reinforced metal-matrix composites. I. Continuous Fibres. *Acta Metall. Mater.* 39(2):153–64, 1991.

GOTO, S., AND M. MCLEAN. Role of interfaces in creep of fibre-reinforced metal-matrix composites. II. Short Fibres. *Acta Metall. Mater.* 39(2):165–77, 1991.

GREIL, P. Opportunities and limits in engineering ceramics. *Powder Met. Int.* 21:40–45, 1989.

HADIANFARD, M. J., J. C. HEALY, AND Y.-W. MAI. Fracture toughness of discontinuously reinforced aluminium 6061 matrix composites. *J. Mater. Sci.* 28(22):6217–21, 1993.

HAINS, R. W., P. L. MORRIS, AND P. W. JEFFREY. Extrusion of aluminum metal matrix composites. *Proc. Int. Symp. Adv. Struct. Mater., Metall. 27th Ann. Conf.,* ed. D. S. Wilkerson, pp. 53–60. Metallurgy Society of the Canadian Institute of Mining and Metallurgy, Montreal, 1989.

HALLSTEDT, B., Z.-K. LIU, AND J. ÅGREN. Reactions in Al_2O_3-Mg metal matrix composites during prolonged heat treatment at 400, 550 and 600°C. *Mater. Sci. Eng.* A169:149–57.

HAN, N., G. POLLARD, AND R. STEVENS. Microstructure and tensile properties of a cast aluminum alloy/SiC composite. *Proc. First Pacific Rim Int. Conf. Adv. Mater. Process.*, Hangzhou, P.R.C., pp. 727–30, June 1992.

HASHIMOTO, K., S. SEKIGUCHI, AND K. YAMADA. Shear properties of MMC by torsion tests. *Proc. Int. Conf. Interfaces Met.-Ceram. Compos., TMS Ann. Meet.,* pp. 551–58, Anaheim, CA, February 1990.

HAUGHT, D., I. TALMY, D. DIVECHA, ET AL. Mullite whisker felt and its application in composites. *Mater. Sci. Eng.* A144:207–14.

HIHARA, L. H., AND P. K. KONDEPUDI. Corrosion of magnesium-matrix composites. *Proc. 6th Jpn.-U.S. Conf. Compos. Mater.*, pp. 511–17, Orlando, FL, June 1992.

HUNT, W. H., JR. Interfacial zones and mechanical properties in continuous Fiber FP/Al-Li metal matrix composites. *Interfaces Compos. Symp.*, pp. 3–26, New Orleans, LA, March 1986.

Japan External Trade Organization. Aluminum-matrix composite material with superplasticity at a high strain rate, 1991. 91-03-001-76.

JIANG, J., B. DODD, E. SELLIER, ET AL. Large cold plastic deformation of metal-matrix composites reinforced by SiC particles. *J. Mater. Sci. Lett.* 12(19):1519–21, 1993.

JOHNSON, W. S. Fatigue in continuous-fiber/metal matrix composites. NASA Technical Brief, p. 82, November 1992. NASA TM-100628.

JOHNSON, W. S. Damage development in titanium metal-matrix composites subjected to cyclic loading. *Composites* 24(3):187–96, 1993.

KAMAT, S. K., AND M. MANOHARAN. Work hardening behaviour of alumina particulate reinforced 2024 aluminium alloy matrix composites. *J. Comp. Mater.* 27(18):1714–21, 1993.

KEHOE, F. P., AND G. A. CHADWICK. Mechanical and physical properties of squeeze-cast aluminium metal matrix composites containing 5%–30% Saffil. *Mater. Sci. Eng.* A135:209–12, 1991.

KIKUCHI, M., M. GENI, T. TOGO, ET AL. Strength evaluation of whisker reinforced aluminum alloys. *ASM Int. Ann. Meet.*, Pittsburgh, PA, October 1993.

KIM, H. G., M. B. HOUSE, R. B. BHAGAT, ET AL. The effect of damage accumulation in whisker reinforced metal matrix composites. *ASM Int. Ann. Meet.*, Pittsburgh, PA, October 1993.

KOBAYASHI, T., M. MURAKAMI, AND H. TODA. Fracture toughness of SiC whisker reinforced aluminum alloy composites. *ASM Int. Ann. Meet.*, Pittsburgh, PA, October 1993.

KRIESCHKE, R. R., AND T. W. CHOU. Development of a diffusion barrier for SiC monofilaments in titanium. *Mater. Sci. Eng.* A135:145–49, 1991.

KWON, H.-M., S.-W. LEE, AND S.-J. PARK. Characteristics of aluminum base composite materials dispersed with Al_2O_3 particles. *Proc. First Pacific Rim Int. Conf. Adv. Mater. Process.*, Hangzhou, P.R.C., pp. 601–6, June 1992.

LASDAY, S. B. Production and properties of continuous silicon carbide reinforced titanium for high temperature applications. *Ind. Heat.* December:19–26, 1990.

LAWSON, L., E. Y. CHEN, M. SASAKI, ET AL. Fracture toughness and mechanism in an Al-SiC_w composite. *ASM Int. Ann. Meet.*, Pittsburgh, PA, October 1993.

LEE, E. W., AND N. J. KIM, EDS. *TMS Nonferrous Met. Comm. Meet.: Light-Weight Alloys Aerosp. Appl.* II, New Orleans, LA, February 1991.

LERCH, B. A., ET AL. Heat treatment study of the SiC/Ti-15-3 composite system, 1990. NASA TP 2970, E-4985.

LERCH, B. A., E. MELIS, AND M. TONG. Deformation behavior of SiC/Ti-15-3 laminates. *ASM Int. Ann. Meet.*, Cincinnati, OH, October 1991.

LERCH, B. A., AND J. F. SALTSMAN. Tensile deformation damage in SiC reinforced Ti-15V-3Cr-3Al-3Sn, 1991. NASA TM-103620, E-5778.

LI, H., ET AL. Functionally gradient coating on fiber for metal matrix composites. *Proc. First Pacific Rim Int. Conf. Adv. Mater. Process.* Hangzhou, P.R.C., pp. 591–94, June 1992.

LI, K., W. LI, X. ZHAO, ET AL. A study on tribological behavior of Al_2O_3/Al composites at elevated temperature. *Proc. First Pacific Rim Int. Conf. Adv. Mater. Process.*, Hangzhou, P.R.C., pp. 607–10, June 1992.

LI, Q. F., AND D. G. MCCARTNEY. A review of reinforcement distribution and its measurement in metal matrix composites. *J. Mater. Proc. Technol.* 41(3):249–62, 1994.

LIN, C. S. Effect of thermal cycling on ductility and toughness of SiC reinforced 2124 aluminum alloy matrix composite. *38th Int. SAMPE Symp. Exhibit.*, Anaheim, CA, May 1993.

LIN, R. Y., R. J. ARSENAULT, G. P. MARTINS, ET AL., EDS. *Proc. Int. Conf. Interfaces Met.-Ceram. Compos., TMS Ann. Meet.*, Anaheim, CA, February 1990.

LIU, H., AND F. H. SAMUEL. Effect of some metallurgical parameters on the properties of a SiC particulate reinforced aluminum composite. *Proc. First Pacific Rim Int. Conf. Adv. Mater. Process.*, Hangzhou, P.R.C., pp. 525–600, June 1992.

LIU, K. C., C. O. STEVENS, AND C. R. BRINKMAN. Tensile and cyclic fatigue properties of SiC whisker-reinforced Al_2O_3. Oak Ridge National Laboratory, 1991.

LLOYD, D. J. Metal matrix composites—An overview. *Proc. Int. Symp. Adv. Struct. Mater., Metall. 27th Ann. Conf.*, ed. D. S. Wilkinson, Metallurgy Society of the Canadian Institute of Mining and Metallurgy, Montreal.

LUCAS, K. A., AND H. CLARKE. *Corrosion of Aluminum-Based Metal Matrix Composites.* John Wiley, New York, 1993.

LUSTER, J. W., M. THUMANN, AND R. BAUMANN. Mechanical properties of aluminium alloy 6061-Al_2O_3 composites. *Mater. Sci. Technol.* 9(10):853–62, 1993.

MAJUMDAR, B. S., G. M. NEWAZ, B. LERCH, ET AL. Fatigue fracture mechanisms in titanium matrix composites. *ASM Int. Ann. Meet.*, Pittsburgh, PA, October 1993.

MALL, S., AND J. J. SCHUBBE. Thermo-mechanical fatigue behavior of a cross-ply SCS-6/Ti-15-3 metal-matrix composite. *Mater. Sci. Technol.* 9(10):49–57, 1993.

MARSH, G. Engineering ceramics. Part 2. *Aerosp. Compos. Mater.* 24–26, 1990.

MASOUNAVE, J., AND F. G. HAMEL, EDS. *ASM Conf. Proc. Nat. Res. Counc. Can.: Fabr. Particulates Reinforced Met. Compos.* 1990.

Metal Matrix Composites. NERAC, Tolland, CT. 1993. PB93-858348/WMS.

MISRA, R. S., AND A. B. PANDEY. Some observations on the high-temperature creep behavior of 6061 Al-SiC composites. *Metall. Trans. A* 21(7):2089–90, 1990.

MITAL, S. K., C. C. CHAMIS, AND P. K. GOTSIS. Micro-fracture in high-temperature metal-matrix laminates. *Compos. Sci. Technol.* 50(1):59–70, 1994.

MITTNICK, M. A., AND J. MCELMAN. Continuous silicon carbide fiber reinforced metal matrix composites. *Proc. Int. Symp. Adv. Struct. Mater., Metall. 27th Ann. Conf.*, ed. D. S. Wilkerson, pp. 61–70. Metallurgy Society of Mining and Metallurgy, Montreal, 1989.

MOREL, M. R., D. A. SARAVANOS. Tailoring of inelastic metal-matrix laminates with simultaneous processing considerations. *Compos. Sci. Technol.* 50(1):109–17, 1994.

MORTENSEN, A. Metal matrix composites: composite systems MIT Industrial Liaison Progress Report 3-30-93, 1993.

Moving towards the non-metallic aero engine. *Metallurgia,* August:370–74, 1988.

NEWAZ, G. W., AND B. S. MAJUMDAR. A comparison of mechanical response of MMC at room and elevated temperatures. *Compos. Sci. Technol.* 50(1):85–90, 1994.

NUTT, S. N. Interfaces and failure mechanisms in Al-SiC composites. *Interfaces Compos. Symp.*, New Orleans, LA, pp. 157–68, March 1986.

OCHIAI, S., AND K. OSAMURA. Effects of interface on strength of metal matrix composites. *Proc. First Pacific Rim Int. Conf. Adv. Mater. Process.*, pp. 585–90, Hangzhou, P.R.C., June 1992.

OKURA, M. Fabrication and properties of fiber reinforced metal matrix composites. In *Advances in Metal Matrix Composites,* ed. M. A. Taha and N. A. El-Mahallawy, 255–64. Trans Tech, Aedermannsdorf, Switzerland, 1993.

PANDEY, A. B., R. S. MISHRA, AND Y. R. MAHAJAN. Creep fracture in Al-Si metal matrix composites. *J. Mater. Sci.* 28(11):2943–49, 1993.

PUTCHA, N. S. FEA eases composites design. *Adv. Mater. Process.* 138(3):49–53, 1990.

RAY, S. Review—Synthesis of cast metal matrix particulate composites. *J. Mater Sci.* 28(20):5397–413, 1993.

RHYNE, E., D. A. KOSS, J. HELLMAN, ET AL. Interfacial failure in SCS-6 fiber-reinforced Ti-24-11. *ASM Int. Ann. Meet.*, Pittsburgh, PA, October 1993.

RICHTER, D. Commercial alternatives in metal matrix composites. *Adv. Mater. Technol. Int.* 60:57–58, 1992.

RITLAND, M. A., AND D. W. READEY. Processing and properties of Al_2O_3-Cu composites. *ASM Int. Ann. Meet.*, Pittsburgh, PA, October 1993.

ROBI, P. S., R. C. PRASAD, AND P. RAMAKRISHNAM. Fracture behavior of discontinuous carbon fiber reinforced 6061 aluminum alloy composites. *ASM Int. Ann. Meet.*, Pittsburgh, PA, October 1993.

ROHATGI, P. K., AND R. ASTHANA. Solidification processing of metal-matrix composites. *Proc. Int. Symp. Adv. Struct. Mater., Metall. 27th Ann. Conf.*, ed. D. S. Wilkinson, pp. 43–52. Metallurgy Society of the Canadian Institute of Mining and Metallurgy, Montreal, 1989.

SCHRECENGOST, T. R., B. A. SHAW, R. G. WENDT, ET AL. Nonequilibrium alloying of graphite-reinforced aluminum metal matrix composites. *Corrosion Sci.* 49(10):842–49, 1993.

SELVADURAY, G., R. HICKMAN, D. QUINN, ET AL. Relationship between microstructure and physical properties of Al_2O_3 and SiC reinforced aluminum alloys. *Proc. Int. Conf. Interfaces Met.-Ceram. Compos., TMS Ann. Meet.*, pp. 271–80, Anaheim, CA, February 1990.

SHARMA, S. C., AND S. R. ARUN. Fabrication and evaluation of the mechanical properties of aluminium alloy-glass particulate composites. *1st Asia-Pacific Conf. Mater. Process.*, Singapore, February 1993. *J. Mater. Proc. Technol.* 37(1–4):381–86, 1993.

SHEPPARD, L. M. Progress in composites processing. *Ceram. Bull.* 69(4):666–73, 1990.

SHERMAN, R. G. Data base development for selected metal matrix composites. Nevada Engineering Technical Corporation, 1990. FR 8/81-7/87, NSWC Tr 88-110, N60921-82-C-0245.

SINGH, P. M., AND J. J. LEWANDOWSKI. Fracture during tension testing of a SiC_p discontinuously reinforced aluminum alloy. *ASM Int. Ann. Meet.*, Pittsburgh, PA, October 1993.

SINHAROY, A., AND R. B. BHAGAT. Micromechanical modeling toward predicting the local damage events in titanium matrix composites. *ASM Int. Ann. Meet.*, Pittsburgh, PA, October 1993.

SORENSEN, N. J. A planar model study of creep in metal-matrix composites with misaligned short fibers. *Acta Metall. Mater.* 41(10):2973–83, 1993.

SRIVATSAN, T. S., R. ANNIGERI, AND W. H. HUNT, JR. The microstructure, tensile properties and fracture behavior of X2080 aluminum alloy-ceramic particle composites. *ASM Int. Ann. Meet.*, Pittsburgh, PA, October 1993.

SURESH, S., A. MORTENSEN, AND A. NEEDLEMAN. *Fundamentals of Metal-Matrix Composites.* Butterworth-Heinemann, Newton, MA, 1993.

SYN, C. K., D. R. LESUER, AND O. D. SHERBY. Fracture behavior and toughness of multi-layer laminated metal composites of Al 5182 and Al 6061-25 vol% SiC_p. *ASM Int. Ann. Meet.*, Pittsburgh, PA, October 1993.

TELESMAN, J., L. J. GHOSN, AND P. KANTZOS. Methodologies for prediction of fiber bridging effects in composites. *J. Comp. Technol. Res.* 15(3):234–41, 1993.

THOMAS, M. P., AND J. E. KING. Effect of thermal and mechanical processing on tensile properties of powder formed 2124 aluminium and 2124 Al-SiC_p metal matrix composite. *Mater. Sci. Technol.* 9(9):742–53, 1993.

THURSTER, R. J. Tensile and creep properties of a SiC fiber-reinforced beta titanium composite. *ASM Int. Ann. Meet.*, Pittsburgh, PA, October 1993.

TODA, H., AND T. KOBAYASHI. Fatigue crack initiation and growth characteristics of SiC whisker reinforced aluminum alloy composites. *ASM Int. Ann. Meet.,* Pittsburgh, PA, October 1993.

TOURATIER, M., A. BEAKOU, AND J. Y. CHATELLIER. On the mechanical behavior of aluminum alloys reinforced by long or short alumina fibers or SiC whiskers. *Compos. Sci. Technol.* 44(4):369–83, 1992.

TURNBULL, A. Review of corrosion studies on aluminium metal matrix composites. National Physical Laboratory, Teddington, England, 1990.

TURNER, S. P., R. TAYLOR, F. H. GORDON, ET AL. Thermal conductivities of Ti-SiC and Ti-TiB_2 particulate composites. *J. Mater. Sci.* 28(14):3969–76, 1993.

VAIDYA, R. U., AND K. K. CHAWLA. Thermal expansion of metal-matrix composites. *Compos. Sci. Technol.* 50(1):13–22, 1994.

VERMA, R., A. K. GHOSH, T. MUKHERJI, ET AL. Measurement of interfacial shear properties of fiber-reinforced composites of Ti-1100 alloy and SCS-6 SiC monofilament fiber. *ASM Int. Ann. Meet.,* Pittsburgh, PA, October 1993.

VOITURIEZ, C. A., AND I. W. HALL. Strengthening mechanisms in silicon carbide whisker reinforced aluminum composites, University of Delaware, Newark, CCM 90-32, 1990, *J. Mater. Sci.* 26(15):4241–49, 1991.

VYLETEL, G. M., J. E. ALLISON, AND D. C. VAN AKEN. 1993. The influence of matrix microstructure and TiC reinforcement on the cyclic response and fatigue behavior of 2219 Al. *Metall. Trans. A* 24(11):2545–57, 1993.

WANG, H. F., C. L. LIN, J. C. NELSON, ET AL. Interfacial stability and mechanical properties of the Al_2O_3 fiber reinforced Ti matrix composite. *ASM Int. Ann. Meet.,* Pittsburgh, PA, October 1993.

WILKS, T. E. The development of a cost-effective particulate reinforced magnesium composite. *Int. Congr. Exhibit.,* Detroit, MI, February 1992. SAE 920457.

WILKS, T. E. Lightweight magnesium composites for automotive applications. *Soc. Automot. Eng. Int. Congr. Exhibit.,* pp. 49–54, Detroit, MI, February/March 1994. SAE 940846.

WOOD, J. V., P. DAVIES, AND J. L. F. KELLIE. Properties of reactively cast aluminium-TiB_2 alloys. *Mater. Sci. Technol.* 9(10):833–40, 1993.

WRIGHT, P. K., R. NIMMER, G. SMITH, ET AL. The influence of the interface on mechanical behavior of Ti-6Al-4V/SCS-6 composites. *Proc. Int. Conf. Interfaces Met.-Ceram. Compos., TMS Ann. Meet.,* pp. 559–81, Anaheim, CA, February 1990.

YIH, P., AND D. D. L. CHUNG. Copper-matrix composites fabricated by using copper coated silicon carbide whiskers. *ASM Int. Ann. Meet.,* Pittsburgh, PA, October 1993.

ZAHL, D. B., S. SCHMAUDER, AND R. M. MCMEEKING. Elastic behaviour of discontinuously reinforced composites. *Z. Metallk.* 84(11):802–5, 1993.

ZHANG, J., R. J. PEREZ, AND E. J. LAVERNIA. Damping behavior of 6061 Al/SiC graphite hybrid metal matrix composites processed by spray deposition. *ASM Int. Ann. Meet.,* Pittsburgh, PA, October, 1993.

ZHANG, J., R. J. PEREZ, AND E. J. LAVERNIA. Effect of SiC and graphite particulates on the damping behavior of metal matrix composites. *Acta Metall. Mater.* 42(2):395–409, 1994.

ZOK, F., S. JANSSON, A. G. EVANS, ET AL. The mechanical behaviour of a hybrid metal matrix composite. *Metall. Trans. A* 22(9):2107–17.

ZWEBEN, C. Metal matrix composites. *ASM Int. Ann. Meet.,* Pittsburgh, PA, October 1993.

6

CMC Properties

6.1 INTRODUCTION

Ceramic matrix composites offer design, materials, and process engineers further opportunities to design performance properties to meet particular end-use specifications. The main thrust in advanced composite structural ceramics has been driven by the potential opportunities in heat engine applications, where greater fuel efficiency and reduced exhaust emissions are the targets. The technology is now at an advanced stage of development, and so innovative products are beginning to reach the marketplace.

The term CMC covers a wide range of materials with different properties, different fabrication routes, and thus different advantages, disadvantages, and applications. Most generally, the composite consists of a matrix material in which an embedded second phase produces an advantageous property such as increased strength or toughness. The second phase may be grown in situ during fabrication of the material or it may be manufactured separately, for example, as a fiber or whisker, and mixed into the matrix material during production of the composite. The second phase may have a variety of shapes such as spherical, platelike, or rodlike, depending on the properties sought. Improved properties over those of the matrix or competing materials can include higher strength, greater toughness, creep resistance or fatigue resistance, and better thermal, electrical, and electromagnetic properties. Some CMCs may be of interest primarily for their electrical and microelectronic properties.

Table 6.1 lists the different types of CMCs and includes carbon-carbon composites which in some instances, along with ceramic coatings and infiltrants, is considered a CMC.

TABLE 6.1. Generic Types of Ceramic and Glass Composites and Carbon-Carbon[1]

Nanocomposites
In situ composites
Particulate composites
Whisker-toughened ceramics
Glass and glass-ceramic matrix composites
Gas- and liquid-phase infiltration composites
Reaction-bonded fiber composites
Lanxide composites
Carbon-carbon composites

6.1.1 Matrix-Fiber Compatibility

Chemical and thermal expansion compatibility between the reinforcing phase and the matrix are important, and this severely restricts the phases that can be combined successfully. For fiber composites, in general, a fiber with the same CTE as the matrix, or a high one, is preferred. High-temperature chemical reactions between the fiber and the matrix during fabrication can have significant effects on the properties of the composite: the most severe are either degradation of the fibers or production of too strong a fiber-matrix bond, resulting in a brittle, low-strength composite. To control the bond strength and improve toughness, an interlayer between the fiber and the matrix may be necessary; for example, fibers may be coated with graphite before composite manufacture or, as for Nicalon fibers in a glass-ceramic, an appropriate interface may be developed by appropriate heat treatment.[1]

6.1.1.1 Fracture and sliding and failure processes. Evidence has been provided to illustrate that fracture and sliding at the fiber-matrix interface is not a simple process that can be described by a single parameter. Moreover, while progress has been made in identifying and measuring some of the governing parameters, it is evident that some remain that have been only superficially examined. While it is not necessary to understand all the nuances of a material system in order to use it in practice, the probability of success and of avoiding spectacular disasters increases strongly with fundamental understanding.

The fracture and sliding at the fiber-matrix interface are also complicated processes dependent on several variables. There are numerous issues regarding measurement and interpretation of interface properties and elucidation of the roles of these variables. An analytical model of pullout-pushout tests that allows separation of many of these factors has been developed.

6.1.2 Continuous and Discontinuous Additives

In view of the limitations of monolithic ceramics, it is not surprising that the genesis of structural ceramics has included addition of the aforementioned reinforcing second phase. Adding a second ceramic phase with an optimized interface improves fracture toughness, decreases the sensitivity of the brittle matrix to the aforementioned microscopic flaws, and even improves strength. It has been demonstrated experimentally that dispersing whiskers in a brittle matrix mitigates crack growth. The presence of whiskers near the crack tip modifies fracture behavior by effectively increasing the required crack driving force through several mechanisms. These mechanisms include crack deflection, crack pinning, whisker bridging, and whisker pullout. In addition, lowered creep rates and varied success in improving resistance to thermal shock have been reported. Whisker reinforcement may also be combined with other toughening mechanisms, such as continuous-fiber reinforcement, which results in a material system known as a hybrid composite, or employment of a matrix material capable of undergoing stress-induced transformation toughening. In addition to capturing the inherent scatter in strength, the reliability analysis of components fabricated from composite ceramics must account for the material symmetry imposed by the reinforcement. For example, the whisker orientations encountered in hot-pressed and injection-molded whisker-toughened ceramics usually impart a local transversely isotropic material symmetry. Whether the second phase imparts orthotropic, transversely isotropic, or isotropic material symmetry, structural reliability models must account for material orientation in a rational manner.

Composite ceramics (e.g., continuous-fiber, laminated, and woven ceramics) offer significant potential for raising the thrust-to-weight ratio of gas turbine engines by tailoring directions of high specific reliability. In general, ceramic continuous-fiber composites exhibit an increase in fracture toughness, which allows for graceful rather than catastrophic failure. When loaded in the fiber direction, these composites retain substantial strength capacity beyond the initiation of transverse matrix cracking, although neither constituent exhibits such behavior when tested alone. For most applications the design failure stress is taken to coincide with the first matrix-cracking stress. The reason for this is that matrix cracking usually indicates a loss of component integrity and allows high-temperature oxidation of the fibers, which leads to embrittlement of the composite.

Composite reinforcement offers an opportunity to increase the toughness and hence critical flaw size of ceramic materials (the critical flaw size increases with the square of the toughness). The presence of particulates, whiskers, or continuous fibers enhances toughness by at least three mechanisms:

- Deflection of the propagating crack tip
- Energy absorption by pullout of the fibers (implying a weak bonding between the fiber and the matrix)
- Bridging of the crack (i.e., holding the two faces of the crack together).

6.1.2.1 Particulate composites. A considerable amount of research has been carried out on brittle particle-reinforced matrices, and the science and technology arguably are mature. The second-phase particle can produce small but significant increases in toughness and consequent increases in strength through crack deflection processes. Important parameters include thermal expansion mismatch, elastic modulus ratio, and degree of interfacial coherence.[2]

Compared with whisker-reinforced systems, particle-reinforced systems present less processing difficulty and should permit higher volume fractions of the reinforcing phase. They should be superior to ZrO_2-toughened systems at elevated temperatures because the efficiency of ZrO_2 transformation toughening decreases as temperature increases. A good understanding exists of the basic science of crack pinning and deflection and consequent toughening.[3–5]

6.1.3 Toughening Mechanisms

As discussed previously, three different forms of secondary phases are being incorporated in CMCs: particulates, whiskers, and continuous fibers. The mechanisms by which these secondary phases affect reinforcement of ceramic composites represent an area of intense interest and remain controversial, although a number of theories have been proposed.[6]

The major toughening mechanism that has been exploited in CMCs is crack deflection by interaction of an advancing track with the dispersed phase of hard refractory particulates or whiskers. When a crack approaches a hard secondary phase with a weakened interface, it is deflected from its original plane. This mechanism is associated with the stress field surrounding the dispersed secondary phase caused by thermal expansion and/or elastic modulus mismatch. As noted earlier, it is preferred that the thermal expansion coefficient of the dispersed phase be higher than that of the matrix; this produces a residual radial tension and tangential compression strain state in the matrix, which tends to divert the crack around the dispersed particle. In the case of randomly dispersed whiskers, the crack front twists; and this twist deflection is more important than the tilt deflection in determining the toughness enhancement. Thus, it has been suggested that tilting and twisting of the advancing crack front is an effective toughening mechanism.[7–12] The major conclusions of various researchers[1,8,13,14] can be summarized as follows.

- Rods (whiskers) are more effective toughening agents than disk-shaped particles, which are more effective than spherical particles.
- The degree of toughening increases with the fiber aspect ratio up to a practical limiting ratio of about 10, above which there is little further improvement.
- The degree of toughening is independent of particle size.
- The degree of toughening reaches an asymptotic value of about 4 for high aspect ratio rods at a volume fraction of ≥ 0.15.

The absolute size of the rods does not affect the toughness calculations. However, the maximum benefit in strength is achieved with finer rods.

It is widely recognized that reinforcement with continuous fibers represents the most attractive option for the enhancement of fracture toughness, but the fabrication of such composites presents special difficulties. The theoretical analysis of composites in which the fibers are unidirectionally aligned is well developed. Crack growth in the matrix phase requires that energy be supplied from the elastic stored energy near the crack tip. In the presence of fibers, energy is absorbed as the crack grows, and the supply of energy to the crack tip growth region is limited. In addition, as a crack propagates through the matrix, the bonds between the matrix and the fiber are broken and the subsequent separation of the matrix causes the fibers to be pulled out.

The total work of fracture is the sum of two terms: one representing fiber-matrix debonding and the other fiber pullout, which is the dominant term. Both terms are directly proportional to the volume fraction V_f of fiber reinforcement; therefore high volume fractions of fine-diameter fibers provide the highest degree of toughness benefit. The balance between strength and toughness is critically dependent on the magnitude of the interfacial shear strength between the fiber and the matrix; a high value causes suppression of matrix cracking and therefore favors high strength, whereas a low value is desirable for high toughness. In contrast to continuous-fiber-reinforced ceramics, whisker pullout is limited by the short lengths of the whiskers (typically less than 100 μm).

Fiber surface treatment is significant as a route in tailoring the interface such that fiber pullout is encouraged as the matrix begins to crack. The application of a 1-μm-thickness coating of carbon on SiC whiskers has been shown to enhance the fracture toughness of whisker-reinforced SiC composites, with values of 10–12 MPa $m^{1/2}$ being reported.

In addition to crack deflection and fiber pullout, it is also suggested that fibers, and to some extent whiskers, can prevent crack propagation by bridging a crack and thereby holding the two faces together. Obviously, the highest mechanical properties are obtained when the applied stress is parallel to the fiber alignment. The strength of the composite is at a minimum when the stress is perpendicular to the fiber direction. For most engineering applications, more complex stressing situations apply and multidirectional reinforced composites are required. Composites based on layered laminated structures, with different fiber directions in each layer, two-dimensional woven mats, and three-dimensional braided fibers have been extensively evaluated in composites. The theoretical analysis of these composites is being developed, although the analytical techniques developed for plastic composites are also being adopted. As will be appreciated, in plastic composites strain transfer between the matrix and the reinforcement is critical, and the matrix can be treated as a homogeneous continuum. This is not the case in ceramic composites. Not only is prediction of the performance of CMCs difficult, but it is also often difficult to produce composites in which the fibers are selectively placed to provide the desired stiffness and load-bearing properties. The actual performance of CMCs often deviates significantly from theoretical predictions, and this is commonly attributed to the introduction of flaws during fabrication and to changes in microstructure associated with the presence of the second-phase material.

6.1.4 Whisker-Toughened Ceramics

In terms of mechanical properties, whisker-toughened ceramics are the most successful category of discontinuous-fiber-reinforced ceramics. In principle, discontinuous-fiber-reinforced ceramics are the simplest fiber-reinforced CMCs to manufacture. Short fibers or whiskers are mixed intimately with powdered matrix material and then either hot-pressed uniaxially or hot-pressed isostatically after isostatic cold-pressing. Temperatures in excess of 1500°C and pressures in excess of 100 MPa are required for consolidation of ceramic matrices such as alumina. Following this process the reinforcing phase in the resulting composite is randomly oriented in the pressing plane. Agglomeration of the fibers or whiskers makes it difficult to produce homogeneous materials with reinforcing volume fractions greater than ~0.3.

Composites of this type containing short random fibers, as opposed to whiskers, have tended to be of lower strength than the unreinforced matrix but can have substantially greater toughness: techniques exist for aligning the fibers before consolidation, and composites produced in this way can have higher strengths than the matrix. Whisker-toughened ceramics containing randomly oriented whiskers can be substantially stronger than the unreinforced matrix and have greater toughness. Successful systems of this type are SiC whiskers in Si_3N_4 and SiC whiskers in Al_2O_3. As an example of the improvement that can be obtained by whisker toughening, an unreinforced Al_2O_3 had strength of 500 MPa and K_{IC} of 4.5 MPa $m^{1/2}$; the addition of 30 vol% of SiC whiskers increased these properties to 650 MPa and 9.0 MPa $m^{1/2}$, respectively.

The toughening of whisker-toughened ceramics arises partly from crack deflection processes and partially from conventional fiber-toughening processes such as debonding and pullout.

A competing method of toughening ceramics uses the martensitic transformation of ZrO_2. There is promising research aimed at combining whisker toughening and ZrO_2 toughening, and recent results appear to indicate synergistic effects, the combined toughening effect being greater than the sum of the individual effects. The advantages of whisker toughening over ZrO_2 toughening include the lower densities of whisker-toughened materials and retention of the toughening mechanisms up to high temperatures—a major disadvantage of ZrO_2 toughening being that the contribution to toughness from the martensitic transformation decreases as temperature increases.

A major problem facing the future development and exploitation of whisker-toughened ceramics that must be overcome is the toxicity and carcinogenic properties of whiskers.

6.1.5 Continuous-Fiber-Reinforced Ceramic Composites

Although whisker-toughened ceramics have enhanced toughness and reliability, they do not substantially lessen the possibility of catastrophic failure, a problem that restricts their use in certain applications. Continuous-fiber-reinforced ceramic composites, however, can provide significant increases in fracture toughness along with ability to fail in a noncatastrophic manner. Prewo and Brennan[15–17] have demonstrated that incorporating fibers with high strength and stiffness into brittle matrices with similar CTEs yields ceramic composites with the potential of meeting high-temperature performance requirements. Typical stress-strain curves of unidirectional systems are bilinear when loaded along the fiber direction, with a distinct breakpoint that usually corresponds to transverse matrix cracking. Since monolithic ceramics are much stronger in compression than in tension, fibers are incorporated to mitigate failure by bridging inherent matrix flaws. Yet one should be mindful that the failure characteristics of these composites are controlled by a number of local phenomena including matrix cracking, debonding and slipping between matrix and fibers, delamination, and fiber breakage.

6.1.6 Fabric-Reinforced CMCs

Advancements in textile weaving technology have resulted in significant new opportunities for utilizing high-performance two- and three-dimensional fabric-reinforced CMCs in high-temperature structural applications (see Fareed et al.[18] and Ko et al.[19]).

Attractive features include improvements in damage tolerance and reliability, flexibility in fiber placement and fabric architecture, and the capability of near net shape fabrication. This last-mentioned feature is of particular interest because applications where these materials can have a significant effect often require complex geometric shapes. However, designing structural components fabricated from materials incorporating ceramic fiber architectures also represents new and distinct challenges in analysis and characterization. Preforms, which serve as the composite skeleton, are produced by weaving, knitting, and braiding techniques[20]. Woven fabrics (i.e., two-dimensional configurations) exhibit good stability in the mutually orthogonal warp and fill directions. Triaxially woven fabrics, made from three sets of yarns interlacing at 60° angles, offer nearly isotropic behavior and higher in-plane shearing stiffness. A three-dimensional fabric, consisting of three or more yarn diameters in the thickness direction, is a network in which yarns pass from fabric surface to fabric surface. These three-dimensional systems can assume complex shapes and provide good transverse shear strength, impact resistance, and through-thickness tensile strength. Furthermore, the problem of interlaminar failure is totally eliminated.[20–23]

Complex textile configurations and complicated yarn-matrix interface behavior represent a challenge in determining the properties of these composites. Considerable effort has been devoted to evaluating the effectiveness of various reinforcement architectures based on approximate geometric idealizations. Chou and Yang[24] have summarized the results of extensive studies on modeling the thermoelastic behavior of woven two-dimensional fabrics and braided three-dimensional configurations.

As processing methods improve, it is likely that the popularity of fabric-reinforced CMCs will increase. Fiber preform design, constitutive modeling, and material processing will require the combined talents and efforts of material scientists and structural engineers.

6.2 CMC PROPERTIES

Each of the two types (continuous and discontinuous) of CMCs offers potential advantages over the other as well as over monolithic materials, the key points being increased fracture toughness, increased strength, and composite-like properties (Table 6.2).

For CMCs, three general temperature ranges are of interest:

Low to medium temperature <1204°C
High temperature 1204–1649°C
Ultrahigh temperature >1649°C

The specific strength of fiber CMCs in particular is shown in Figure 6.1.

In selecting a reinforcement for use in a CMC, many factors must be considered (Table 6.3). First, the type of reinforcement must be determined, whether it be a particulate, a continuous-fiber, or a three-dimensional fiber preform.

An ideal reinforcement for a structural CMC in a very demanding environment could have the following properties.

- Continuous fiber Can be woven, braided, and so on
- Small diameter <20 μm (but application-dependent)
- High strength >2068-MPA tensile strength in air and other hostile environments
- Creep resistance
- CTE ≈Matrix
- High elastic modulus >45 MPa

TABLE 6.2. Potential Advantages of Fiber-Reinforced Ceramic Composites

Specific properties
High strength and strength-to-density ratio
High strength at temperature
Low density
High stiffness-to-density ratio
Toughness (impact and thermal shock)
Improved fatigue strength
Improved creep strength
Improved stress rupture life
Composite-like properties
Controlled thermal expansion and conductivity
Improved hardness and erosion resistance
Ability to tailor-make specific properties
Multiple combinations of above properties
Ability to fabricate complex components to near net shape

6.2.1 Attractiveness of CMCs

CMCs with fibers have attractive performance advantages:

- Low density
- High strength (high specific strength) and high modulus (stiffness)
- High hardness

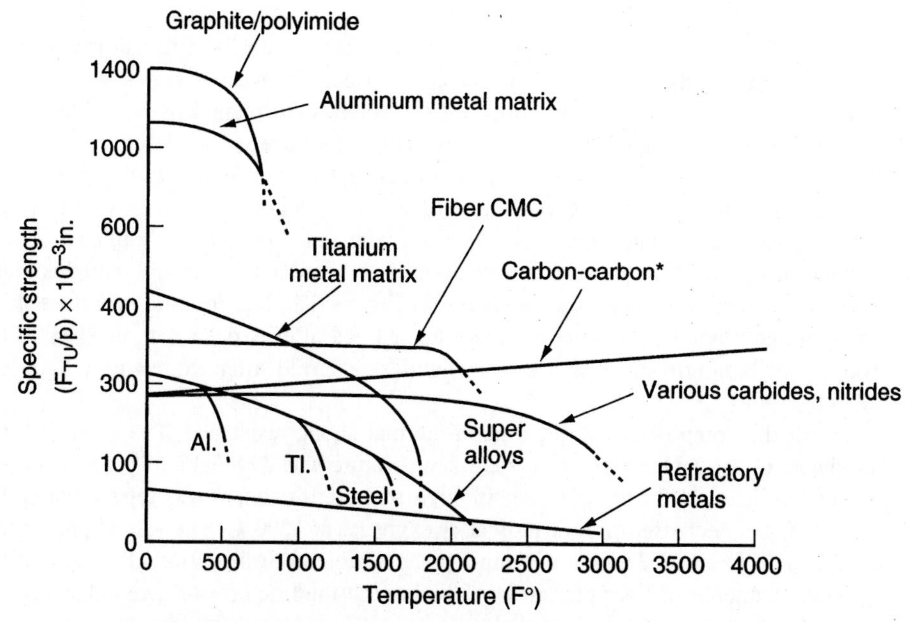

Figure 6.1. Specific strength comparisons of advanced materials. °F = 1.8°C + 32.

TABLE 6.3. A Key Factor: Reinforcement for CMC

Type
Particulate
Platelet
Whisker
Fiber
Filament versus tow
Mat preform
Three-dimensional woven or braided preform
Selection Criteria
Chemical compatibility
Coefficient of thermal expansion
Strength
Modulus
Diameter
Length
Density
Thermal limit
Electromagnetic properties

- High fracture toughness (compared with monolithic ceramics) (see Table 6.4)
- Low thermal expansion (can be adjusted)
- High-temperature stability in oxidative and chemical environments
- Ability to tailor properties for end-use specification.

6.2.2 Fibers, Woven Fabrics, and Preforms

6.2.2.1 SiC CMCs. Recent studies have focused on the effect of preform architecture on the densification and mechanical properties of composites where a three-dimensional braided Nicalon preform, a two-dimensional plain woven SiC fabric laminate, and a unidirectional SiC laminate were compared. It was found that the 3-D braided structure achieved the highest density, an effect that was attributed to the more uniform pore network that prevents early sealing of the pores. Nevertheless, relatively large pores resulting from inadequate matrix infiltration were not completely eliminated. The 3-D braided composite exhibited a considerably higher flexural strength and modulus than the 2-D woven laminate composite as shown in Figure 6.2. The room-temperature flexural strength and modulus remained virtually unchanged after exposure in air at 1000°C for 100 h, but a significant decrease in properties occurred after exposure at 1200°C for 100 h.

All the composites exhibited good thermal shock resistance. The average fracture toughness of the 3-D composite at room temperature was 29.8 MPa $m^{1/2}$, and for the 2-D woven laminate, 16.2 MPa $m^{1/2}$. Again, high fracture toughness was retained after 100 h at 1000°C but decreased significantly after exposure at 1200°C. The loss of properties after exposure in an oxidizing environment was attributed to fiber strength degradation. A significant amount of fiber pullout was observed at the fracture surface, indicating that a weak fiber-matrix bond was formed.

TABLE 6.4. Fracture Toughness of Structural Ceramics: Monolithic, Transformation-Toughened, and Composites

Material	Fracture Toughness (MPa $m^{1/2}$)	Critical Flaw Size (μm)
Monolithic ceramics		
Alumina	3–4.5	15–35
Silicon nitride	4–6	33–75
Siicon carbide	4.5–6	40–70
Zirconia	7–12	75–250
Glass-ceramic	1–2	
Transformation-toughened		
Zirconia (MgO)	9–12	160–300
Zirconia (Y_2O_3)	6–9	75–165
Alumina (ZrO_2)	6.5–15	85–450
Particulate Reinforcement		
Alumina-TiC_p	4.2–4.5	35–40
Silicon nitride-TiC_p	3.0–4.5	35–40
Silicon nitride-SiC_p	3.5–5.0	35–40
Whisker reinforcement		
Alumina-SiC_w	7–9	130–200
Silicon nitride-SiC_w	5–8	
Zirconia-SiC_w	15–20	
Fiber reinforcement		
LAS/SiC_f	15–25	
SiC/SiC_f	8–15	

The recent attention of several U.S. firms has recently been focused on three-dimensional braided SiC preforms infiltrated with lithium aluminosilicate (LAS). The 3-D braided composites, containing 40 vol% fiber, produced a 30% increase in fracture toughness when compared with a conventional one-dimensional (1-D) laminated composite. The fracture toughness was somewhat less than the 26 MPa $m^{1/2}$ obtained by researchers in the United Kingdom. The 3-D reinforcement, as expected, was reported to improve the resistance of the composite to shear stresses and to enhance fatigue performance.

Others have evaluated SiC matrix composites prepared by CVI using methyl dichlorosilane as the reactant and Nicalon fibers and Nextel fibers as the reinforcement. In a recent series of experiments, it was reported that Nicalon-reinforced composites exhibited the highest strength and the highest fiber pullout. The researchers attributed the differences in performance to differences in fiber-matrix interfacial bonding and to differences in fiber geometry because Nextel fabric was used in one case and unidirectional Nicalon in the other.

Other U.S. firms have evaluated SiC/SiC composites reinforced with one-dimensional and two-dimensional braided fiber configurations prepared by CVI for applications in vacuum-tight power plant plasma containers. The composite comprised 50 vol% fiber reinforcement, had a density of 2.5 g/cm^3, and had a Young's modulus of 240 GPa. The unidirectional composite exhibited a bending strength of 620 MPa, whereas the 2-D braided composite had a bending strength of 350 MPa.

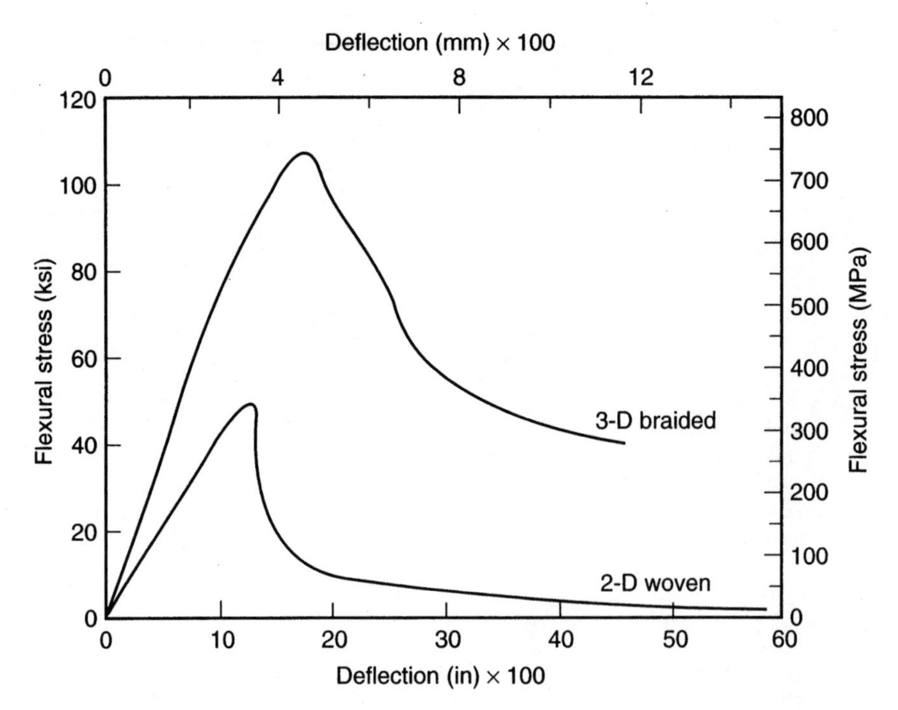

Figure 6.2. Flexural stress-deflection behavior of 2-D woven laminate and 3-D braided Nicalon-SiC composites.

Riccitiello and associates[25] conducted an evaluation of 2-D CMCs in aeroconvective environments and measured the tensile strength, tensile modulus, and Iosipescu shear strength, at room temperature, at 1093°C, and at 1538°C, for the CMCs in both the orthotropic and quasi-isotropic configurations.

Their evaluation used CMCs fabricated via the CVI/CVD process with Nicalon fabric, Nextel 440, and carbon fabric in 2-D orthotropic and quasi-isotropic configurations with a SiC matrix.

The tensile strength of the C/SiC composites remained relatively constant over the temperature tests (Figure 6.3), whereas the Nicalon- and Nextel-reinforced composites decreased in strength for test temperatures above 1093°C (Figures 6.3b and c, respectively). This behavior at elevated temperatures is the same as that reported for the reinforcing fiber alone.[25]

The corresponding tensile modulus shows that the modulus for the Nicalon-reinforced material in the orthotropic configuration was higher than that for both the Nextel- and the carbon-reinforced materials at room temperature (Figure 6.4a–c). Based on the rule of mixtures, the Nicalon-reinforced material should have been similar to the other composites. A plausible explanation for the higher modulus obtained for the Nicalon-reinforced composite could be related to the interphase used to prevent bonding between the fiber and matrix which could have been compromised during the processing.

In summary, from testing the continuous fiber-reinforced ceramic composites, they found the following.

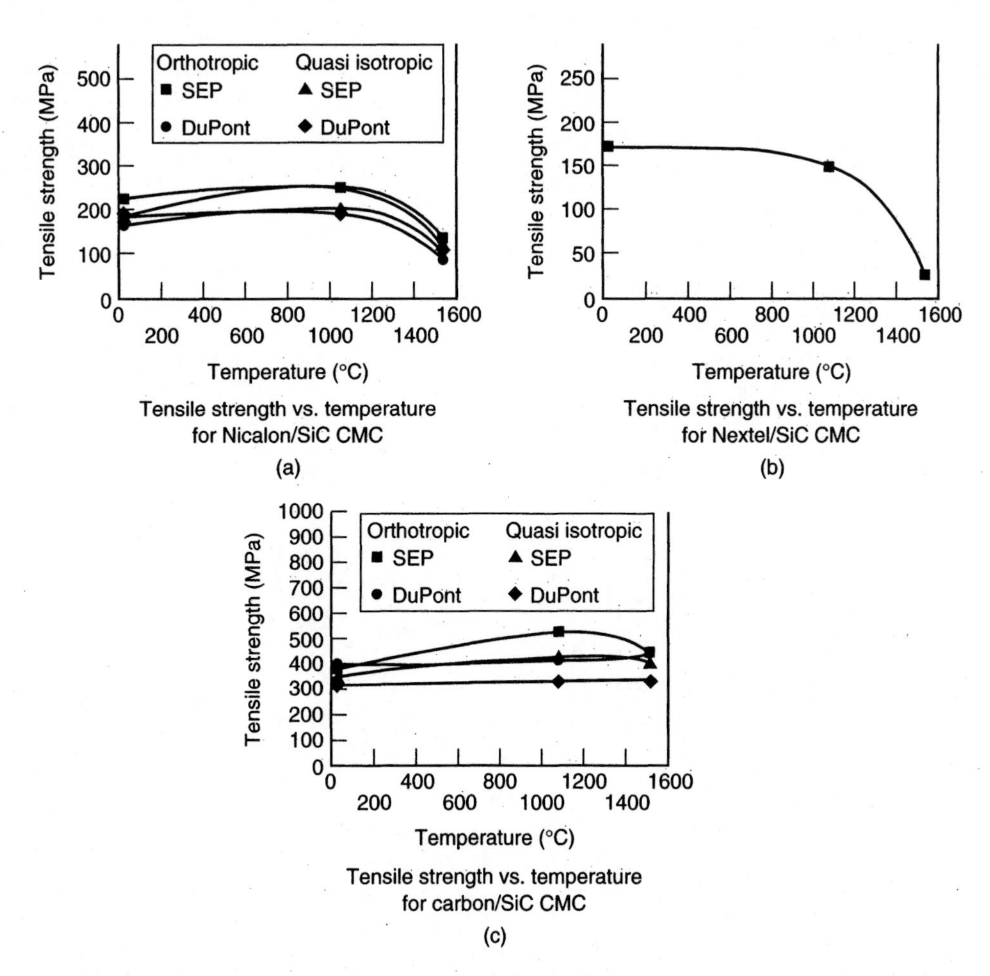

Figure 6.3. (*a*) Tensile strength versus temperature for Nicalon-SiC CMC. (*b*) Tensile strength versus temperature for Nextel-SiC CMC. (*c*) Tensile strength versus temperature for carbon-SiC CMC.

For the Nicalon-SiC orthotropic and quasi-isotropic configuration:

- After 100 min of aeroconvective exposure at a surface temperature of 1500°C, the composite showed a slight mass gain and no surface recession.
- After 100 min of convective exposure at a surface temperature of 1100°C, the composite showed no mass change or surface recession in the orthotropic configuration. The same result would be expected for the quasi-isotropic configuration.
- The physical property data suggested that the carbonlike interphase material was affected during aeroconvective exposure, particularly after exposure to the 1500°C environment.
- The physical property data indicated that regardless of the method used to obtain a temperature of 1500°C, the material properties were severely compromised.

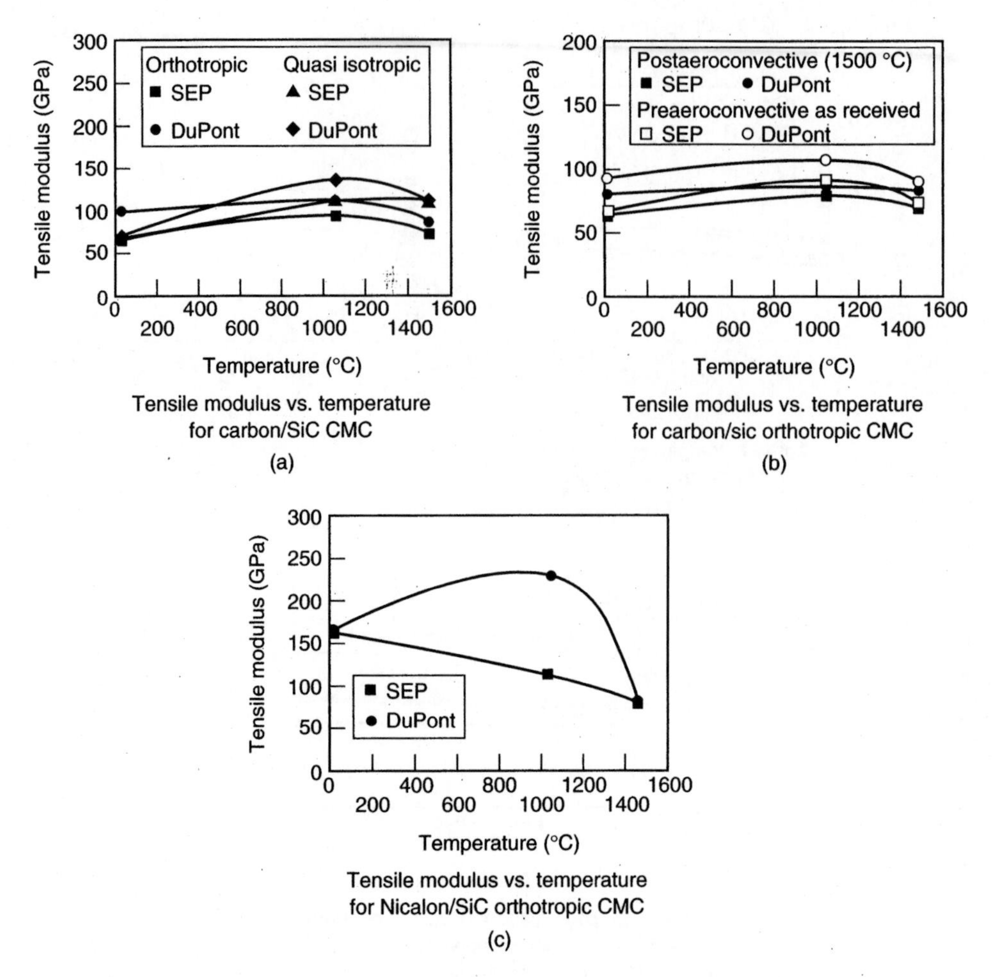

Figure 6.4. (*a*) Tensile modulus versus temperature for carbon-SiC CMC. (*b*) Tensile modulus versus temperature for carbon-SiC orthotropic CMC. (*c*) Tensile modulus versus temperature for Nicalon-SiC orthotropic CMC.

For the Nextel-SiC orthotropic configuration:

- After 100 min of aeroconvective exposure at a surface temperature of 1500°C, the composite showed a mass loss and minor spallation with no measurable overall surface recession.
- After 100 min of aeroconvective exposure at a surface temperature of 1100°C, a mass loss rate of 2.04×10^{-7} g/cm·s was measured with no spallation or surface recession. This rate was higher than that experienced by the carbon fiber-reinforced composites when exposed to a 1500°C surface temperature for the same length of time.
- The 1500°C aeroconvective exposure data suggest that the measured mass loss may have been influenced by the loss of boria from the fiber as well as by contributions from subsurface oxidation and from minor spallation.

- The physical property data indicated that the material was severely affected by the aeroconvective exposure as well as testing at a temperature of 1538°C.

For the carbon-SiC orthotropic configuration:

- After 100 min of aeroconvective exposure at a surface temperature of 1500°C, the composite showed a slight mass loss or surface recession.
- After 100 min of aeroconvective exposure at a surface temperature of 1100°C, the composite showed no mass loss or surface recession.
- The physical property data showed that the composite experienced no change after aeroconvective exposure.

Khandelwal and Johnson[26] measured tensile strength and creep behavior of 2-D SiC/SiC CMC material from room temperature to 1371°C. The average room-temperature strength was 191 MPa, which decreased to 92.4 MPa at 1371°C. The creep rate of the material was between 300 and 750 mm/mm·h at 982°C and increased to 400–1500 mm/mm·h at 1204°C.

The mechanical behavior of 2-D woven and 3-D two-step and four-step braided (2-S B and 4-S B) composites of a SiC-based fiber in a SiC matrix (SiC/SiC) was studied by Karandikar, Parvizi-Majidi, and Chou.[27]

They found that under tensile loading, 2-D woven, 3-D two-step braided, and 3-D four-step braided SiC/SiC composites showed gradual damage development and non-catastrophic failure. Damage initiated in the form of cracks from the intertow pores. Cracks then developed in the transverse and longitudinal bundles, giving rise to nonlinear stress-strain curves and modulus degradation.

The 2-D woven composite exhibited the lowest tensile failure strain and also exhibited the lowest tensile strength because only 50% of the fibers are oriented in the loading direction. For the 3-D composites, multiple cracking developed in the longitudinal bundles and gave rise to a higher strain.

Societe Europeenne de Propulsion (SEP), a French manufacturer of SiC matrix composite made by CVI, has been very successful in processing covalent matrix composites.[28] The mechanical properties of both SiC Nicalon-SiC carbon (T300)/SiC composites are given in Tables 6.5 and 6.6.[28]

The tensile strengths of 200 and 350 MPa at 1000°C are most impressive. In recent work on the environmental behavior of these composites, performed on uncoated SiC/SiC composites, a strength loss together with some structural modifications of the fiber-matrix interface after long-term holds (more than 100 h) in air at temperatures ranging between 800 and 1300°C was observed.

The observations show a progressive consuming of the carbon interphase layer by oxidation and its replacement by a silica layer. From a mechanical point of view, a decrease in the Young's modulus is first observed, which corresponds to removal of the carbon layer followed by a sharp increase corresponding to formation of the silica layer which achieves a strong fiber-matrix bond.[29] In parallel, a strong increase in the fiber matrix shear strength measured by microindentation has been observed[30]; then the composite exhibits a brittle failure.

As far as fatigue is concerned, tests performed in tension-tension on 2-D SiC/SiC

TABLE 6.5. Mechanical Properties of 2D C/SiC Composites[28]

Property	Unit	23°C	1000°C	1400°C
Fiber volume fraction	%		40	
Density			2.5	
Open porosity	%		10	
Tensile strength	MPa	200	200	150
Elongation to rupture	%	0.3	0.4	0.5
Tensile Young's modulus	GPa	230	200	170
Flexural strength	MPa	300	400	280
Compressive strength parallel	MPa	580	480	300
Compressive strength perpendicular	MPa	420	380	250
Interlaminar shear strength	MPa	40	35	25
Toughness, K_{IR}	MPa $m^{1/2}$	30	30	30

processed by CVI illustrates that the fatigue limit of the material appears to be quite high, 140 MPa as compared to the elastic proportional limit of about 160 MPa.[31]

Rouby and Reynaud reported at a recent Agard meeting[32] that the fracture properties of CMCs are controlled by two main microstructure parameters: first, the average fiber strength and the strength distribution (described currently by Weibull's statistics), and second, the interface properties, particularly the interfacial shear stress associated with friction or debonding and controlled by adequate fiber coating and also residual thermal stresses.

To verify their beliefs they carried out tests at room temperature under tension-tension sinusoidal cycling, between zero and controlled maximum stress *S*, at a frequency of 1 Hz. The material was a cross-weave composite of SiC fiber (Nicalon) bundles embedded in a SiC matrix obtained by CVI (2-D SiC/SiC). The composite was processed by SEP and is characterized by a failure strain of about 0.2–0.3%. Similar results were obtained for a cross-plied SiC (Nicalon) glass-ceramic composite processed by Aerospatiale. The MAS-L matrix was obtained by using the sol-gel route and hot-pressing.

Extensive work on the mechanical behavior of CMCs,[33] crack growth resistance,

TABLE 6.6. Mechanical Properties of 2D SiC/SiC Composites[28]

Property	Unit	23°C	1000°C	1400°C
Fiber volume fraction	%		45	
Density			2.1	
Open porosity	%		10	
Tensile strength	MPa	350	350	350
Elongation to rupture	%	0.9	0.9	
Tensile Young's modulus	GPa	90	100	100
Flexural strength	MPa	500	700	700
Compressive strength parallel	MPa	580	600	700
Compressive strength perpendicular	MPa	420	450	500
Interlaminar shear strength	MPa	35	35	35
Toughness, K_{IR}	MPa $m^{1/2}$	32	32	32

and creep properties of 2-D woven Nicalon fiber cloth SiC/C/SiC composite materials, indicates that the shape of the resistance curve depends essentially on the predominant toughening mechanism (extensive matrix microcracking or fiber-matrix debonding), whereas the frontal process zone size is the same for the materials.

Lamicq and Boury[34] made a concerted effort to determine the potential for thermostructure design of SiC/SiC and C/SiC composite materials (Figure 6.4).

These two materials are currently the most used for designing thermostructure-oriented parts or subsystems in an oxidizing environment.

C/SiC composites are made up of a carbon fiber reinforcement, a pyrocarbon interphase, and a SiC matrix, whereas the SiC/SiC composites are made up of a SiC fiber reinforcement, a pyrocarbon interphase, and a SiC matrix.

Fiber reinforcements can be either bidirectional—consisting of a mere fabric stacking (2-D) or made up of layers bound together to form a multilayer (2.5-D)—or three-dimensional as in the case of carbon reinforcement prepared by using the needling technique. The interphase and then the matrix are deposited by CVI of a carbon or SiC precursor.

In other tests the researchers found that the strength potential of the materials at room temperature appeared relatively low. On the other hand, the low evolutions with temperature led to remarkable results at high temperatures for unaged material. Average rupture values were higher than

200 MPa for a 2-D SiC/SiC up to 1500°C
300 MPa for a 2.5-D C/SiC up to 1500°C
230 MPa for a 3-D C/SiC up to 1500°C.

Figure 6.5 can be used to compare these strength performances with those of some common metallic materials.

It can be noted that, from 675°C upward, some specific features (reduced to density for isomass comparison) are comparable. At 975°C, the mechanical advantage becomes obvious because of the CMC low mean density.

Additional testing by Lamicq and associates[34] covered rupture criteria (tension,

Property	INCO 718 at 600°C	Haynes 188 at 833°C	C/SiC 2.5-D up to 1166°C	SiC/SiC 2-D up to 977°C
Density (kg/m^3)	8200	8980	2100	2300
Ultimate tensile strenth • (MPa) (minimum at 3 standard deviations)	1150	240	210	140
Young's modulus (GPa) (initial or elastic value)	165	150	65	220
Specific rupture strength ($MPa/kg \cdot m^3$)	140	27	100	61
Specific stiffness ($GPa/kg \cdot m^3$)	20	17	31	95

Figure 6.5. CMC versus metallic mechanical characteristics.

compression, interlaminar shear, and translaminar shear); vibratory, thermal, mechanical, and thermomechanical calculations; and fatigue (Table 6.7).

Thus, from the various resultant property tests information about CMC design is beginning to be formalized. This allows better appreciation of the real potential and application possibilities of materials that are still very new (Table 6.8).[35–40]

Glass matrix and SiC continuous fibers have attracted attention because of their ability to achieve high strength and controlled failure. Prewo and Brennan[41] used a slurry infiltration processing technique to produce a SiC/7740 borosilicate glass matrix composite using monofilaments of SCS-6 SiC. They conducted studies on composites with two levels of fiber: 0.35 and 0.65. A flexure strength of 830 MPa at 22°C was reported. This value increased to 930 MPa at 350°C and to 1240 MPa at 600°C. This trend is due to softening of the matrix analogous to the behavior of carbon fiber composites. Weaker strengths were exhibited by the 35% fiber specimens which had a value of 650 MPa at room temperature.

Additional testing revealed a fracture toughness of 18.9 MPa $m^{1/2}$, corresponding to approximately a 140-fold increase in strength and a 30-fold increase in toughness.

Similar strength-versus-temperature behavior was observed in additional experiments[41] involving Nicalon fibers in 7740 glass and in 7930 high-silica glass. Higher flexure strengths, with temperature patterns comparable to those of the SCS-6 SiC/7740 system were shown by the Nicalon-7740 glass composites after normalizing the data for fiber fraction. The strength increase in this system was attributed to the greater surface area and finer microstructure of the SiC yarn composites. Strength values for the higher silica glass composites ranged from 400 to 800 MPa at temperatures of 22–1100°C.

In all cases, the ultimate strength of the composite at high temperatures was limited by the softening point of the glass. However, the benefit of the increase in viscoelasticity was soon lost when the specimen began to bend.

The limited temperature capability of glass as a matrix material led to the use of glass-ceramics. This matrix material offered the ease of vitreous preparation combined with the high-temperature capability of a crystallized ceramic. Brennan and Prewo[42] formulated composites using a lithium aluminosilicate glass-ceramic and SiC fibers. A fiber

TABLE 6.7. Properties of CMC Materials Manufactured by the Si-Polymer Route[35]

Property		C/SiC[a]			SiC/SiC[a]		
Infiltration cycles		1	1	3	1	1	3
Fiber coating			PyC	PyC		PyC	PyC
Porosity	(%)	20–25	20–25	10–15	20–25	20–25	10–15
Density	(g/cm^3)	1.6–1.7	1.6–1.7	1.9–2.0	1.9–2.0	1.9–2.0	2.3–2.4
Tensile strength	(MPa)	100–120	180–260	250–300	30–40	140–160	180–200
Elongation	(%)	0.2	0.4–0.6	0.4–0.6	0.1	0.4	0.2
Young's Modulus	(GPa)	60–80	60–80	70–80	80	80	90
Interlaminar shear strength	(MPa)	4	2	12	6	6	20
Bending strength	(MPa)	100–120	80–100	190–210	150–160	160–190	280–300
Bending strength, 1000°C	(MPa)				150–160	160–190	280–300
Bending strength, 1100°C	(MPa)				150–160		

[a] Lamination: 0/90° UD prepregs; fiber content 45–55%.

TABLE 6.8. Typical Mechanical Properties of 2D-CMC in Comparison to Monolithic SiC as Published in the Literature[35]

Properties	CVI Route				Polymer Route[a]			Monolithic Ceramic
	Isothermal		p,T-Gradient		Si-Polymer		Si Infiltration	
	C/SiC	SiC/SiC	C/SiC	SiC/SiC	C/SiC	SiC/SiC	C/C-SiC	SSiC
Tensile strength (MPa)	350	150–200	270–330	300–350	100–120	30–40	90–110	
Elongation at break (%)	0.9	0.3–0.5	0.6–0.9	0.5–0.8	0.2	0.1	0.23	
Young's modulus (GPa)	90–100	170–230	90–100	180–220	60–80	80	60–70	405
Compression strength (MPa)	580–700	300–580	450–570	440			300	2900
Flexure strength (MPa)	500–700	280–400	450–500	500–600	100–120	150–160	160–200	350–400
Shear strength (MPa)	35	25–40	45–55	65–75	4	6	55–60	
Open porosity (%)	10	10	10–15	10–15	20–25	20–25	3–7	
Fiber content (vol%)	45	40	42–47	40–50	45–55	45–55	55–65	
Density (g/cm^3)	2.1	2.5	2.1–2.2	2.3–2.5	1.6–1.7	1.9–2.0	1.8–1.95	3.1
CTE								
Parallel (10^{-6} 1/K)	3^b	3^b	2^c	4^c	3	3	$1–2^d$	4.1
Perpendicular	5^b	$1.7–3.4^b$	5^c	4^c	5	2.5	4.6^d	
Thermal conductivity								
Parallel (W/m·K)	$14.3–20.6^b$	$15–19^b$	14	20			$10–15^e$	105
Perpendicular	$6.5–5.9^b$	$5.7–9.5^b$	6	10			$6–8^e$	
Specific heat (J/kg·K)	620–1400	620–1200	600				1150–1850	660
Manufacturer[f]	SEP	SEP	MAN	MAN	Dornier	Dornier	DLR	Sintec

[a] Values without additional fiber coating and after one infiltration cycle.
[b] RT–1000°C.
[c] 100–1000°C.
[d] RT–1500°C.
[e] 200–1700°C.
[f] SEP, References 28 and 37; MAN, reference 38; Dornier, reference 39; DLR, Deutsche Luft- und Raumfahrt; Sintec, reference 40.

fraction of 50% was studied. Flexure strengths were determined by three-point bending, resulting in values of 620 and 370 MPa for unidirectional and 0/90° cross-ply composites, respectively. The unreinforced LAS material showed a strength of 190 MPa. At 1000°C, the strength of the unidirectional composite increased to 900 MPa again as a result of the viscoelastic behavior of the glass-ceramic.

The development of a composite test methodology for ceramic fiber composites takes into consideration the anisotropic nature of continuous ceramic fiber composites and makes the need for tensile, compressive, and shear data abundantly clear. A typical Nicalon-SiC fiber-reinforced glass-ceramic has the nominal properties listed in Table 6.9.

Typical tensile and compressive stress-strain curves of one particular SiC-glass-ceramic composite[43] are provided in Figure 6.6. It is clear that with the degree of anisotropy exhibited by ceramic fiber composites, very specialized testing involving both in-plane as well as off-axis tests is required.

In general, toughness values for SiC fiber-reinforced ceramic composites are much higher than for the unreinforced matrix.[44,45] However, large discrepancies in toughness have been reported in the literature. Differences in processing that affect the interfacial area may be the prime reason for these inconsistencies. Prewo observed values of 18.9 $MN/m^{1.5}$ for a monofilament SiC-reinforced 7740 glass composite as measured by the notched beam technique. This value was higher than that measured for the continuous yarn composites, which had observed values of 11.5 $MN/m^{1.5}$.

High fracture toughness can be attributed to the fiber pullout mechanism. High values of toughness were recorded in LAS/SiC fiber compositions that showed extensive pullout on the fracture surface.[42]

According to Prewo and Tressler,[46] the advent of SIC fibers in CMCs has allowed higher performance temperatures to be achieved. SiC fibers show good corrosion and oxidation resistance while maintaining their strength and toughness at elevated temperatures.

Room-temperature aging studies conducted on a 7740/SiC-reinforced composite showed no degradation after exposure to 540°C for 500 h.[47]

The mechanical properties of several mature Nicalon SiC-reinforced glass and glass-ceramic composites are shown in Tables 6.10–12. Table 6.10 illustrates the properties of Corning SiC/1723 glass matrix composite. Code 1723 glass is an alkaline earth aluminosilicate and is readily reinforced with SiC fibers with close thermal expansion matching (ma-

TABLE 6.9. Typical Properties of Nicalon SiC Fiber-Reinforced Glass-Ceramic Composites[a]

Tensile strength (MPa)	
0°	600
90°	20
In-plane shear strength (MPa)	40
Interlaminar shear strength (MPa)	15
Interply tensile strength (MPa)	15
Compression strength, 0°	1000

[a] SiC $V_f \sim 0.4$; all properties at 25°C; properties typical for LAS, BMAS, and CAS glass-ceramic matrix composites.

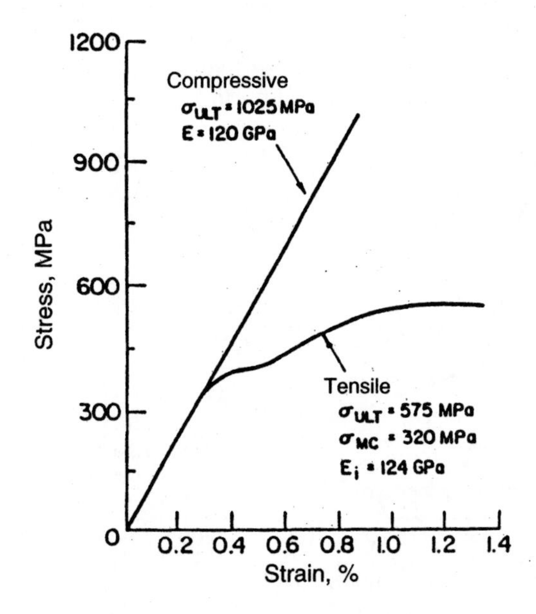

Figure 6.6. Tensile and compressive stress-strain response for SiC/LAS glass-ceramic composites.[42]

trix 5.4×10^{-6} C^{-1}). Its high-temperature behavior is, like that of graphite-glass composites, governed by the anneal point of the glass (710°C for Corning 1723). The high-temperature flexure properties degrade rapidly after 700–800°C as shown in Figure 6.7. The refractoriness of such glass matrix composites is thus matrix-limited. Table 6.10 illustrates the excellent strength and toughness attained by SiC/1723 glass composites—a unidirectional tensile strength of 734 MPa at room temperature with 1.2% ultimate failure strain.

SiC/LAS and SiC/CAS (calcium aluminosilicate) glass-ceramic matrix composite

TABLE 6.10. Properties[a] of Corning SiC/1723 Composites[48]

Longitudinal (0°) tensile properties	
Ultimate strength (MPa)	734
Failure strain (%)	1.2
Elastic modulus (GPa)	128
Poisson's ratio	0.24
Proportional limit stress (MPa)	
Proportional limit strain (%)	
Transverse (90°) tensile properties	
Strength (MPa)	27
Failure strain (%)	0.035
Elastic modulus (GPa)	101
Off-axis properties	
In-plane shear strength (MPa)	
Shear modulus (GPa)	
Interlaminar shear strength (MPa)	73[b]
Interply tensile strength (MPa)	

[a] All properties at ambient room temperature. Nicalon SiC, $V_f = 0.35$–0.40. Corning Code 1723 alkaline earth aluminosilicate glass.
[b] Short-beam shear.

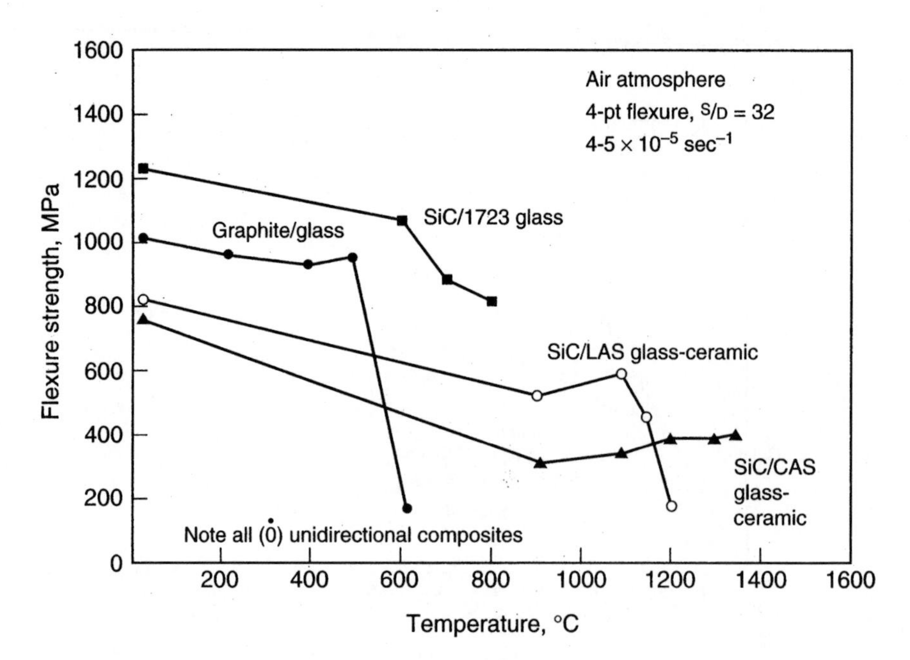

Figure 6.7. High-temperature strength of various Corning ceramic fiber composites.

behavior is listed in Tables 6.11 and 6.12. These glass-ceramic matrix composites are more refractory than glass matrix composites, as shown in Figure 6.7.

An interesting comparison of these materials shows where the off-axis behavior of the CAS and 1723 matrices appears to have significantly higher values of transverse elastic modulus E_{22}, as well as out-of-plane properties (in-plane shear and interlaminar shear strengths, and interply tensile strength) than do LAS composites. It would be expected that this desirable trait would be accomplished by the achievement of a stronger fiber-matrix interfacial bond. This, in fact, is supported by the interfacial shear resistance (τ_{debond}) data published by Mandell.[48,49]

This interrelation between in-plane and off-axis properties is essential property data for designers contemplating the use of such CMCs.

TABLE 6.11. Properties of SiC/LAS-II Composites[a]

Property	Data Source	
	Prewo[47]	Larsen et al.[43]
Longitudinal, 0°		
Tensile properties		
Ultimate strength (MPa)	641–686	586
Failure strain (%)	0.78–1.08	0.93
Elastic modulus (GPa)	124–140	121
Poisson's ratio		
Proportional limit stress (MPa)	367–464	290
Proportional limit strain (%)	0.26–0.34	0.25

[a] UTRC SiC/LAS-II Compglas.

TABLE 6.12. Properties of Corning SiC/CAS-II Glass-Ceramic Composites

Density	2.7 g/cm^3
Longitudinal (0°) tensile at 25°C	
Ultimate strength (MPa)	450
Ultimate strain (%)	0.95
Elastic limit stress (MPa)	207
Elastic limit strain (%)	0.15
Elastic modulus (GPa)	124
Longitudinal (0°) compression at 25°C (MPa)	1075
Off-axis properties at 25°C	
Transverse (90°) tension strength (MPa)	28
In-plane shear strength (±45°) (MPa)	65
Interlaminar shear strength (SBS) (MPa)[a]	68
Interlaminar shear strength (DNC) (MPa)[b]	30
Transverse elastic modulus (GPa)	117
Shear modulus (±45°) (GPa)	48
Thermal expansion, 25–1000°C	
Longitudinal	4.3×10^{-6}/°C
Transverse	4.5×10^{-6}/°C

Source: Corning Glass Works.
[a] SBS, Short-beam shear.
[b] DNC, Double-notch compression.

Prewo and Brennan[50] have reported that carbon fiber-reinforced borosilicate glass retains its flexural strength up to 600°C. Above this temperature, the composite strength is reduced as a result of matrix softening. Thus, although glass matrix is attractive for its lower processing temperature, its application temperature is limited to 600°C. An alternative is to replace glass with a glass-ceramic, such as LAS, which increases the use temperature to 1000°C or higher. Because carbon fibers are prone to oxidation above 300°C, they are replaced by SiC fibers which show no appreciable reduction in either strength or modulus up to 800°C.

Thermomechanical properties of SiC-reinforced glass-ceramics have been studied by a number of investigators.[51,52] The results of investigations of Nicalon (SiC) fiber-reinforced LAS are summarized as follows.

1. Nicalon-LAS composites retain their flexural strength up to 800°C when tested in air, and up to 1100°C when tested in an inert atmosphere such as argon. The fracture mode of these composites depends strongly on the test environment.
2. A room-temperature tension test in air of unidirectional Nicalon-LAS composites shows multiple matrix cracking and fiber pullout. The composite fails gradually after the maximum load is reached. When the same composite is tested at 900°C and above, fiber failure occurs after a single matrix crack has formed, resulting in a catastrophic failure with a sudden load drop. This change in failure mode is attributed to fiber strength degradation as well as increased fiber-matrix bonding in the oxidative atmosphere. The initial carbon-rich layer between the SiC fiber and the LAS matrix is replaced by an amorphous silicate as a result of oxidation at 650–1000°C. This creates a strong interfacial bond and transforms the composite from a relatively tough material to a brittle material.[52]

3. The presence of a carbon-rich interface is very important in obtaining a strong, tough SiC fiber-reinforced LAS composite because this interface is strong enough to transfer the load from the matrix to the fibers, yet weak enough to debond prior to fiber failure. This enables the composite to accumulate a significant amount of local damage without failing in a brittle mode.

Glass and ceramic matrix composites with continuous fibers were tested with instrumented impact apparatuses by Hasson and Fishman.[53,54] The composite architecture was cross-ply (0/90°). The glass matrix composite had a borosilicate matrix with untreated high-modulus carbon fibers, and the CMCs had a ceramed modified LAS matrix (LAS III) with SiC fibers or prewoven SiC 2-D fabric. All the 0/90° components showed that the edge-on orientation had a lower dynamic rupture work per unit area than the interlaminar orientation, with the exception of glass matrix composite materials cut at 45° to the final ply axis. For the CMCs the continuous SiC fiber material was tougher than the composite with the 2-D prewoven fabric. For all the composites the percentage of dynamic rupture work attributed to crack initiation was less than 20%, and thus most of the rupture work was attributed to fracture of the fibers and fiber bundles after fiber-matrix interfacial bond failure with the attendant frictional sliding mechanism toughness contribution.

Metcalfe, Donald, and Bradley[55] studied and tested a magnesium aluminosilicate glass-ceramic matrix composite (MAS I and II) reinforced with SiC fibers (Tyranno). Their study demonstrated that the MAS composite was manufacturable at a much lower temperature (810°C) than is possible for CMCs. Historically glass-ceramics have been consolidated at about the liquidus temperature, but this study proved that it is possible to consolidate at a temperature only slightly above the glass transition and simultaneously achieve a low level of porosity. Some degree of optimization of thermal expansion is possible by making a suitable choice of heat treatment schedule, thereby minimizing residual matrix stress caused by differential shrinkage. This composite with its good mechanical properties ($\sigma \approx 1$ GPa and $E > 75$ GPa up to 500°C) and high work of fracture (>30 kJ/m^2 at ambient temperature rising to 45 kJ/m^2 at 500°C) could fill a gap in the materials field, that is, provide a strong, tough material with higher temperature capability than polymeric composites. The possibility of being able to increase the toughness of a composite further by postfabrication heat treatment is suggested by this work.

Jarman, Layden, McCluskey, et al.[56] participated in a government-sponsored program to evaluate Nicalon braid, braid-SCS-6 SiC monofilament hybrid, and Nicalon 3-D woven fabric-reinforced glass CMCs in terms of their microstructure and mechanical properties. Some aspects relating to glass powder particle size distribution, slurry formulation, and slurry infiltration techniques were addressed.

Additionally they found that

1. Glass CMCs reinforced with SiC monofilament fiber had significantly higher elastic moduli, tensile strengths, and proportional limits than similar composites reinforced with only Nicalon yarn.
2. The addition of SiC-type yarn fibers to SiC monofilament-reinforced glass-ceramics resulted in improved structural integrity, increased elastic modulus, and increased tailorability of composite mechanical properties.

3. The carbon-rich coating on the SCS-6 monofilament was prone to oxidative degradation in a glass-ceramic composite.
4. Long-term oxidative stability of SiC monofilament-reinforced glass-ceramics required the development of fiber coatings designed specifically for ceramic reinforcement (Table 6.13).

Flexural strengths measured by four-point bending at room temperature for the 2-D Nicalon cloth–SiC material ranged from 83 to 517 MPa for individual samples according to Mazdiyasni.[57] Strength generally correlated with density and was typically 180 and 450 MPa for 65% and 90% theoretical density, respectively. The material exhibited high strain tolerance and toughness, provided the extent of fiber-matrix bonding was controlled. Fracture toughness values measured with the single-edge notch beam technique were in the range 7–10 MPa $m^{1/2}$ depending on the orientation of the cloth. The flexural strength at 1000°C in air was ~170 MPa.

Other researchers obtained a room-temperature flexural strength of 950 MPa for a unidirectional composite containing 60 vol% of large-diameter (140-μm) SiC monofilaments. Process times were very long (~1 month) and brittle failure was observed.

The fiber-reinforced SiC/SiC composite made by SEP is usually referred to as Cerasep. It contains the Nicalon polycarbosilane-derived SiC fiber. Room-temperature flexural strength is reported to be 450–800 MPa for about 40 vol% fiber. Most of the strength is retained at 1100°C, and about half is retained at 1400°C for short times in vacuum. The elastic modulus is 230 MPa.

Whereas the strength of SiC/SiC composites is comparable to that of monolithic SiC made by sintering or hot-pressing, the toughness of the CVI fiber-reinforced materials is much superior. For 2-D Cerasep the fracture toughness is reported to be in excess of 25 MPa $m^{1/2}$ for room temperatures up to 1000°C. The thermal shock resistance of the fiber-reinforced material is also much superior to that of monolithic SiC.

6.2.2.2 SiC/Si_3N_4. Si_3N_4 composites with SiC fibers and mats have been successfully manufactured. Rupture tensile strengths above 600 MPa have been reported for materials with 30 vol% of SCS-6 fibers.[57] However, aging of these composites at temperatures ranging between 1260 and 1370°C for 100 h leads to a twofold decrease in their rupture strength, and so the aged matrix and composite exhibit the same residual strength.[57] In fatigue at high temperatures (1000°C) the behavior of these materials resembles that of glass IMCs: the fatigue limit of the composite is nearly equal to the elastic proportional limit, that is, 200 MPa for a rupture strength of 400 MPa.[58]

SiC monofilament-reinforced Si_3N_4 matrices have been explored as higher-temperature CMCs by Lewis, Chamberlain, Daniel, et al.[59] A novel preform preparation method using matrix tape-casting was used in conjunction with reaction-bonded Si_3N_4 matrix chemistry and processing. Composite response and moderate fracture energy may be achieved utilizing a partially sacrificial thick carbon-rich interface, and despite the oxidation problem creep and stress rupture properties exhibit useful increments over those of turbine superalloys.

NASA Lewis, under the HITEMP program,[60] has examined an SiC/Si_3N_4 composite system. The composite consists of nearly 30 vol% aligned SiC fibers in a relatively porous Si_3N_4 matrix. Large-diameter (142-μm) (CVD) SiC fibers containing carbon-rich surface

TABLE 6.13. Flexural Properties of Dual SiC Fiber and SiC Monofilament-Reinforced Composites[56]

	SiC Monofilament Plus		Calculated Maximum Flexural Stress at Failure (MPa)[a]				Flexural Elastic Modulus (GPa)			
Composite Number	Matrix	Yarn	As-pressed	Ceramed	Aspressed, Oxidized[b]	Ceramed, Oxidized[b]	As-pressed	Ceramed	Aspressed, Oxidized[b]	Ceramed, Oxidized[b]
0°/90° Composites										
377-88	LAS (a)		1250 S	574 S	226 S	168 T,S	145	107	102	117
229-88	LAS	Nicalon	766 S	385 S	> 620 S	206 T	151	120	134	118
13-88	LAS	Nicalon	538 S	484 S			166	152		
228-88	LAS	Nicalon	1070 S	446 T,S	470 T,S	578 S	181	151	144	153
230-88	LAS	Nicalon	898 T,C,S	639 S			169	164		
293-88	LAS	Nicalon	1110 C	309 T		388		162		
317-88	LAS	Nicalon		431 S,T				154		
360-88	LAS	Nicalon		453 C		300 T,S		165		165
361-88	LAS	Nicalon		312 S				135		
379-88	LAS	Nicalon		898 S,C		397 S		174		166
438-88[c]	LAS	Nicalon		714 S		420 T		169		149
166-88	LAS (b)		320	Delaminated	> 238 S		162	Delaminated	137	
67-88	LAS	Nicalon	756 S	720 S,C	393 T,S	366 T	203	139	148	151
99-88	MAS		1140 T,S	933 T,S	496 T,S	611 S	224	173	161	168
100-88	MAS	Nicalon	1060 S,T	498 S	462 S	253 S	210	181	151	139
191-88	CAS	Nicalon	657 S	493 S	> 387 S	> 307 T	181	128	151	134
10-88	Al-silicate		1280 S,C				149			
0° Composites										
423-89	LAS (a)	Nicalon	1139 T,S				212			
536-89	LAS	Nicalon	789 T,S		601 S		207		208	
537-89	LAS	Nicalon	994 T,S,C		500 S		196		137	

[a] Failure mode: S, shear; C, compression; T, tensile.
[b] Standard exposure—flowing oxygen at 800°C for 64 h.
[c] Ceramed under pressure.

coating were used as reinforcement, and because SCS-6 fibers cannot be woven or bent to a radius of less than 12 mm, the fabrication potential of the first-generation RBSN composites using these fibers was limited to 2-D laminated structures (Table 6.14).

Unidirectional and 2-D SCS-6/RBSN laminates exhibited high specific stiffness and strength, high toughness, notch sensitivity, and thermal shock resistance up to 1400°C.[61,62] The composite also showed excellent strength and graceful failure at temperatures up to 1550°C after 15 min exposure in air.

6.2.2.3 Al_2O_3-glass matrices. Alumina fibers have become attractive because of their increased oxidative stability as compared to that of SiC fibers. However, the low strength of these fibers remains a drawback and may limit the performance of the composites. Alumina fibers possess a CTE of approximately 7.0 ppm/°C compared to SiC fibers which have a coefficient of about 3.0 ppm/°C. Bacon[63] calculated the stresses in various matrices using Lamé equations and found the level of stress in borosilicate glass 7740 and silica glass to be acceptable. The strength observed in the FP alumina fiber-7723 glass composites was 277 MPa which is much less than that calculated by the rule of mixtures (ROM) for a 30% fiber composite (544 MPa). Composites produced with alumina fibers showed a considerable increase in strength over that of the unreinforced matrix. Results produced during three-point bending of a silica glass–alumina fiber composite resulted in a strength of 187 MPa for a 37% fiber fraction, which is approximately four times the strength of the matrix alone.

Further tests by Bacon showed the ability of the alumina fiber to retain strength at high temperature. Very small losses in strength were recorded over the temperature range 22–1000°C.

In a recent study Michalske and Hellmann[64] fabricated alumina continuous fibers in four different silicate matrices. Observed strengths ranged from 156 MPa for barium sealing glass (9013) to 305 MPa for borosilicate glass (7740). These values were lower than expected. Composite strengths were determined to be limited by the fiber strength and the residual stresses produced as a result of the thermal expansion mismatch. The ultimate strength of a composite may be determined by a simple ROM for composite strength. Ultimate strength occurs where the fiber load at composite failure is equal to the original fiber strength. However, fiber degradation occurring in processing-induced flaws in the composites can decrease the actual strengths. High residual stresses calculated in the barium glasses (130 MPa) played an important role in the overall strength of these compos-

TABLE 6.14. Room-Temperature Tensile Property Data for SiC Fiber- and SiC Whisker-Reinforced RBSN Composites

Property	Unreinforced RBSN	RBSN + 10% SiC_w	30% SiC/RBSN	30% SiC_f + RBSN + 10% SiC_w
Elastic modulus (GPa)	110 ± 14	120	193 ± 7	151 ± 5
Tensile strength (MPa)			227 ± 41	206± 30
Matrix Ultimate	84 ± 26	93	682 ± 150	511 ± 28
Strain (%)	0.08	0.08	0.12	0.14
Matrix Ultimate	0.08	0.08	0.45	0.40

ites. Matrix stress in the 7740 composites was much less (10 MPa); however, a reduction in strength was still observed.

Chevron notch tests to determine the toughness of alumina fiber-glass matrix composites were conducted by Michalske and Hellmann.[64] Results showed that although no fiber pullout occurred in these systems because of the strong chemical bonding, toughening increases were still experienced as a result of crack shielding. The mismatch in elastic modulus between the fibers and the matrix shielded the fiber from matrix crack extension. This action increased toughness without fiber pullout. Toughness values for the alumina fiber-7740 glass composites were 3.7 MPa $m^{1/2}$.

6.2.2.4 Al_2O_3/MgO. Coblenz[65] examined the behavior of an alumina fiber-reinforced MgO matrix. Composites were produced by slurry infiltration, and mechanical properties were evaluated. Low strengths, approximately 100 MPa, with no evidence of toughening were observed. These poor results were caused by the strongly bonded interface and fiber degradation produced during processing. However, FP fibers impregnated in a $SiO_2 + B_2O_3$ matrix produced higher strengths, approximately 200 MPa. These samples exhibited delamination during testing with fracture toughness values of 4.4 MN/$m^{1.5}$.

The potential for alumina fibers to reinforce ceramic matrices is highly dependent on the incidence of bonding at the interface. Overall, the oxidative stability and corrosion resistance of these fibers is good. However, the formation of strong bonds in matrices such as MgO and SiO_2 greatly reduces the strength and toughness. Furthermore, heating to temperatures greater than 1200°C degrades the fiber by enhancing grain growth and crystallization. This action limits the high-temperature performance of composites made with these fibers.

Coatings applied on the fibers before processing may provide the solution to strong bonding and lead to an increase in strength and toughness. Rice et al[66] demonstrated the use of barrier coatings to inhibit bonding and promote toughness. They noted improved fracture behavior of coated alumina (fibers in matrices such as silica, cordierite, mullite, and zirconia). A fourfold increase in strength over that of uncoated fiber composites was observed, and a change in fracture behavior from brittle to tough took place because of the weakly bonded interface provided by the coating. Considerable increases in fracture toughness due to the coating have been reported.

6.2.2.5 SiC/ZrB_2. ZrB_2 with an addition of SiC was demonstrated in earlier studies to be a material with superior properties for leading edges of hypersonic atmospheric flight and reentry vehicles. Continuous-fiber-reinforced composites have been prepared with ZrB_2/SiC matrix materials and reported by Stuffle, Lougher, and Chanat[67] in a recent study to improve reliability.

Pure ZrB_2 and ZrB_2 with additions of SiC suffer from inherent brittleness and thus demonstrate flaw sensitivity, thermal shock sensitivity, and catastrophic failure. The addition of carbon reduces brittleness, although carbon is somewhat deleterious to the ablation response of the material[68] and does not change the catastrophic nature of failure. Discontinuous reinforcement of the ZrB_2/SiC system with carbon fiber and SiC platelets has also been investigated in attempts to improve the failure characteristics but has not succeeded. Continuous-fiber reinforcement, on the other hand, can drastically improve failure behavior. Mechanisms involving crack deflection, crack bridging, fiber pullout, load transfer, and stress delocalization are operative. Continuous-fiber-reinforced brittle matrix composites can demonstrate flaw sensitivity, increasing strain to failure beyond matrix cracking, and large strain to failure. Failure does not occur in a catastrophic fashion and hence the reliability of continuous-fiber-reinforced composites is greatly improved over that of the unreinforced or discrete reinforced material.

Stuffle, Lougher, and Chanat[67] reinforced the ZrB_2/SiC material with Textron SCS-6 fiber. The SCS-6 fiber, a 140-μm-diameter monofilament, was prepared by CVD of SiC onto a carbon core. The fiber had an amorphous carbon coating, approximately 2–3 μm thick, which served as a debonding interface.

The SCS-6/20 vol% RBSiC/ZrB_2 composite system demonstrated good composite-type mechanical properties even though the reaction-bonded matrix material and SCS-6 fiber have considerably differing CTEs. This was done after a series of arc jet ablation tests to evaluate the material's potential as a leading edge reentry component for future spacecraft.

The measured weight loss of the composite was approximately four times greater than that of the traditional material in the arc jet ablation testing. This was probably due to oxidation of the SCS-6 fiber in the surface region. It is expected that the mass loss rate will level off to a comparable rate after the oxide layer is fully formed.

As a result, the investigators concluded that the SCS-6/20 vol% RBSiC/ZrB_2 system appears very promising as leading edge material. The material demonstrates dramatically improved properties from the standpoint of reliability over those of the traditional unreinforced material and performs comparably under simulated reentry conditions.

6.3 WHISKERS

6.3.1 Introduction

In recent years, ceramic materials have begun to emerge as serious candidates for use in load-bearing structural applications in severe environments. Monolithic ceramics such as Si_3N_4, SiC, Al_2O_3, and ZrO_2 possess excellent room- and elevated-temperature strength and chemical resistance to oxidizing environments. However, these monolithic ceramics typically exhibit low values of fracture toughness and are susceptible to catastrophic brittle fracture at both low and high temperatures. It is the brittle nature of these materials that has been the major impediment to their widespread use as structural components.

The introduction of a fibrous reinforcement phase into monolithic ceramics has been observed to significantly increase the fracture toughness of the material. These ceramic composites are much tougher and in many cases stronger than their corresponding monolithic counterparts. Much success has been achieved with ceramic composites in glass, glass-ceramic, and polycrystalline ceramics such as Al_2O_3, ZrO_2, Si_3N_4, and SiC systems. Not only are low-temperature mechanical properties improved, but in many cases elevated temperature properties are also enhanced.

The major key to the success of ceramic composites rests with the reinforcement material. SiC whiskers possess much greater thermal stability than the currently available SiC fibers. In addition, their mechanical properties significantly surpass those of SiC fibers. However, at present SiC whiskers cannot be synthesized in continuous lengths.

6.3.2 Whisker Composites

Whisker composites have a more limited relationship to traditional ceramics than do particulate composites. However, some relationship does exist. Mullite crystallites may often take on a tabular or needle (whiskerlike) character, particularly in the presence of the glassy phases frequently found in some traditional ceramics, for example, some porcelain enamels. Crystalline phases formed in some crystallized glasses may also have a whiskerlike character. However, purposeful design and processing of whisker composites, especially artificial whisker composites (where the whiskers are added in the processing as opposed to being formed in situ) are relatively recent. While some work on whisker composites started about 20–30 years ago, most activity has occurred in the last 10–15 years.

The overall understanding of existing whisker composites is reasonable to good, appearing to be a more favorable cause of crack deflection by rod-shaped particles. However, better understanding is needed of (1) whisker strength, that is, the effects of changes in both the average strength and the strength distribution of whiskers on resultant composite strength, (2) the effect of the Young's modulus of the whiskers versus that of the matrix on the performance of resultant composites, for example, to determine the relative benefits, if any, of using SiC whiskers as opposed to Si_3N_4 whiskers, (3) whisker size, and (4) whisker-matrix interfaces. The whisker-matrix bonding, that is, the interfacial strengths, can in principle be varied substantially by coatings or other surface treatments of the whiskers and is clearly known to be a major factor in continuous-fiber composites.

Whisker and Particulate Reinforcements. Compared with continuous-fiber-reinforced ceramics, particulate- and whisker-reinforced composites offer the potential for low-cost processing because they can be produced by the standard fabrication techniques used for monolithic materials. Their relative ease of fabrication and the availability of a range of second-phase materials have encouraged a concentration of research attention on these materials even though it is recognized that their performance is lower than what can be achieved by continuous-fiber reinforcement.

Although similar fabrication techniques can be adopted, particulate- and whisker-reinforced materials are more difficult to process than monolithic materials for a number of reasons. The following principal differences and problems are associated with the introduction of a second phase material.

- Achievement of a homogeneous dispersion of matrix powder and particulate and whisker reinforcement
- Achievement of a random orientation of whiskers
- Occurrence of damage to whiskers during powder processing and dispersion
- Effects of a second phase on flaw size and sintering.

Most of the fabrication processes employed for monolithic materials, such as uniaxial pressing, isostatic pressing, injection molding, extrusion, and slip-casting, can be used for particulate- and whisker-reinforced composites; but those involving fluid-processing techniques are generally preferred because they minimize a number of the problems noted earlier. Wet processing is adopted to ensure the best homogeneity in the matrix ceramic powder-particulate or whisker mixture. The choice of wet-processing conditions is to some extent determined by the particular process to be adopted to form the component, and the mixing of the materials may be carried out in either a dispersed or a flocced state.[69–74]

For simple slip-casting of CMCs, which is a process suited to the production of complex-shaped components, high-solids, low-viscosity slurries can be obtained by choosing a pH where both the matrix and the whiskers have a high zeta potential of the same charge. For the more common methods of fabrication, such as hot-pressing, flocced mixing is generally adopted. Flocculation occurs when the whiskers and the matrix powder are either differently charged or have zeta potentials near zero. To achieve the desired level of homogeneity, high-intensity shear mixing of the flocs is necessary. Researchers are investigating coacervation as a process for obtaining intimate mixtures of ceramic powders and particulate or whisker reinforcements. Coprecipitation of the components from the suspension in the liquid phase is reported to have produced significant improvements in the mechanical properties of some materials.

In whisker-reinforced composites, local shear forces, which occur during mixing and forming operations, can induce whisker alignment, causing uneven shrinkage during sintering and possible component fracture and also producing anisotropic properties in the composite. Some manufacturers claim to have developed a proprietary process that produces homogeneous and randomly oriented whisker-reinforced composites.

According to theoretical predictions,[69] the degree of toughening increase is dependent on the morphology of the dispersoids. Rod-shaped particles are predicted to be more effective in toughening than disk-shaped particles, which are more effective than spheres. Based on these considerations, a lot of work has been done in the field of whisker reinforcement.[70,71] Whisker additions up to 50 vol% to an alumina matrix increase the toughness values up to nearly 4 times to ~9 MN $m^{-3/2}$ starting from 2.5 MN $m^{3/2}$ for the monolithic alumina. Strength is increased simultaneously up to a factor of 2.5 to a maximum strength of 900 MPa.[72,73] As toughening mechanisms, crack bridging is most effective[74] accompanied by crack deflecting and branching effects.

6.3.2.1 Silicon carbide–silicon nitride (SiC/Si_3N_4). In considering high-temperature applications, the LTV Missiles and Electronics Group[75] examined three composite materials. These included SiC_w/Si_3N_4, $SiC_{platelet}/MoSi_2$, and SiC_w/Y_2O_3. Y_2O_3 has many desirable properties such as a high melting point, good oxidation resistance, chemical stability with many reinforcements, low volatility, and an elastic limit greater than 1300°C. Disadvantages include poor thermal shock resistance during rapid heating (partially due to low strength) and low fracture toughness. Improvements in fracture toughness and strength have been demonstrated with SiC whisker reinforcement, but poor creep resistance has deterred further study on this composite.

SiC_w/Si_3N_4, however, does show promise. Flexure strengths of 700 MPa at room temperature and 504 MPa at 1371°C have been recorded. This composite also demonstrated excellent thermal and pressure cycling durability. High-temperature mechanical fasteners such as bolts and stand-off posts are immediate applications being investigated for this composite.

Yust and DeVore[76] measured the friction and wear of a $SiC/Si_3N_4/20_w$ matrix composite at room temperature in both lubricated and unlubricated sliding. SiC and

Si_3N_4 spheres were tested against the composite in both unidirectional and reciprocating sliding. The range of experimental parameters included normal forces from 1 to 45 N, velocities of 0.1 and 0.3 m/s, and test durations of 5 min to 3 h. The lubricant used in the room-temperature tests was a commercial synthetic, fully formulated engine oil (5W30). In unlubricated sliding, Si_3N_4 produced more severe wear of the composite than did SiC.

Sliding against a SiC sphere causes measurable wear. However, the wear rate is one to two orders of magnitude lower. Lubricated reciprocating sliding and unidirectional lubricated sliding result in no measurable wear on the composite disk and only very mild wear on spheres of either Si_3N_4 or SiC. For reciprocating sliding the friction coefficient is less than 0.05, whereas the unidirectional sliding is less than 0.02.

The flexural strength and fracture toughness of SiC/Si_3N_4/30w material were determined as a function of temperature from 25 to 1400°C in an air environment by Choi and Salem.[77] They found that both strength and toughness of the composite material were almost the same as those of the monolithic counterpart. The room-temperature strength was retained up to 1100°C; however, appreciable strength degradation started at 1200°C and reached a maximum at 1400°C as a result of stable crack growth. In contrast, the fracture toughness of the two materials was independent of temperature, with an average value of 5.66 MPa $m^{1/2}$. It was also observed that the composite material exhibited no rising *R*-curve behavior at room temperature, as was the case for the monolithic material. These results indicate that SiC whisker addition to the Si_3N_4 matrix did not have any favorable effects on strength, toughness, or *R*-curve behavior.

Ohji and Yamauchi[78] investigated the tensile creep and creep rupture behavior of Si_3N_4 at 1200–1350°C using hot-pressed materials with and without SiC_w. Stable steady-state creep was observed under low applied stress at 1200°C. Accelerated creep regimes, which were absent below 1300°C, were identified above that temperature. The appearance of accelerated creep at the higher temperatures was attributable to formation of microcracks throughout a specimen. The whisker-reinforced material exhibited better creep resistance than the monolith at 1200°C; however, the superiority disappeared above 1300°C. Considerably high values, 3–5, were obtained for the creep exponent in the overall temperature range. The exponent tended to decrease with decreasing applied stress at 1200°C. The primary mechanism was considered to be cavitation-enhanced creep.

Zheng et al.[79] studied reinforcement of Si_3N_4 with SiC_w-platelet. Specimens containing 0–30% whiskers were produced.

Initial densities were increased by platelet additions but reduced by whisker additions. Sintered densities decreased appreciably with increasing whisker content. Density with 30 vol% whiskers was only about 50% that of the matrix material alone. Density differences caused by different sintering temperatures were small relative to those resulting from differences in whisker content. Density also decreased with increasing platelet content, but the decrease was smaller. With the alumina addition in combination with the magnesium oxide–neodymium oxide addition, density at a 20 vol% platelet level was still 95% of that of the matrix without any whisker or platelet content. The differences were attributed to more efficient particle packing and less inhibition of shrinkage during sintering with the platelets. In addition, whiskers reacted more with the liquid sintering phase, causing more mass loss. This was speculated to be due to the lower purity of the whiskers.

As might be expected from the very low densities of the whisker-containing composites, fracture toughness was not significantly improved over the matrix material, which has a toughness of about 4.5 MPa $m^{1/2}$. However, platelet-containing composites had increased fracture toughness. With 20 vol% platelets, toughness was about 6.5 MPa $m^{1/2}$.

Ohji et al.[80] investigated the high-temperature toughness and tensile strength of SiC_w/Si_3N_4. Hot-pressed Si_3N_4 specimens containing 20 wt% SiC_w were produced. Mechanical testing was carried out from room temperature to 1300°C. Fracture toughness was determined using the chevron notched beam technique.

The tensile test results as a function of temperature were not reported in detail, but it was mentioned that tensile strength showed substantial degradation above 1000°C. Tensile strength at 1300°C was 290 MPa (using a displacement rate of 0.1 mm/min), which was less than one-half that at room temperature.

6.3.2.2 Silicon nitride–silicon nitride (Si_3N_4/Si_3N_4). Chu and Singh[81] prepared $Si_3N_4/Si_3N_4/35_w$ and determined mechanical properties and microstructural details. UBE SN-WB whiskers and UBE-SN-E10 powder were used.

Young's modulus was determined by the sonic method, and hardness by the Vicker's indentation method. Flexural strength (four-point) and fracture toughness (indentation method) were also determined. Young's modulus and hardness were independent of whisker content. Fracture toughness increased from 6.5 MPa $m^{1/2}$ to 8.8 MPa $m^{1/2}$ with 5 vol% whiskers, decreased with additional whisker content up to 20%, and then increased again. This was hypothesized to be due to competing factors of increased toughening and increasing whisker content (attributable to whisker pullout and crack deflection mechanisms) but decreased toughening as the result of a decrease in matrix grain size with increasing whisker content. The decrease in grain size with increasing whisker content could be seen in the microstructural analysis.

Flexural strength variation with whisker content was similar to that for fracture toughness. Strength increased from 673 MPa to 802 MPa with 5 vol% whiskers but decreased and then increased with increasing whisker content.

The microstructure and mechanical properties of SiC_w/Si_3N_4 matrix composites with different sintering additives was examined by several researchers. Results of one group showed that the addition of SiC_w could significantly improve the flexural strength and reliability of Si_3N_4. The microstructural development and the mechanical properties of the SiC_w/Si_3N_4 composites were influenced by sintering additives and whisker loading. The stability of the SiC_w was also strongly dependent on the amount of sintering additive.

Several automotive companies in Japan are currently investigating toughening of Si_3N_4 with the addition of SiC_w. Starting with amorphous Si-C-N powders CVD from a $[Si(CH)_3]_2$ NH-NH_3-N_2, followed by hot-pressing (N_2 atmosphere for 3 h at 1700–1800°C), Japanese researchers obtained room-temperature flexural strengths of 1000 MPa. The toughness was found to peak at about 15 vol% SiC with a K_{IC} of 7–8 MPa $m^{1/2}$ as shown in Figure 6.8. The SiC_p were found to be dispersed not only in the grain boundaries but also within the Si_3N_4 grains. Below 10 vol% the SiC dispersions accelerated the growth of elongated Si_3N_4 grains, which is believed to be responsible for the improved toughness. There was a degradation in strength as a function of temperature.

Others used another innovative approach to reinforcing Si_3N_4 with SiC_w whereby

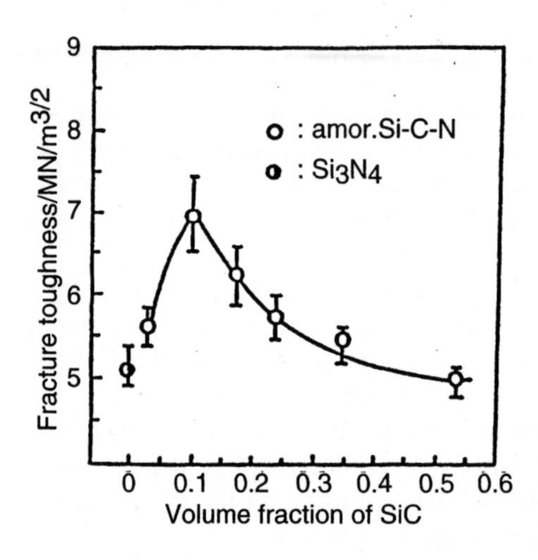

Figure 6.8. Effects of SiC dispersions on the fracture toughness of composites prepared from fine, amorphous Si-C-N powder.[115]

the whiskers were formed in situ. Carbon black, α-Si_3N_4, silica, $CoCl_2$ (catalyzer), and NaCl (space-forming agent) were mixed and heated at 1600°C for 1 h, sintering aids were added, and the mixture was hot-pressed.

Compared to that of physically mixed powders, the bulk density was improved but the bend strength was lower, most probably because of residual carbon. The advantages of this process were its economy and the reduced health hazards associated with the handling of very fine SiC_w.

At Osaka University, SiC_w-reinforced Si_3N_4 was hot isostatically pressed without additives. Material with 20 vol% SiC_w exhibited a four-point flexural strength of about 500 MPa, whereas the same material with 30 vol% SiC_w showed a higher flexural strength at room temperature but a lower flexural strength at 1400°C.

Other Japanese researchers, working at GIRI Osaka, fabricated SiC_w/Si_3N_4 composites with a 30 wt% SiC_w content by heating at 1700–2000°C for 1 h under 1 MPa of N_2. Fully dense composites were obtained at 2000°C. The optimum contents of the sintering aids, Y_2O_3 and La_2O_3, were 20 and 30 mol% for whisker contents of 10 and 20 wt%, respectively. Room-temperature bending strengths were 596 and 560 MPa for 10 and 20 wt% additions of whisker, respectively. Moreover, these composites had strengths that were more than 80% of the room-temperature value at 1300°C.

The fracture toughness for $SiC/Si_3N_4/0$–30_w, which was measured by the IM method, varied from 6.3 MPa $m^{1/2}$ for the Si_3N_4 matrix to 7.2 MPa $m^{1/2}$ for the composite containing 15 wt% whisker. On the other hand, K_{IC} obtained from the CN method varied from 5.1 MPa $m^{1/2}$ for the Si_3N_4 to 6.3 MPa $m^{1/2}$ for the composite having 30 wt% whisker.

6.3.2.3 Silicon carbide–aluminum oxide (SiC_w/Al_2O_3). Majidi and Chou[82] conducted experimental studies on the fracture toughness and toughening mechanisms of SiC_w/Al_2O_3 composites. Three composites with V_f values of 17, 23, and 29 were examined. Figure 6.9 presents the fracture toughness measured as a function of temperature. There appears to be little change in the toughness up to a temperature of about 1000°C. The composite with 23% whiskers was a lower-grade material produced at an early stage

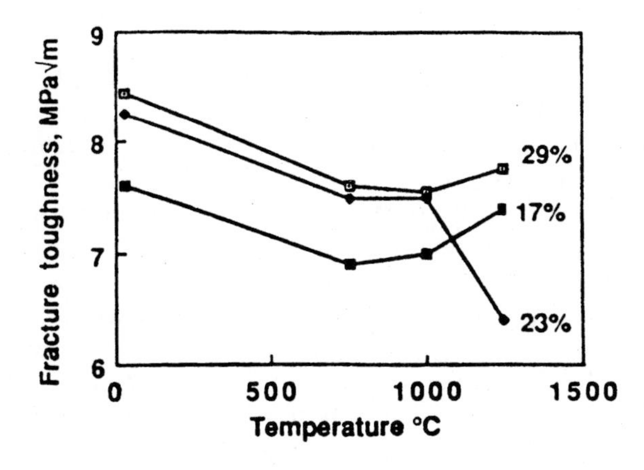

Figure 6.9. Fracture toughness as a function of temperature for SiC_w-reinforced Al_2O_3 composites with various whisker volume fractions (V_f).[82]

of the development of SiC_w/Al_2O_3 composites. The fracture toughnesses measured for this material are lower than those of the other two composites.

The fracture surfaces at room temperature showed the evidence of crack deflection and whisker pullout. Crack deflection was clearly the dominant mechanism, whereas whisker pullout was limited to a few whiskers and involved pullout lengths of only about twice the whisker diameter. A significant number of whiskers broke in or near the plane of crack propagation and therefore did not contribute to the pullout energy.

It was observed that with increasing temperature crack deflection gradually decreased, whereas whisker pullout became more significant. Up to 1000°C, crack deflection was still the major toughening mechanism. At 1250°C, however, whisker pullout appeared to be much more extensive than crack deflection. At this temperature, crack deflection was significantly reduced.

Kragness, Amateau, and Messing[83] examined laminated SiC_w/Al_2O_3 composites fabricated by tape-casting and hot-pressing. They produced whisker loadings of 10 and 20 V_f with a predominance of fiber orientation in the direction of casting. Elastic, thermal, and mechanical properties of the composites were measured in the longitudinal and transverse directions. The elastic moduli for 10 and 20 vol% whisker Al_2O_3 composites were approximately the same as for unreinforced alumina. The CTE was reduced by addition of the whiskers. The fracture toughness of the composites increased with whisker content from 3.8 MPa $m^{1/2}$ for pure alumina to 8.0 MPa $m^{1/2}$ for 20 vol% whisker in the longitudinal direction, which is more than twice that of the monolithic laminate.

Becher, Angelini, Warwick, et al.[84] exposed $Al_2O_3/SiC/20_w$ to applied stresses in four-point flexure at temperatures of 1000°C, 1100°C, and 1200°C in air for periods of up to 14 weeks. At 1000 and 1100°C, an "apparent" fatigue limit was established at stresses of ~75% of the fast fracture strength. However, after long-term (>6-week) tests at 1100°C, some evidence of crack generation as a result of creep cavitation was detected. At 1200°C applied stresses as low as 38% of the 1200°C fracture strength were sufficient to promote creep deformation with accompanying cavitation and crack generation and growth resulting in failures in times of <250 h.

SiC whiskers and alumina powder could be mixed in water medium when the whiskers were flocculated and the powder deflocculated, and the composite slip could be cast to high green density. However, sintering of the composite to fully dense material was found to be difficult. The reasons were the formation of whisker networks and the development of differential sintering stresses in the matrix around the whiskers.

As a result, hot-pressing was believed to be necessary to densify the composites. The MgO additive, according to Yang and Stevens,[85] can effectively promote densification of the composite and inhibit abnormal grain growth. High pressure is required to achieve substantial densification prior to grain growth.

The improved fracture resistance of whisker-reinforced CMCs involves more than one energy-absorbing mechanism. The mechanisms involve microcracking, load transfer, bridging, and crack deflection. Giannakopoulos and Breder[86] developed a model for the individual toughening mechanisms in a rigorous way: microcracking is induced through a nonlinear mechanical constitutive equation, bridging is connected with pullout, locking, and residual stresses, and crack deflection appears as in a statistical way. Therefore, all the microproperties of the composite become involved in the final macroscopic response. Finally, because of the nature of the crack tip stress fields, a synergistic toughening result is reached through an energy release rate balance equation.

Lin and associates[87] showed that the flexure strength and fracture toughness of the Al_2O_3/SiC_w composites were increased by increasing the SiC_w content. The flexure strength of Al_2O_3 + 20 vol% ZrO_2 (2 mol% Y_2O_3) + SiC_w increased with SiC_w content only to 20 vol% SiC_w, after which the strength decreased as a result of excessive microcracking, but the fracture toughness increased monotonically with the SiC_w content.

Second, the flexural strength and fracture toughness of the Al_2O_3 + 20 vol% ZrO_2 (2 mol% Y_2O_3) + SiC_w composite were always higher than those of the Al_2O_3 + SiC_w composite. The toughening effect of both SiC_w and the ZrO_2 component had an obvious additive effect, at least for the Al_2O_3 matrix. The addition of ZrO_2 decreased the strengthening effect of SiC_w to some effect.

Finally, the main mechanisms of strengthening and toughening by SiC whiskers were whisker pullout and bridging (and in some cases the whiskers were ruptured) and crack deflection. The grain refinement and the dynamic T_m phase transformation of ZrO_2 (2 mol% Y_2O_3) were very important in the toughening of the composites.

Baek and Kim[88] carried out a series of investigations on the effect of SiC_w length on Al_2O_3 matrix composites. Both flexural strength (four-point bending) and fracture toughness (single-edge notched-beam test and chevron notch test) increased with increasing whisker length. They compared strength with that predicted using a mathematical model based on the rule-of-mixtures modified to account for variations in fiber length and orientation. The model tended to overestimate composite strength at lower whisker lengths.

Early studies[89] on whisker-reinforced ceramics revealed that greater than twofold increases in fracture toughness were achieved in alumina[90–92] with the incorporation of strong small-diameter (<1 μm) SiC_w. Toughness values of 8 to 10+ MPa $m^{1/2}$ for large crack extensions were obtained with alumina reinforced with 20 vol% SiC_w. These are values compared to toughness values, for similar crack extensions, of 2.5+ to 4.5 MPa $m^{1/2}$ for aluminas, 4.5 to 6+ MPa $m^{1/2}$ for dense Si_3N_4s, and 6 to 13+ MPa $m^{1/2}$ for transformation-toughened zirconia ceramics at room temperature.

Becher et al.[89] examined the effects of a number of factors on properties of SiC_w oxide matrix composites.

Fracture toughness measurements were made using the applied moment double-cantilever beam method. Toughness results were reported in terms of a toughening increment, dK_{wr}, in the inherent matrix toughness. A variety of techniques were used to characterize the microstructure and composition of the composites and whiskers. Theoretical predictions of toughening effects from various factors were made using both the stress intensity in the presence of a bridging zone on the strain energy change associated with a bridging zone.

The influence of the SiC_w characteristics on toughening was also investigated by Becher et al.[90] Alumina composites were fabricated using SiC_w from various sources. Although the goal was to examine the effect of whisker diameter, this was complicated by possible differences in other factors such as surface chemistry, which could modify the whisker-matrix interface.

They found that composite toughening increased with increasing whisker diameter and that composites containing whiskers having low surface oxygen contents were tougher. The increase in toughness with increasing whisker diameter was predicted by the theoretical model of Becher et al.[91,92] The improvement in toughness with whiskers with low surface oxygen content is consistent with the formation of weaker interfaces during composite fabrication (indicating that chemical bonding at an interface is enhanced by the presence of a thin oxide layer on a whisker). This was consistent with a dramatic decrease in whisker pullout length during fracture with decreased whisker surface oxygen content.

In agreement with other predictions of the theoretical model, overall toughness of the composites increased with matrix grain size and with increasing matrix Young's modulus.

Shih et al.[93] studied the effect of whisker variations on alumina matrix composites. Composites containing 30 vol% whiskers were consolidated both by hot-pressing and by hot-pressing followed by HIP.

Mechanical properties are shown in Table 6.15. The characteristic strength listed is that at which the probability of failure was 0.63. After hot-pressing, the mechanical properties were best with type I whiskers, followed by type II and type III. Differences were not large with regard to strength, but Weibull modulus was appreciably higher with type I whiskers than with the other two types. There were also appreciable differences in fracture toughness. The use of HIP after hot-pressing raised strength but had a detrimental effect on toughness. Weibull modulus was adversely affected by HIP with type I whiskers.

Possible reasons for the differences in properties resulting from the use of different whiskers were considered to be whisker diameter, aspect ratio, surface chemistry, surface morphology, and stability during processing. The increase in flexure strength resulting

TABLE 6.15. Mechanical Properties of SiC_w-Reinforced Al_2O_3 as a Function of Sintering Variables[93]

Whisker	Process	Flexural Strength (MPa)	Weibull Modulus	Fracture Toughness (MPa $m^{1/2}$)
I	HP	651	15.2	9.4 ± 0.4
II	HP	600	6.7	8.0 ± 0.4
III	HP	590	6.4	6.6 ± 0.8
I	HP + HIP	681	9.9	8.1 ± 0.6
II	HP + HIP	718	16.0	6.5 ± 0.5

from HIP after hot-pressing was attributed to elimination of porosity in the composite, whereas the reduction in fracture toughness was speculated to be due to increased bonding between the whiskers and matrix caused by HIP.

Fracture toughness of very fine-grain alumina increases from less than 3 MPa $m^{1/2}$ to more than 9 MPa $m^{1/2}$ with the addition of SiC whiskers. Comparable increases in toughness were obtained in other whisker-reinforced ceramics. Tables 6.4 and 6.16 are illustrative comparisons of the fracture toughness of structural ceramics and the properties of SiC_w/Al_2O_3 composites.[94]

Liu, Fine, and Cheng[95] initiated a lubricated rolling wear study of SiC_w-reinforced Al_2O_3 composite and monolithic Al_2O_3 versus M2 tool steel using a cylinder-on-cylinder apparatus. They noted that the composites wore considerably less than Al_2O_3. The wear of the tool steel against the composites was also considerably less than against Al_2O_3. Microfracture occurred on a smaller scale in the composites than in the Al_2O_3, and this was attributed to the differences in microstructure and fracture toughness.

Kunz, Chia, McMurty, et al.[96] conducted a study to investigate the interface region of Al_2O_3 composites with SiC_w and SiC_{pl}. They fabricated almost dense (>98% theoretical density) $Al_2O_3/SiC/30_w$ and $Al_2O_3/SiC/30_{pl}$ composites by uniaxial hot-pressing at 1550°C and 12.5 MPa.

They also determined that the effect of surface treatments on the fracture toughness at low loadings is not obvious. However, addition of oxidized platelets or whiskers of SiC to the alumina matrix resulted in an increase in the single-edge notched-beam testing of fracture toughness at 30 vol% loading.

SEM fractographic study showed that the use of oxidized and carbon-coated SiC_{pl} favored intergranular fracture of the reinforcement, whereas noncoated platelets failed in a transgranular mode.

The crack propagation resistance of alumina composites is greatly enhanced by the addition of SiC_{pl} or SiC_w. The effects of these two reinforcements are similar (Tables 6.17–19).

Lin and Becher[97] investigated the creep resistance of SiC_w/Al_2O_3 between 1200 and 1300°C in air and found that Al_2O_3 composites reinforced with 10 and 20 vol% whiskers

TABLE 6.16. Properties of Al_2O_3/SiC_w Composites

Whisker Type	Whisker Volume (%)	Theoretical Density (%)[a]	Strength (psi)[b]	Fracture Toughness (MPa $m^{1/2}$)[c]
Arco Chem	15	99.7	77,000	7.0
Arco Chem	30	99.3	54,000	6.8
Tateho	15	99.2	63,500	5.7
Tateho	30	98.5	54,000	4.8
Tokai Carbon	15	99.2	65,500	5.1
Tokai Carbon	30	98.5	57,000	6.8
Alumina[d]	0	99.0	55,000	4.0

[a] Theoretial density: 15 vol% whiskers, 3.86 g/cm³; 30 vol% whiskers, 3.75 g/cm³.
[b] Four-point strength: military standard type B bend bars.
[c] Single-edge notch beam technique.
[d] Average values for A-16 SG alumina with no reinforcement.

TABLE 6.17. Mechanical Properties of Al_2O_3/SiC Platelet Composite, 20 Vol% Reinforcement[96]

Surface Treatment	Density (% TD)[a]	Flexural Strength (ksi)	Modulus of Elasticity (msi)	SENB Toughness (MPa m$^{1/2}$)
SiC Platelets				
As received	98.7	19.6	38.9	4.2
Oxidized	99.5	15.7	37.0	3.6
Acid-treated	100	18.9	35.3	4.1
Carbon-coated	99.3	19.5	34.2	4.3
No reinforcement	100	64.0	57.9	4.6

[a] TD, Total density.

exhibited creep rates that were one to two orders of magnitude lower than those of 30 and 50 V_f whisker-reinforced alumina composites. The degradation in creep resistance due to increasing the whisker contents above 20 vol% was attributed to the extensive creep cavitation and enhanced oxidation of SiC_w. At a whisker content of 10 vol%, a larger matrix grain size was obtained that appeared to decrease the creep rate at 1200°C. The creep resistance of alumina composites was degraded by the presence of glassy phases[98] that promoted creep deformation by grain boundary sliding and cavitation, thus enhancing the creep processes.

The most impressive results have been achieved for Al_2O_3 and mullite with SiC_w. It was reported that the bending strength and the fracture toughness were significantly improved by the additions, as shown in Tables 6.20 and 6.21.[99–101] These composites also showed excellent creep resistance. A thermal shock resistance of T_c = 900°C for an Al_2O_3/20 vol% SiC_w composite was obtained, which is much higher than that for monolithic Al_2O_3 (T_c < 400°C). This composite has found considerable practical use as a cutting tool.

Lei, Zhu, and Zhou[102] investigated the microstructure, mechanical properties, and toughening mechanisms of hot-pressed SiC_w/Al_2O_3 ceramic composites by means of x-ray diffraction and SEM. Experimental results showed that fracture toughness and Vickers hardness of composites increased with increasing SiC content. With 20 wt% whiskers a fracture toughness of 9 MPa m$^{1/2}$ was obtained, which is nearly twice that of the matrix.

TABLE 6.18. Mechanical Properties of Al_2O_3/SiC Reinforcement Composites, 30 Vol% Reinforcement[96]

Reinforcement	Density (% TD)[a]	Flexural Strength (ksi)	Modulus of Elasticity (msi)	SENB Toughness (MPa m$^{1/2}$)
SiC Platelets				
As received	98.4	23.4	38.3	4.3
Oxidized	100	40.3	58.4	6.0
SiC Whiskers				
As received	98.9	55.0	58.0	7.3
Oxidized	98.0	62.2	57.4	8.0
No Reinforcement	100	64.0	57.9	4.6

[a] TD, Total density.

TABLE 6.19. Mechanical Properties of Al_2O_3/SiC Reinforcement Composites, Effect of V_f Reinforcement[96]

	Volume Fraction (%)				
SiC Platelets[a]	0		10		30
Density (%TD)	100		100		100
MOR (ksi)	64.0		32.8		40.3
MOE (Mpsi)	57.9		58.6		59.1
SENB Toughness (MPa $m^{1/2}$)	4.6		4.5		6.0
SiC Whiskers[a]	0	15		20	30
Density (%TD)	100	100		99.1	98.0
MOR (ksi)	64.0	83.3		67.2	62.2
MOE (msi)	57.9	59.3		57.2	57.4
SENB toughness (MPa $m^{1/2}$)	4.6	5.8		6.7	8.0

[a] Oxidized reinforcements. TD, Total density.

The grain size of the matrix also has an influence on fracture toughness, and an increment of 1–2 MPa $m^{1/2}$ was obtained when the grain size increased from 0.05 to 0.5 mm. It was also observed that no glassy phase existed along the whisker-matrix boundaries but that at certain interfaces small traces of glassy phase containing Al and Si were formed. Whisker pullout and bridging, as well as crack deflection, are considered to be the main toughening mechanisms. The existence of a glassy phase decreased the interfacial bonding strength and thus facilitated the pullout of whiskers.

6.3.2.4 Silicon carbide–aluminum oxide (SiC_w/Al_2O_3-ZrO_2). Ricoult[103] found that during high-temperature oxidation in air, SiC whisker-reinforced Al_2O_3/ZrO_2 composites degraded by the formation of a whisker-depleted mullite zirconia scale. The reaction kinetics have been studied as a function of time and temperature for composites with whiskers preoxidized for different lengths of time. A model compatible with observations on Al_2O_3/ZrO_2/SiC and the results reported in the literature for Al_2O_3/SiC_w composites was proposed: the oxidation occurs at an internal reaction front. Oxygen diffuses along dislocations and grain boundaries through the mullite scale to react at this front with SiC, thereby forming amorphous silica and graphite. Silica penetrates grain boundaries and further reacts with alumina and zirconia to form mullite and zircon, and the second reaction product, graphite, is oxidized into carbon monoxide when the reaction front moves deeper into the composite.

TABLE 6.20. Properties of Al_2O_3/20 Vol% SiC, Mullite/20 Vol% SiC_w-Reinforced Composites[71]

Material	K_{IC} at 25°C (MPa $m^{1/2}$)	Bending strength (MPa) At 25°C	Bending strength (MPa) At 1200°C
CR-10 Al_2O_3 composite	9.0	805	520
Linde-A Al_2O_3 composite	8.6	600	450
Al_2O_3	4.6		
Mullite composite	4.6	438	
Mullite	2.2		

TABLE 6.21. Properties of Whisker-Reinforced Composites

Whisker Content (%)	Temperature (°C)	Density (% Total)	Flexure Strength (MPa)	K_{IC} (MPa m$^{1/2}$)	Source[a]
Alumina, SiC whisker					
0	25	99.7	649	4–5	1
5	25		445	3.6	2
20[b]	25		620–807	8.8	3
30	25		620–669	4–4.8	1
Silicon nitride, SiC whisker					
0	25	99.7	758	4.6	4
	1000		586	4.9	
20	25		758	4.8	4
	1000		620	6.2	4
30	25		441	6.0	4
	1000		827	7.6	4

[a] *Sources:* (1) Becher, P. F., and G. C. Wei, Toughening behavior in $SiC_{(w)}$ reinforced alumina, *J. Am. Ceram. Soc.*, 67(12):C267–C269 (1984). (2) Homeny, J., W. L. Vaughn, and M. K. Ferber, Processing and mechanical properties of $SiC_{(w)}$-alumina matrix composites, *Am. Ceram. Soc. Bull.* 67(2), 333–338 (1987). (3) Greenleaf Corporation, WG-300 Engineering Data, May 1989. (4) Bulgan, S. J., J. G. Baldoni, and M. L. Huckabee, Si_3N_4-SiC Composites, *Am. Ceram. Soc. Bull.* 66(2), 347–52 (1987).

[b] Used presently as cutting tool.

Yasuda et al.[104] studied the effect of diameter, length, and aspect ratio of SiC whiskers on the bend strength and fracture toughness of 20 vol% SiC_w-alumina composites. The whiskers were deagglomerated ultrasonically, mixed with the alumina powder in butanol, and hot-pressed (1700–1800°C, 33 MPa, 1 h). From the initial data, toughness appeared to increase linearly with both the length and diameter of the whiskers; however, newer data showed no relation between whisker length and fracture toughness. Bend strength, however, decreased. The degradation in strength was found to correspond well with whisker length, leading to the conclusion that the critical flaw size scaled with the length of the reinforcing whisker. Furthermore, the addition of whiskers seemed to inhibit grain growth in alumina as well as enhance the properties at higher temperatures.

Niihara et al.[105] investigated the effect of 2- and 0.3-μm SiC particle additions to alumina, and in general it was found that fracture toughness and strength improved with the dispersions. The samples were made by hot-pressing (1500–1800°C, N_2 atmosphere, 28 MPa) the powder mixtures. The best results were obtained for a 5 vol% dispersion of 0.3-μm β-SiC, where a fracture toughness of 1100 MPa and a toughness of 4.7 MPa m$^{1/2}$ were achieved. This remarkable improvement is believed to be due to dispersion of the SiC particles within the alumina grains and the concomitant compressive residual microstresses that result in the matrix.

6.3.2.5 Silicon carbide–aluminum oxide and silicon carbide–tetragonal zirconia polycrystals (SiC_w/Al_2O_3 and SiC_w/TZP). Akimune et al.[106] and Yang and Stevens[107] investigated and characterized the microstructure of alumina and TZP composites reinforced with SiC_w. They observed that the Al_2O_3/SiC system was stable under fabrication conditions, and that the amorphous phase at the whisker-matrix interface was derived mainly from the SiO_2-rich layer present on the surface of the as-received whiskers.

A reaction between SiC and ZrO_2 occurred at 1650°C. This reaction appeared to be related to the presence of a SiO_2 impurity present in the TZP matrix.

Second, the experimental elastic modulus of the composite was correlated with predictions from the rule of mixtures when the modulus of the whiskers was assumed to be equivalent to that of polycrystalline SiC.

Third, the strength of the alumina composite showed an increasing trend as whisker content increased. In contrast, the strength of the TZP composite diminished as whisker content increased. This was explained in terms of grain size and residual stress present in the composites.[108,109]

Finally, high-temperature annealing at 1000°C showed that the TZP composite had undergone a breakdown in structure with the occurrence of extensive cracking throughout the composite. The alumina composite retained its flexural strength up to 1200°C, above which it deteriorated rapidly.

6.3.2.6 Silicon carbide–zirconia (6 mol% yttria) (SiC_w/ZrO_2 (6 mol% Y_2O_3). Microstructure, mechanical properties, and toughening mechanisms of hot-pressed SiC_w/ZrO_2 (6 mol% Y_2O_3) ceramic composites were investigated by means of x-ray diffraction and SEM by Zhou, Zhu, and Lei.[110] Their results showed that the high density (98%) of the composites was obtained and that Vickers hardness and fracture toughness increased with increasing SiC_w content. With 10 wt% SiC_w the hardness and fracture toughness increased from 14.6 GPa and 3.7 MPa $m^{1/2}$ for the ZrO_2 (Y_2O_3) matrix to 17.2 GPa and 5.5 MPa $m^{1/2}$ for the composite, respectively. SEM observations on crack propagation revealed the following toughening mechanisms: (1) some whiskers were pulled out, bridging the two sides of the cracks which propagated in a straight line; (2) some cracks were deflected on meeting with the whiskers; (3) the whiskers fractured together with the matrix, and the cracks propagated in a straight line; (4) the cracks were branched, and microcracks formed in front of the main crack tip.

6.3.2.7 Silicon carbide–partially stabilized zirconia (SiC_w/PSZ). Yasuda[111,112] reported on his work as summarized in Figures 6.9 and 6.10. It should be noted that higher strength was observed at elevated temperatures in the 30-vol% SiC_w test material than at room temperature (Figure 6.11).

At sintering temperatures of 1400 and 1500°C, no increase in strength or fracture toughness could be observed because of insufficient densification.[113–115]

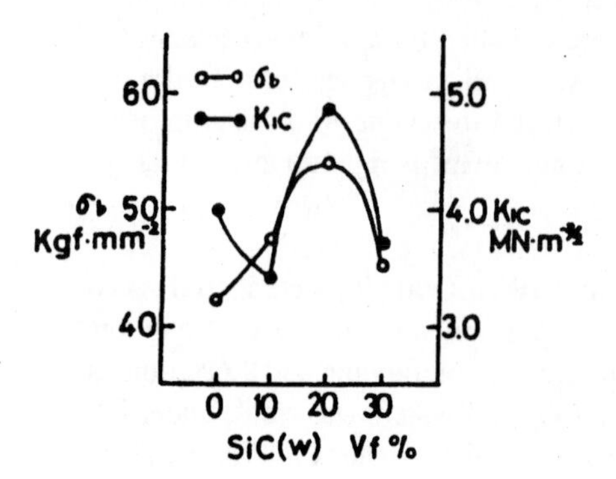

Figure 6.10. Flexural strength and fracture toughness at room temperature.

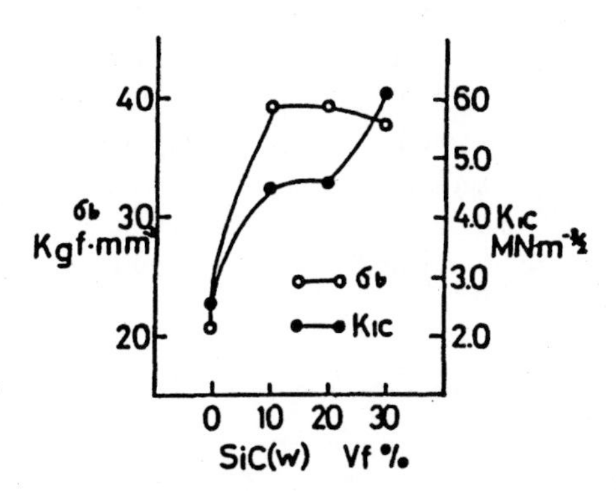

Figure 6.11. Flexural strength and fracture toughness at 1300°C.

6.3.2.8 Silicon carbide–carbon (SiC_w/C). SiC_w/C ceramics have been made by hot-pressing at 1950–2100°C in an argon atmosphere. The microstructure of SiC_w/C ceramics showed a unique distribution of SiC_w in the carbon ceramic matrix. Short SiC_w were distributed and spread in matrix grains, but long SiC_w were distributed at the boundaries. Short SiC_w play a strengthening role in the matrix grains as second-phase particles, and long SiC_w play a toughening role in the matrix because whiskers prevent the cracks from propagating. This reinforcing mechanism is identified as the two-directional reinforcing mechanism of the SiC_w in carbon ceramics.

These research results by Tian, Tong, Dong, et al.[116] show that in carbon ceramics added SiC_w increase the bend strength and fracture toughness. Optimal bend strength and fracture toughness are 400–450 MPa and 4–6 MPa $m^{1/2}$, respectively—about four to six times the strength of graphite. The experimental results indicate that the composite can withstand 600°C thermal shock and 800°C oxidation.

6.3.2.9 Silicon carbide–mullite (SiC_w/Mullite). Becher, Tiegs, and Ruh[117,118] found in their recent studies that whisker reinforcement could be combined with transformation and microcrack toughening associated with ZrO_2 particles in ceramic composites to achieve further improvements in fracture toughness. For example, the addition of 20 V_f SiC_w to mullite yielded a toughness of 4.5–5 MPa $m^{1/2}$. However, with the same whisker content, toughness values of ~7 MPa $m^{1/2}$ were obtained for a mullite-20 V_f monoclinic ZrO_2 composite where microcrack toughening can contribute, and 10–11 MPa $m^{1/2}$ in a mullite-20 V_f tetragonal ZrO_2 composite where transformation toughening can be triggered.[117,118]

Mullite + 35.4% ZrO_2 + 1.5% Y_2O_3 and mullite + 30.4% ZrO_2 + 3% Y_2O_3 test samples with and without 20% SiC_w and mullite + 32.4% ZrO_2 + 2.2% MgO test samples with and without 30% SiC_w were subjected to air exposure at 1000°C for 260 h by Ruh and Mazdiyasni[119] and Mah, Mendiratta, Katz, et al.[120] Weight gains were observed only for samples containing both Y_2O_3 and SiC_w. For room-temperature strength measurements after exposure, no significant change was noted for samples containing MgO with and without SiC_w, but significant strength degradation was seen in samples containing Y_2O_3 with and without SiC_w.

When mullite was used as matrix material, very promising toughness values were obtained by the introduction of SiC_w and ZrO_{2p} simultaneously. The highest values of fracture toughness were reported[117] as 10.5 MPa $m^{1/2}$ for a mullite matrix containing 20 V_f% SiC_w and 20 V_f% tetragonal ZrO_2 (see Table 6.23).

Substantial improvement in flexure strength has been seen in 0–40 V_f composites. By increasing the SiC_w V_f in the mullite composite from 0 to 40 V_f%, increases in flexure strengths from 320 MPa to 940 MPa in T-type samples have been observed. Fracture toughness was also calculated for the SiC_w-mullite composites. The laminated composites were significantly tougher than monolithic mullite, having a maximum toughness of 6.85 MPa $m^{1/2}$ for the L-type 40 V_f% SiC_w samples. This toughness is 2.5 times that of monolithic mullite. The results of other work on the thermomechanical properties of monolithic and 20 V_f% SiC_w/mullite composite for both longitudinal and transverse directions can be seen in Table 6.22.

6.3.2.10. Silicon carbide–lithium aluminosilicate (SiC_w/LAS I and II). Two types of SiC_w have been evaluated at UTRC as potential reinforcements for LAS matrices in order to provide increased toughness and strength. From the results of this study,[121] both whiskers, which are essentially stoichiometric SiC with a small amount of oxygen present, gave composites that exhibited linear stress-strain behavior and fractured in a brittle fashion. The whiskers strengthened the matrices but did not alter the brittle nature of fracture appreciably. Unreinforced LAS-I and LAS-II exhibited a room-temperature flexural strength of ~70–90 MPa. A β-SiC_w-reinforced LAS III matrix composite exhibited a flexural strength of 276 MPa, whereas an α-SiC_w-reinforced LAS I matrix composite exhibited a room-temperature flexural strength of 450 MPa.

Xue and Chen[122] reported on a new LAS/SiC composite that was rich in alumina (78 wt%) and lean in silica (21 wt%) and lithia (1 wt%). This formulation offers a new option of converting the glass-ceramic matrix to a mullite-alumina matrix on annealing above 1400°C, and hence better creep resistance and other high-temperature mechanical properties. Using a transient-phase processing method, they were able to hot-press a composite containing 30 V_f% SiC_w at ~1350°C to achieve full density. Flexural strength measurements up to 1400°C confirmed the improved high-temperature strength and creep resistance over that of conventional LAS. The fracture toughness was also higher than that

TABLE 6.22. Thermomechanical Properties of Monolithic and 20 Vol% SiC_w/Mullite Composite for Both Longitudinal and Transverse Directions[75]

	Mullite	Composite[a]
Young's modulus (GPa)	225	284 (L) 276 (T)
Poisson's ratio	0.25	0.24 (L) 0.24 (T)
Fracture strength (MPa)	290	550 (L) 500 (T)
Fracture toughness (MPa $m^{1/2}$)	1.8	4.3 (L) 3.7 (T)
Coefficient of thermal expansion ($\times 10^{-6}$/°C)	5.2	4.80 (L) 4.83 (T)
Thermal conductivity (W/(m · K))	4.9	7.8 (L) 5.4 (T)

Source: Center for Advanced Materials, Pennsylvania State University.
[a] (L), Longitudinal; (T), transverse.

TABLE 6.23. Examples for Oxide-Based Composites with Various Reinforcements[177]

Matrix	Reinforcing Component: Material	Shape	Strength (MPa)	Toughness ($MN\ m^{-3/2}$)
Al_2O_3	SiC	Particle	530	
	SiC	Whisker	800	9
	SiC	Fiber		10.5
	ZrO_2	Particle	1080	12
	ZrO_2/SiC	Particle + whisker	1000	13.5
	TiC	Particle	690	4.3
	TiN	Particle	430	4.7
Mullite	SiC	Whisker	450	5.0
	SiC	Fiber	650–850	
	ZrO_2/SiC	Particle + whisker	580–720	6.7–11
ZrO_2	Mullite	Whisker		15
	SiC	Whisker	600	11
	Al_2O_3	Platelet	735	9.5
Spinel	SiC	Whisker	415	

of LAS. The results suggested that the new composition may be chosen as a better candidate matrix for SiC_f composite.

McMahon, Wang, Quon, et al.[123] studied LAS glass-ceramics reinforced with SiC_w and fabricated by hot-pressing. Room-temperature modulus of rupture (MOR) was evaluated through biaxial flexure tests. The MOR for the unreinforced material was 69 MPa. Minor strengthening effects were observed for composites containing 2.5 and 5.0 wt% whiskers, while a decrease in the MOR occurred for the composite containing 20 wt% whiskers. The increase in porosity with increasing whisker content is believed to be the major reason for this behavior.

6.3.2.11 Silicon carbide–calcium aluminosilicate (SiC_w/CAS). Brennan and Nutt[124] incorporated different types of SiC_w into LAS and CAS glass-ceramic matrices. Certain types of SiC_w were characterized by a carbon-rich near-surface chemistry that became more carbon-rich after composite fabrication. In these materials, the flexural strength at 20°C increased by up to 400% and the fracture toughness increased by up to 500%. Crack propagation modes were characterized by crack deflection, whisker-matrix debonding, and crack bridging. In contrast, SiC_w with stoichiometric near-surface chemistry generally did not form carbon-rich interfaces during composite fabrication, resulting in composites with low strength and fracture toughness.

6.3.2.12 Silicon carbide–glass (SiC_w/Gl). Researchers at Corning Glass have also been successful in strengthening various glasses and glass-ceramics by the addition of SiC_w.[125] Fracture toughness values for these composites were also significantly higher than those for the matrices alone; however, most of this increase can be accounted for by the increase in strength of the composite.

Investigations on whisker-reinforced glass and glass-ceramics were reported by

Gadkaree and Chyung.[125] Several of the whisker types previously described were used in the investigation. Glass-ceramic compositions were based on barium "stuffed" cordierite (MgO-$4Al_2O_3$-$10SiO_2$) and a barium osumilite ($2BaO$-$4MgO$-$6Al_2O_3$-$18SiO_2$).

Flexural strengths and fracture toughnesses were determined, and results were in general agreement with many of those obtained with whisker-reinforced crystalline matrix materials. Several-fold improvements in strength and fracture toughness were obtained by incorporating SiC_w into the glasses and glass-ceramics. Incorporation of whiskers into the glass particle slurries produced a large viscosity increase, necessitating much higher composite processing temperatures than for the matrix alone. The commercially available whiskers differed greatly in their ability to reinforce glasses and glass-ceramics. An increase in thermal expansion mismatch and the presence of alkali metal oxides in the matrix appeared to have an adverse effect on composite quality. For glass-ceramics, eliminating the glassy phase and increasing the softening point of the matrix resulted in better high-temperature properties.[126]

6.3.2.13 Silicon carbide–zircon ($ZrSiO_4$) (SiC_w/Zircon). The strength of a monolithic zircon was increased by the addition of whisker reinforcement. The strength was increased from 281 MPa to 342 MPa by the addition of SiC_w. The whisker reinforcement also increased the critical stress for first matrix cracking in SiC monofilament-reinforced composites. An average critical stress for first matrix cracking of 441 MPa was obtained in composites reinforced with SiC_w and SiC filaments. In comparison, composites reinforced with monofilaments alone has an average matrix cracking strength of 287 MPa.

Second, Singh[127] found that the ultimate strength of composites remained unaffected (between 647 and 700 MPa) as a result of the incorporation of SiC_w, which suggested that the ultimate composite strength is controlled by the properties of continuous SiC monofilaments. Third, an average work-of-fracture (WOF) value of 35 kJm^{-2} was obtained for composites reinforced with SiC and filaments, which is much higher than an average WOF value of 16 kJm^{-2} for composites reinforced with SiC filaments alone. This behavior was found to be the result of a lower fiber-matrix interfacial shear strength in whisker- and filament-reinforced composites.

Finally, the results from Singh's study suggested that whisker reinforcement of the zircon matrix phase can be used to increase first matrix cracking strength in this class of composites reinforced with SiC monofilaments. At the same time, composites with enhanced toughness can also be produced via tailored interfacial properties.[127]

6.3.2.14 Silicon carbide–aluminum oxide–mullite—cordierite–zirconia (SiC_w/Al_2O_3/mullite/cordierite /ZrO_2). The simultaneous use of ZrO_2 dispersion reinforcement and SiC_w addition has also been studied in the case of Al_2O_3, mullite, and cordierite matrices, with increases in fracture toughness being reported.[128,129] The results of this work are shown in Tables 6.24–6.26.

The ZrO_2/SiC_w system has also been studied. Densification up to at least 99% of theoretical by hot-pressing with up to 30 V_f SiC_w is possible, and while there was a drop in room-temperature strength, fracture toughness improved remarkably from the 6 MPa $m^{1/2}$ value for the matrix to 12 MPa $m^{1/2}$ with 30 V_f SiC_w. Flexural strength at 1000°C was reported to be 400 MPa, twice the value for the matrix.[130] Results are shown in Figures 6.12 and 6.13.

TABLE 6.24. Strength and Toughness of ZrO_2-Al_2O_3/SiC_w Composites[128]

	Composition (vol%)			
No.	Matrix	Whisker	K_{IC} (MPa m$^{1/2}$)	Strength (MPa)
1	Al_2O_3		4.7	520
2	Al_2O_3	+20 SiC	8.5[a]	650[a]
3	Al_2O_3 + 15 t-ZrO_2		6.2	1080[b]
4	(Al_2O_3 + 15 t-ZrO_2	+20 SiC	13.5[c]	~700
5	Al_2O_3 + 40 m-ZrO_2		2	
6	(Al_2O_3 + 40 m-ZrO_2)	+ 20 SiC	8.5	880
7	m-ZrO_2		<1	
8	m-ZrO_2	+ 20 SiC	9.0[d]	ND

[a] Data from Ref. 129.
[b] Sample hot isostatic pressed at 1600°C for 10 min.
[c] Annealed at 1500°C in Ar for 24 h.
[d] Determined from indentation crack lengths.

SiC_w composites have been studied relative to the effect of hot isostatic pressing on an injection-molded and a N_2 pressure-presintered body of SiC_w/Si_3N_4, and a fracture toughness of 9 MPa m$^{1/2}$ has been reported.[131,132] This improvement is thought to be due to whisker alignment by injection molding and densification by hot isostatic pressing.

6.3.3 Zirconium Oxide

6.3.3.1 Zirconium oxide–aluminum oxide (ZTA_w). Zirconia-toughened alumina (ZTA) make up a promising group of ceramic composites. The toughness values reached are up to about a factor of 4 higher compared with commercial grades of Al_2O_3 (K_{IC} values of about 3–4 MN m$^{3/2}$). The mechanisms are based on a volume expansion of about 4% and the same shear strain of about 6% connected with a tetragonal or monolithic transformation during cooldown from sintering to room temperature. The effects can be described in the case of unstabilized ZrO_2 as stress-induced transformation toughening leading to crack shielding and microcrack toughening by deflecting the propagating

TABLE 6.25. Strength and Toughness of ZrO_2-Mullite/SiC_w Composites[128]

	Composition (vol%)			
No.	Matrix	Whisker	K_{IC} (MPa m$^{1/2}$)	Strength (MPa)
1	Mullite		2.8	244
2	Mullite + 10 ZrO_2	3.5	4.4	
3	Mullite	+20 SiC	4.4	452
4	Mullite + 10 ZrO_2	+20 SiC	5.4	580
5	Mullite + 10 ZrO_2	+30 SiC	6.7	551
6	Mullite[a]	+10 SiC	4.1	277
7	Mullite[a]	+20 SiC	4.6	407

[a] Extruded.

TABLE 6.26. Strength and Toughness of ZrO_2-Cordierite/SiC_w Composites (ZrO_2 + 3 mol% Y_2O_3)[128]

No.	Composition (vol%) Matrix	Whisker	K_{IC} (MPa m$^{1/2}$) RT	Strength (MPa) RT	1000°C
1[a]	Cordierite		2.2	180	170
2	Cordierite + 20 ZrO_2		ND	190	160
3[a]	Cordierite	+20 SiC	3.7	260	245
4	Cordierite + 20 ZrO_2	+20 SiC	ND	380	300

[a] Hot-pressed at 1250°C for 30 min.

primary crack. In the case of a combination of ZrO_2 transformation toughening and whisker reinforcement, very high toughness values of about 13.5 MN $m^{3/2}$ were reported accompanied by strength values of ~700 MPa.[133] The addition of tetragonal ZrO_2 particles to Al_2O_3 seems to be more effective in increasing strength, and the whisker addition resulted in the improvement of both toughness and strength.[117,132–133]

6.3.4 Boron Carbide

6.3.4.1 Boron carbide–aluminum oxide (B_4C/Al_2O_3). High strength and toughness values have been achieved at low loadings of B_4C_w in Al_2O_3. Toughness values of 7.5 MPa m have been achieved at 10–15 V_f B_4C_w. To achieve the same toughness, a loading of nearly double the amount of SiC_w was required.

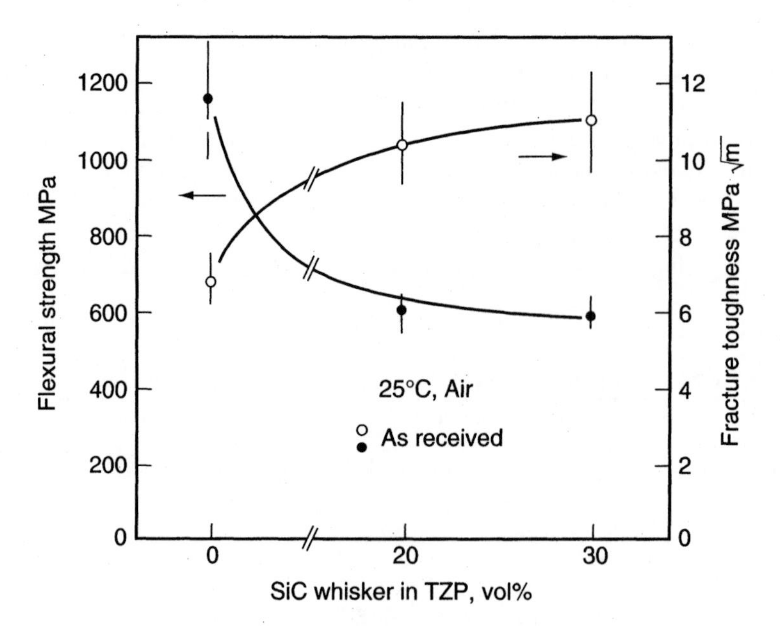

Figure 6.12. Room-temperature flexural strength and fracture toughness of TZP-SiC_w composites as a function of whisker content.[130]

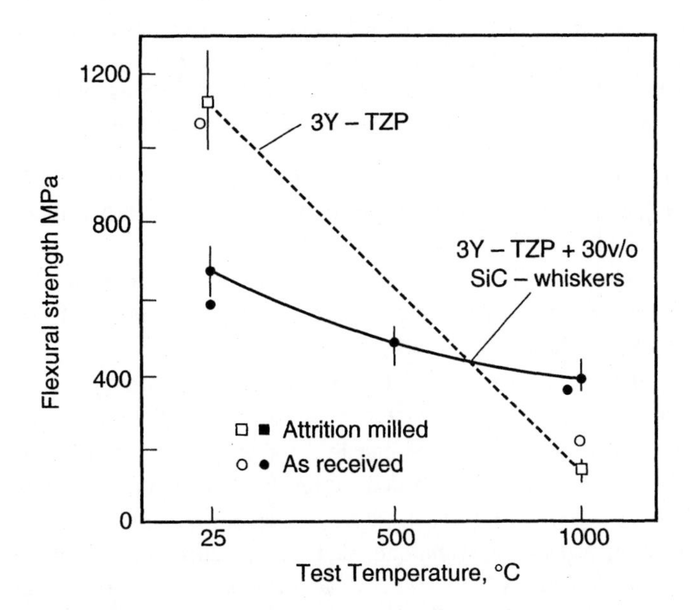

Figure 6.13. Flexural strength of TZP with and without SiC_w reinforcement. Thermal mismatch stresses reduce strength at room temperature, whereas modulus load transfer becomes effective at 1000°C.[130]

The B_4C_w are much larger than currently manufactured SiC_w, with the objective of making the B_4C_w in a range of sizes that are nonrespirable, thereby reducing handling problems.

6.3.5 Silicon Nitride

6.3.5.1 Silicon nitride-yttria-magnesia-silica-calcia ($Si_3N_{4w}/Y_2O_3/MgO/SiO_2/CaO/Si_3N_4$). Pyzik and Beaman[134] reported on the manufacturing and mechanical properties obtained with β-Si_3N_{4w}. They developed self-reinforced Si_3N_4 with a fracture toughness (K_{IC}) of >8 MPa $m^{1/2}$ and a flexure strength of >1000 MPa. The material was based on the $Si_3N_4/Y_2O_3/MgO/SiO_2/CaO$ system in which elongated grains of β-Si_3N_4 were formed in situ from oxynitride glass. This technology is based on controlling the chemistry of the starting powder mixture, nucleation and growth of the β-SiN, and crystallization of the glassy phase.

Second, the morphology of the β-Si_3N_4 grains was tailored by controlling the glass-phase chemistry, glass content, and processing conditions.

Third, they found that in $Y_2O_3/MgO/SiO_2/CaO$-based self-reinforced Si_3N_4, cracks propagated mainly through the glass phase. The crack path was primarily affected by the grain morphology and glass chemistry. The crack traveled preferentially through calcium-rich regions. The primary toughening mechanisms were crack bridging and deflection.

Finally, the presence of elongated grains in Si_3N_4 materials is a necessary but not sufficient condition for improved properties. Other workers have obtained high-aspect-ratio Si_3N_4 materials but not with the combination of high strength and toughness. The key factors were a small grain size coupled with the appropriate glass chemistry, which weakens the Si_3N_4-glass interface. In the investigated system, the optimum combination of mechanical properties was obtained with a Y_2O_3/MgO ratio of 3:1–1:2, a CaO range of 0.1–0.5 wt%, and a Si_3N_4 content between 90 and 96 wt%.

Dusza, Sajgalik, and Reece[135] conducted an analysis of Si_3N_4/β-Si_3N_{4w} ceramic composites with Al_2O_3 and Y_2O_3 additions and found the following.

1. Si_3N_4/β-Si_3N_{4w} composites with high relative densities (>0.97) were successfully prepared with small additions of Al_2O_3 + Y_2O_3 by hot-pressing at 1750–1850°C for 1 and 2 h under a pressure of 27–32 MPa.
2. The addition of β-Si_3N_{4w} up to 20 wt% had no influence on the densification behavior of the composites, but at higher whisker contents the maximum attainable density was reduced significantly.
3. An increase in the fracture toughness of Si_3N_4/β-Si_3N_{4w} composites was observed with increasing whisker content at up to 10 wt% addition.
4. The main mechanisms responsible for toughening were crack deflection, crack branching, whisker-matrix debonding, and to a lesser extent whisker pullout.
5. Future research on Si_3N_4/β-Si_3N_{4w} composites should concentrate on the development of processing techniques that will produce defect-free systems with whisker contents of about 30 wt%. It is also important to study the influence of whisker characteristics (size and shape) and whisker orientation on interfacial debonding and fracture toughness.

6.3.6 Molybdenum Disilicide ($MoSi_2$)

6.3.6.1 Molybdenum disilicide–silicon carbide ($MoSi_2/SiC_w$). The need for increasing operating temperatures of gas turbines to improve their thermodynamic efficiency has provided a large impetus in recent years for the development of high-temperature materials such as aluminides and silicides. Among these, $MoSi_2$ is receiving increased attention as a promising structural material because of its high melting temperature (2020°C) and its superior oxidation and corrosion resistance. The formation of a thin layer of SiO_2 when exposed to high temperatures and aggressive environments helps to improve its oxidation and corrosion resistance. Despite the brittleness of the material at low temperatures, several attempts[136,137] have been made to improve its ductility and strength using SiC as a reinforcement. These composites were found to have more oxidation resistance than aluminide-based intermetallics and to have relatively more ductility than conventional structural ceramics.[138]

Mechanical properties of $MoSi_2$ in single crystalline and polycrystalline forms, as well as those of SiC-reinforced materials, have been studied under compression or flexure loads. $MoSi_2$, with its tetragonal crystal structure, has a limited number of available slip systems and is therefore brittle below 1000°C. Reinforcements with 20 V_f of SiC improve not only its strength but also its ductility.[139] A recent investigation[140] showed that additional improvements could be made by alloying the matrix with WSi_2 which forms an extended solid solution with $MoSi_2$. Sadananda et al.[141] conducted a systematic investigation of creep behavior under uniaxial compression of both monolithic $MoSi_2$- and SiC-reinforced $MoSi_2$ composites in the temperature range 1100–1400°C.

They also found that creep rates were significantly reduced by the presence of SiC_w reinforcement. The reduction was attributed to increased resistance to plastic flow as well as inhibition of grain boundary sliding.

Other efforts by Cook et al.[142] involved the use of a variety of reinforcements to improve the oxidation behavior of $MoSi_2$ composites. The oxidation behavior of $MoSi_2$ composites prepared by the XD process showed that composites of $MoSi_2$ and SiC had the best oxidation resistance and that the oxide that formed on $MoSi_2$/SiC XD composite consisted of a continuous stable layer of SiO_2.

6.4 PARTICULATES

Many early and current ceramics are particulate composites, empirically derived based on processing-property developments. More recently, composites have often resulted from practical processing technology (SiC from "siliconizing" preforms of SiC and C, leaving excess Si and usually some C). Using additives to sinter or hot-press Si_3N_4 results in toughness which generally increases with additive levels.[143] Particulate composites, whether made by design or as a result of processing, have been the most amenable to traditional ceramic processing—primarily powder-based processing and pressureless sintering. They are closest to traditional ceramics in their mechanical behavior, exhibiting catastrophic failure generally described by conventional fracture mechanics. Thus, in one sense the needs and opportunities for improving particulate composites parallel those for improving conventional ceramics: refining the understanding of mechanism-microstructure-processing relationships.

A central need in understanding the mechanisms of toughening particulate composites is to better address the multiple mechanisms of toughening and strengthening. Clearly, the fact that toughening and strengthening may be separate entities as opposed to one directly predicting the other is indicated by the frequent rise, and frequent subsequent decrease, of the σ/K_{IC} values as a function of composite composition.

The expression σ/K_{IC} is the ratio of tensile (typically flexure) strength (σ) to fracture toughness (K_{IC}) as a measure of the mechanical performance against a primarily composite composition parameter (V_f of the additive phase). Because this σ/K_{IC} parameter is apparently new, a brief discussion is in order. If all compositions of a given composite type were uniformly processed, exhibited the same strengthening in the same proportion to toughening, and had the same response to machining, this ratio would then simply be a horizontal band reflecting the range of the reciprocal of flaw size independent of composition. However, if changes occurred as a function of composite composition (e.g., of processing defects, machining flaws), the mechanisms of toughening or strengthening would then change the σ/K_{IC} ratio as a function of composition.

6.4.1 Silicon Carbide

6.4.1.1 Silicon carbide–silicon nitride (SiC_p/Si_3N_4). Tokai Carbon Company[144] has developed new SiC_p to be added to Si_3N_4 ceramics and has improved the toughening properties of the composites considerably by the following factors.

1. Diameter, morphology, content, and distribution of the reinforcement
2. Effects on the stress and strain between reinforcement and matrix, like thermal expansion coefficient and elastic modulus
3. Effects on the interface between the reinforcement and the matrix.

Various diameters of SiC_p did not affect the fracture toughness of the composites but did affect the toughening mechanism. Presumably, pullout mechanisms helped improve fracture toughness mainly in composites containing relatively small-diameter, large-aspect-ratio SiC. Deflection mechanisms probably contributed to improving fracture toughness mainly in composites containing relatively large-diameter, small-aspect-ratio SiC. Only the flexural strength of the composite containing 2.5-µm-diameter particles with a 2.5-mm diameter and an aspect ratio of 6 was close to the optimum morphology for reinforcement.

Table 6.27 shows the resultant improved mechanical properties of Si_3N_4 obtained by distributing SiC_p in it by CVD hot-pressing Si/C/N composite powder and a Y_2O_3/Al_2O_3 promoter.

6.4.1.2 Silicon carbide–sialon (SiC_p/Sialon). Akimune, Hirosaki, and Ogasawara[145] reported on a study where SiC_p were used to strengthen sialon and SiC_p-sialon composites were produced and their mechanical properties examined.

SiC_p-sialon composites produced at a hot isostatic pressing temperature of 1850°C under 100 MPa nitrogen pressure exhibited the highest strength of the specimens at room temperature. The Weibull modulus increased with increasing hot isostatic pressing temperature and nitrogen gas pressure. This is attributed to the fact that the SiC_p inhibited sialon grain growth, which was also controlled by the hot isostatic pressing conditions. In this sintering process SiC_p were trapped in the sialon grains, resulting in higher bending strength (Table 6.28).

6.4.1.3 Silicon carbide–yttria-stabilized tetragonal zirconia polycrystal–mullite (SiC_p/Y-TZP/mullite). Mullite matrix composites reinforced by both SiC_p and Y-TZP were fabricated by hot-pressing. The effect of the combined SiC_p and Y-TZP on improving the mechanical properties appeared to be additive or synergetic. The reinforcing mechanisms in the composites are mainly crack deflection and crack branching caused by SiC_p, and microcrack toughening caused by the addition of Y-TZP. The strength of the composites stay almost constant from room temperature to 1000°C, and the thermal shock resistance of the composites was significantly improved.

TABLE 6.27. Mechanical Properties of Si_3N_4 with SiC_p Distributed in It

Composite	Toughness (MPa m$^{1/2}$)	Strength (MPa)	Thermal Expansion Coefficient (°C^{-1})	Young's Modulus E (GPa)^a	ΔT_c (°C)^b	ΔT_c (°C)
Si_3N_4	5.5	1100	3.4×10^{-6}	305	1050	1050
Si_3N_4/10 vol% SiC	5.7	1130	3.7×10^{-6}	315	857	920
Si_3N_4/25 vol% SiC	6.5	1550	4.1×10^{-6}	330	759	800
Si_3N_4/32 vol% SiC	6.1	1260	4.2×10^{-6}	340	486	530

^{a} Calculated using a linear relation between $E = 305$ GPa for Si_3N_4 and $E = 440$ GPa for SiC.
^{b} Calculated from strength, thermal conductivity, Young's modulus, and thermal expansion coefficient. Poisson's ratio was assumed to be 0.25 for all the composites.

TABLE 6.28. Mechanical Properties of Sialon and SiC_p-Sialon Composites[145]

	Hot-pressed	Hot Isostatic Pressed		
	A	B	C	D
Sintering conditions				
Temperature (°C)	1720	1850	1850	1950
Pressure (MPa)	24.5			
Nitrogen pressure (MPa)	0.1	10	100	100
Duration (h)	0.5	1	1	1
Crystal phase	α-Sialon	←	←	←
	β-Sialon	←	←	←
		β-SiC	β-SiC	β-SiC
Density (Mg/m^3)	3.20	3.27	3.27	3.27
Young's modulus (GPa)	323 ± 2	332 ± 2	330 ± 2	3.30 ± 2
Bending strength (MPa)	762 ± 109	781 ± 83	824 ± 65	746 ± 49
Weibull modulus	8.2	10.9	14.3	16.5
Fracture toughness (MPa $m^{1/2}$) (SEPB)[a]	4.6 ± 0.1	5.0 ± 0.1	4.7 ± 0.1	5.0 ± 0.1

[a] SEPB, Single-edge precracked beam.

The flexural strength and fracture toughness of the mullite matrix composites are significantly improved by introducing SiC_p and Y-TZP together and are greater than those of the mullite matrix composite reinforced by only SiC_p or Y-TZP. The maximum flexural strength and fracture toughness of the composites are 600 MPa and 6.7 MPa $m^{1/2}$, respectively, with the addition of 20 V_f of Y-TZP and 35 V_f of SiC_p. The strength stays almost constant to a temperature of about 1000°C.

Adiabatic engine piston caps made of SiC_p/Y-TZP/mullite composites are potential candidates as thermal insulating structural components for heat engines.[146]

6.4.1.4 Silicon carbide–aluminum oxide (SiC_p/Al_2O_3). Pickard and his colleagues at the University of California at Santa Barbara[147] measured the mechanical properties of a range of direct melt oxidation (DMOX) materials (see Table 6.29).[148–151]

The rate of reaction growth front and microstructure are influenced by molten alloy composition and element dopants. As mentioned earlier, filler, filler size, and residual metal in the matrix influence composite properties. The residual aluminum alloy, for example, significantly contributes to the increasing room-temperature strength and toughness of SiC_p-reinforced Al_2O_3 composites (see Figure 6.14). As the aluminum alloy increases in temperature to the melting point (~600°C), the strength of the composite with 7-μm filler and a larger volume of residual metal decreases because of the reduced load-bearing-material cross section. Also, for the same composite, the toughness diminishes with temperature increase and softening of the aluminum alloy because of its decreasing contribution to crack bridging. As the curve levels out at higher temperatures, relatively high strength and toughness value are retained as a result of the reinforced ceramic matrix.

Creep rates[151] are generally low for these composites, making them attractive for high-temperature industrial heating applications (see Figure 6.15). Creep has not been observed in SiC_p/Al_2O_3 composites at or below 1200°C.

The best strengths were obtained when SiC_p preforms were used. The use of such

TABLE 6.29. Physical Properties of SiC_p/Al_2O_3 Composites[148]

Property[a]	Units		25°C	(73°F)	1000°C	(1832°F)	1550°C	(2822°F)
Density	b/cm³	lb/in³	3.4–3.5	(0.12–0.13)				
Hardness	Rockwell A		80–90					
Flexural strength[b]	MPa	(ksi)	400–500	(65–73)	200–250	(29–36)	175–225	(26–33)
Modulus	GPa	(msi)	310–330	(45–48)				
Poisson ratio			0.25–0.29					
Shear modulus[c]	GPa	(msi)	120–130	(17–19)				
Fracture toughness[d]	MPa $m^{1/2}$	(ksi/$in.^{1/2}$)	7.0–7.5	(6.4–6.8)	3.0–4.0	(2.7–3.6)	2.5–3.5	(2.3–3.2)
Coefficient of thermal expansion[e]	ppm/°C	(ppm/°F)	7.8	(3.9–4.5)				
Thermal conductivity	W/m·K	(BTU in./h·ft^2·°F)	60–70	(485–555)	15–25	(100–175)	5.5	(38)

[a] Particulate loading of 55 vol%; 5 to 20 μm particle diameter. Loadings can be tailored up to ~75 vol%.
[b] Four-point bend.
[c] Sonic method.
[d] Chevron notch beam.
[e] Average value from 25 to 1400°C.

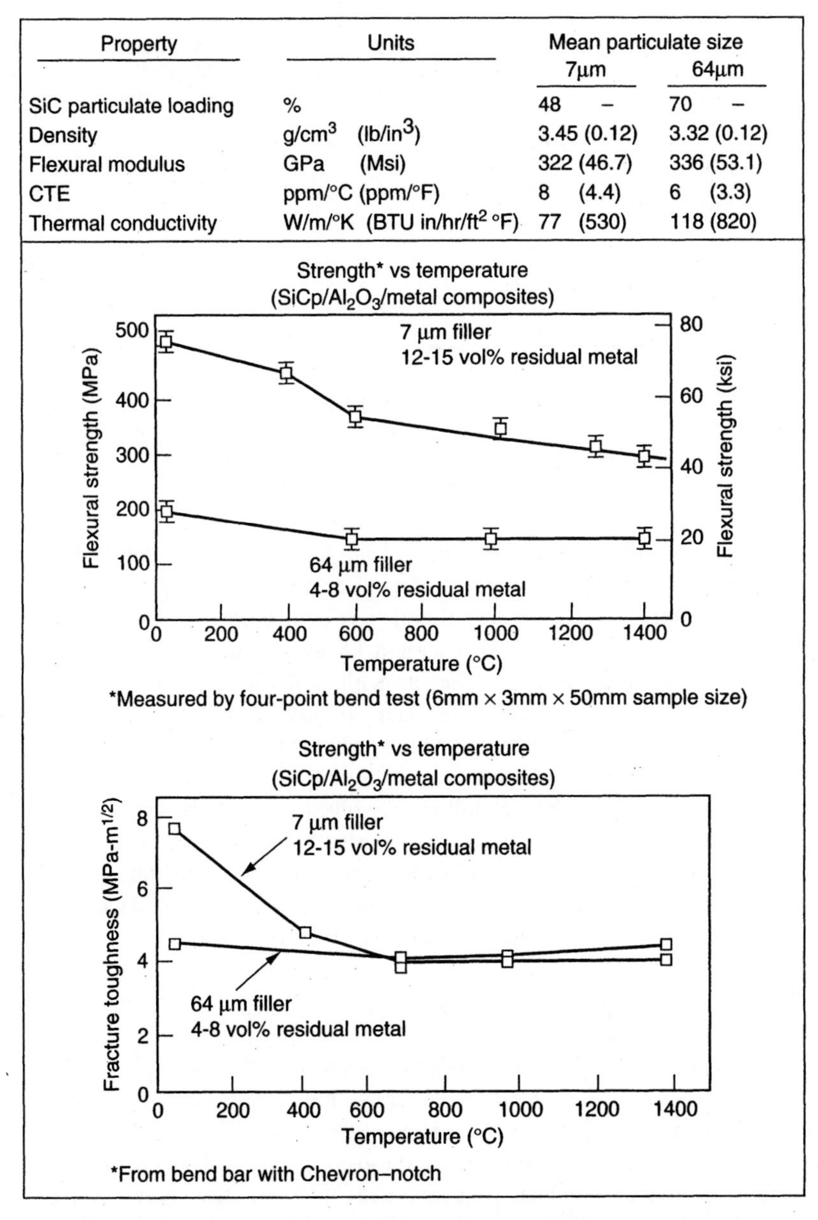

Property	Units	Mean particulate size 7μm	64μm
SiC particulate loading	%	48 –	70 –
Density	g/cm³ (lb/in³)	3.45 (0.12)	3.32 (0.12)
Flexural modulus	GPa (Msi)	322 (46.7)	336 (53.1)
CTE	ppm/°C (ppm/°F)	8 (4.4)	6 (3.3)
Thermal conductivity	W/m/°K (BTU in/hr/ft² °F)	77 (530)	118 (820)

Figure 6.14. Comparison of properties for two types of SiC_p/Al_2O_3 composites. Finer filler size and increased volume percent of residual metal favor strength and toughness. As the residual aluminum alloy softens with increase to melting temperature (about 600°C), strength and toughness decrease to a relatively stable value suitable for many high-temperature applications.[148]

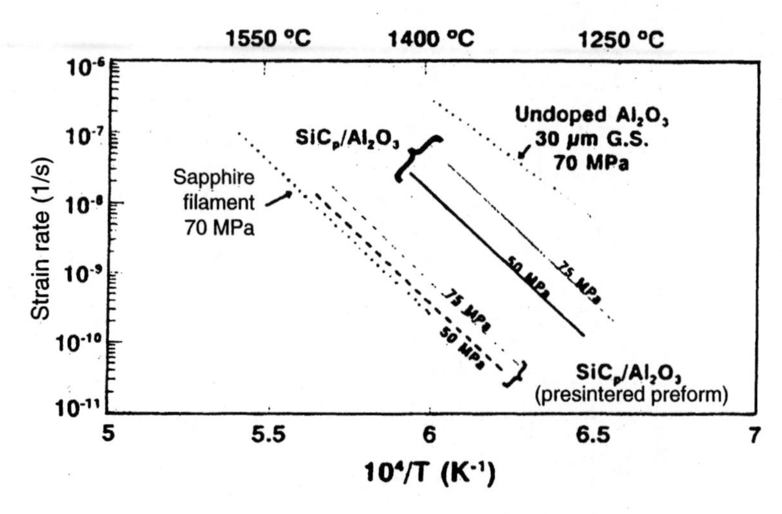

Figure 6.15. Creep behavior of SiC_p/Al_2O_3 composites.[148]

preforms appeared to dictate the oxide growth process and prevent the formation of flaws present in particulate-free materials. However, the thermal expansion mismatch between SiC and Al_2O_3 caused strength to diminish as particulate size increased. The implication for high strength is that preforms should be used that have small particulates closely matched in thermal expansion coefficient with the DMOX matrix.

The attainment of both high strength and high toughness was subject to the dilemma that the phenomena involved have characteristics that cause strength to decrease but toughness to increase as the size of the reinforcement increases. However, an optimization should be possible. Notably, the contribution of the alloy to the toughness can be enhanced by precipitation hardening without compromising the strength. Also, the use of a preform with high-aspect-ratio rods, rather than particulates, should provide good toughness without degrading strength. It remains to be ascertained whether such preforms need to be coated to ensure sufficient interface failure.[147]

6.4.2 Aluminum Oxide

6.4.2.1 Silicon nitride–yttria–aluminum oxide ($Si_3N_4/Y_2O_3/Al_2O_3p$). $Si_3N_4/Y_2O_3/Al_2O_3$ and $Si_3N_4/Y_2O_3/Al_2O_3$/AlN have had a large impact on the development of Si_3N_4 as an engineering ceramic. As is now well known, a more elongated grain structure and more complete densification are promoted by the addition of Y_2O_3, which leads to higher strength and toughness. Tsuge et al.[152,153] and Komeya[154] obtained the highest fracture strength for Si_3N_4 with Y_2O_3/Al_2O_3 additions, for which three-point bending strengths of 1460 MPa at room temperature and 1260 MPa at 1200°C were achieved. The improved materials from these combinations had high strength from room temperature to 1000°C and have recently been put to practical use in engine components.

Improvement of the fracture toughness of Si_3N_4 has been achieved by causing anisotropic grain growth. Kawashima et al.[155] reported on a toughness of 11.3 MPa $m^{1/2}$

and a three-point bending strength of 774 MPa which were obtained as a result of the formation of large, elongated grains. Higher strength (1147 MPa) was attained with microstructures composed of homogeneous small grains.

6.4.2.2 Aluminum oxide–metal carbides (5 V_f SiC, 5 V_f B_4C, and 5 V_f TiC) and Al_2O_3/ZrO_2 (Y_2O_3-stabilized) 10 V_f. The effect of rapid-rate furnace sintering and HCD-plasma sintering on density, microstructure, and mechanical properties of three $Al_2O_3/5V_f$ metal carbide composites and $Al_2O_3/10$ V_f ZrO_2 was studied by Bengisu and Inal,[156] and they compared their results to those for conventionally sintered composites.

The following conclusions were extracted from their study.

1. Rapid-rate sintering in a furnace increases the final fractional density of conventionally sintered Al_2O_3/5% SiC composites from 76% to 86% in a 20-min sintering period. Similarly, higher densities are achieved with the HCD-plasma sintering method in shorter sintering times. Final densities of Al_2O_3/5% B_4C composites are also increased by rapid-rate furnace sintering or by plasma sintering.
2. Mechanical properties are improved in 20-min rapid-rate-sintered Al_2O_3/5% SiC and Al_2O_3/5% B_4C as compared to 20-min conventionally sintered composites. These improvements are attributed to increased fractional densities.
3. It is possible to achieve high densities and mechanical properties similar to those of rapid-rate-sintered Al_2O_3/5% TiC by conventional sintering in alumina embedding powder. Density and mechanical property differences are minimal in rapid-rate and conventionally sintered Al_2O_3/10% ZrO_2 composites because ZrO_2 is easily sintered.
4. HCD-plasma sintering provides the finest microstructure, but high densities are achieved only in Al_2O_3/ZrO_2 composites.
5. The finest microstructure in particulate CMCs is achieved by plasma sintering.

6.4.3 Zirconium Oxide

6.4.3.1 Zirconium oxide–aluminum oxide (ZrO_{2p}/Al_2O_3). Zirconia-toughened alumina, in which ZrO_2 is present as a secondary dispersed phase, has attracted much scientific and technological interest in recent years because of its enhanced toughness and strength. The mechanical properties of ZTA depend mainly on the amount of ZrO_2 added, the size of the ZrO_2 grains, and the relative content of t-ZrO_2.

Dense ZTA ceramics with homogeneously dispersed ZrO_{2p} show excellent mechanical properties according to Matsumoto, Hirota, Yamaguchi, and associates.[157] They found that the average bending strength with 5 mol% ZrO_2 was 420 MPa. It increased to the maximum of 650 MPa with 15 mol% ZrO_2. The value was the same as that for ZTA prepared from sinterable CVD powders.[158] In the range of up to a 15 mol% ZrO_2 addition, the increase in strength was almost proportional to that of K_{IC}. The ZTA with 15–30 mol% ZrO_2 did not show any increase in strength. The t-ZrO_2 particles did not increase the strength as much as they enhanced the toughness. This may be because of the lower relative density (<99%) in the ZTA above 15 mol% ZrO_2.

6.4.3.2 Zirconium oxide-mullite. High-temperature strength, creep resistance, and good thermal stability are the main characteristics of pure mullite ceramics, and according to Ismail, Shiga, Katayama, et al.[159] by dispersing ZrO_2 in mullite, fracture toughness can be increased. The methods evaluated for dispersion of ZrO_2 in mullite were

1. Reaction of zircon ($ZrSiO_4$) with Al_2O_3
2. Mechanical mixing of mullite powder with ZrO_2
3. Sol-gel method for obtaining a homogeneous dispersion of ZrO_2 in mullite.

They found that the maximum amount of ZrO_2 that could be dispersed in mullite without causing a decrease in strength was 15 V_f. The flexural strength and fracture toughness of 15 V_f ZrO_2 dispersed mullite-ZrO_2 ceramic at room temperature were 500 MPa and 4.3 MPa $m^{1/2}$, respectively. The flexural strength of the same composite at 1400°C was 300 MPa.

6.4.3.3 Zirconium oxide–molybdenum disilicide ($ZrO_{2p}/MoSi_2$). Petrovic, Honnell, Mitchell, et al.[160] fabricated $ZrO_{2p}/MoSi_2$ matrix composites by wet-processing or hot-pressing, using high-quality unstabilized, partially stabilized, and fully stabilized ZrO_2 powders. Composite room-temperature indentation fracture toughness increased with increasing V_f of ZrO_2 reinforcement. Unstabilized ZrO_2 produced the highest composite fracture toughness, 7.8 MPa $m^{1/2}$ as compared to 2.6 MPa $m^{1/2}$ for pure $MoSi_2$. Unstabilized ZrO_2 composites exhibited matrix microcracking, and the spontaneous tetragonal-to-monoclinic ZrO_2 phase transformation induced significant plastic deformation in the $MoSi_2$ matrix.

6.4.4 Titanium Boride

6.4.4.1 Titanium boride–silicon carbide (TiB_{2p}/SiC). SiC/TiB_{2p} ceramic composite material exhibits enhanced mechanical and fracture properties over those of the monolithic α-SiC at temperatures in excess of 1000°C as seen in Table 6.30.[161]

SiC-based composites with transition metal boride particulates have been developed for electroconductive applications such as heating elements and igniters[162,163] and also as wear-resistant structural parts for high temperatures such as valve train components and rocker arm pads in super-hot-running engines.[164,165] These composites combine the high thermal and electric conductivity of, for example, TiB_2 and ZrB_2 with the oxidation resistance of SiC. Additionally, because of thermal mismatch stresses on the order of 2 GPa, toughening mechanisms such as crack deflection and stress-induced microcracking with a pronounced process zone, as well as crack flank friction, have been proven to occur.

An optimum V_f for reinforcing particulates of 25–30 V_f has been reported, yielding a flexural strength of 710 MPa and a fracture toughness of 5.0–5.7 MPa $m^{1/2}$.[166] Composites with a lower TiB_2 content of 15 V_f exhibit a mean strength of 485 MPa combined with a K_{IC} of 4.5 MPa $m^{1/2}$. The strength of SiC-based materials with 50 V_f ZrB_2, HfB_2, NbB_2, or TaB_2 particles also ranges between 400 and 500 MPa.[162] Similar strength values (480 MPa) combined with much higher fracture toughness of 7–9 MPa $m^{1/2}$ have been reported for large-scale lots of 16 V_f TiB_2 composites sintered without pressure.[163] Because the sintering was carried out with temperatures exceeding 2000°C, yielding 98–99% of

TABLE 6.30. Typical Properties of Nonreinforced and TiB_2-Reinforced Sintered α-SiC[161]

Property	Sintered α-SiC	SiC/TiB_2 Composite
Density (gm/cm^3)	3.21	3.40
Young's modulus (GPa)	400	440
Flexural strength, 25°C (MPa)	380	450
Fracture toughness, 25°C (SENB) (MPa $m^{0.5}$)	4–5	6–8
Hardness (Vickers) (GPa)	28	29
Coefficient of thermal expansion (10^{-6}/°C)	4.02	4.60
Thermal conductivity (W/m·K)	125	65

the theoretical density and an average TiB_2 particle size of 2.0 μm, it is obvious that the reinforcing phase also acts as a grain growth inhibitor for SiC. The high-temperature strength of SiC/TiB_2 and SiC/ZrB_2 composites was found to remain nearly constant at 480 MPa up to 1200°C, and is hence superior to that of many sialons.[162,163]

The addition of TiB_2 to a SiC matrix causes a significant reduction in the grain size of the matrix in the sintered microstructure compared to that in the sintered α-SiC. It is thought that this decrease is related to the pinning of the grain boundaries by the TiB_{2p} during densification and grain growth. The differential thermal expansion between SiC and TiB_2 causes weakening of the interface during cooling, and when the particle size is large enough, some limited separation occurs. Consequently, this microstructure results in enhanced toughening because of crack deflection and modifications in fracture crack advance as well as chemical compatibility. Most of the properties of this composite are generally similar to SiC as seen in Table 6.30.[161]

Hexoloy ST, an α-phase sintered SiC/TiB_2 is 50–75% tougher than the unreinforced Hexoloy SA grade. Both materials are made by the same process of pressing, injection molding, or extrusion; therefore, the material can be obtained in production volumes.

The unreinforced grade retains most of its excellent abrasion, corrosion, and oxidation resistance; however, when the added fracture toughness of the reinforced material is considered, the new material is expected to make it suitable for applications where the unreinforced material would fracture, such as gas turbine engine rotors and reciprocating-engine valve trains, where it could potentially be 30–40% less expensive than Si_3N_4 based on raw material costs. Unlike Si_3N_4, the composite does not require hot isostatic pressing or surface grinding to attain maximum strength and toughness.

Dense SiC/TiB_{2p} composites were obtained by Tani and Wada[167] through pressureless sintering of $SiC/TiO_2/B_4C/C$ powder compacts. During the process, TiO_2, B_4C, and C reacted to form TiB_2, followed by the consolidation of SiC matrix with the aid of excess B_4C and C. The sintered body with additional hot isostatic pressing at 1900°C exhibited the average four-point flexural strength of more than 700 MPa at both 20 and 1400°C.

6.4.4.2 Titanium boride–aluminum oxide (TiB_{2p}/Al_2O_3). Liu and Ownby[168] investigated the mechanical properties of composite ceramics composed of 0–20 V_f TiB_{2p} dispersed in an α-Al_2O_3 matrix. The Al_2O_3/TiB_2 composite powder was hot-pressed at 1470°C for 20 min to achieve over 98.8% of the theoretical composite density. The strength and fracture toughness of the two-phase, hot-pressed composite were both significantly improved compared to those of the single-phase Al_2O_3. Thus, the addition of TiB_{2p} improved the mechanical properties. Both fracture toughness and flexural strength were significantly improved with only 5 V_f TiB_2 in the Al_2O_3 matrix. The resulting increase in fracture toughness of Al_2O_3 was shown to be caused by additions of TiB_2 in comparison to other hard particle additions.[168]

6.4.5 Titanium Carbide

6.4.5.1 Titanium carbide–silicon carbide (TiC_p/β SiC). Endo, Ueki, and Kubo[169] hot-pressed SiC/TiC composite ceramics with a 0–100 wt% TiC to determine the effect of composition (amount of TiC) on elastic modulus, hardness, flexural strength, and fracture toughness. The composites exhibited superior mechanical properties compared to monolithic SiC and TiC, especially in fracture toughness, K_{IC}, for values of 30–50 wt% TiC composite. The maximum values of K_{IC} and room-temperature flexural strength were 6 MPa $m^{1/2}$ for a 50 wt% TiC and 750 MPa for a 30 wt% TiC composite, respectively. The observed toughening could be attributed to the deflection of cracks as a result of dispersion of the different particles. Although no third phases were detected by both transmission electron microscopy (TEM) and x-ray diffraction studies, an EDS study and resistivity measurements indicated some possibility of solid solutions being present. Composites containing more than 30 wt% TiC exhibited resistivity lower than 10 Ω-cm, which is favorable for electrodischarge machining (EDM) of ceramics.

6.4.6 Boron Carbide

6.4.6.1 Boron carbide–titanium boride (B_4C_p/TiB_2). Kang and Kim[170] investigated the improvement of TiB_2 with a dispersion of B_4C_p. Using 1 wt% Fe as a reactive additive, hot-pressing at 1700°C for 60 min at 35 MPa resulted in 99% dense composites with a clear maximum in strength of 700 MPa for 10 V_f B_4C and in a K_{IC} of 7.6 MPa $m^{1/2}$ for 20 V_f B_4C. This optimizing effect was attributed to both grain growth inhibition and a change in the fracture mode from transgranular to intergranular caused by the B_4C addition. Because studies on the B_4C-rich side of this system also indicate optimum properties at approximately 60–70 V_f B_4C, a change in strengthening and toughening mechanisms must occur at a composition between 40 and 50 V_f B_4C.

6.4.7 Borides and Carbides

6.4.7.1 Molybdenum disilicide ($MoSi_2$)/TiB_2-HfB_2-ZrB_2-SiC. $MoSi_2$ was reinforced in situ with TiB_2, HfB_2, ZrB_2, or SiC-30_p by the XD process. The compressive yield strengths of these composites, along with that of an unreinforced alloy, were investigated as a function of temperature from room temperature to 1300°C by Aikin.[171] Although particle type had some effect, the major difference in high-temperature yield stress among the alloys was related to reinforcement size and shape.

The SiC-reinforced alloy exhibited a 0.2% offset yield at temperatures of 600°C and above. At 800 and 1000°C the SiC-reinforced and base alloys exhibited similar strengths, whereas the SiC-reinforced alloy was significantly stronger than the base alloy at 1200°C and above. The three diboride-reinforced alloys all exhibited enough low-temperature plasticity for a 0.2% offset yield stress to be determined at room temperature. At low temperatures the strengths were high, approximately 1900 MPa for the TiB_2- and HfB_2-reinforced alloys and 2300 MPa for the ZrB_2-reinforced alloy. At 1000°C the diborides have nearly identical strengths, whereas at 1200 and 1300°C TiB_2 and ZrB_2 have similar strengths and the HfB_2-reinforced alloy is significantly stronger.

As noted previously, incorporating micron-sized particles of TiB_2, HfB_2, ZrB_2, or SiC substantially improves the mechanical properties of the composites. The high-temperature strength of the reinforced alloys is significantly higher than that of the unreinforced alloy, with diboride-reinforced alloys showing the greatest improvement and the SiC-reinforced alloy located between the base alloy and the diboride-reinforced alloys. The size and shape of the particles appear to be the primary factors controlling alloy strength. The smallest particles, and hence the smallest interparticle spacing, lead to the greatest improvement in strength. Among the diboride-reinforced alloys, the chemical composition of the particles appears to have a secondary effect on strength. This may be due to differences in the solid solution strengthening contributions of the various constituent elements.

Finally, low-temperature plasticity is improved by reinforcement with micron-sized particles.

6.4.7.2 Molybdenum disilicide–silicon nitride ($MoSi_{2p}/Si_3N_4$). Kao[172] studied the physical and mechanical properties of hot-pressed $Si_3N_4/MoSi_2/15\text{-}30_p$. The average room-temperature four-point bend strength, fracture toughness, and electrical resistivity were 522 MPa, 3.6 MPa $m^{1/2}$, and 6.3×10^5 Ω-cm for the 15 V_f $MoSi_2$ composite, and 487 MPa, 5.3 MPa $m^{1/2}$, and 0.31 Ω-cm for the 30 V_f $MoSi_2$ composite. The mechanical properties of the composites were very close to those of hot-pressed Si_3N_4 ceramics. The high electric conductivity of the 30 V_f $MoSi_2$ composite was attributed to the percolation effect of $MoSi_{2p}$.

The effect of adding $MoSi_2$p on the strength of the Si_3N_4 matrix is difficult to quantify because of the absence of the base Si_3N_4 ceramic; therefore, more developmental work will be required to determine this effect.[172]

6.5 PLATELETS

The mechanical properties of brittle ceramics can be improved considerably by dispersing high-strength reinforcing particles. In recent years most experiments have concentrated on whisker reinforcement, as this type of composite material can be produced by means of conventional powder processing just like monolithic materials. The reinforcement with continuous long fibers in 2-D and 3-D techniques, on the contrary, requires the application of cost-intensive production processes.

In recent times a shadow has been cast over the application of whisker-reinforced materials because of concerns about health-damaging influences when they are inhaled. Therefore, alternative, less expensive reinforcement components with uncritical geometry, such as platelet-shaped crystals, are being tested. At present, SiC and Al_2O_3 are available in the form of monocrystalline platelets. They have a reinforcement potential comparable to that of whiskers and offer certain advantages such as

- Higher thermal stability and a lower range of defects
- Simple P/M preparation with a homogeneous dispersion of platelets
- Low cost.

However, the strength of the composites can be reduced strongly when platelets are randomly dispersed, as the critical flaw size rises.

Work performed by Janssen and Heussner[173] concluded the following.

- In ceramic materials randomly embedded platelets improve fracture toughness considerably by crack deflection. The influence of pullout and crack bridging is negligible in this case. These mechanisms can be activated only if platelets are oriented parallel to the tensile stress direction.
- The strength of the composites is reduced if platelets are dispersed at random; however, the critical defect size can be minimized by using platelets with small diameters or by orienting them parallel to the tensile stress plane.
- Platelet-reinforced composites can be produced by conventional, inexpensive processes. When composites are fabricated by sintering, platelets retard the densification as a result of hydrostatic tension and skeleton formation.
- First results show that platelets have a reinforcing potential similar to that of whiskers. Above all, the *R*-curve behavior and the increased creep resistance are of special importance for structural applications.

6.5.1 Silicon Carbide

6.5.1.1 Silicon carbide–silicon nitride and molybdenum disilicide (SiC_{pl}/Si_3N_4) ($SiC_{pl}/MoSi_2$). Baril, Tremblay, and Fiset[174] used different grades of high-aspect-ratio SiC_{pl} to reinforce Si_3N_4. Dispersion of additives (4 wt% Y_2O_3 and 3 wt% Al_2O_3) was achieved by ball milling in ethanol using Al_2O_3 balls, whereas dispersion of platelets was obtained by ball milling using plastic balls. Consolidation of the composites was carried out by uniaxial hot-pressing. They found a slight decrease in flexural strength, whereas significant increases in elastic properties, fracture toughness, and Weibull modulus were noted irrespective of platelet size. Grain growth was unaffected by the addition of platelets. Even though both matrix grains and platelets were aligned, only the alignment of matrix grains had a significant effect on the material's anisotropic behavior.

The best combination of room-temperature mechanical properties was achieved with 30 V_f superfine (SF)-grade SiC composite, where flexural strength of 726 MPa, fracture toughness of 8.5 MPa $m^{1/2}$, Weibull modulus of 28, and Young's modulus of 350 GPa were obtained.

The mechanical properties of $MoSi_2$ were found by Richardson and Freitag[175] to increase significantly (up to 98% higher flexure strength) through the addition of SiC_{pl}. The ultimate properties achieved at elevated properties were influenced by the intergranular phases formed during fabrication, whereas those achieved at room temperature were influenced by residual porosity.

At room temperature the best properties measured were for a composition using 99.95% pure $MoSi_2$ (flexure strength of 297 MPa, fracture toughness of 7.50 MPa $m^{1/2}$, pin shear strength of 54.4 MPa). At elevated temperature the best properties measured were for a composition using 99.5% pure $MoSi_2$ (flexure strength of 340 MPa at 1093°C).

When compared to $SiC_w/MoSi_2$ composites of comparable purity and levels of reinforcement, SiC_{pl}-reinforced $MoSi_2$ demonstrated similar mechanical properties. Platelet reinforcement does, however, offer environmental and cost advantages over whisker reinforcement. A combination of reinforcement type (e.g., whiskers plus platelets), as suggested in the work of Carter,[176] may provide some additional increase in mechanical properties. However, significant improvements in properties above the brittle-to-ductile transformation point require continuous fiber reinforcement.

6.5.2 Aluminum Oxide

6.5.2.1 Aluminum oxide–zirconium oxide (Al_2O_{3pl}/ZrO_2). Some investigators have found platelets to be a promising alternative to whiskers. When Lehmann and Ziegler[177] introduced Al_2O_{3pl} (5 V_f) in partially Y_2O_3- and CeO_2-stabilized ZrO_2 matrices, the K_{IC} value increased from 8.2 $MN/m^{3/2}$ to 9.5 MN $m^{3/2}$. However, the fracture strength decreased from 1430 MPa to 735 MPa. This relationship between toughness and strength is observed in other systems, too, if the relatively large-sized alumina platelets, the only ones now available to the investigators, are incorporated.

Huang and Nicholson[178] reported on the varying degrees of success that have been achieved by reinforcing Y-PSZ or Y-TZP with AlO_{3p}, AlO_{3w}, or Al_2O_{3pl}. The bend strength of Y-PSZ at 1000°C was increased by two to four times with 10–40 wt% Al_2O_3 addition. A bend strength of 188 MPa was obtained at 1230°C for Y-PSZ/15 vol% Na-β-Al_2O_3, compared with ~80 MPa for Y-PSZ. A total resistance to fracture of ~700 J/m^2 was reported for Y-PSZ/20 vol% Na-β-Al_2O_3 composites at 1300°C, double the value for Y-PSZ.

Platelets as reinforcement have improved environmental safety, production economics, and thermal stability, compared with whiskers. The room-temperature fracture toughness of Al_2O_3 with SiC_{pl} is comparable to that achieved with SiC_w (6.6 MPa $m^{1/2}$ at 7 vol%) and better than the values for particulate reinforcement (e.g., 5.3 versus 4.3 MPa $m^{1/2}$ at 10 vol%). Fracture toughness is increased at the expense of strength. Microstructure and mechanical properties of ≤20 vol% Al_2O_{3pl} reinforced with Y-TZP were studied at room temperature. It was found that the fracture toughness (by the chevron notch beam method) increased from 8.2 MPa $m^{1/2}$ to 9.5 MPa $m^{1/2}$ with optimum (5 vol%) platelet addition. This was accompanied by a strength reduction from 1555 MPa to 645 MPa. The toughening mechanisms associated with platelet inclusion were identified as crack deflection and modulus load transfer. The transformation toughening became less effective at higher platelet loadings.

Furthermore, Huang and Nicholson examined the influence of platelet reinforcement on the high-temperature mechanical properties. They fabricated Y-PSZ/Al_2O_{3pl} composites by conventional and tape-casting techniques followed by sintering and hot isostatic pressing. The room-temperature fracture toughness increased from 4.9 MPa $m^{1/2}$ for Y-PSZ to 7.9 MPa $m^{1/2}$ (by the indentation strength-in-bending method) for 25 vol% Al_2O_{3pl} with aspect ratio 12. The room-temperature flexural strength decreased 21% and 30% (from 935 MPa for Y-PSZ) for platelet contents of 25 vol% and 40 vol%, respectively. Al_2O_{3pl} improved high-temperature strength (by 110% over Y-PSZ with 25 vol% platelets at 800°C and by 40% with 40 vol% platelets at 1300°C) and fracture toughness (by 90% at 800°C and 61% at 1300°C with 40% platelets). An amorphous phase at the Al_2O_{3pl}/Y-PSZ interface limited mechanical property improvement at 1300°C.

Cutler, Mayhew, Prettyman, et al.[179] examined the high-toughness Ce-TZP/Al_2O_3

ceramics. The addition of SrO to Ce-TZP led to a decrease in ZrO_2 grain size with an increase in hardness. Toughness decreased as strength increased in accordance with conventional ZrO_2 ceramics. The simultaneous addition of SrO and Al_2O_3 to Ce-TZP, however, resulted in the in situ formation of $SrO{\cdot}6Al_2O_{3pl}$. Ce-TZP/Al_2O_3/$SrO{\cdot}6Al_2O_3$ ceramics have the strength (500–700 MPa) of Ce-TZP/Al_2O_3, the hardness (13–14 GPa) of Y-TZP/Al_2O_3, and the high toughness (14–15 MPa $m^{1/2}$) of Ce-TZP. Optimum toughness was obtained at a SrO/Al_2O_3 molar ratio of ~0.1 for compositions containing 15 and 30 vol% Al_2O_3, whereas the maximum toughness occurred at a SrO/Al_2O_3 molar ratio of ~0.03 for compositions with 60 vol% Al_2O_3. The in situ formation of platelets allowed the achievement of high platelet loadings. High toughness was obtained with compositions up to 60 vol% Al_2O_3.

The benefit of simultaneous additions of Al_2O_3 and SrO to Ce-TZP is that tough ceramics can be produced with good hardness and strength. The retention of strength at high toughness values in these Ce-TZP matrix ceramics suggests that *R*-curve behavior does not limit strength to the same extent as in Ce-TZP. Further work is needed to explore elevated-temperature properties including strength, toughness, and creep resistance. The role of $SrO{\cdot}6Al_2O_{3pl}$ in toughening Ce-TZP deserves further study.

6.5.3 Zirconium Boride

6.5.3.1 Zirconium boride–zirconium carbide (ZrB_{2pl}/ZrC). A relatively new composite material consisting of a ZrC matrix reinforced by ZrB_{2pl} and containing residual Zr exhibits a combination of high strength, fracture toughness, and specific stiffness. In preliminary tests, this material compared favorably to the commonly used implant alloys Ti-6Al-4V and Co-Cr-Mo. In addition, the results of wear tests on this material compared to ultrahigh-molecular-weight polyethylene indicate promising results for use in orthopedic devices.[180]

ZrB_2-reinforced ZrC matrix composites fabricated by the DIMOX-directed metal oxidation process[181–192] have exhibited the mechanical properties mentioned before, especially high fracture toughness. A comparison of the properties of the ZrB_2-reinforced composite containing 6 vol% residual Zr with Co-Cr-Mo and Ti-6Al-4V alloys, Al_2O_3, and ZrO_2 is shown in Table 6.31. As indicated in the table, the flexural strength of the 6 vol% Zr material is comparable to that of Zr, and both these materials have higher strengths than Al_2O_3. The composite material also compares favorably with the metallic

TABLE 6.31. Properties of a ZrB_2/ZrC/Zr Composite and Other Implant Materials

Property	ZrB_2/ZrC/6Zr[a]	Al_2O_3	ZrO_2	Ti-6Al-4V	Co-28Cr-6Mo
Density (kg/m^3)	6200	3900	6100	4429	8474[b]
Strength (MPa)[c]	FS = 850	FS = 400	FS = 900	YS = 860 UTS = 985	YS = 448–517 UTS = 655–889
Fracture toughness, K_{IC} (MPa $m^{1/2}$)	12	5–6	9–10	53	Not available
Young's modulus (GPa)	390	380	200	115	241
Specific stiffness (10^3 GPa-m^3/kg)	62.9	97.4	32.8	25.9	28.4

[a] The amount of residual zirconium is given in volume percent.
[b] This value was calculated using the rule of mixtures.
[c] FS, Flexural strength; YS, yield strength; UTS, ultimate tensile strength.

systems (whose yield strengths are reported). The fracture toughness of the 6 vol% Zr material is comparable to that of Zr and is higher than that of Al_2O_3. The specific stiffness of the ZrB_2/ZrC/Zr material is higher than that of Zr and the metallic systems and is lower than that of Al_2O_3.

6.6 NANOCOMPOSITES

Ceramic nanocomposites have been vigorously investigated by Niihara and workers.[193–195] Ceramic composites can be divided into two types, microcomposites and nanocomposites. In microcomposites, micrometer-sized materials such as particles, platelets, whiskers, and fibers are dispersed at the grain boundaries of the matrix. The main purpose of the composites is to improve fracture toughness. On the other hand, nanocomposites can be grouped into three types; intergranular, intragranular, and nano/nanocomposites, as shown in Figure 6.16.[194] In intra- and intergranular nanocomposites (Figure 6.16) nano-sized particles are dispersed mainly within the matrix grains or at the grain boundaries of the matrix, respectively, and these systems lead toward improvement in the mechanical properties such as hardness, fast-fracture strength, fracture toughness, creep resistance, and fatigue strength, from room temperature to high temperatures. On the other hand, nano/nanocomposites are composed of nanometer-sized dispersoids and matrix grains. The primary purpose of nano/nanocomposites is to add new properties to ceramics, such as machinability and superplasticity, that are usually associated with metals.

At the initial stage, ceramic nanocomposites were successfully prepared by CVD. It is, however, recognized that the P/M processes are more promising for engineering ceramics because components of complex shape are required. Many kinds of nanocomposites, such as Al_2O_3/SiC, Al_2O_3/Si_3N_4, Al_2O_3/TiC, mullite/SiC, B_4C/TiB_2, SiC/amorphous SiC, and Si_3N_4/SiC, have been successfully prepared by the usual P/M techniques such as

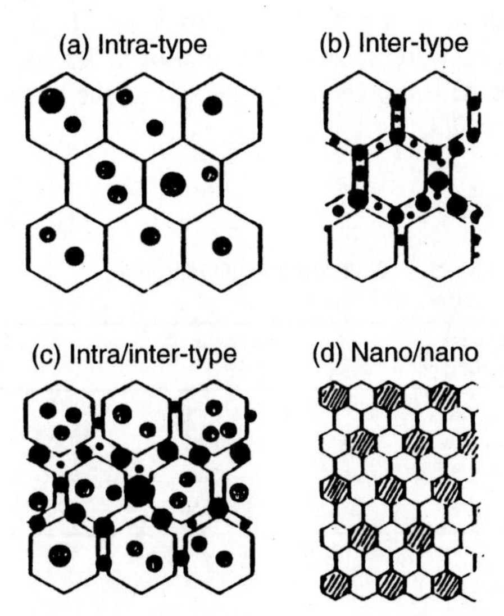

Figure 6.16. The classification of ceramic nanocomposites.[194]

pressureless sintering, hot-pressing, and hot isostatic pressing.[196] Niihara showed in examples of Al_2O_3/SiC/5_p nanocomposites that most of the finer SiC_p, typically less than 0.2 μm, were predominantly dispersed within the matrix grains with some larger SiC_p at the grain boundaries. Similar observations of second-phase dispersion were also made for the MgO/SiC, Al_2O_3/Si_3N_4, and natural mullite-SiC composites. These observations were confined to intragranular-type nanocomposites in which the nano-sized particles were predominantly dispersed within the matrix grains. Intra- and intertype nanocomposites have also been fabricated by controlling the sintering conditions. Mechanical properties such as fracture strength and toughness were greatly improved by the addition of nano-sized SiC and Si_3N_4 dispersions to the matrix grains. Niihara and Nakahira[195] found that there were variations of fracture strength with SiC content for the Al_2O_3/SiC nanocomposite. The strength of Al_2O_3 was increased almost three times by the dispersion of only 5 vol% of nano-sized SiC_p. A further improvement in the strength of up to 1550 MPa was also obtained by annealing at 1300°C for 1 h in either air or an inert atmosphere. Significant improvements in the fracture strength were also found for other nanocomposites, as summarized in Table 6.32.[197] The extreme increase in fracture toughness is also apparent from the table.

The high-temperature strength was also significantly improved by the nano-sized particle dispersion. Niihara[198] found that there was variation in fracture strength with temperature for Al_2O_3- and MgO-based nanocomposites. Monolithic Al_2O_3 and MgO exhibited low strength at room and high temperatures, whereas the Al_2O_3/SiC and MgO/SiC nanocomposites showed an impressive improvement in strength to well over 1000°C. The greatest improvement in high-temperature strength was observed for the MgO/SiC nanocomposite. Even in the temperature range 1000–1400°C, the MgO/30 vol% SiC nanocomposite exhibited higher strength than the monolithic MgO ceramic. High-temperature strength degradation in oxide ceramics occurs through grain boundary sliding, cavitation formation, and/or diffusional creep. However, the nano-sized particle dispersion within the matrix grains promotes transgranular rather than intergranular fracture.[199] This phenomenon was also observed at temperatures above 1000°C. Thus, it may be concluded that the increase in high-temperature strength is mainly due to the prohibition of grain boundary sliding, cavitation, and creep deformation by dislocation pinning, which is produced by the dispersion within the matrix grains. In summary, mechanical properties of nanocomposites were substantially improved by particle dispersion. Nanoparticle dispersion has an excellent potential for increasing the fracture strength and toughness of a wide range of ceramics.

TABLE 6.32. Improvements in the Mechanical Properties Observed for Ceramic Nanocomposites[197]

Composite System	Toughness (MPa $m^{1/2}$)	Strength (MPa)	Maximum Operating Temperature (°C)
Al_2O_3/SiC	3.5 → 4.8	350 → 1520	800 → 1200
Al_2O_3/Si_3N_4	3.5 → 4.7	350 → 850	800 → 1300
MgO/SiC	1.2 → 4.5	340 → 700	600 → 1400
Si_3N_4/SiC	4.5 → 7.5	850 → 1550	1200 → 1400

6.7 FUTURE TRENDS

A better understanding of the detailed micromechanisms of fatigue degradation in advanced materials, together with more precise specification of material property requirements, will ultimately allow the development of composite microstructures optimized for specific applications. However, compromises among the different property sets will be required. In fact, tradeoffs have already been made between strength and toughness in transformation-toughened ceramics.[200] Compromises such as these are common in the microstructural design of alloys. In steels, for example, high strength favors increased resistance to fatigue crack initiation but can be deleterious to crack growth. Similarly, microstructural features may have inverse effects on toughness and subcritical crack growth.

Another important area of advanced materials research is the problem of weakness under nonaxial loading and its ramifications for component design, fabrication, and reliability. Some of the most highly acclaimed high-temperature materials are continuous-fiber-reinforced CMCs and IMCs. These materials have high strength, toughness, and fatigue resistance when loaded in the axial direction, thanks to the development of techniques for controlling the properties of the fiber-matrix interface. Although the focus on axial properties was justifiable in the early days of composite development, now the major limitation of these materials is weakness under nonaxial loading. The low-bond-strength interfaces that are so advantageous for axial properties make these composites unusually vulnerable to delamination under shear or impact loading or to tensile failure under transverse loading.

There are parallels between this situation and the development of PMCs for critical, damage-tolerant applications. Axial strength in PMCs is now often taken for granted. However, these materials are not yet widely used in commercial airframes, for example, because they are vulnerable to delamination, especially following impact damage. The solution to this problem may be found in multidimensional reinforcements—braided, knitted, woven, and stitched fiber preforms—in which a significant V_f of fibers supports loads not only in both in-plane directions but also in the through-thickness direction.[201]

Early test results for these three-dimensional PMCs are very encouraging. Axial properties suffer somewhat, but substantial gains are made in transverse strength, shear strength, and residual strength after impact. The same or similar reinforcement architectures will likely be introduced in IMCs and CMCs along with a new set of fatigue problems. Again referring to the PMC analog, the mechanics of failure of brittle three-dimensional composites will likely be qualitatively different from that of unidirectional materials.[202] Instead of a single dominant crack or a few cracks growing normal to the fibers, there may be complex distributed damage involving highly heterogeneous, local stress fields and the interactions of many microcracks, failed fibers, and, at high temperatures, plasticity.

REFERENCES

1. Phillips, D. C. 1991. Ceramic composites: Their current status and some requirements for future development. *Comp. Sci. Technol.* 40:1–17.
2. Suzuki, H. 1987. A perspective on new ceramics and ceramic composites. *Phil. Trans. Roy. Soc. Lond.* A322:457–78.

3. Kochendörfer, R. 1990. Potential and design aspects of ceramic matrix composites. *4th Eur. Conf. Compos. Mater.*, pp. 43–51, September 1990, Stuttgart.
4. Kochendörfer, R. 1988. Monolithic and fiberceramic components for turboengines and rockets. *AGARD Conf. Proc. No. 449, 72nd Specialist Meet.*, October 1988, Bath, England.
5. Heraud, L. 1990. Main characteristics and domains of application of C/SiC and SiC/SiC ceramic matrix composites. *Proc. Verbundwerk,* Wiesbaden, Germany.
6. Kerans, R. J., T. A. Parthasarathy, P. D. Jero, et al. 1993. Fracture and sliding in the fiber/matrix interface and failure processes in ceramic composites. *Brit. Ceram. Trans.* 92(5): 181–96.
7. Courtright, E. L. 1991. Engineering property limitations of structural ceramics and ceramic composites above 1600°C. *15th Ann. Conf. Compos. Adv. Mater.,* January 1991. PNL SA 18908, Conf. 910162-1, DE-AC06-76RLO1830.
8. Duffy, S. F. 1991. Structural design methodologies for ceramic-based material systems. NASA TM 103097, E 5418.
9. Evans, A. G. 1972. The strength of brittle materials containing second phase dispersions. *Phil. Mag.* 26(6):1327–44.
10. Green, D. J. 1983. Fracture toughness predictions for crack bowing in brittle particulate composites. *J. Am. Ceram. Soc.* 66(1):C4–C5.
11. Faber, K. T., and A. G. Evans. 1983. Crack deflection processes. I. Theory. *Acta Metall.* 31(4):565–76.
12. Angelini, P., and P. F. Becher. 1987. In situ fracture of SiC whisker reinforced Al_2O_3. *Proc. Ann. Meet. Electron Microsc. Soc. Am.* 45:148–49.
13. Singh, R. N. 1991. Influence of testing methods on mechanical properties of ceramic matrix composites. *J. Mater. Sci.* 26:6341–51.
14. Becher, P. F., C. Hsueh, and P. Angelini. 1988. Toughening behavior in whisker reinforced ceramic matrix composites. *J. Am. Ceram. Soc.* 71:1050–61.
15. Brennan, J. J., and K. Prewo. 1982. Silicon carbide-fiber-reinforced-glass-ceramic matrix composites exhibiting high strength and toughness. *J. Mater. Sci.* 17(8):2371–83.
16. Prewo, K., and J. J. Brennan. 1980. High-strength silicon carbide fiber-reinforced glass-matrix composites. *J. Mater. Sci.* 15(2):463–68.
17. Prewo, K., and J. J. Brennan. 1982. Silicon-carbide-yarn-reinforced glass-matrix composites. *J. Mater. Sci.* 17(4):1201–06.
18. Fareed, A. S., et al. 1988. Fracture of silicon carbide/lithium ceramic composites. *Advances in Ceramics,* vol. 22: *Fractography of Glasses and Ceramics,* 261–78.
19. Ko, F., M. Koczak, and G. Layden. 1987. Structural toughening of glass matrix composites by 3-D fiber architecture. *Ceram. Eng. Sci. Proc.* 8:822–31.
20. Ko, F. 1989. Preform fiber architecture for ceramic matrix composites. *Am. Ceram. Soc. Bull.* 68(2):401–14.
21. Rolincik, P. G. 1992. Autoweave-technique for automated 3D weaving technology. Textron Specialty Materials, Lowell, MA.
22. Dharan, C. 1980. Pultruded braided hybrid composites. *12th Nat. SAMPE Tech. Conf.,* October 1980, Seattle, WA.
23. Drozda, T. J. 1989. *Composites Applications—The Future Is Now.* SME, Dearborn, MI.
24. Chou, T. W., and J. M. Yang. 1986. Structure-performance maps of polymeric, metal, and ceramic matrix composites. *Metall. Trans.* 17A:1547–59.

25. Riccitiello, S. R., W. L. Love, and A. B. Peterson. 1992. Evaluation of 2D ceramic matrix composites in aeroconvective environments. *24th Int. SAMPE Tech. Conf.,* pp. 1107–22, October 1992.

26. Khandelwal, P. K., and B. P. Johnson. 1992. Tensile and creep behavior of 2-D SiC/SiC composite. Advanced Turbine Technology Applications Project and Internal High Pressure Turbine Engine Technology Program, Allison Gas Turbine and Southern Research Institute Reports TE-1227, 1228, and 735, pp. 127–133.

27. Karandikar, P. G., A. Parvizi-Majidi, and T. W. Chou. 1992. Tensile, compressive and shear behaviors of 2D woven and 3D braided SiC/SiC composites. *Proc. Am. Soc. Compos. 7th Tech. Conf. Compos. Mater.,* pp. 755–82, Orlando, FL. Technomic, Lancaster, PA.

28. Cavalier, J. C., A. Lacombe, and J. P. Rouges. 1989. Ceramic-matrix composites, new materials with very high performances. *3rd Eur. Conf. Compos. Mater.,* ed. A. R. Bunsell, P. Lamicq, and A. Massian, pp. 99–110.

29. Cuttard, T., D. Fargeot, and C. Gault. 1990. Evolution of SiC/SiC and SiC/MASL mechanical properties in relation with the interface structure. Revue des composites et des matériaux avancés. *4th Eur. Conf. Compos. Mater.,* September 1990, Stuttgart.

30. Passilly, B., O. Sudre, and M. Parlier. 1990. Caractérisation mécanique des interfaces dans les composites céramiques. *4th Eur. Conf. Compos. Mater.,* September 1990, Stuttgart.

31. Reynaud, P., D. Rouby, and G. Fantozzi. 1991. A model describing the changes in ceramic-fibre composite under cyclic loading. *5th Eur. Conf. Compos. Mater.,* September 1991, Stuttgart.

32. Rouby, D., and P. Reynaud. 1993. Role of interfaces on the cyclic fatigue behavior of ceramic matrix composites. *AGARD Workshop Introd. Ceram. Aerosp. Struct. Compos.,* pp. 6-1 to 6-12, April 1993, Antalya, Turkey.

33. Gomina, M., and J. L. Chermant. 1993. Mechanical behavior of CMCs: Crack growth resistance and creep aspects. *AGARD Workshop Introd. Ceram. Aerosp. Struct. Compos.,* pp. 8-1 to 8-9, April 1993, Antalya, Turkey.

34. Lamicq, P., and D. Boury. 1993. Ceramic matrix composite parts design. *AGARD Workshop Introd. Ceram. Aerosp. Struct. Compos.,* pp. 12-1 to 12-10, April 1993, Antalya, Turkey.

35. Krenkel, W. 1993. CMC design consequences. *AGARD Workshop Introd. Ceram. Aerosp. Struct. Compos.,* pp. 13-1 to 13-14, April 1993, Antalya, Turkey.

36. Sygulla, D., A. Mühlratzer, and P. Agatonovic. 1993. Integrated approach in modelling, testing and design of gradient-CVI derived CMC components. *AGARD Workshop Introd. Ceram. Aerosp. Struct. Compos.,* pp. 14-1 to 14-10, April 1993, Antalya, Turkey.

37. Desnoyer, D., A. Lacombe, and J. M. Rouges. 1991. Large thin composite thermostructural parts. *Proc. Int. Conf. Spacecr. Struct. Mech. Test.,* April 1991, Noordwijk, Netherlands.

38. Mühlratzer, A. 1992. CVI-verfahren zur herstellung faserverstärker keramik. *CCG-Sem. Faserkeramikwerkst. Eigenschaften Anwendungen,* November 1992, Stuttgart.

39. Ostertag, R., and U. Trabandt. 1992. Fibre-reinforced ceramic components by filament winding for intermediate temperature applications. *Proc. Int. Symp. Adv. Mater. Lightweight Struct.,* March 1992. ESTEC, Noordwijk, Netherlands.

40. Sintec Keramik. 1991. Ceramic sintering and ceramic properties. Data Sheet.

41. Prewo, K., and J. J. Brennan. 1980. High strength silicon carbide fiber reinforced glass-matrix composites. *J. Mater. Sci.* 15(2):463–68.

42. Prewo, K., G. K. Layden, E. J. Minford, et al. 1985. Advanced characterization of silicon carbide fiber reinforced glass-ceramic matrix composites. UTRC Report R85-916629-1, ONR Contract NOOO14-81-C-0571.

43. Larsen, D. C., S. L. Stuchly, and S. A. Bortz. 1985. *Met. Matrix Carbon Ceram. Matrix Compos.*, pp. 313–34. NASA Conf. Pub. 2406.

44. Gac, F. D., J. V. Milewski, J. J. Petrovic, et al. 1985. *Met. Matrix Carbon Ceram. Matrix Compos.*, pp. 260–76. NASA Conf. Pub. 2406.

45. Lehman, R. L. 1988. Ceramic matrix fiber composites. In *Treatise on Material Science and Technology,* vol. 29, 229–92. Academic Press, New York.

46. Prewo, K., R. E. Tressler, et al. eds. 1985. *Tailoring Multiphase and Composite Ceramics,* 529–47. Plenum Press, New York.

47. Prewo, K. 1986. Tension and flexural strength of silicon carbide fiber-reinforced glass-ceramics. *J. Mater. Sci.* 21:3590–3600.

48. Mandell, J. F., D. H. Grande, and K. A. Dannemann. 1968. High temperature testing of glass/ceramic matrix composites. *ASTM 2nd Symp. Test Meth. Des. Allowables Fibrous Compos.*, November 1986, Phoenix, AZ.

49. Mandell, J. F., K. C. C. Hong, and D. M. Grande. 1987. *Ceram. Eng. Soc. Proc.* 8:937–39.

50. Prewo, K., and J. J. Brennan. 1989. Fiber reinforced glasses and glass ceramics for high performance applications. In *Reference Book for Composites Technology,* vol. 1, ed. S. M. Lee. Technomic, Lancaster, PA.

51. Rice, R. W., and D. Lewis III. 1989. Ceramic fiber composites based upon refractory polycrystalline ceramic matrices. In *Reference Book for Composites Technology,* vol. 1, ed. S. M. Lee. Technomic, Lancaster, PA.

52. Mah, T., M. G. Mendiratta, A. P. Katz, et al. 1987. Recent developments in fiber-reinforced high temperature ceramic composites. *Ceram. Bull.* 66:304.

53. Hasson, D. F., and S. G. Fishman. 1988. Impact behavior of fiber reinforced glass and ceramic matrix composites. *Proc. Int. Symp. Adv. Struct. Mater.*, ed. D. S. Wilkinson, August 1988, pp. 187–93, Montreal.

54. Hasson, D. F., and S.G. Fishman. 1991. Impact behavior of ceramic matrix composite materials. *6th Int. Conf. Compos. Mater.*, vol 2, ed. F. L. Matthews, N. C. R. Buskell, J. M. Hodgkinson, et al., 2.40–2.47. Elsevier, London.

55. Metcalfe, B. L., I. W. Donald, and D. J. Bradley. 1993. Preparation and properties of a SiC fibre-reinforced glass-ceramic matrix composite. *Brit. Ceram. Trans.* 92(1):13–20.

56. Jarman, D. C., G. K. Layden, P. H. McCluskey, et al. 1990. Advanced fabrication and characterization of fiber reinforced ceramic matrix composites. UTRC Report R-90-917548-4, Contract NOOO14-86-C-0649.

57. AGARD *Workshop Introd. Ceram. Aerosp. Struct. Compos.* pp. 1-1 to 16-12, April 1993.

58. Yamamura, T., T. Ishikawa, M. Shiouya, et al. 1989. An approach to manufacture a new type of ceramic composite. *Proc. 1st Jpn. Int. SAMPE Symp.*, pp. 1084–89, Osaka.

59. *Proc. 21st Univ. Conf. Ceram. Sci., Penn State Univ.*, July 1985. Plenum Press, New York.

60. Bhatt, R. T. 1991. Status and current directions for SiC/Si_3N_4 composites. *4th Ann. HITEMP Rev.*, pp. 67-1 to 67-4, October 1991. NASA Conf. Pub. 10082.

61. Bhatt, R. T. 1988. The properties of silicon carbide fiber-reinforced silicon nitride composites. In *Whisker- and Fiber-Toughened Ceramics,* ed. R. A. Bradley et al., 199–208. ASM International, Chicago, IL. NASA TM-101356.

62. Bhatt, R. T., and R. E. Phillips. 1990. Laminate behavior for SiC fiber-reinforced reaction-bonded silicon nitride matrix composites. *J. Compos. Technol. Res.* 12(1):13–23.

63. Bacon, J. F., K. Prewo, and R. D. Veltri. 1978. Glass matrix composites. II. Alumina reinforced glass. *Proc. 2nd Int. Conf. Compos. Ceram. Mater.*, pp. 753–69.

64. Michalske, T. A., and J. R. Hellmann. 1988. Strength and toughness of continuous-alumina-fiber-reinforced glass-matrix composites. *J. Am. Ceram. Soc.* 71(9):725–31.

65. Coblenz, W. S., R. W. Rice, D. Lewis, et al. 1984. Progress in ceramic refractory fiber composites. *Proc. NASA/DOD Conf. Met. Matrix Carbon Ceram. Matrix Compos.*, pp. 191–216.

66. Rice, R. W., J. R. Spann, D. Lewis, et al. 1984. The effect of ceramic fiber coatings on the room temperature mechanical behavior of ceramic-fiber composites. *Ceram. Eng. Sci. Proc.* 5(7/8):614–24.

67. Stuffle, K., W. Lougher, and S. Chanat. 1992. Preparation and characterization of continuous fiber reinforced zirconium diboride matrix composites for a leading edge material. *24th Int. SAMPE Tech. Conf.*, pp. T935–T949, October 1992.

68. Bull, J., D. Rasky, and J. Karika. 1992. Characterization of selected diboride composites. *16th Ann. Conf. Compos. Mater. Struct.*, January 1992, Cocoa Beach, FL.

69. Mazdiyasni, K. S. 1991. Ceramic composite matrices from metal organic precursors. *Mater. Sci. Eng.* A144:83–90.

70. Faber, K. T. 1982. Thesis, University of California, Berkeley.

71. Becher, P. F., and G. C. Wei. 1984. Toughening behavior in SiC_w reinforced alumina. *J. Am. Ceram. Soc.* 67(12):267–69.

72. Tiegs, T. N., and P. F. Becher. 1987. Sintered Al_2O_3-SiC-whisker composites. *Am. Ceram. Soc. Bull.* 66(2):339–42.

73. Becher, P. F., and T. N. Tiegs. 1987. *J. Am. Ceram. Soc.* 70(9):651–54.

74. *Fifth Eur. Conf. Compos. Mater.*, September 1991, Stuttgart.

75. Geiger, G. 1991. Progress continues in composite technology. *Ceram. Bull.* 70(2):212–19.

76. Yust, C. S., and C. E. DeVore. 1990. The friction and wear of lubricated Si_3N_4/SiC_w composites. Contract DE-AC05-84OR21400, Conf.-901071-7, DE90015790.

77. Choi, S. R., and J. A. Salem. 1992. Strength, toughness and *R*-curve behavior of SiC whisker-reinforced composite Si_3N_4 with reference to monolithic Si_3N_4. *J. Mater. Sci.* 27(6):1491–98.

78. Ohji, T., and Y. Yamauchi. 1993. Tensile creep and creep rupture behavior of monolithic and SiC-whisker-reinforced silicon nitride ceramics. *J. Am. Ceram. Soc.* 76(12):3105–12.

79. Zheng, X. Y., F. P. Zeng, M. J. Pomeroy, et al. 1990. Reinforcement of silicon nitride ceramics with whiskers and platelets. In *Fabrication Technology,* British Ceramic Proceedings No. 45, ed. R. W. Davidge and D. P. Thompson, 187–98. Institute of Ceramics, Stokes-on-Trent, U.K.

80. Ohji, T., Y. Goto, and A. Tsuge. 1991. High-temperature toughness and tensile strength of whisker-reinforced silicon nitride. *J. Am. Ceram. Soc.* 74(4):739–45.

81. Chu, C.-Y., and J. P. Singh. 1990. Mechanical properties and microstructures of Si_3N_4-whisker-reinforced Si_3N_4 matrix composites. *Ceram. Eng. Sci. Proc.* 11(7–8):709–20.

82. Majidi, A. P., and T.-W. Chou. 1989. Elevated temperature studies of continuous and discontinuous fiber reinforced ceramic matrix composites. University of Delaware, Newark. ASME 89 GT 124.

83. Kragness, E. D., M. F. Amateau, and G. L. Messing. 1991. Processing and characterization of laminated SiC whisker reinforced Al_2O_3. *J. Compos. Mater.* 25:416–32.

84. Becher, P. F., P. Angelini, W. H. Warwick, et al. 1990. Elevated-temperature-delayed failure of alumina reinforced with 20 vol% silicon carbide whiskers. *J. Am. Ceram. Soc.* 73(1):91–96.

85. Yang, M., and R. Stevens. 1991. Microstructure and properties of SiC whisker reinforced ceramic composites. *J. Mater. Sci.* 26(3):726–36.

86. Giannakopoulos, A. E., and K. Breder. 1991. Synergism of toughening mechanisms in whisker-reinforced ceramic-matrix composites. *J. Am. Ceram. Soc.* 74(1):194–202.

87. Lin, G. Y., T. C. Lei, Y. Zhou, et al. 1993. Mechanical properties of Al_2O_3 and $Al_2O_3 + ZrO_2$ ceramics reinforced by SiC whiskers. *J. Mater. Sci.* 28(10):2745–49.

88. Baek, Y. K., and C. H. Kim. 1989. The effect of whisker length on the mechanical properties of alumina-SiC whisker composites. *J. Mater. Sci.* 24:1589–93.

89. Becher, P. F., and T. N. Tiegs. 1988. Temperature dependence of strengthening by whisker reinforcement: SiC whisker reinforced alumina in air. *Adv. Ceram. Mater.* 3(2):148–54.

90. Wei, G. C., and P. F. Becher. 1985. Development of SiC-whisker-reinforced ceramics. *Am. Ceram. Soc. Bull.* 64:289–304.

91. Tiegs, T. N., and P. F. Becher. 1987. Thermal shock behavior of an alumina-SiC whisker composite. *J. Am. Ceram. Soc.* 70:C-109 to C-111.

92. Becher, P. F., T. N. Tiegs, J. C. Ogle, et al. 1986. In *Fracture Mechanics of Ceramics,* vol. 7, ed. R. C. Bradt, A. G. Evans, D. P. H. Hasselman, and F. F. Lange, pp. 61–73. Plenum Press, New York.

93. Shih, C. J., J.-M. Yang, and A. Ezis. 1990. Processing and performance of several SiC whisker-reinforced Al_2O_3 matrix composites. *Mater. Manuf. Process.* 5(1):35–49.

94. Bray, D. J. 1988. Reinforced alumina composites. *SME Fabric. Compos. 1988 Conf.,* pp. 3–24, September 1988, Dearborn, MI.

95. Liu, H., M. E. Fine, and H. S. Cheng. 1994. Lubricated rolling wear of SiC-whisker-reinforced Al_2O_3 composites against M2 tool steel. *J. Am. Ceram. Soc.* 77(1):179–85.

96. Kunz, S. M., K. Chia, C. H. McMurtry, et al. 1988. Interface considerations in ceramic matrix composites. *SME Fabric. Compos. 1988 Conf.,* pp. 36–57, September 1988, Dearborn, MI.

97. Lin, H.-T., and P. F. Becher. 1991. High-temperature creep deformation of alumina-SiC-whisker composites. *J. Am. Ceram. Soc.* 74(8):1886–93.

98. Weiderhorn, S. M., and B. J. Hockey. 1991. High temperature degradation of structural composites. *Ceram. Int.* 17(4):243–52.

99. Lin, H. T., and P. F. Becher. 1990. Creep behavior of a SiC whisker reinforced alumina. *J. Am. Ceram. Soc.* 73(5):1378–81.

100. Niihara, K., A. Nakahira, T. Uchiyama, et al. 1986. High-temperature mechanical properties of Al_2O_3-SiC composites. In *Fracture Mechanics of Ceramics,* vol. 7, ed. R. C. Bradt, A. G. Evans, D. P. H. Hasselman, and F. F. Lange, pp. 103–16. Plenum Press, New York.

101. Lipetzky, P., S. R. Nutt, and P. F. Becher. 1988. Creep behavior of an Al_2O_3-SiC composite. *Mater. Res. Soc. Symp. Proc.* 120:271–77.

102. Lei, T. C., W. Z. Zhu, and Y. Zhou. 1991. Mechanical properties and toughening mechanisms of SiC_w/Al_2O_3 ceramic composites. *Mater. Chem. Phys.* 28(1):89–97.

103. Ricoult, M. B. 1991. Oxidation behavior of SiC-whisker-reinforced alumina-zirconia composites. *J. Am. Ceram. Soc.* 74(8):1793–802.

104. Yasuda, E., T. Akatsu, Y. Tanabe, et al. 1988. Development of SiC whisker reinforced alumina. *Proc. Int. Congr. Adv. Compos. New Syst. Technol.,* September 1988, Weisbaden, Germany.

105. Niihara, K., A. Nakahira, G. Sasaki, et al. 1988. Development of strong Al_2O_3/SiC composites. *Proc. Int. Meet. Adv. Mater.* MRS, Pittsburgh, PA.

106. Akimune, Y., Y. Katano, and Y. Shichi. 1987. Microstructure and mechanical properties of SiC-whisker/Y-TZP composites. *Sintering,* Tokyo.

107. Yang, M., and R. Stevens. 1990. Fabrication of SiC whisker reinforced Al_2O_3 composites. *J. Mater. Sci.* 25:4658–66.

108. Homeny, J., W. L. Vaughn, and M. K. Ferber. 1987. *Am. Ceram. Soc. Bull.* 66(2):681–86.

109. Akimune, Y., Y. Katano, and Y. Shichi. 1988. Mechanical properties and microstructure of air-annealed SiC-whisker/Y-TZP composites. *Adv. Ceram. Mater.* 3(3):138–42.

110. Zhou, Y., W.-Z. Zhu, and T.-C. Lei. 1992. Mechanical properties and toughening mechanisms of SiC_w/ZrO_2 (6% mol Y_2O_3) ceramic composites. *Ceram. Int.* 18(3):141–45.

111. Yasuda, E., et al. 1987. Mechanical properties of carbon-coated SiC-whiskers/Al_2O_3 composites. *Proc. 25th Basic Discuss. Meet. Ceram.*, p. 41. Ceramic Society of Japan, Tokyo.

112. Yasuda, E., et al. 1986. Mechanical properties of SiC whisker reinforced PSZ. *Proc. 3rd Int. Conf. Sci. Technol. Zirconia,* p. 358. Ceramic Society of Japan, Tokyo.

113. Kondo, I., and N. Tamari. 1987. Sintering and mechanical properties of SiC whisker-reinforced zirconia. *Proc. Ann. Meet. Ceram. Soc. Jpn.* Ceramic Society of Japan, Tokyo.

114. Office of Naval Research Far East, vol. 12(4):1987.

115. Koczak, M. J., K. Prewo, A. Mortensen, et al. 1991. Inorganic composite materials in Japan: Status and trends. Office of Naval Research Far East M7.

116. Tian, J., X. Tong, L. Dong, et al. 1994. Reinforcing mechanism of SiC whiskers in SiC_w/ carbon ceramic composites. *4th Int. Symp. Ceram. Mater. Components Eng.*, ed. R. Carlsson, T. Johansson, and L. Kahlman, 804–12. Elsevier, London.

117. Tiegs, T. N., and P. F. Becher. 1987. *Am. Ceram. Soc. Bull.* 66:339–42.

118. Ruh, R., and K. S. Mazdiyasni. 1990. Mullite-SiC-whisker-reinforced ceramic composites: Characterization and properties. In *Fiber Reinforced Ceramic Composites,* ed. K. S. Mazdiyasni. Noyes, Park Ridge, NJ.

119. Ruh, R., and K. S. Mazdiyasni. 1985. Fabrication of mullite-SiC whisker composites and mullite partially stabilized ZrO_2-SiC whisker composites. NASA-DOD Adv. Compos. Work. Group, January 1985. NASA Lewis Research Center, Cleveland, OH.

120. Mah, T., M. G. Mendiratta, A. P. Katz, et al. 1985. *J. Am. Ceram. Soc.* 68(9):C-248.

121. Prewo, K., J. J. Brennan, and G. K. Layden. 1986. *Am. Ceram. Soc. Bull.* 65:305–13.

122. Xue, L. A., and I.-W. Chen. 1993. A new SiC-whisker-reinforced lithium aluminosilicate composite. *J. Am. Ceram. Soc.* 76(11):2785–89.

123. McMahon, G., S. S. B. Wang, D. H. H. Quon, et al. 1991. A study on SiC whisker reinforced lithium aluminosilicate composites. *MPIF Proc. Powder Metall. Aerosp. Def. Technol.,* ed. F. H. Froes, March 1991, Tampa, FL.

124. Brennan, J. J., and S. R. Nutt. 1992. SiC-whisker-reinforced glass-ceramic composites: Interfaces and properties. *J. Am. Ceram. Soc.* 75(5):1205–16.

125. Gadkaree, K., and K. Chyung. 1986. Silicon carbide whisker reinforced glass and glass-ceramic composites. *Am. Ceram. Soc. Bull.* 65:370–76.

126. Vaidya, R. U., and K. N. Subramanian. 1991. Elevated temperature mechanical properties of continuous metallic glass ribbon-reinforced glass-ceramic matrix composites. *J. Mater. Sci.* 26(5):1391–94.

127. Singh, R. N. 1991. Mechanical properties of a zircon matrix composite reinforced with silicon carbide whiskers and filaments. *J. Mater. Sci.* 26(7):1839–46.

128. Claussen, N., and G. Petzow. 1986. Whisker-reinforced oxide ceramics. *J. Phys.* 47(2):Cl-693–702.

129. Tiegs, T. N., and P. F. Becher. 1985. Whisker reinforced ceramic composites. In *Tailoring Multiphase and Ceramic Composites,* ed. N. Claussen and G. Petzow. Plenum Press, New York.

130. Claussen, N., K. L. Weisskopf, and M. Ruhle. 1986. Tetragonal zirconia polycrystals reinforced with SiC whiskers. *J. Am. Ceram. Soc.* 69(3):288–92.

131. Niwano, K. 1991. Properties and applications of SiC whiskers. In *Silicon Carbide Ceramics 2: Gas Phase Reactions, Fibers and Whisker Joining,* ed. S. Somiya and Y. Inomata, 99–116, Ceramic Research and Development in Japan Series. Elsevier, Amsterdam.

132. Claussen, N., and G. Petzow. 1985. Processing and microstructure of SiC whisker reinforced Z_rO_2-toughened ceramics (ZTC). *Proc. Conf. Tailoring Multiphase Compos. Ceram.,* ed. R. E. Tressler et al. Plenum Press, New York.

133. Lehmann, J., B. Muller, and G. Ziegler. 1989. *Euro-Ceramics,* vol. 1, 196–200.

134. Pyzik, A. J., and D. R. Beaman. 1993. Microstructure and properties of self-reinforced silicon nitride. *J. Am. Ceram. Soc.* 76(11):2737–44.

135. Dusza, J., D. Sajgalik, and M. Reece. 1991. Analysis of Si_3N_4 + β-Si_3N_4 whisker ceramics. *J. Mater. Sci.* 26(24):6782–88.

136. Gac, F. D., and J. J. Petrovic. 1985. Feasibility of a composite of SiC whiskers in an $MoSi_2$ matrix. *J. Am. Ceram. Soc.* 68:C-200.

137. Gibbs, W. S., J. J. Petrovic, and R. E. Honell. 1987. SiC whisker-$MoSi_2$ matrix composites. *Ceram. Eng. Sci. Proc.* 8:645.

138. Meschter, P. J., and D. S. Schwartz. 1989. Silicide-matrix materials for high temperature applications. *J. Met.* November:52.

139. Carter, D. H., W. S. Gibbs, and J. J. Petrovic. 1989. Mechanical characterization of SiC-whisker-reinforced $MoSi_2$. *Proc. 3rd Int. Symp. Ceram. Mater. Components Heat Eng.,* p. 977. American Ceramic Society, Westerville, OH.

140. Petrovic, J. J., and R. E. Honell. 1990. SiC reinforced $MoSi_2/WSi_2$ alloy matrix composites. *Ceram. Eng. Sci. Proc.* 11:734.

141. Sadananda, K., H. Jones, and J. Feng. 1991. Creep of monolithic and SiC whisker-reinforced $MoSi_2$. *Ceram. Eng. Sci. Proc.* 12(9–10):1671–78.

142. Cook, J., R. Mahapatra, E. W. Lee, et al. 1991. Oxidation behavior of $MoSi_2$ composites. *Ceram. Eng. Sci. Proc.* 12(9–10):1656–70.

143. Rice, R. W., K. R. McKinney, C. Cm. Wu, et al. 1985. *J. Mater. Sci.* 20:1392–1406.

144. Japan External Trade Organization. 1992. New SiC particles for reinforcing ceramics, 92-03-001-02, March 1992, p. 18.

145. Akimune, Y., N. Hirosaki, and T. Ogasawara. 1991. Mechanical properties of SiC-particle/sialon composites. *J. Mater. Sci. Lett.* 10:223–26.

146. Huang, X. X., J. S. Hong, and J. Guo. 1994. SiC particle and T-TZP reinforced mullite matrix composites. *4th Int. Symp. Ceram. Mater. Components Eng.,* ed. R. Carlsson, T. Johansson, and L. Kahlman, 795–803. Elsevier, London.

147. Pickard, S. M., E. Manor, H. Ni, et al. 1992. The mechanical properties of ceramic composites produced by melt oxidation. *Acta Metall. Mater.* 40(1):177–84.

148. Lasday, S. B. 1993. Production of ceramic matrix composites by CVI and DMO for industrial heating industry. *Ind. Heat.* April:31–35.

149. Antolin, P. B., and J. Weinstein. 1991. Improved corrosion resistance of ceramic-matrix composites. *Ceram. Bull.* 70(3):336–40.

150. Henderson, T. J., C. A. Anderson, and J. D. Stachiw. 1991. Novel ceramic matrix composites for deep submergence pressure vessel applications. Naval Ocean Systems Center, San Diego, CA. AD-A242 740, NOSC TD 2222.

151. Fareed, A. S., D. J. Landini, T. A. Johnson, et al. 1993. High-temperature ceramic matrix composites by the directed metal oxidation process. SME Technical Paper EM92-216, SME, Dearborn, MI; *Ind. Heat.* April:31–35.

152. Tsuge, A., K. Nishida, and M. Komatsu. 1975. *J. Am. Ceram. Soc.* 58:323.

153. Tsuge, A., H. Inoue, and K. Komeya. 1979. *81st Ann. Meet. Am. Ceram. Soc.*, April 1979, Cincinnati, OH.

154. Komeya, K., and H. Inoue. 1984. Development of nitrogen ceramics. *Am. Ceram. Soc. Bull.* 63(9):1158–64.

155. Kawashima, K., H. Okamoto, et al. 1991. Grain size dependence of the fracture toughness of silicon nitride ceramics. *J. Am. Ceram. Soc.* 99:320–23.

156. Bengisu, M., and O. T. Inal. 1991. Rapidly sintered particulate ceramic matrix composites. *Ceram. Int.* 17(3):187–98.

157. Matsumoto, Y., K. Hirota, and O. Yamaguchi. 1993. Mechanical properties of hot isostatically pressed zirconia-toughened alumina ceramics prepared from coprecipitated powders. *J. Am. Ceram. Soc.* 76(10):2677–80.

158. Hori, S., M. Yoshimura, S. Somiya, et al. 1985. Mechanical properties of ZrO_2-toughened Al_2O_3 ceramics from CVD powders. *J. Mater. Sci. Lett.* 4(4):413–16.

159. Ismail, M. G. M. U., H. Shiga, K. Katayama, et al. 1994. Microstructure and mechanical properties of ZrO_2 toughened mullite synthesized by sol-gel method. *4th Int. Symp. Ceram. Mater. Components Eng.*, ed. R. Carlsson, T. Johansson, and L. Kahlman, 381–92. Elsevier, London.

160. Petrovic, J. J., R. E. Honnell, T. E. Mitchell, et al. 1991. ZrO_2-reinforced $MoSi_2$ matrix composites. *Ceram. Eng. Sci. Proc.* 12(9–10):1633–42.

161. Seshadri, S. G., M. Srinivasan, J. W. MacBeth, et al. 1990. Fabrication and mechanical reliability of SiC/TiB_2 composites. In *Ceramic Materials and Components for Engines,* ed. V. Tennery, 1419–37. Las Vegas, NV.

162. Jimbou, R., K. Takahashi, and Y. Matsushita. 1986. SiC-ZrB_2 electroconductive ceramic composite. *Adv. Ceram. Mater.* 1:341.

163. McMurtry, C. H., W. D. Boecker, G. Seshadri, et al. 1987. Microstructural and material properties of SiC-TiB_2 particulate composites. *Am. Ceram. Soc. Bull.* 66:325–329.

164. Janney, M. A. 1986. Microstructural development and mechanical properties of SiC and of SiC-TiC composites. *Am. Ceram. Soc. Bull.* 64(5):357.

165. Janney, M. A. 1987. Mechanical properties and oxidation behavior of a hot-pressed SiC-15 vol% TiB_2 composite. *Am. Ceram. Soc. Bull.* 66(2):322.

166. Ly Ngoc, D. 1989. Gefugeverstarkung von SiC-keramiken. Doctoral Thesis, University of Stuttgart.

167. Tani, T., and S. Wada. 1991. Pressureless-sintered and HIPed SiC-TiB_2 composites from SiC-TiO_2-B_4C-C powder compacts. *J. Mater. Sci.* 26(13):3491–96.

168. Liu, J., and P. D. Ownby. 1991. Enhanced mechanical properties of alumina by dispersed titanium diboride particulate inclusions. *J. Am. Ceram. Soc.* 74(1):241–43.

169. Endo, H., M. Ueki, and H. Kubo. 1991. Microstructure and mechanical properties of hot-pressed SiC-TiC composites. *J. Mater. Sci.* 26(14):3769–74.

170. Kang, E. S., and C. H. Kim. 1990. *J. Mater. Sci.* 25:580.

171. Aikin, R. M., Jr. 1991. Structure and properties of in situ reinforced $MoSi_2$. *Ceram. Eng. Sci. Proc.* 12(9–10):1643–55.

172. Kao, M.-Y. 1993. Properties of silicon nitride-molybdenum disilicide particulate ceramic composites. *J. Am. Ceram. Soc.* 76(11):2879–83.

173. Janssen, R., and K.-H. Heussner. 1991. Platelet-reinforced ceramic composite materials. *Powder Met. Int.* 23(4):241–45.

174. Baril, D., S. P. Tremblay, and M. Fiset. 1993. Silicon carbide platelet-reinforced silicon nitride composites. *J. Mater. Sci.* 28(20):5486–94.

175. Richardson, K. R., and D. W. Freitag. 1991. Mechanical properties of hot pressed SiC platelet-reinforced $MoSi_2$. *Ceram. Eng. Sci. Proc.* 12(9–10):1679–89.

176. Carter, D. H., W. S. Gibbs, and J. J. Petrovic. 1988. Mechanical characterization of SiC whisker-reinforced $MoSi_2$, LA-UR-883776. LANL, Los Alamos, NM.

177. Lehmann, J., and G. Ziegler. 1990. Oxide-based ceramic composites. *4th Eur. Conf. Compos. Mater.: Dev. Sci. Technol. Compos. Mater.*, pp. 425–33, September 1990, Stuttgart.

178. Huang, X.-N., and P. S. Nicholson. 1993. Mechanical properties and fracture toughness of α-Al_2O_3-platelet-reinforced Y-PSZ composites at room and high temperatures. *J. Am. Ceram. Soc.* 76(5):1294–1301.

179. Cutler, R. A., R. J. Mayhew, K. M. Prettyman, et al. 1991. High-toughness Ce-TZP/Al_2O_3 ceramics with improved hardness and strength. *J. Am. Ceram. Soc.* 74(1):179–86.

180. Ravi, V. A., T. D. Claar, B. Berelsman, et al. 1994. Zirconium-based composites for orthopedic applications. *J. Met.* 46(2):50–2.

181. Johnson, W. B., T. D. Claar, and G. H. Schiroky. 1989. Preparation and processing of platelet reinforced ceramics by the directed reaction of zirconium with boron carbide. *Ceram. Eng. Sci. Proc.* 10(7/8):588–98.

182. Claar, T. D., et al. 1989. Microstructure and properties of platelet reinforced ceramics formed by the directed reaction of zirconium with boron carbide. *Ceram. Eng. Sci. Proc.* 10(7/8):599–609.

183. Johnson, W. B., A. S. Nagelberg, and E. Breval. 1991. Kinetics of formation of a platelet-reinforced ceramic composite prepared by the directed reaction of zirconium with boron carbide. *J. Am. Ceram. Soc.* 74:2093–2101.

184. Ravi, V. A., et al. 1992. Platelet reinforced ceramics for severe thermal shock applications. In *Processing and Fabrication of Advanced Materials for High Temperature Applications,* ed. V. A. Ravi and T. S. Srivatsan, pp. 175–83. TMS, Warrendale, PA.

185. Ravi, V. A., et al. 1993. Zirconium diboride/zirconium carbide/zirconium composites for orthopaedic applications. *Trans. 19th Ann. Meet. Soc. Biomater.* vol. XVI, p. 142. Minneapolis, MN.

186. Heimke, G. 1984. Structural characteristics of metals and ceramics. In *Metal and Ceramic Biomaterials,* vol. I, ed. P. Ducheyne and G. W. Hastings, 46. CRC Press, Boca Raton, FL.

187. Soltesz, U., and H. Richter. 1984. Mechanical behavior of selected ceramics. In *Metal and Ceramic Biomaterials,* vol. II, ed. P. Ducheyne and G. W. Hastings, 46. CRC Press, Boca Raton, FL.

188. Davidson, J. A., and F. S. Georgette. 1989. State of the art materials for orthopedic prosthetic devices. In *Composites Applications—The Future Is Now,* ed. T. J. Drozda, 406. SME, Dearborn, MI.

189. *Structural Alloys Handbook,* vol. 3. 1990. Battelle MCIC, Columbus, OH.

190. Davidson, J. A., A. K. Mishra, and P. Kovacs. 1993. A new low-modulus, high-strength, biocompatible Ti-13Nb-13Zr alloy for orthopaedic implants. *Trans. 19th Ann. Meet. Soc. Biomater.,* vol. XVI, 145, Minneapolis, MN.

191. Christel, P., et al. 1989. Mechanical properties and short-term in-vivo evaluation of yttrium-oxide partially-stabilized zirconia. *J. Biomed. Mater. Res.* 23:49.

192. Mears, D. C. 1977. Metals in medicine and surgery. *Int. Met. Rev.* June:131.

193. Niihara, K. 1991. *Ann. Chim. Paris* 16:479.

194. Niihara, K. 1991. *J. Am. Ceram. Soc. Jpn.* 99(10):974.

195. Niihara, K., and A. Nakahira. 1991. *Advanced Structural Inorganic Composites,* ed. P. Vincenzini, 637. Elsevier, Amsterdam.

196. Niihara, K. 1989. *Proc. MRS Int. Meet. Adv. Mater.,* vol. 4, ed. Y. Hamano, O. Kamigaito, T. Kishi, et al., 129. Materials Research Society, Pittsburgh, PA.

197. Niihara, K. 1990. *J. Jpn. Soc. Powder Metall.* 37:348.

198. Niihara, K. 1989. *Proc. 1st Jpn. Int. SAMPE Symp.,* ed. N. Igata, I. Kinbara, T. Kishi, et al., 1120. Nikkan Kogyo Shinbun, Tokyo.

199. Niihara, K. 1989. *Fine Ceramics,* ed. H. Okuda, T. Izeki, K. Komeya, et al. Japan Standardization Association, Tokyo.

200. Marshall, D. B., and M. V. Swain. 1988. Crack resistance curves in magnesia partially-stabilized zirconia. *J. Am. Ceram. Soc.* 71:399–407.

201. Ko, F. K. 1989. Preform fiber architecture for ceramic-matrix composites. *Ceram. Bull.* 68:401–14.

202. Cox, B. N., et al. 1992. Mechanisms of compressive failure in 3-D composites. *Acta Metall. Mater.* 40:3285–98.

BIBLIOGRAPHY

Adv. Mater. Process. 146(1), 1994.

Brockmeyer, J. W. Ceramic matrix composite applications in advanced liquid fuel rocket engine turbomachinery. *ASME Int. Gas Turbine Aeroeng. Congr. Exposit.,* ASME-92-GT-316, June 1992.

Buckley, J. D. *Proc. 15th NASA/DOD Conf. Met. Matrix Carbon Ceram. Matrix Compos.* Cocoa Beach, FL, 1991. NASA CP 3133.

Butkus, L. M., J. W. Holmes, and T. Nicholas. Thermomechanical fatigue behavior of a silicon carbide fiber-reinforced calcium aluminosilicate composite. *J. Am. Ceram. Soc.* 76(11):2817–25,1993.

Carroll, D. R., and L. R. Dharani. Effect of temperature on the ultimate strength and modulus of whisker-reinforced ceramics. *J. Am. Ceram. Soc.* 75(4):786–94, 1992.

Ceramic Technology International, 1995. Sterling, Great Britain.

Chou, T.-W., and A. P.- Majidi. Elevated temperature behavior of glass and ceramic matrix composites. July 1987–May 1991. Final Report AFOSR-87-0383.

Chu, C.-Y., J. P. Singh, and J. L. Routbort. High-temperature failure mechanisms of hot-pressed Si_3N_4 and Si_3N_4/Si_3N_4-whisker-reinforced composites. *J. Am. Ceram. Soc.* 76(5):1349–55, 1993.

Cutler, R. A., K. M. Rigtrup, and A. V. Virkar. Synthesis, sintering, microstructure, and mechanical properties of ceramics made by exothermic reactions. *J. Am. Ceram. Soc.* 75(1):36–43, 1992.

Danchaivijit, S., and D. K. Shetty. Matrix cracking in ceramic-matrix composites. *J. Am. Ceram. Soc.* 76(10):2497–2504, 1993.

DAUSKARDT, R. H., R. O. RITCHIE, AND B. N. COX. Fatigue of advanced materials. Part I, *Adv. Mater. Process.* 144(1):26–31; Part II, 144(2):30–35.

DICARLO, J. A. CMCs for the long run. *Adv. Mater. Proc.* June:41–44, 1989.

ECKEL, A. J., AND T. P. HERBELL. Ceramic composites survive severe thermal shocks. NASA Technical Brief, December 1993. p. 64.

FANG, N. J.-J., AND T.-W. CHOU. Characterization of interlaminar shear strength of ceramic matrix composites. *J. Am. Ceram. Soc.* 76(10):2539–48, 1993.

FOHEY, W., M. BATTISON, J. HALADA, ET AL. Evaluation of ceramic composites for turbine engine components. WL-TR-92-4019, July 1992; FR-July 1986–January 1992.

GOGOTSI, Y. G. Particulate silicon nitride-based composites. *J. Mater. Sci.* 29(10):2541–56, 1994.

HITEMP Rev. 1991, Cleveland, OH, October 1991. NASA Conf. Pub. 10082.

HOLMES, J. W. Tensile creep behaviour of a fibre-reinforced SiC-Si_3N_4 composite. *J. Mater. Sci.* 26(7):1808–14, 1991.

HOLMES, J. W., Y. H. PARK, AND J. W. JONES. Tensile creep and creep-recovery behavior of a SiC-fiber-Si_3N_4-matrix composite. *J. Am. Ceram. Soc.* 76(5):1281–93, 1993.

HSU, J.-Y., AND R. F. SPEYER. Fabrication and properties of SiC fibre-reinforced $LiO_2 \cdot Al_2O_3 \cdot 6SiO_2$ glass-ceramic composites. *J. Mater. Sci.* 27(2):381–90, 1992.

JAMET, J. F. Fiber-reinforced glass-ceramic matrix composites: New class of materials for space applications. Societe Nationale Industrielle Aerospatiale, France, 1991. ETN 91-99279, 911-430-107.

JIANG, Z. Z., Z. T. ZHANG, AND Y. HUANG. Preparation and high temperature strengthening of tetragonal zirconia-mullite whisker composites. *Brit. Ceram. Trans.* 93(4):154–56.

JONES, R. H., C. H. HENAGER, JR., AND P. F. TORTORELLI. Elevated-temperature effects of oxygen on SiC/SiC composites. *J. Met.* 45(12):11–13, 1993.

LASDAY, S. B. Ceramic/ceramic composite components advance furnace systems and processes. *Ind. Heat.* April:26–29, 1993.

LEONARD, L. Ceramic-matrix composites: Mettle for the nasty jobs. *Adv. Compos.* July/August:37–43, 1990.

LEUNG, S., E. G. MEHRTENS, G. T. STEVENS, ET AL. On high temperature mechanical and fracture properties of an Al_2O_3/SiC_w ceramic matrix composite. *J. Mater. Sci. Lett.* 13(1):817–20, 1994.

LIN, B. W., AND T. ISEKI. Different toughening mechanisms in SiC/TiC composites. *Brit. Ceram. Trans. J.* 91:147–50, 1992.

LIN, G. Y., T. C. LEI, Y. ZHOU, ET AL. Microstructure and mechanical properties of SiC whisker reinforced ZrO_2-6 mol% Y_2O_3 composites. *Mater. Sci. Technol.* 9(8):659–64, 1993.

LIN, R. Y., R. J. ARSENAULT, G. P. MARTINS, ET AL., EDS. *Proc. TMS Ann. Meet.: Interfaces Met.-Ceram. Compos.,* Anaheim, CA, February 1990.

LUCAS, K. A., AND H. CLARKE. *Corrosion of Aluminum-Based Metal Matrix Composites.* John Wiley, New York, 1993.

MAKINO, A., AND C. K. LAW. SHS combustion characteristics of several ceramics and intermetallic compounds. *J. Am. Ceram. Soc.* 77(3):778–86, 1994.

MALLICK, P. K. *Fiber-Reinforced Composites: Materials, Manufacturing, and Design,* 2nd ed., Mechanical Engineering, vol. 83. Marcel Dekker, 1993.

MARSH, G. Engineering ceramics. Part 2. *Aerosp. Compos. Mater.* 24–26, 1990.

METCALFE, B. L., W. DONALD, AND D. J. BRADLEY. Development and properties of a SiC fibre-reinforced magnesium aluminosilicate glass-ceramic matrix composite. *J. Mater. Sci.* 27(11):3075–81, 1992.

MITCHELL, M. R., AND O. BUCK, EDS. *Cyclic Deformation, Fracture, Nondestructive Evaluation of Advanced Materials,* STP 1157. ASTM, Philadelphia.

MORRIS, W. L., B. N. COX, D. B. MARSHALL, ET AL. Fatigue mechanisms in graphite/SiC composites at room and high-temperature. *J. Mater. Sci.* 28:792–800, 1993.

NIEH, T. G., AND J. WADSWORTH. Superplasticity in fine-grained 20% Al_2O_3/YTZ composite. *Acta Metall. Mater.* 39(12):3037–45, 1991.

PADTURE, N. P. In situ-toughened silicon carbide. *J. Am. Ceram. Soc.* 77(2):519–23, 1994.

PORTER, J. R. Reinforcements for ceramic-matrix composites for elevated temperature applications. *Mater. Sci. Eng.* A166:179–84, 1993.

RANJBAR, K., B. T. RAO, T. R. R. MOHAN, ET AL. Effect of chemically added zirconia and yttria on the mechanical properties of zirconia-dispersed alumina. *Am. Ceram. Soc. Bull.* 73(2):63–66, 1994.

REVANKAR, V. V. S., AND V. HLAVACEK. Making ceramic fibers by chemical fibers. Lewis Research Center, NASA Technical Brief, December 1994, p. 82.

RICCITIELLO, S. R., W. L. LOVE, AND W. C. PITTS. A ceramic matrix composite thermal protection system for hypersonic vehicles. *SAMPE Q.* 24(4):10–17, 1993.

Schott Corporation. Fiber reinforcement makes glass ductile. *Adv. Mater. Process.* March:4, 1994.

SHEPPARD, L. M. Toward economical processing of composites. *Ceram. Ind.* March:79–83.

SHIH, C. J., J.-M. YANG, AND A. EZIS. Microstructure and properties of reaction-bonded/hot-pressed SiC_w/Si_3N_4 composites. *Comp. Sci. Technol.* 43:13–23, 1992.

SHIN, D.-W., AND H. TANAKA. Low-temperature processing of ceramic woven fabric/ceramic matrix composites. *J. Am. Ceram. Soc.* 77(1):97–104, 1994.

SINGH, R. N. SiC fiber-reinforced zircon composites. *Ceram. Bull.* 70(1):55–6, 1991.

SINGH, R. N. Interfacial properties and high-temperature mechanical behavior of fiber-reinforced ceramic composites. *Mater. Sci. Eng.* A166:185–98, 1993.

SISKIND, K. S. Glass-ceramic matrix composites for advanced gas turbines, 1990. AIAA-90-2014.

SORNAKUMAR, T., V. E. ANNAMALAI, R. KRISHNAMURTHY, ET AL. Mechanical properties of composites of alumina and partially stabilized zirconia. *J. Mater. Sci. Lett.* 12(16):1283–85, 1993.

SPEARING, S. M., F. W. ZOK, AND A. G. EVANS. Stress corrosion cracking in a unidirectional ceramic-matrix composite. *J. Am. Ceram. Soc.* 77(2):562–70, 1994.

STIEFF, P. S., AND A. TROJNACKI. Bend strength versus tensile strength of fiber-reinforced ceramics. *J. Am. Ceram. Soc.* 77(1):221–29, 1994.

TENNERY, V. J., ED. *Proc. 3rd Int. Symp. Ceram. Mater. Compos. Eng.* Las Vegas, NV, November 1988.

UPADHYAYA, D. D., P. Y. DALVI, AND G. K. DEY. Processing and properties of Y-TZP/Al_2O_3 composites. *J. Mater. Sci.* 28:6103–06, 1993.

URQUHART, A. W. Novel reinforced ceramics and metals: A review of Lanxide's composite technologies. *Mater. Sci. Eng.* A144:75–82.

VAIDA, R. U., AND K. N. SUBRAMANIAN. Metallic glass-ribbon-reinforced glass-ceramic matrix composites. *J. Mater. Sci.* 25:3291–96, 1990.

WACHTMAN, J. B., JR., R. C. BRADT, R. F. DAVIS, ET AL. Japanese structural ceramics research and development. FASAC Technical Assessment Report, SAIC, July 1989.

WAGNER, R., AND G. ZIEGLER. *J. Eur. Ceram. Soc.* 1991.

WHITEHEAD, A. J., AND T. F. PAGE. Fabrication and characterization of some novel reaction-bonded silicon carbide materials. *J. Mater. Sci.* 27(3):839–52, 1992.

WILSON, M. Ceramic matrix composites. *Ceram. Int.*, 93–99, 1992.

WU, X., AND J. W. HOLMES. Tensile creep and creep-strain recovery behavior of silicon carbide fiber/calcium aluminosilicate matrix ceramic composites. *J. Am. Ceram. Soc.* 76(10):2695–2700, 1993.

YANG, J.-M., S. T. J. CHEN, S. M. JENG, ET AL. Processing and mechanical behavior of SiC fiber-reinforced Si_3N_4 composites. *J. Mater. Res.* 6(9):1926–36.

YASUTOMI, Y., A. CHIBA, AND M. SOBUE. Development of reaction-bonded electroconductive silicon nitride-titanium nitride and resistive silicon nitride-aluminum oxide composites. *J. Am. Ceram. Soc.* 74(5):950–57, 1991.

YEH, J. R. Evaluation of interfacial frictional properties for fiber-reinforced ceramic composites. *J. Comp. Mater.* 25:1158–70.

ZHANG, J., R. J. PEREZ, AND E. J. LAVERNIA. Effect of SiC and graphite particulates on the damping behavior of metal matrix composites. *Acta Metall. Mater.* 42(2):395–409, 1994.

7

Repair of Composite Structures

7.0 INTRODUCTION

The repair of structures fabricated from composite materials is a many-faceted problem. The research includes identifying defective areas, determining whether the damage is detrimental to structural performance, developing efficient repair procedures, and performing tests to demonstrate that the repair will allow the component to be put back into service for the remainder of its design life.

The need for composite material repair has been demonstrated in a variety of government and industry flight service programs.[1–20]

In the early 1970s NASA and Air Force Materials Laboratory (AFML) recognized the need to increase confidence in the long-term durability of advanced composites that would encourage aircraft manufacturers and operators to make production commitments for the use of composite structures on commercial aircraft. Thus, programs were initiated systematically for the design, fabrication, testing, and flight service evaluation of numerous composite components to help increase the necessary confidence. Major emphasis was placed on the evaluation of advanced composites used in commercial transport aircraft because of their high utilization rates, exposure to worldwide environmental conditions, and systematic maintenance procedures. Additionally, the commercial transport aircraft companies themselves conducted exhaustive studies and real-time testing to evaluate their designed aircraft and aircraft components.

The increased usage of advanced composite structures in commercial aircraft is anticipated and will require the development of generic repair techniques and processes for various types of structures so that airline operators can maintain the components.

Existing technology applicable to the repair of structural composite hardware emphasizes the utilization of sophisticated materials and processes to restore damaged components to a fully functional state. Many of the repair techniques for structural composites were developed in conjunction with the materials of construction and reflect the state of the art at the time of development. In practice, under actual application conditions, the repair processes initially developed have proven to be overly complicated, time-consuming, and in many cases impractical because of the unadaptability of the specialized equipment and materials required to produce a structural repair.

R&D has continued to investigate the potential for simplified processing, new materials availability, and heating and joining methods, and as a result, structural repairs and proposed structural repair concepts continue to be viable.

7.1 WHY REPAIR?

An important consideration is whether a detected flaw is detrimental to the safe use of the structure and whether a repair can be made. This problem is addressed in considering the variety of flaws that occur. Interlayer delaminations, surface cracks, delaminations between the resin matrix composite and metal inserts, and impact damage have been used in various studies to determine their effect on the performance of stiffened elements and honeycomb-core panels representative of current and near-term aerospace applications. Figure 7.1 shows a typical set of data for a laminate, a hat-stiffened panel, and a honeycomb-core panel that has simulated delaminations and is subjected to compression. The data shown for the laminate are for Gr/Ep T300/5208, and the data shown for the hat-stiffened and honeycomb-core panels are for Gr/PI Celanese Celion 6000/PMR-15. The compression strength, normalized as a percentage of unflawed strength, is plotted as a function of the ratio of delamination width to panel width. For a 5-cm delamination in the laminate, a 30% reduction in the compression strength was observed, whereas for the same size of delamination in the hat-stiffened and honeycomb-core panels, 40% and 60% reductions in the compression strength were observed, respectively.

The data shown are for one part of an ongoing effort to determine the effects of various types of damage on composite structural performance.[21,22] Test results suggest that lightly loaded panels, which may be designed based on stiffness requirements, may achieve their design strength in the presence of impact damage. Heavily loaded, strength-critical panels, however, experience significant strength reductions because of impact damage that is not visually detectable.

Most experts affirm the general view that performance properties of advanced composites are not considered the problem in aircraft repair. Restoring them to original performance criteria is the first goal of repair.

Most repair relates to two types of damage: environmental factors (such as hail, lightning and bird strikes, and debris kicked up on takeoff or landing) and hangar rash (mishandling of aircraft or components on the ground). Puncture-type damage and microcracking of composites are common and can damage face skins and allow moisture ingression in the underlying honeycomb. Large-part or large-area repair can require replacement of the part or its removal for remanufacturing. For damage over a smaller area,

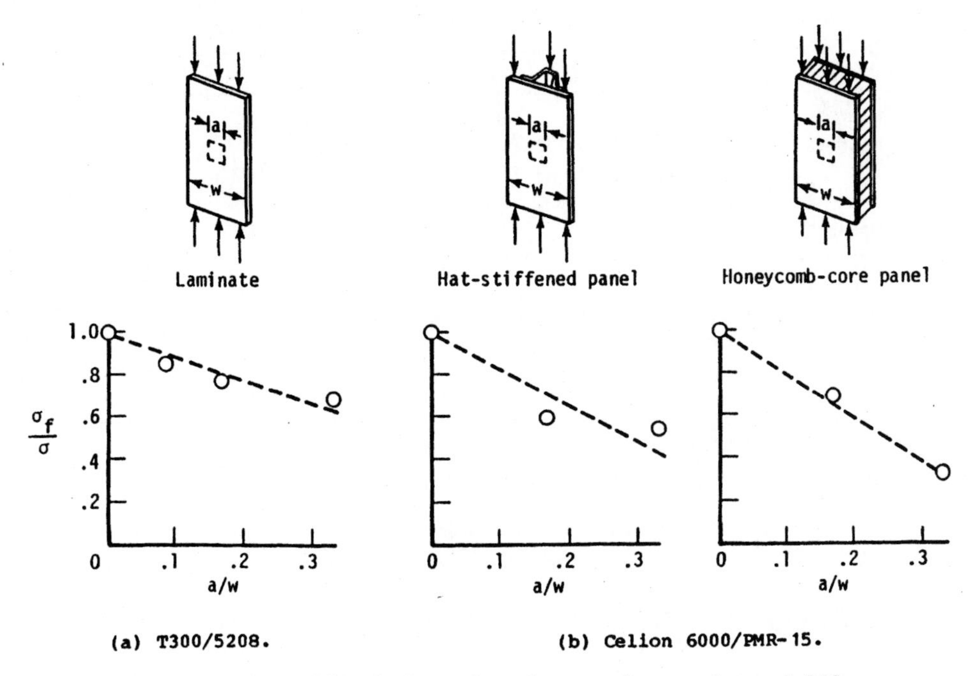

Figure 7.1. Effects of delamination on the performance of compression panels 0.15 m wide and 0.30 m long.

operators use prepregs and wet lay-up techniques in bonded, bolted, or flush aerodynamic patching.[23]

To accomplish safe, effective repair one must rely on original equipment manufacturer (OEM) maintenance manuals, as well as on more detailed structure manuals and the supplements to both. The Federal Aviation Administration (FAA) reviews these documents during aircraft certification but does not approve them per se. Instructions for continued airworthiness must include a limitations section that specifies mandatory replacement times, structure inspection intervals, and related structure inspection procedures as specified by the applicable certification rule. This rule applies to maintenance instructions for standard repairs but obviously doesn't cover all repair situations.

Some instructors claim that the basic problem is that structure repair manuals are written for basic composite repairs. When damage is more extensive, the best advice is to consult the manufacturer.

Manuals should be available with the history (traceability) for each composite component, along with a rating of its importance to the flight of the aircraft. One needs to know ribbon direction in a honeycomb, fiber orientation in fiber parts, component thickness documentation, the number of plies per laminate, especially in radomes where the thickness of the skin cannot be changed, as well as technique standardization. Figure 7.2 is an example from a manual showing how repair information can be charted.

Item	Skin Material	Repair Method and Cure Temperature				
		Wet Lay-up/Postcure			Prepreg	
		150°F	200°F	250°F	250°F	350°F
Flap track fairings	Glass-epoxy, 250°F cured	5.0 (2XPLS)*	5.0 (2XPLS)*	NA	No size limit	Na
	Graph-glass epoxy, 350°F cured	2.0 Interim only	NA (2XPLS)*	300°F only 5.0 (2XPLS)*	NA	No size limit
Wing tip Wing panels Aileron tab	Glass-epoxy, 250°F cured	5.0 (2XPLS)*	5.0 (2XPLS)*	NA	No size limit	NA
Aileron	Graph-glass epoxy, 350°F cured	3.0 Interim only	NA	5.0 (2XPLS)	5.0 (2XPLS)	No size limit
Spoiler	Graph-epoxy, 350°F cured	3.0 Interim only	NA	6.0 (2XPLS)	6.0 (2XPLS)	No size limit

Primary structure Secondary structure

* Two extra piles (per face sheet damaged) required on repair.

Figure 7.2. Maximum repairable damage size for 737-300 aircraft. 1 in = 25.4 mm; 150°F = 66°C, 200°F = 93°C, 250°F = 121°C, 350°F = 177°C.

7.2 TYPES OF REPAIR AND TYPES OF DAMAGE

7.2.1 Classification

The initial step in the repair process is classification of the damage type and an assessment of the extent of the damage. The extent of the damage can be determined by visual inspection in combination with an ultrasonic C-scan. Damage or other laminate anomalies are classified as follows.

1. *Negligible.* Negligible damage requires no rework other than possible cosmetic surface refinishing. This damage may consist of minor dents, scratches, or voids that will result in no degradation of structural performance, based on damage tolerance studies.
2. *Repairable.* Repairable damage requires rework of various degrees to return the component to full flight status. Repairable damage types are defined in Table 7.1 and classified according to damage cause. For example, voids and blisters are classified as intrinsic defects that are built in during the manufacturing process. Other defects can be caused by mechanical damage during service or by in-service deterioration. Mechanical damage can result from excessive loads, ballistic penetration, impact during handling, overheating, and numerous other causes. In-service deterioration can be caused by environmental damage such as lightning exposure, weathering, moisture, and rain erosion.
3. *Nonrepairable.* Nonrepairable damage is damage in which the structure cannot be restored to a satisfactory condition, whether because of cost considerations or technical limitations. This class of damage requires removal and replacement of the damaged parts.

TABLE 7.1. Defects Common to Composite Structures[24]

Defect	Definition and Probable Cause	Defect[a]
Void	An open area contained within the laminate; no apppareṅt changes in laminate thickness; generally small. Caused by volatile entrapment during cure.	1
Blister	A raised area, pillowlike, soft, can be depressed by light pressure, close to the laminate outer surface. Caused by a partial entrapment of volatiles during cure.	1
Delamination	Separation of adjacent composite plies. Caused by rough handling, poor interlaminar adhesion, or improper machining procedures.	1,2,3
Debond	Separation at an adhesive bond line. Caused by improper processing, marginal adhesive, built-in bonding stresses, and, for honeycomb structures, the intrusion and freezing of moisture.	1,2,3
Core Damage	Core cell walls wrinkled or buckled. Core cell walls ruptured or split. Debond of core ribbon at the node bond caused by improper processing, design loads exceeded, impact damage.	1,2,3
Fracture or splintering	Cracks in laminate matrix or matrix and fiber; combined cracking and delaminating of laminate outer fibers caused by improper handling or mechanical minor impact.	2,3
Abrasion	A wearing away of the laminate surface; may extend into fiber. Caused by environmental conditions (rain, dust, and so on) mechanical (misfit, oversanding, and so on)	1,2,3
Recess	Molded in depression in a laminate. Caused by foreign object on mold surface.	1
Scratch	Elongated surface discontinuity that is small in width compared to length. Caused by contact with sharp-edged object.	1,2,3
Erosion	Surface abrasion. Caused by natural environment (e.g., rain, dust, dirt, hail).	3
Porosity	Similar to voids only microscopic in size.	1
Gouge	Elongated surface discontinuity extending through one or several laminate plies removing matrix and fiber. Caused by tool damage in machining or contact with a sharp object.	1,2,3
Penetration	Breaking through of a laminate surface or bonded honeycomb panel (top skin into core or totally through the panel). Caused by impact.	1,2,3
Matrix degradation	Breakdown of the matrix material caused by chemical attack or excessive and local exposure to high heat.	2,3
Matrix-Fiber degradation	Breakdown of the composite structure due to sonic and/or mechanical fatigue.	3

[a] 1, Intrinsic; 2, mechanical; 3, degradation.

Jones and Graves[24] reported on an extensive study involving the development, demonstration, and verification of repair techniques and processes for Celion 6000/LARC-160 Gr/PI composite structures.

During this study, repair concepts were developed and evaluated for three general structural categories: flat laminates, honeycomb sandwich panels, and hat-stiffened panels. The repairs developed for flat laminates were also applicable to the sandwich panel skin and to the hat section. Honeycomb sandwich panels were subject to variable damage situations: core crushing or buckling, node bond separation, skin-to-core debending, top skin penetration and core breakage, and total penetration through the sandwich structure. The prevalent damage types associated with the three basic structures are given in Table 7.2 along with candidate repair concepts. A particular repair type may apply to many types of damage. The repair concepts for a fracture, gouge, penetration, or splinter of a laminate are identical. Damage to advanced composites may also appear as a combina-

TABLE 7.2. Damage Modes Versus Candidate Repair Methods[24]

△, Thick laminate only; ●, feasible repair technique; □, no plies fractured; ▲, face sheet only inolved (honeycomb).

			Candidate Repair Concepts									
		Damage Mode	Surface Coat or Refinish	Partial Penetration Patch (Flush Surface)	Partial Penetration Patch (Nonflush Surface)	Full Penetration Patch (Two Flush Surfaces)	Full Penetration Patch (One Flush Surface)	Full Penetration Patch (No Flush Surface)	Partial Core Replacement	Full-Depth Core Replacement	Sandwich Panel Section Replacement	Stringer Rebond
Skin or Stringer		Stringer debond										●
Skin or Stringer		Bond line porosity										●
Skin or Stringer	Sandwich Panel, Laminate	Abrasion	●									
Skin or Stringer	Sandwich Panel, Laminate	Erosion	●									
Skin or Stringer	Sandwich Panel, Laminate	Scratch	●									
Skin or Stringer	Sandwich Panel, Laminate	Delamination				▲ △	▲ ●	●				
Skin or Stringer	Sandwich Panel, Laminate	Porosity or void				▲ △	▲ ●	●				
Skin or Stringer	Sandwich Panel, Laminate	Recess	▲ □	▲ ●	▲ ●	▲ △	▲ ●	●				
Skin or Stringer	Sandwich Panel, Laminate	Fracture	▲			▲ △	▲ ●	●				
Skin or Stringer	Sandwich Panel, Laminate	Gouge		▲ △	▲ △	▲ △	▲ ●	●	●			
Skin or Stringer	Sandwich Panel, Laminate	Penetration				▲ △	▲ ●	●	●			
Skin or Stringer	Sandwich Panel, Laminate	Splintering			▲ △	▲ △	▲ ●	●				
Skin or Stringer	Sandwich Panel, Laminate	Matrix degradation				▲ △	▲ ●	●				
	Sandwich Panel	Core debond							●	●	●	
	Sandwich Panel	Core penetration							●	●	●	
	Sandwich Panel	Core degradation								●	●	
	Sandwich Panel	Bond line porosity							●	●	●	

tion of several distinct damage modes. For example, laminate penetration is often accompanied by splintering, cracking, or delamination. Note that the laminate repairs shown in Table 7.2 also apply to the honeycomb panels and to the hat or skin and stringer panels.[25]

They verified the repair methodology through compression testing at room temperature and at 316°C. The repairs restored 70–95% of the undamaged control component ultimate strength at room temperature and at elevated temperatures.[24]

Goering and Griffiths[26] conducted an exhaustive study on the design and analysis techniques of aircraft battle damage repair (ABDR) of advanced composite structures. In their work they defined the analytical methodology that can be used by repair personnel to determine the response of a composite wing structure in the undamaged, damaged, and repaired conditions. They defined the type of damage that can be expected from a ballistic impact and described how to collect experimental data from composite structures that have been ballistically damaged.

They reported that the majority of the repair techniques reported in the literature address skin repairs. There is a much smaller base of information on composite stiffener repairs, and information specifically addressing composite substructure repairs is virtually nonexistent. Most references that do not mention substructure describe the applicability of proposed skin repairs in the vicinity of substructure members rather than actual repair of the substructure. References 27–29 present the only bonded substructure repair identified by the survey. The types of repair considered in each of the references reviewed are summarized in Table 7.3.

According to the referenced literature skin repairs are generally reduced to either a metal or composite patch bolted, bonded, or bolted and bonded to the outer mold line (OML) of the structure. Composite patches have tailorable stiffnesses and are able to conform to more complex surfaces than metal patches, but their application is typically more complicated than that of metal patches. Some common OML repair concepts are shown in Figure 7.3. Although Figure 7.3 shows bolted repairs, similar concepts are applicable to bonded repairs. Several flush repair concepts have also been proposed, as shown in Figure 7.4. Most of the flush repair concepts use bonded patches, but some flush bolted repairs have been made. Other proposed repairs are shown in Figures 7.5–7.7.[30–56]

7.3 DESIGN CRITERIA

The composite community has given a great deal of emphasis to the design, materials selection, analysis, and manufacture of composite structures but has paid limited attention to the repair of these structures.

Numerous studies have been conducted to analyze metallic structures, components, and so on, by FEA methods and others and then attempt to extend the analytical methods to cover repairs to fiber composite materials. Other concerns may be due to a lack of appreciation of the difficulties associated with the repair of composite structures.

Reisdorfer[57] considered two repair design concepts for the new V-22 Osprey tiltrotor helicopter: high-strain structure and stability-critical structure. Because the V-22, B-2 bomber aircraft, and the developmental RAH-66 Comanche helicopter make extensive use of composite materials, repairability was a major consideration during the air-

TABLE 7.3. Summary of Composite Repair References[26]

Reference[a]	30	31	32	33	34	35	36	37	38	39	40	41	27	28	42	29	43	44	45 & 46	47 & 48	49	50	51	52	53	54	55	56
Skin repair	x	x	x	x	x	x	x	x	x	x	x	x	x	x	x	x	x	x	x	x	x	x	x	x	x	x	x	x
Stiffener repair		x																	x	x	x	x	x	x	x			
Substructure repair																x			x	x			T					
Bolted repair	T	T	T		D	T		D		D	T		T	T	T		A	T	T	D	D	D	T			T	T	
Bonded repair		T		T	D	T	T	D	D	D	T	T	T		D	T								T	T	T	T	T
≤ 4 in. damage	x			x	x	x	x	x			x		x	x							x			x	x	x		x
4–8 in. damage	x						x					x			x	x	x	x			x	x					x	x
> 8 in. damage		x	x																x				x					

[a]D, Discussion of repair technique only; A, discussion and application of repair technique; T, discussion, application, and test of repair technique.

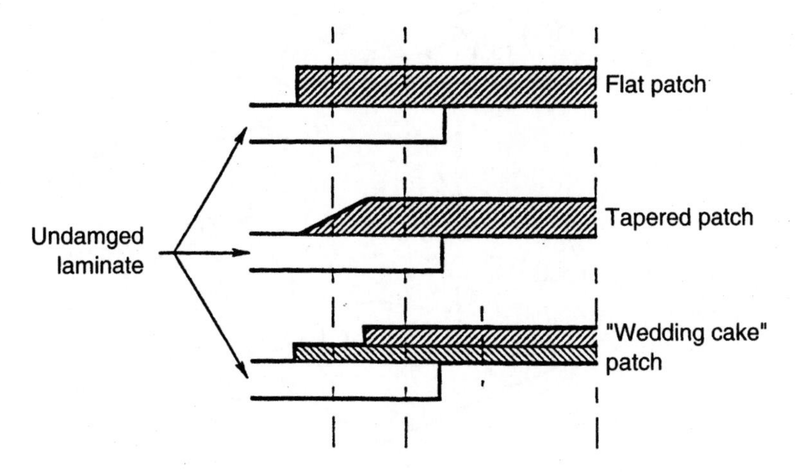

Figure 7.3. Typical outer mold line skin repair concepts.[26]

craft's development and eventual introduction into the field. To support these systems, the structures research community must begin to increase its appreciation of the difficulties associated with repair and to concentrate more resources on the design and processing of repairs for composite structures. Table 7.4 illustrates the technical difficulties associated with the repair of composite structures, as contrasted to the original design of the structure, and suggests some of the challenges involved in designing a repair. To further complicate the challenges, the different logistic philosophies and requirements of the military services must be satisfied, especially for systems with multiservice use.

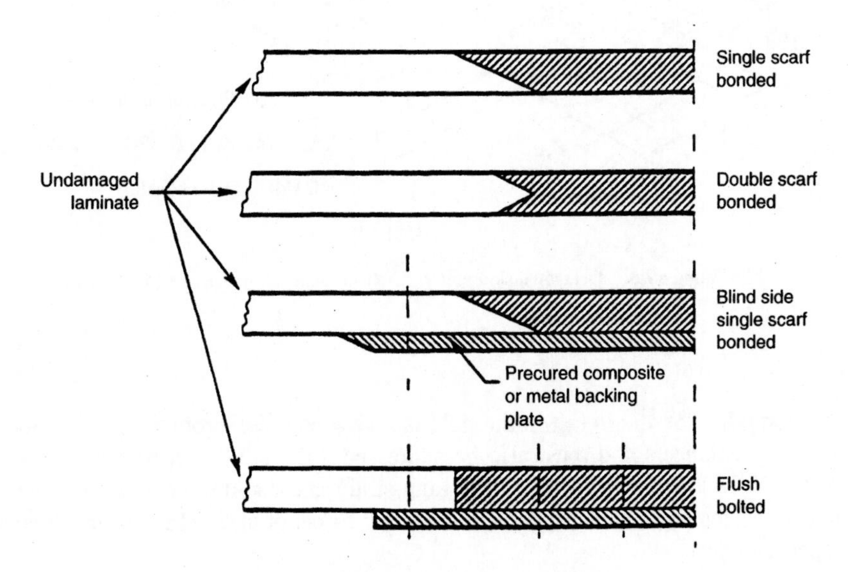

Figure 7.4. Flush repair concepts.[26]

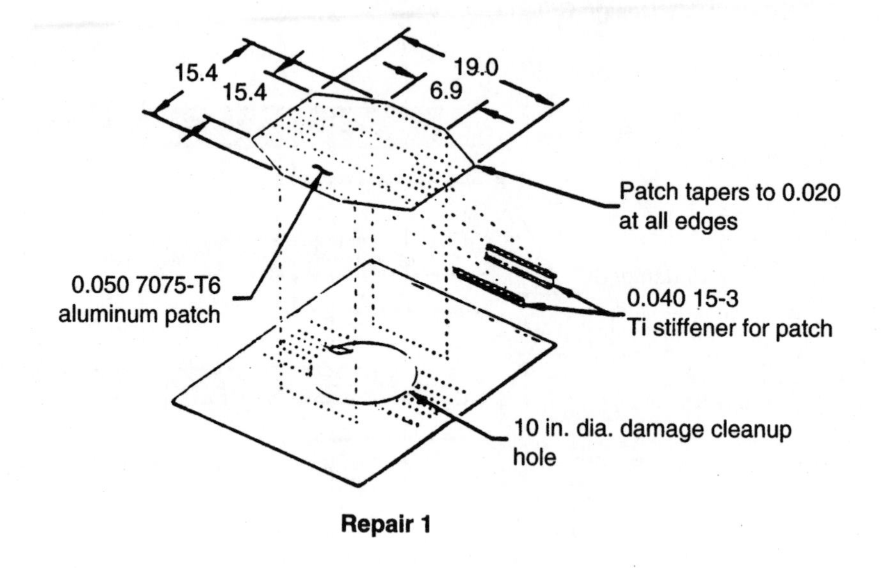

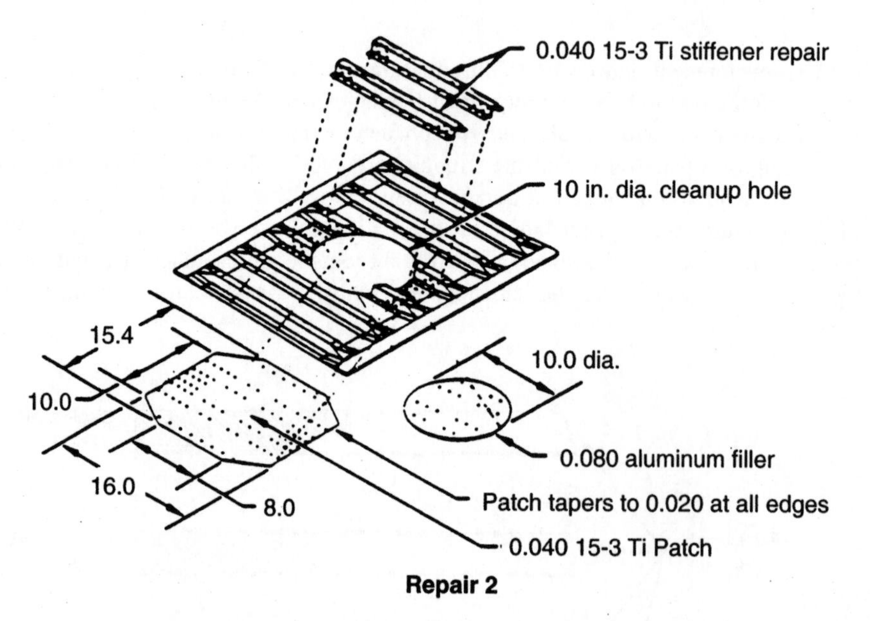

Figure 7.5. Octagonal skin repairs for 10-in-diameter damage.[26] 1 in = 25.4 mm.

Supportability of an aircraft in the field is a very complex problem that is seldom understood by engineers and must also be addressed. Figure 7.8 reflects the proposed repairs of the V-22 helicopter based on the high-strain and stability-critical structures—wings, flaperons, fuselage, and empennage—relative to depot and field skill levels available to repair damage.

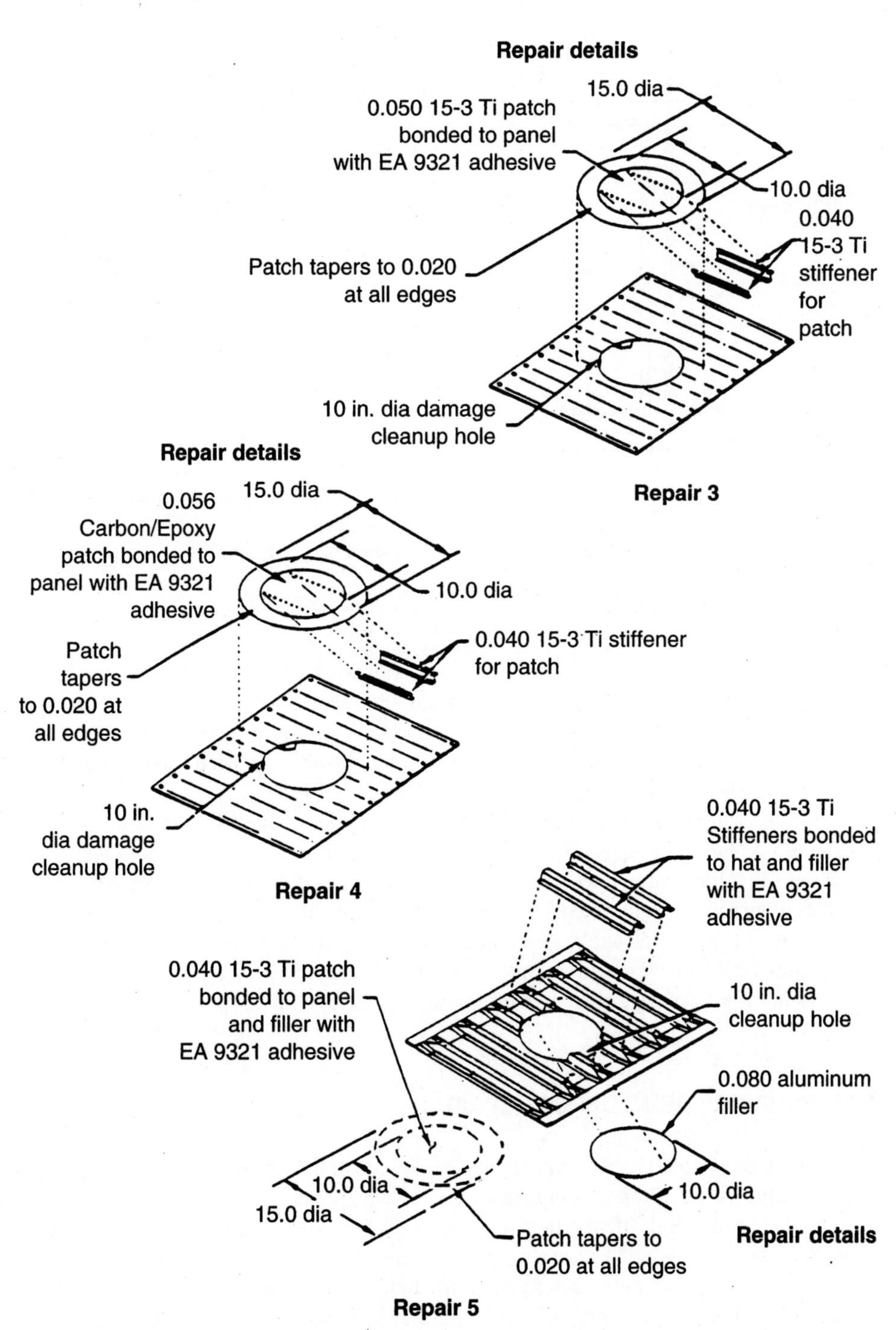

Figure 7.6. Circular skin repairs for 10-in-diameter damage.[26] 1 in = 25.4 mm.

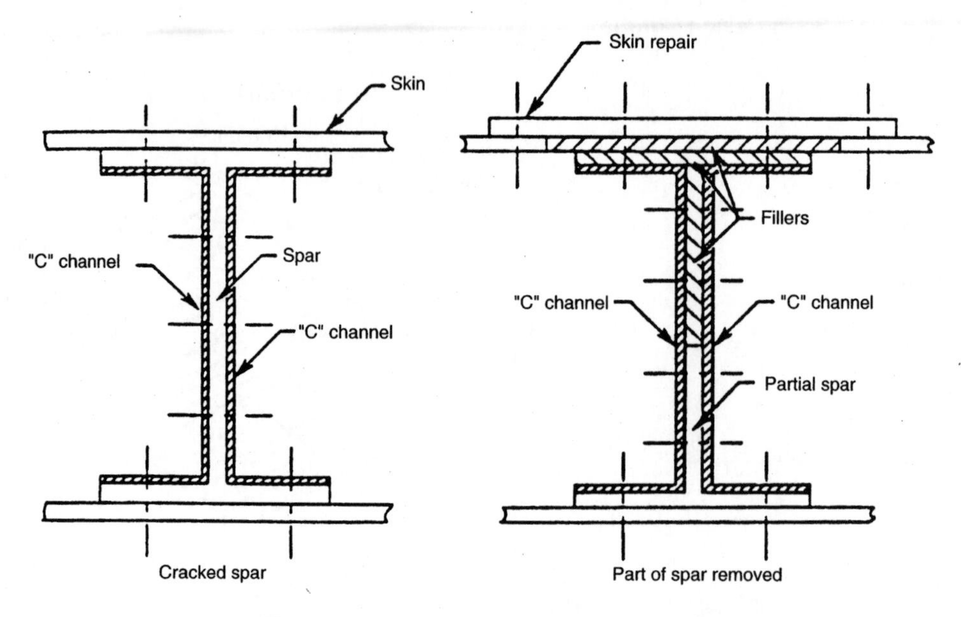

Figure 7.7. Primary substructure repair concepts.[26]

7.4 GENERIC DESIGN CONCEPTS

There are a number of candidate repair procedures, and the approach usually taken is to evaluate them and perform tests on damaged and repaired specimens representing a wide range of advanced composite structures found in a variety of military aircraft. The three structural types most often considered are monolithic, sandwich, and stiffened panels subdivided by design concept and variation (Figure 7.8).

The generic design concepts for the basic types of panel structures are presented in Table 7.5. For each type of structure, the table lists construction techniques, material choices involved, and generic applications. References 58–67 reflect a variety of components, component requirements for repair, and test results for various components.

As a result of all the studies, analyses, repair programs, and lessons learned from a logistics or depot viewpoint,[6] finite element methods[8] have been refined to assist engineers in the task of repairing advanced composite structural components.

7.5 FIELD REPAIR VERSUS DEPOT REPAIR

Field repair of composites is carried out under constraints when compared with depot facility capabilities. Proper equipment such as autoclaves and freezers are normally not available in the field. Most repairs are accomplished with the aid of heating blankets and vacuum processing.[68–70] Processing is further discussed in Section 7.6.

Other requirements, for example, for field repair of graphite structural composites in naval aircraft include resin systems that have long-term (at least 6-month) stability at ambient temperatures. Second, a cure in 1 h at 121–149°C using a heating blanket and

TABLE 7.4. Difficulties in Repair of Composites

Original Design	Repair Design
Straight load paths	Discontinuities or eccentricities
Tailored designs	Material availability
More readily inspectable	More difficult to inspect
Extensive tooling	Minimum tooling
Highly skilled labor	Lower level of labor skill

mechanical pump provides a performance (e.g., high-temperature properties and water resistance) approaching that of the resin used in the original advanced composite material.[71]

For depot repairs the choices include bolted versus bonded[72] and precured patches versus cocured in-place patches.[73–75]

Joint configurations are generic in nature: an external patch, a precured external bonded doubler, a near flush patch, and a flush scarf cocured patch.[72–75]

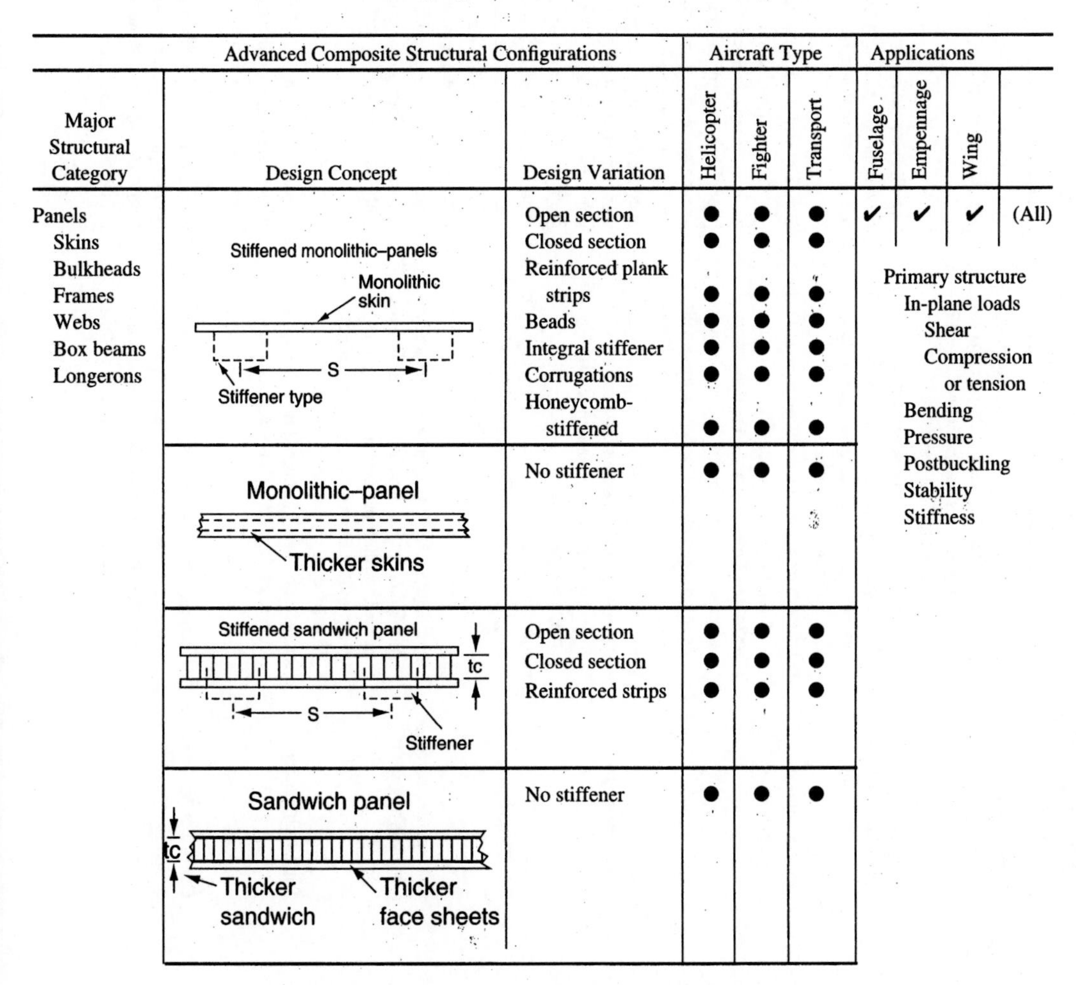

Figure 7.8. Aircraft advanced composite structural types.

TABLE 7.5. Generic Design Concepts for Advance Composite Airframe Structures

Type of Structure	Construction	Materials	Potential Applications
Bead-stiffened panels	Laminated fiber-epoxy skin with integrally molded stiffening beads. Flat patterns laid up in cauls or dies and autoclave-cured or press-molded to final form.	Woven Kevlar (3- or 4-ply) locally reinforced with unidirectional and/or woven graphite.	Floors and decks, interior bulkheads, interior skin panels
Skeleton-stiffened panels	Single-piece laminated fiber-epoxy outer skin. Single-piece molded fiber-epoxy inner skeleton. Skin and skeleton separately cured and postbonded together.	Woven Kevlar outer skin (5-ply). Woven Kevlar inner skeleton (4-ply). Both locally reinforced with unidirectional graphite.	Crew doors, cabin doors, exterior skin panels
Honeycomb-stiffened panels	Laminated or filament wound fiber-epoxy skin with transverse or longitudinal stiffening members. Open, square-edge honeycomb stiffeners with laminated or filament wound fiber-epoxy caps. Stiffeners cocured with skin.	Woven Kevlar, unidirectional fiberglass, woven graphite or graphite filament-wound skins (4–6 ply). Nomex or NFT (fiberglass-phenolic) honeycomb core (1/2–1 in. thick). Woven graphite, unidirectional graphite or graphite filament wound caps (1- to 3-ply).	Exterior skin panels, floors and decks
Open-section stiffened panels	Laminated fiber-epoxy skin with transverse or longitudinal J-section, Z-section, or C-section stiffening members. Stiffeners cocured with skin.	Woven Kevlar, unidirectional fiberglass and/or woven graphite skins (3- to 6-ply). Stiffeners same material as skin with unidirectional graphite reinforcement in caps.	Interior and exterior skin panels, floors and decks, interior bulkheads
Closed-section stiffened panels	Laminated fiber-epoxy skin with transverse or longitudinal hat section stiffening members. Stiffeners cocured with skin.	Woven Kevlar, unidirectional fiberglass and/or woven graphite skins (3- to 6-ply). Stiffeners same material as skin with unidirectional graphite reinforcement in caps.	Floors and decks, interior bulkheads, interior and exterior skin panels
Corrugation-stiffened panels and webs	Laminated fiber-epoxy skin with integrally molded corrugations. Flat patterns laid up in cauls or dies and autoclave-cured or press-molded to final form. Corrugations formed with square or tapered ends.	Woven graphite (2-ply) with woven Kevlar spacer ply. Corrugation caps reinforced with unidirectional graphite.	Interior bulkheads, floors and decks, work platforms, frame and beam webs. ACAP, CRF.
Sandwich panels	Honeycomb core. Laminated fiber-epoxy face sheets. Facings adhesively bonded to core.	Nomex or aluminum core (3/8–1 in. thick). Woven Kevlar, unidirectional fiberglass, or woven or unidirectional graphite facings (2- to 8-ply).	Exterior skin panels, fuel cell covers, floors and decks, work platforms, interior bulkheads, frames. ACAP, CRF, Night Hawk.
Open-section frames and ribs	Z-section, E-section, or C-section transverse framing members. Flat pattern fiber-epoxy laminates laid up in cauls or dies and autoclave-cured or press-molded to final form.	Woven Kevlar and/or woven graphite webs (4- to 9-ply). Frame caps and fastener flanges reinforced with woven and/or unidirectional graphite (3- to 8-ply).	Main landing gear frame, lower fuselage frames, roof structure frames, aft transmission support frame, cabin doorjamb, longitudinal beams. ACAP, CRF.

Open-section beams and longerons	Z-section, J-section, or C-section longitudinal framing members. Flat, beaded, or corrugated webs. Flat fiber-epoxy laminates laid up in cauls or dies and autoclave-cured or press-molded to final form.	Woven Kevlar or woven graphite webs (4- to 5-ply). Beam caps and flanges reinforced with unidirectional graphite (4- to 8-ply). Woven and/or unidirectional graphite doublers in areas of high load introduction.	Lower fuselage beams and longerons, roof structure outboard longerons, inboard transmission support, longitudinal beams. ACAP, CRF.
Closed-section beams and longerons	Hat-section longitudinal framing members. Flat pattern fiber-epoxy laminates cocured with the skin panels over removable mandrels.	Woven graphite webs (5-ply). Beam caps and flanges reinforced with unidirectional graphite (4- to 14-ply).	Keel beams. ACAP, CRF, ESSS, Night Hawk.
Sandwich structure beams and longerons	I-section or C-section longitudinal framing members. Flat pattern fiber-epoxy laminates or filament windings over honeycomb core. Autoclave-cured or press-molded to final form.	Nomex, HFT, or aluminum honeycomb web with woven Kevlar-, woven graphite-, or graphite filament-wound facings (2- to 3-ply). Beam caps and fastener flanges reinforced with woven and/or unidirectional graphite.	Transmission support beams, inboard transmission support, rear fuselage keel beam, drag beam. ACAP.

These repairs are applicable to a wide variety of light to moderately bonded (up to 0.17 MPa) stiffened and honeycomb sandwich structures sustaining damage over a reasonably large area (up to 645 cm^2).

Finally, studies by Ong and Chen[76] showed that the use of composite patch repairs on cracked metallic aircraft structures was an accepted technique for improving fatigue life and maintaining high structural efficiency. Both B/Ep and Gr/Ep composite patches, bonded with either room-temperature-cured or high-temperature-cured adhesives, attained sufficiently high fatigue lives to meet the damage tolerance requirement. However, the B/Ep patch with high-temperature adhesives gave better fatigue lives than the Gr/Ep patch with room-temperature adhesives.[77–79]

7.5.1 Materials

As new materials have become available, the potential for simplified processes by which repair materials are applied and converted to consolidated structural forms has become unlimited. For all structural repair concepts that have been proposed, the following four basic requirements are universal.

1. Survey, assessment, and definition of advanced repair material technology.
2. Definition of damage-affected areas and preparation of damaged and adjacent area application of the repair.
3. Application of pressure to provide full consolidation of the materials of repair, consistent with the selected method of chemical conversion.
4. Conversion of the selected material to a fully cured structural repair.

All repair procedures begin with an assessment of damage to the structure requiring repair. In all cases it is assumed that the damage area to be repaired is at least partially accessible from both sides; that is, both surfaces of the composite component can be prepared to produce surfaces able to receive the available material. Assessment of damage is conducted utilizing currently available techniques consisting of visual and manual low-frequency ultrasonic (coin tap) methods. After the damaged area has been mapped and its physical boundaries marked, materials and a method of repair are defined by a process specification.

Most engineers believe that matrix preimpregnation yields the composite integrity necessary to effect optimum, permanent repairs. Although field impregnation of dry reinforcement with single-component or multicomponent matrices somewhat simplifies the issue of materials inventory and speed of repair, quality risks are high. The mixing proportion, mix uniformity, pot life, realizable level of impregnation, and inherent slowness of the mixing and impregnation processes are limiting variables.

In assessing matrix chemistries, epoxy(Ep), bismaleimide, and engineering thermoplastic backbones are primary candidates. Although acrylic, urethane, and polyester matrices offer the potential advantages of rapid cure and field stability, the background data that exist are limited to substantiating adequate resistance to temperature and moisture. Potential difficulties in fully curing thick sections and the requirement for a higher level of operator sophistication if alternative energy (radiation) curing mechanisms are utilized are also negatives.

Although many reinforcement compositions and physical forms have been considered, in most cases graphite fiber has been selected as the primary reinforcement. In addition to inherent high strength and stiffness it offers the advantage of a degree of property "selection" when the availability of a range of fiber types is considered. It is possible to design increased stiffness into damaged areas by selecting a 30% stiffer fiber (without sacrificing strain) as the repair fiber. Stress concentrations thus induced may be compensated for by hybrid fibers or by incorporating higher strain matrices at interfaces. Although five or eight harness satin weave balanced fabrics are the primary reinforcement physical forms chosen, it is possible to selectively introduce highly directional (90%+) fibers in special structural applications. Graphitic mat and Kevlar fabric are other reinforcements to be considered as well as the materials shown in Tables 7.6–7.9. These repair materials are being designated for a new RAH-66 Comanche utility and scout helicopter.[77]

The physical application of composite repair materials and process materials and equipment presupposes their availability in acceptable sizes, shapes, and forms. Two approaches to providing these requirements involve "prekitted" sized and shaped materials of repair and "bulk" form materials of repair. Although presized and shaped kits minimize the time and sophistication of field repair, their universal use for all damage types is limited.

In reviewing and surveying applicable materials technologies the following list reflects the several forms, types, and so on, that are potentially and conceptually usable.

1. *Staged prepreg or optional film bond.* This concept utilizes conventional multifunctional Ep-aromatic amine chemistry common to the parent composite material. The graphite-reinforced prepreg in this case, however, is vacuum- or high-pressure-processed by the supplier at the impregnation level. The resultant prepreg is essentially free of air and dissolved or absorbed volatile components and requires only reflow and cross-linking to achieve integrous composite properties. The advanced state of cure required presupposes the ability to heat rapidly (10–20°C/min) to realize a usable minimum melt viscosity. It is probable that some form of external heat will be required to preform this material. An optional or mandatory compatible flexibilized film adhesive is provided to achieve interlaminar strength at prepreg-to-parent and/or prepreg layer-to-layer interfaces. These films, such as the American Cyanamid FM 250-330 series, are currently available.

TABLE 7.6. Repair Materials (Resins)

Application	Material Identification	Mix Ratio, A/B	Cure Cycle	
Laminating resin	EA9396, A/B	100/30	(1) 77°F	for 3–5 days
			(2) 150°F	for 60 min
	EA956, A/B	100/58	(1) 75°F	for 24 h
			(2) 150°F	for 2 h
			(3) 190°F	for 1 h
Paste adhesive	Magnobond 6380, A/B	100/27	(1) 77°F	for 5–7 days
			(2) 150°F	for 120 min
Liquid shim	Magnobond 6380, A/B	100/27	(1) 77°F	for 5–7 days
			(2) 150°F	for 120 min
Potting compound	Epocast 1652, A/B	100/12	(1) 150°F	for 120 min
	Epocast 1614	1 part patty	(1) 260°F	for 60 min

TABLE 7.7. Repair Materials and Hardware

Application	Material Identification	Thickness
Dry graphite fabric	SGP-196 PW IM7 Graphite	0.1905-mm (0.0075-in.) ply
Dry glass fabric (veil)	E-glass 108 Style	0.0508-mm (0.002-in.) ply
Expanded copper foil	Dry pre-preg	
Graphite shims		0.03, 0.06, 0.09 in.
Metal shims	7075-T6 aluminum	Standard sheet stock
	Ti-6Al-4V titanium	Standard sheet stock
	301 stainless steel	Standard sheet stock
Fasteners	Hi-Lites	Protruding shear head
		HST10YV-x-y
		HST1071TAWT-x
		Protruding tension head
		HST12YW-x-y
		HST1072TAWT-x
		130° countersunk shear head
		HST315YV-x-y
		HST1071TAWT-x
		Swivel collar
		HST1096TA-x
	Hi-Loks	Protruding shear head
		HL10V-x-y
		HL40-x-y
		HL1094W-x
		Protruding tension head
		HL12V-x-y
		H48-x-y
		HL1087APBW-x
		130° countersunk shear head
		HL1094W-x
		Swivel collar
		HL75-xAW
	Big-Foot	Protruding head
		MBF3003-x-y
		MBF2110-x-y
		MBF2120-x-y
		130° countersunk shear head
		MBF3006-x-y
		MBF2113-x-y
		MBF2123-x-y

2. *"Conventional" multifunctional epoxy prepreg.* In this concept, blocked (stabilized) tertiary amine and/or dicyandiamide chemistry is needed to yield epoxy cross-link properties similar in elevated-temperature wet performance to those of aromatic amine-Ep cross-linking. The material in this case is inherently more similar to parent composite prepreg in terms of handling and hence more universally useful for any damage geometries encountered. Additionally, no secondarily introduced adhesive is required to promote interfacial properties.

TABLE 7.8. Determination of Number of Wet Lay-up Replacement Plies

Original Material Type	Nominal Ply Thickness	Required Number of Plies
Plain-weave graphite fabric	0.1905 mm (0.0075 in.)	2
8 HS graphite fabric	0.381 mm (0.015 in.)	4
Unidirectional graphite tape	0.1524 mm (0.006 in.)	3

3. *Single-side adhesive-coated prepreg.* This concept utilizes a two-stage preimpregnated material with different properties on opposing surfaces. The more advanced stage surface inherently provides optimal composite properties of Ep-aromatic amine or BMI chemistry. The adhesive (secondarily coated) surface produces required interfacial properties when adequately cured.
4. *Layered prepreg.* A preimpregnated isolation layer is selected and processed to provide an integral barrier for transport of matrix and curative but is thermally or chemically degradable so as not to affect mixing and cross-linking. The single-component material should be inherently storable, and cross-linking chemistry will result in acceptable elevated-temperature wet properties. Reactant stoichiometry is inherent, as no mixing is required.
5. *Inhibited prepreg.* This technology is similar to item 4 but different in that the curative or backbone reaction is blocked chemically by coordination or complexation or physically by encapsulation to affect stability. Activation can be thermal or photo-chemical.
6. *Multilayered epoxy prepreg and/or film.* This approach utilizes sequentially applied stoichiometrically formulated layers of backbone and curative. These layers may be applied as either prepreg or prepreg and film. The requirements of excellent preimpregnation, elimination of premixing, and speed of repair are inherent. Typically the reactants require a mechanical and thermal assist in ensuring adequate molecular proximity to ensure the degree of cross-linking expected. Reactants are selected to ensure maximum mobility under minimum heat and pressure conditions, however. Optionally the curative can be secondarily applied to the backbone prepreg at the layer level by spraying, dipping, rolling, and so on, but the chemistry must be carefully developed such that the stoichiometry is insensitive to an excess of either reactant.
7. *BMI prepreg.* Conceptually this method implies a superior degree of preimpregnation such that melt under minimum mechanical pressure is adequate to generate interlaminar properties. BMI or (Ep-triazene) modified BMI film materials as pro-

TABLE 7.9. Expendable Materials Used in the Repair of Composite Structures

Material Identification
Fluoropeel
Fluoropeel tape
FEP
N-10
Bagging film

posed in item 1 are similarly applicable. This approach presupposes availability of a minimal 204°C heat source and time at a temperature required to promote the maleimide homopolymerization reaction.

8. *Acrylic-polyester-urethane prepregs.* These chemistries offer the potential advantage of very rapid alternative energy curing (as compared to thermosetting epoxies) but are comparatively unproven as composite matrices. Problems to be overcome include compatability of surface chemistries (with graphite fiber and/or parent composite), typically high rates of shrinkage and therefore prestress, safety in handling, activation energy penetration (in thick graphite-reinforced sections), and maintenance of cured properties at elevated-temperature (121°C) wet conditions.
9. *Solvated thermoplastic prepregs.* It is proposed that a solvated thermoplastic prepreg be sequentially layered, devolatilized, and ultimately reflowed. The selection of engineering thermoplastic is directly driven by its solubility in low-boiling or high-vapor-pressure carriers and its resultant chemical resistance and elevated-temperature wet mechanical properties. It is conceivable that postrepair sealing can permit use of a lower melting (less moisture or chemical-resistant) thermoplastic. If current technology engineering thermoplastics are required, however, this repair technique may apply to only BMI parent structures. It is possible that the reflow temperatures and times required for these materials may induce damage in adjacent Ep/Gr parent composite structures. This approach offers a real advantage in that its implicit degree of prepreg flexibility is consistent with repairs of damaged areas on highly contoured components.
10. *Thermoplastic prepregs.* Conceptually this approach is similar to item 9, but heat is required to ply and form this typically inflexible prepreg. The possibility of carrier solvent entrapment and the necessity for sequential layer treatment are eliminated, but high reflow temperature requirements remain. Primary candidate engineering thermoplastics include Torlon (PAI), Ryton (PES), Ultem (PEI), and PEEK.
11. *Interpenetrating network and pseudo interpenetrating network (IPN and PDIPN) prepregs and adhesives.* This comparatively immature technology offers a real advantage in the sense that the best properties of dissimilar thermosets and thermosets/thermoplastics exist simultaneously at a molecular level. No processability advantage is perceived, however, when compared to either generic material type.
12. *Bonded and bolted-bonded repair.* Both concepts presume that preformed laminates or metallics are available in shapes and sizes consistent with all types of damage encountered. In addition, they imply the use of current technology adhesives and assume their viability after extended 60°C storage and being tested under 121°C wet conditions.
13. *Tape-flexible or rigid.* Graphite fabric-reinforced prepreg with pressure-sensitive, but heat activated, adhesive sequentially layered.

7.5.2 Processing Procedures

The means of introducing energy or pressure conditions other than ambient to the immediate repair area are normally dictated by services available at the site of repair. Secondary means of converting energy to a required form or degree must be addressed coincidentally with the repair matrix ability to consolidate, fuse, or cure. The following list is intended to suggest viable means of meeting such requirements.

ENERGY

1. *Radiant or convective heating.* Lamps, heat guns, engine gases, and so on, are convenient and simple means of introducing local heat to the repair area. Control and degree are critical variables that must be addressed in the context of the candidate matrix curing requirements. These energy sources offer both geometry and size-of-repair latitude (Figure 7.9).
2. *Blanket heating.* An insulating blanket (Figure 7.10) encapsulating strip heaters offers the advantage of simple local heating of flat and mildly contoured sections. Additionally, control is enhanced via the known time-at-temperature characteristics of the heating element. Mehrkam and Cochran[78] performed an evaluation on a range of potential materials for elevated-temperature repair. Their efforts were directed at the development of patch materials that could be used for depot and field-level repairs of 204°C composite structures. Initial work focused on BMI and other elevated-temperature systems (dicyanate ester and high-temperature Ep). Initial patch-processing concepts were developed for these materials, and field-usable repair materials were storable at ambient temperatures, formable to cover contoured surfaces, and processable using heating blanket or vacuum bag procedures.

   This work led to the development of field repair concepts for Ep, BMI, and polyimide composites which were demonstrated to restore structural integrity to damaged composite laminates.
3. *Resistance heating (sacrificial or integral).* The advantage of sacrificial resistance heating in the form of a metallic or graphitic surface-positioned flexible grid over which is placed a flexible insulating blanket is in its conformability (Figure 7.11). Conceivably a changing contour with small radii can be locally heated and the grid stripped after completion of the repair. Integral resistance heating can be accomplished by impregnating partially metal-coated graphite fiber with the repair matrices. The level and location of metallization at the electrical connection points is dictated by the size of the repair area. Resistive layers can be sequentially positioned via a predetermined number and position formula for various repair thick-

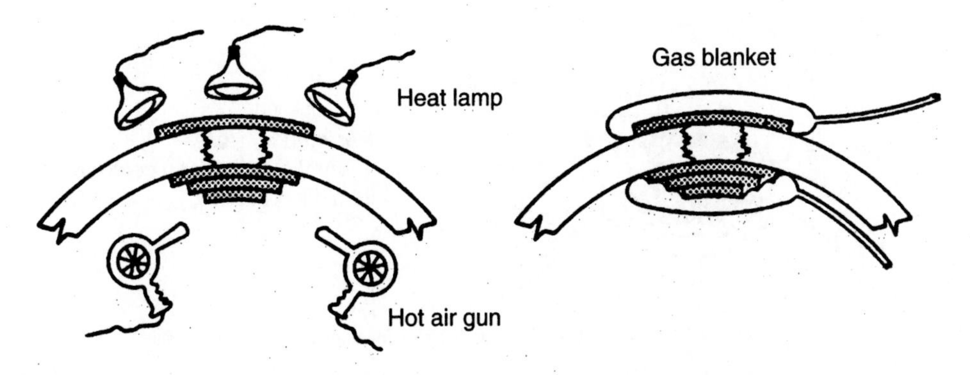

Figure 7.9. Lamps, air guns, and gas blanket heating systems.

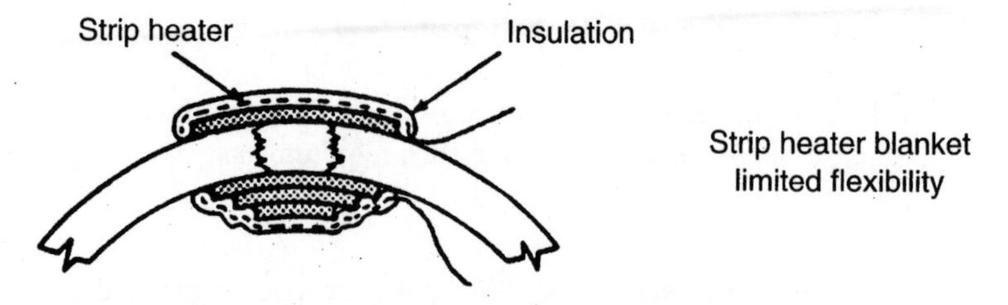

Figure 7.10. Heating blankets.

nesses. Internal heating provides a uniform, potentially high-temperature means of curing or fusing. Xiao, Hoa, and Street[79] examined two fusion bonding methods as repair processes for thermoplastic composites. They believe that these two processes, resistance heating and induction heating, offer significant advantages over conventional adhesive bonding and mechanical fastening.

Their work and studies were concerned with bonding APC-2 composite, and the PEEK prepreg itself was used as a heating element. The ends of a single ply of prepreg were treated with chromic-sulfuric acid to remove the matrix and expose the fibers. The exposed fibers were then connected to a power supply by copper clamps. The fibers were coated with conductive paint to provide a good contact between the heating element and the power supply, which is crucial for resistance heating. This improved the clamping efficiency about 15–20%, and the current distribution was more uniform across the width of the prepreg sheet.

The bonding quality was found to be a function of current density, heating time, and pressure. Current density determines the maximum temperature at the bonding surfaces. Insufficient heating left gaps at the laminate-resin film interface, thus resulting in a low bond strength.

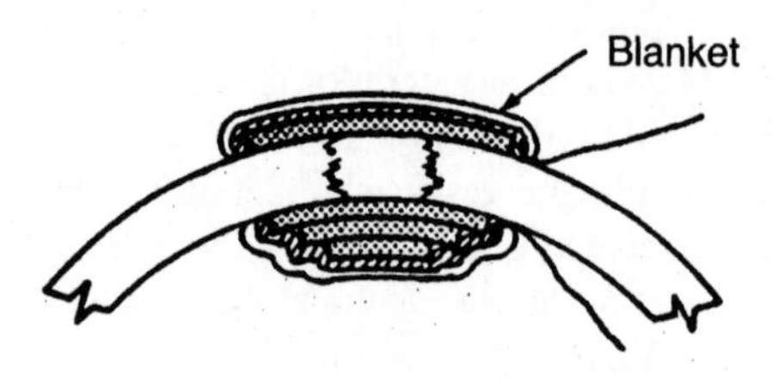

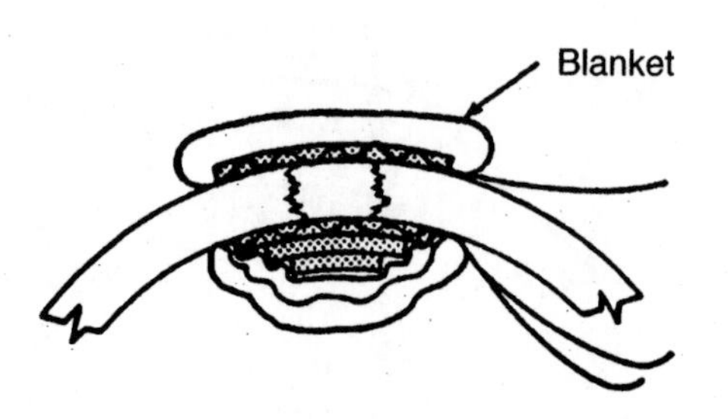

Figure 7.11. Resistance heating. (*a*) Sacrificial grid; (*b*) integral graphite.

The orientation of fibers in the surface ply also plays a role in fusion bonding by resistance heating. For cross-ply laminates, the effect of surface fibers being either perpendicular [(0)/(90) bond] or parallel [(90)/(90) bond] to the fibers in the heating ply was examined. This was found to affect both processing parameters and failure mode. The critical current density for a (90)/(90) bond is lower than that for a (0)/(90) bond, and the former is more sensitive to pressure than the latter. A resin-rich layer is present in the (0)/(90) bond but not in the (90)/(90) bond. As a consequence, the (0)/(90) bond tended to fail within the heating ply. Interlaminar shear strength variations between the different fiber orientations and residual stresses due to the anisotropic thermal conductivity of the individual plies are the most likely causes of the change in failure mode. A (0) ply at the surface of the fusion bond conducts heat away 10 times more quickly than a (90) ply.

Resistance heating bonds in APC-2 produced a lap shear strength above 20 MPa. This value is higher than that for adhesive bonds made using common surface preparation techniques.[80] When considering the added environmental resistance, fusion bonding appears to be the more attractive technique.

Resistance heating has the following advantages: fast process (about 1 min), modest consolidation pressure (<0.6 MPa), easy surface preparation (slight sanding), simple device (dc power supply), and potential to be applied in patching repair. However, at the present time, it also has several disadvantages. Overheating occurs at the edges of the bond and at free edges and fibers slip out, thus creating a slope. However, from the viewpoint of structural continuity, the latter may be a beneficial effect if it reduces stress concentrations. Also, resistance heating produces mainly rectangular bond lines.

Xiao and associates performed induction heating bond trials at 120 kHz with a 2-kW input power, and through heating was achieved on cross-ply APC-2 laminate within 30 s. The bonding at the interface was good, and there was no structural discontinuity throughout the interface. Lap shear specimens showed an adherend failure mode, and thus induction heating appears to be a promising method for healing-type repairs.

By placing a more conductive susceptor at the bond line, induction heating could also be used to produce a more localized interfacial heating.

Because of the "skin effect," heating a small-sized object requires a high frequency, and thus the coupling frequency for a graphite fiber is very high. Fortunately, the distribution of fibers is not uniform but rather clusterlike, and a number of fibers contact each other. Therefore, the actual dimensions of a conductive path in a composite are far larger than the fiber dimensions. Multidirectional laminates also create fiber networks and are easier to heat than unidirectional ones.

Induction heating represents a quite sophisticated method and has advantages such as precisely controlled processing parameters, easy handling, flexible operation, and potential light devices for field use.

Border, Salas, and Black[81] examined induction bonding as a fast heating method for repair, especially for field repair of aircraft. They developed a flexible coil, whose design included an inductor to create the magnetic field and a high-permeability core to focus the field, and the new coil was based on a flexible strip inductor design with a segmented core. By changing the induction coil from a loop to a paired line, the geometry of the heat zone is changed from a circle to a strip. The flexible cable conductor allows the heat zone conform to any necessary repair geometry.

For most bonded repairs a patch is applied over the damaged area and bonded in place. For these types of repairs the bond area can be approximated by a strip. Under these conditions the strip heat zone of the flexible coil is well adapted to the repair geometry, as the coil can simply be bent to fit the shape of the repair. In repairs done on aircraft skins, heating the outer edge of the repair often produces a substantially uniform temperature over the entire repair area because the major heat loss is radially outward from the repair area. Consequently, the strip heat zone is useful for many repair geometries. Most importantly, the heating can be done without scanning the coil over the repair area and so the operator skill required is reduced.

The coil produces a heat zone that is approximately 0.63 cm, and a 2.5-kW power unit has been used to test the flexible coil in the range 80–200 kHz, which is well-suited for heating graphite fabric materials.

4. *Induction heating.* As in resistance heating, the capability of internally heating an integral graphite repair fiber has also been exploited. An externally introduced magnetic flux excites metallic or graphitic substances to provide substantial heat for melting, curing, or fusing layered repair precursor materials (Figure 7.12). The process is rapid and controllable within limits. Although thickness is a limiting variable, geometry is not.
5. *Chemical heating.* Insulating a localized area that has been heated by an external source of chemical heat (prekitted chemistry) or by controlled self-propagating heat of reaction (exothermic cross-linking) can be used to effect a matrix cure (Figure 7.13). In the second case a secondary thermal exposure (postcure) may be necessary to generate acceptable elevated-temperature wet properties. Again, the advantage of indifference to shape and size can be exploited.
6. *Ultrasonic.* Molecular excitement resulting in the generation of internal heat is a potential means of introducing curing energy (Figure 7.14). Developmental efforts in terms of material and equipment have been explored, but the potential advantage of application to repairs of various sizes, shapes, or thicknesses make it attractive.

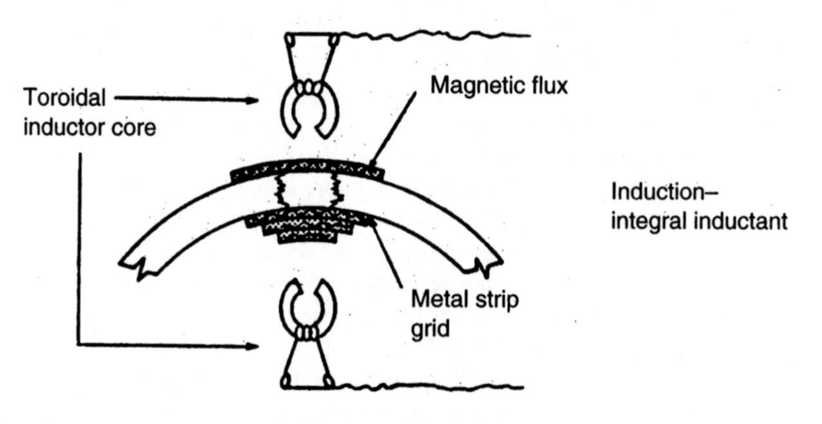

Figure 7.12. Induction heaters.

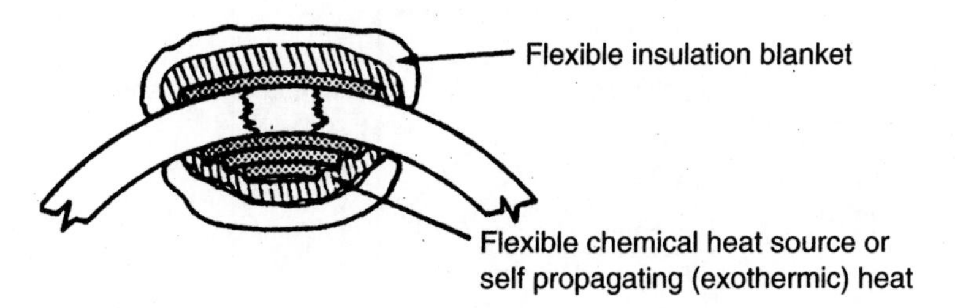

Figure 7.13. Chemical heating technique.

7. *Radiation energy.* Ultraviolet, infrared, electron beam, microwave, and so on, are means of introducing cross-linking that are viable radiation energy sources (Figure 7.15). The state of the art in terms of controlling the reaction and initiating cross-linking in thick sections has been evaluated, and several investigations into the unique molecular structure and actuation schemes for exploiting these potentially very rapid curing concepts have been undertaken.

In addition to the need for energy and pressure (discussed later in the chapter), which normally can be supplied by systems within the confines of a plant, and so on, there are other systems that must exhibit several important characteristics, as shown in Figure 7.16, in order to provide effective field repair methods. The repairs must be executed in field service environments, including on-board-ship environments and outdoor conditions (e.g., sunlight and humidity). Because of the field service environment, it is desirable to avoid any need for complex equipment as this limits the range of applicability of the procedure. Training of personnel in the field for complicated repair procedures is both impractical and quite costly.

Several methods have shown themselves to be quite adaptable to field repair environments in terms of the criteria previously outlined. The first of these methods is magnetic heat induction. It has been shown to be quite adaptable to field repair both by itself and in conjunction with adhesive bonding approaches. Magnetic heat induction involves one or more ferrous magnetic toroids, used to induce a powerful magnetic field. When a steel screen of 3- to 5-mm thickness is placed in the bond line, heat is quickly developed adjacent to the screen, producing a localized melt of the selected material (thermoplastic and/or thermoset).

One of two methods may be adopted. If the bond is to be an overlap joint, the screen may be left in place and the magnetic field stopped. This stops all heating, and a suitable bond is formed almost immediately. The other approach, where a flush joint is

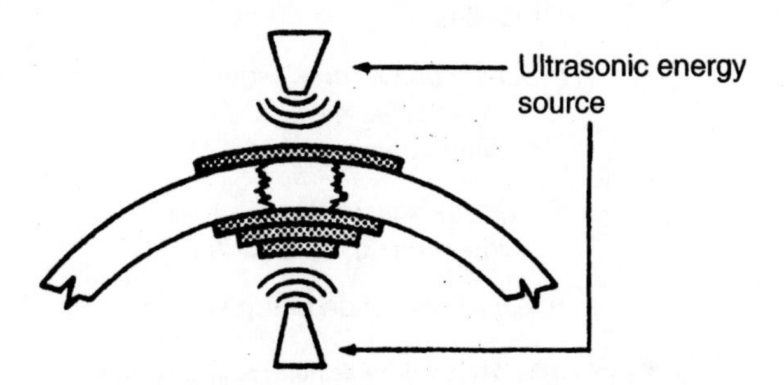

Figure 7.14. Ultrasonics method of generating curing energy.

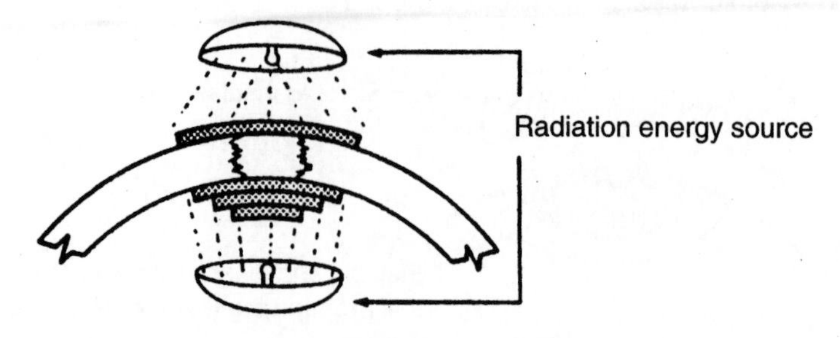

Figure 7.15. Radiation energy sources.

desired, involves the same heating sequence as the first, but when a suitable melt has been achieved, the screen is withdrawn from between the two bonding surfaces, allowing them to contact immediately.

In either case, a strong, reliable bond is produced in a minimum amount of time. A hand-held unit with a portable power supply has been developed to serve the field repair requirement for this procedure.

Adhesive bonding is a commonly used approach in repairing many diverse systems. However, its use in field repair of thermoplastics presents many new considerations. Ep adhesives have shown themselves to be very compatible with thermoplastic substrates in bonding processes, although only a few are suitable for conventional field repair because of the curing cycle required for most epoxies. Most Ep adhesives require either an elevated-temperature cure cycle, or a room-temperature cure of 24 h to 7 days, neither of which is suitable for field repair. Other adhesives, such as synthetic resins and cyanoacrylates, cure quite easily in short times at room temperature but do not show sufficient high-temperature properties or chemical resistance for compatibility in thermoplastic applications. Several options for the use of adhesive systems remain, however.

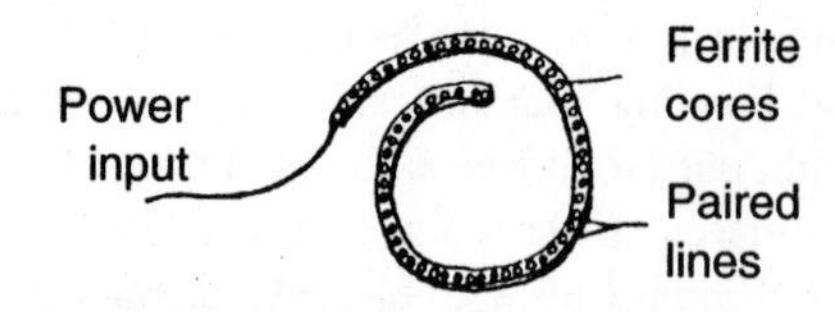

- Low cost
- No complex equipment or facilities
- Simple procedures requiring no special skills or previous training
- Repairability under noncontrolled conditions
- Quick execution of repair

Figure 7.16. Schematic of flexible paired induction coil.

There are several two-part Ep adhesives specially formulated to form tough, strong bonds in less than 1 h. Several of these adhesives, notably a Bostik product (7575 B/A), have been evaluated and show excellent mechanical properties within 1/2 h after application at room temperature. Another Bostik adhesive (7000) has also shown superb characteristics in short cure times. This two-part acrylic adhesive develops extremely strong bonds even to chemically inert thermoplastics such as Torlon and Ryton, with approximately 50% of ultimate physical properties achieved within 5 min after application. This acrylic adhesive is nearly impervious to most solvents and has excellent high-temperature properties.

Another option in the use of adhesives can be realized through magnetic induction heating for accelerated cure of Ep adhesives. For example, a frequently used adhesive, AF 163, has a recommended cure cycle of 60 min at 121°C. Using a 2-min magnetic heat induction cycle as a substitute curing procedure, bond strengths in excess of those achieved through conventional cures are possible.

A third candidate for field repair technology utilizes ultrasonic welding (UW). When employing UW, it is best to use unreinforced resin for the patch material, as the unreinforced material flows more easily than reinforced grades, especially if one is dealing with the same thermoplastic. Hand-held units are available for UW and are quite effective. Most ultrasonic hand-held units draw approximately 450 W, and it is necessary to have an adequate power supply at hand. For the types of repair adaptable to UW, it is necessary to have both a drill capable of clearing out the damage to fairly rounded dimensions as well as a set of patch "plugs" of suitable size. This apparatus can easily be supplied at field locations in a repair package form but is very likely to be slightly more expensive than the previous two methods.

Similar to UW methods in practice is spin or vibrational welding. Spin or vibrational welding is used for repair of the same type of damage as UW and also operates on the principle of mechanical energy. In spin welding, however, the melt is initiated by very rapidly spinning the repair plug relative to the main part. The spinning motion is continued until frictional energy causes the resin to melt and flow slightly around the bond line. As soon as the melt occurs, generally signaled by the small amount of melt adjacent to the bond area, the spinning motion is immediately stopped. The plastic resets almost immediately to a strong, complete bond. Field repair using this process shows it to be extremely well suited to this application. Modifications of existing equipment show a promise of being exceptionally effective in terms of both cost and the mechanical properties of this welding system. Only a minimal power source is required to achieve a high-quality bond.

Finally, inert-gas welding has been examined as an adaptation of metals repair technology applied to thermoplastics systems. Thermoplastics are quite adaptable to modifications of metals welding techniques because of their high-temperature physical characteristics. Thermoplastics melt under heats such as those encountered during gas welding but undergo no chemical change or degradation. The molten plastic flows slightly along the bond line and freezes in much the same fashion as in metals. Care should be taken not to overheat the thermoplastic when welding in this fashion. Investigations continue on the application of inert-gas welding to thermoplastics.

Wegman et al.,[82,83] Holcomb,[84] and Reinhart[85] have prepared and updated manuals, handbooks, and reports covering all aspects of the structural repair of composite components and systems through bonding and curing.

PRESSURE

Contact, weights, and clamps are viable means of consolidation for a number of matrix options that require only melting and cross-linking or "fusing." The decision to utilize little or no external pressure depends on the repair material selected and the shape, size, and accessibility of the damaged area.

Vacuum is another technique where location, shape, size, and accessibility determine the viability of using self-contained vacuum blankets. Film-type vacuum blankets, while having the advantage of conformal application to diverse contours and surfaces, require a reasonable degree of operator proficiency to effect a continuous seal.

Hydraulic or pneumatic bags represent another method where the same factors as indicated for vacuum apply as well. The use of soft positive pressure requires the establishment of a fixed rigid surface to react to the pressure. Adequate sources of pressurized gas and/or fluid must be accessible from the aircraft power train.

Thermal expansion is a viable means of introducing controlled positive pressure. This is accomplished via thermal expansion of a sacrificial external or (in the case of a damaged sandwich) an integral internal organic medium. The concept of introducing local pressure by preforming the expandable material, forming and positioning removable hard backs, and thermally expanding and curing simultaneously is potentially a simple, rapid technique. Expandable elastomerics, foams, or pastes are potential materials that expand and can be made sufficiently thermally conductive to transfer curing temperatures adequately (a secondary posture may possibly be required after initial consolidation). More thermally stable organics are required to expansion mold thermoplastic or BMI matrices.

Shrink Film. The simultaneous identification of a film material and a means of locating and directing the resultant forces induced by thermal shrinkage is another viable technique. The advantage of applicability to positive contoured surfaces offers a real advantage in its application.

7.6 CASE HISTORIES

Repairs to aircraft structures may be made to varying extents ranging from those needed to survive only one flight to those made to last for as long as the aircraft. The extent of the repair depends on the conditions under which it is made; equipment, materials, and expertise available; and required flight length.

Field repairs may have to be made outdoors under diverse noncontrolled conditions, perhaps in a jungle, in a desert, on a mountain, or on-board ship. Available equipment, tools, and materials may be limited, and there will very likely be no electric power for heat-curing of the adhesive bonds and no refrigeration for preservation of adhesives. Simple procedures for application of the repair are therefore needed, ideally requiring no special skills and only limited previous training of the operatives.[88] By their nature, field repairs are usually temporary and more frequently applied to military aircraft. They are usually made to enable the aircraft to make just one journey back to its base where a permanent repair can then be carried out.

Similarly, if a damaged aircraft lands at a foreign base, a quick temporary repair can be made to enable the plane to fly back to its own base for permanent repair.

Temporary repairs can be made to last for longer than one flight. Fast, cheap repairs can be made either in the field or at the aircraft base and designed to last until the next servicing of the aircraft when it can be permanently repaired. The idea of this type of repair is to reduce the downtime of the aircraft between servicing to a minimum.

Permanent repairs are nearly always made at the aircraft base under ideal workshop conditions. All the required materials, tools, and expertise are available to produce a repair that is expected to last for the remaining life of the aircraft.

The following case study examples are used to illustrate repair design technology, and based on these experiences engineers have been able to generate general repair design rules.

Lightening Hole Failure. A shear web failed with two cracks propagating at 45° from a lightening hole in an EH-101 helicopter fuselage constructed of I- and C-section frames and beams machined from light alloy forgings and plate.

The requirement was for a repair that could be easily installed in situ, restore the load-carrying capability of the web, prevent the adjacent holes from failing, and stop the cracks from propagating. Composite materials were selected to be used for the repair of fatigue failures.

The repairs used a wet lay-up of ±45° carbon fabric for shear webs and precured, bonded-on, unidirectional C/Ep strips for the flanges. Where it was necessary to use fasteners in the flanges, ±45° carbon fabric was introduced. In this way, complex frame shapes were accommodated. Such repairs could be generated rapidly and applied in situ.

Fatigue testing of the repaired airframe ensued (Figure 7.17). The oldest repair has experienced the equivalent of 6400 flying hours of low-frequency (maneuver) loading plus 0.86×10^6 cycles of high-frequency (rotor-generated) loading. To date, none of the repairs has failed.

Curved and Tapered Flange Failures. Figure 7.18 shows a failure from the edge of a curved inner flange where it joins with a vertical stiffener and the repair which has survived approximately 14,800 h mean life without applying full composite factors.[86] Figure 7.19 shows the repair resulting from a failure at the thinnest point of a tapered flange that has survived the equivalent of approximately 9800 h mean life without applying full composite factors.

MD-11 Composite Tailcone. Using dry carbon cloth and a liquid resin matrix that can be cured at under 93°C, Mitsubishi Heavy Industries tests[87] found better processing and mechanical properties than the 121°C curing prepregs and film adhesives. The moisture was the main cause of voids in the adhesive layer during the 121°C-vacuum pressure cure cycle. The lower processing temperature (wet lay-up) showed better results than the higher processing temperature (prepreg–adhesive lay-up) for composite repair. See Figure 7.20*a* and *b*.

C-141 Wing Crack Failure. Wing cracks in the U.S. Air Force C-141 have been repaired with a technique, called Bortex patching, in which a boron composite fiber is bonded onto the metal. This is the latest in a series of repair techniques utilizing boron fiber and reinforced thermoset composites.

Boron fiber composite is believed to have several advantages over carbon fiber composite:

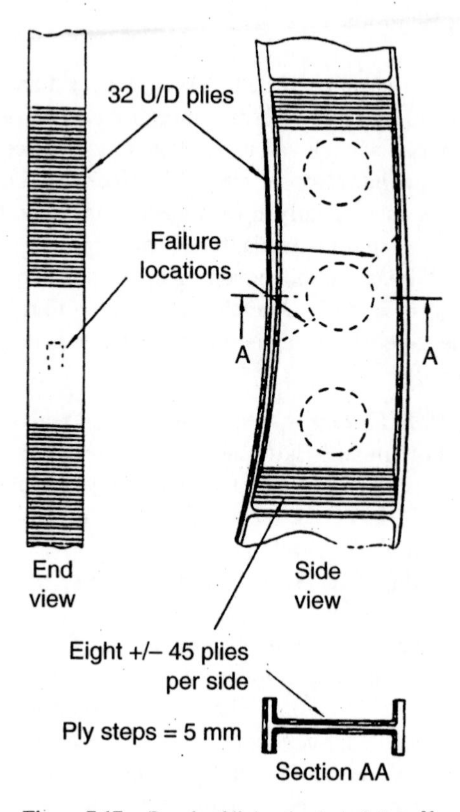

Figure 7.17. Repair of lightening hole failure.[86]

1. Boron is intrinsically stiffer than carbon.
2. Boron is closer to aluminum in the electromotive series than carbon, and therefore the chance of galvanic corrosion is decreased.
3. When using eddy currents to inspect a crack in aluminum below the composite, it is an advantage if the composite has low electric conductivity so that the signals being received from the aluminum are not distorted by signals from the composite. Boron is less conductive than carbon.
4. The difference in CTE between boron and aluminum is smaller than that between carbon and aluminum. Therefore, there is less stress buildup in a boron-to-aluminum bond due to thermal expansion during heat curing.[88]

Boron-epoxy (B/Ep) composite patching techniques offer solutions to repair problems for the commercial aircraft sector.

Installing a composite patch is straightforward. First, the surface to be repaired is prepared according to methods approved by the airframe manufacturer. Second, thermosetting film adhesives are laid down to match the coming composite sheet. Third, alternating layers of composite prepregs and adhesive films are laid down to build a patch of adequate strength and stiffness. Last, heat and vacuum are applied to the patch to "co-cure" the adhesives and composites. Portable equipment provides the required heat and pressure, and so repairs may not require disassembly or special facilities.

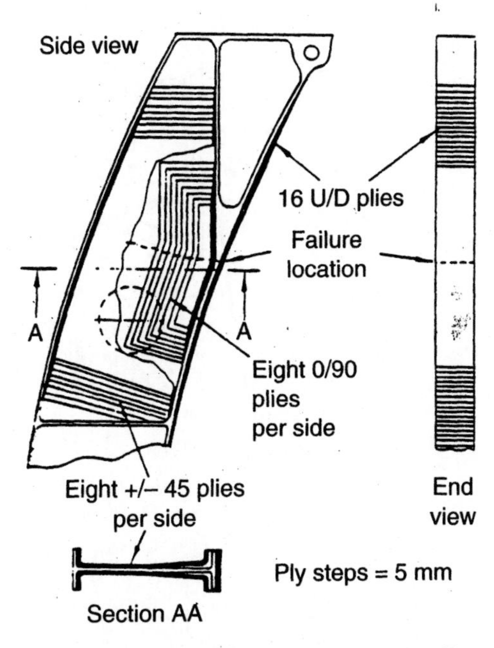

Figure 7.18. Repair of curved flange failure.[86]

Engineers have found that when composites are compared to aluminum patches:

- Fatigue life increases 2.5 times.
- Inspection interval increases 5 times.
- Aircraft availability increases 2.5% (the equivalent of having five more aircraft in a 270-plane fleet).

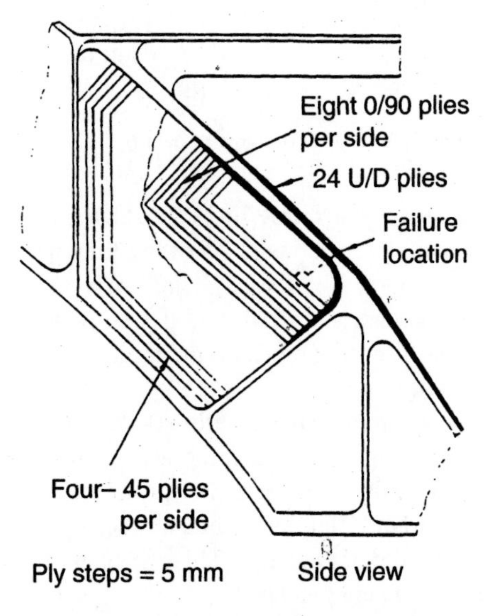

Figure 7.19. Repair of tapered flange failure.[86]

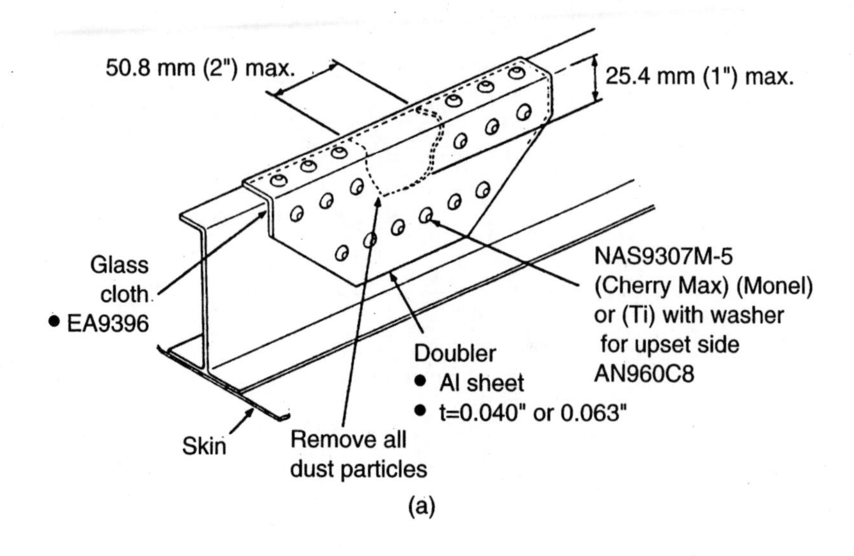

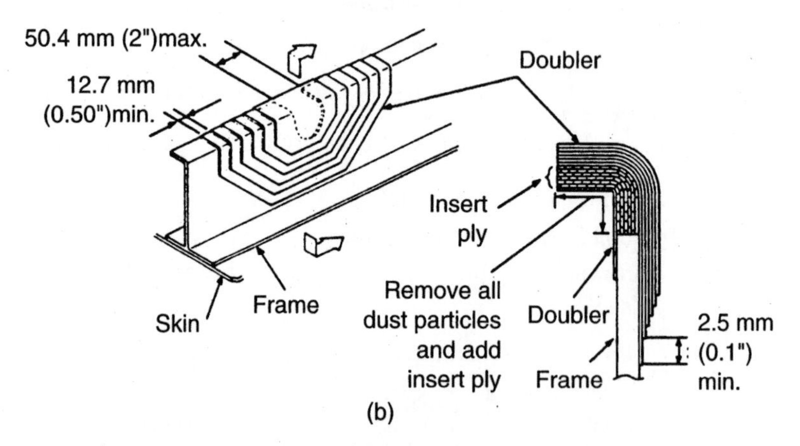

Figure 7.20. Repair of frame by old (*a*) and new (*b*) methods.[87]

In addition to extending airframe life, B/Ep patching saves money in other ways. Composite repairs typically require just 20% of the time necessary for conventional metalworking techniques. In one case, a keel-beam repair on a 767 aircraft would have required 25 days using conventional methods. It was done in 2 days using composites. Moreover, for B/Ep repairs, curing takes up the largest portion of the repair time, freeing technicians for other work.

B/Ep material comes in thin sheets with 4.0- or 5.6-mil boron fibers oriented in one direction. That way, the designer can develop a patch with minimum weight and thickness and maximum strength in the direction of principal stress. Typically, the material is three times stiffer than aircraft aluminum, and so these patches are one-third to one-half as thick as comparable aluminum patches. This means they add less weight to a structure and produce less drag than metal patches.

Bonding a boron patch in place instead of riveting it prevents several major problems: bonding adds no new stress-concentrating fasteners to the airframe. It eliminates blind drilling into airframes that might damage subsurface electrical or hydraulic lines. It avoids the problem of loose or incorrectly installed rivets that may come out and cause foreign object damage to engines. And because B/Ep is transparent to eddy current sensors, it allows postinstallation monitoring of covered cracks.

Beyond repairing cracked aircraft skins and other structures, manufacturers use the material to reinforce load-bearing elements in anticipation of future cracking problems. In one example of prophylactic reinforcement, B/Ep splices on C-141 wing beams reduced local stresses by more than 12%. In this highly cyclic load application, fatigue life increased by a factor of 2.4.

It should be pointed out that this technology is new only to the civilian aircraft industry. Twenty years ago, in its pioneering application, B/Ep patches were used to repair cracks in wing pivots on FB-111s, and during those 20 years, not one of the patches has debonded or failed.[89–94]

7.7 FUTURE OF REPAIR

The value of composites to aircraft today is a reality. Each new generation of aircraft has higher weight fractions, and they are quickly approaching the maximum structure weight possible. In some cases they have already arrived at that point. These amounts vary depending on the type of aircraft and where the user group is in acceptance of composites.

Another reality is that these composite structures will be damaged in their lifetimes. When they are damaged, they need to be fixed. The value of the composite structure is extremely high. The methods by which these structures will be repaired continues to be developed in parallel with composite structure development.[84,85,91,95]

The most important characteristic a future repair material requires is the ability to replace the properties that exist in the original composite material. This means that the ideal would be to replace the original material ply for ply with the same number and orientation. In this situation the transition or interface between the undamaged material and the repair must have sufficient shear strength to transfer the required loads. One technique for doing this involves a scarf joint. An experiment was conducted to determine the tensile strength of a scarf joint with a 20:1 taper. The complete strength of the original material was not obtained in tests, but the possibility of using a slightly longer scarf joint should give the original strength. This is only one technique.

The characteristics of the repair materials of the future are continually under development to meet the critical needs of tomorrow. There is yet another need, and that is for these new materials to gain acceptance. Unless some mechanism or procedure is implemented, current repairs will be very cumbersome and costly in dollars and labor. One idea is to produce common repair materials that can be used on all aircraft.

There is a vital need for a generic repair material that can be used on all aircraft and all parts no matter what the original material. This material must be evaluated fully and then used everywhere a repair is required.

Efforts must continue to address the repair problem, and efforts must be accelerated if composite repair techniques are to meet the needs of the composite aircraft of the future.

REFERENCES

1. Dexter, H. B., and A. J. Chapman. 1980. NASA service experience with composite components. In *Materials 1980,* vol. 12, comp. H. B. Dexter and A. J. Chapman, 77–99. National SAMPE Technical Conference Series. SAMPE, Anaheim, CA.
2. Phelps, M. L. 1979. Assessment of state-of-the-art of in-service inspection methods for graphite epoxy composite structures on commercial transport aircraft. NASA CR-158969.
3. Phelps, M. L. 1981. In-service inspection methods for graphite-epoxy structures on commercial transport aircraft. NASA CR-165746.
4. Stone, R. H. 1980. Development of repair procedures for graphite/epoxy structures on commercial transports. In *Materials 1980,* vol. 12, comp. H. B. Dexter and A. J. Chapman, 688–701. National SAMPE Technical Conference Series. SAMPE, Anaheim, CA.
5. Beck, C. E. 1982. *Advanced Composite Structure Repair Guide.* Northrup Corporation, Aircraft Division, Hawthorne, CA. NOR 82-60, AFWAL/FIBC, F-33615-79-C-3217.
6. Schweinberg, W., P. Manning, L. Ragan, et al. 1983. Depot level repairability, maintainability, and supportability of advanced composites, October 1983, Ft. Worth, TX. AIAA-83-2516.
7. Overd, M. L. 1993. Carbon composite repairs of helicopter metallic primary structures. *Compos. Struct.* 25(1–4):557–65.
8. Jones, R., R. J. Callinan, and K. C. Aggarwal. 1993. Analysis of bonded repairs to damaged fibre composite structures. *Eng. Fract. Mech.* 17(1):37–46.
9. Baker, A. A. 1984. Repair of cracked or defective metallic aircraft components with advanced fiber composites—An overview of Australian work. *Compos. Struct.* 6:153–81.
10. Baker, A. A., R. J. Callinan, M. J. Davis, et al. 1984. Repair of Mirage III aircraft using the BFRP crackpatching technique. *Theor. Appl. Fract. Mech.* 2:1–15.
11. Baker, A. A. 1987. Fibre composite repair of cracked metallic aircraft components—Practical and basic aspects. *Composites* 18(4):293–308.
12. Ratwani, M. M., and H. P. Kan. 1980. Development of composite patches to repair complex cracked metallic structures, NADC-80161-60. NADC, Warminster, PA.
13. Sandow, F. A., and R. K. Cannon. 1987. Composite repair of cracked aluminum alloy aircraft structure. AFWAL-TR-87-3072.
14. Baker, A. A., and R. Jones. 1988. *Bonded Repair of Aircraft Structures.* Martinus Nijhoff, Dordrecht, Netherlands.
15. Molent, L., R. J. Callinan, and R. Jones. 1989. Design of an all boron/epoxy doubler reinforcement for the F-111C wing pivot fitting structural aspects. *Compos. Struct.* 11:57–83.
16. *Advanced Composites Design Guide,* vols. I–V, 1976. Structures Division, AFFDL/AFSC, Wright-Patterson Air Force Base, OH.
17. *Adhesive Bonded Aerospace Structures Standardized Repair Handbook.* AFMLTR-77-206/AFML-TR-77-139.
18. Inspection and repair of advanced composite airframe structure for helicopters. USAAVRADCOM TR-82-D-20.
19. Kevlar epoxy skin panel replaces aluminum in last two bays on bottom of CH-53 cargo ramp. Contract NASA NAS 1-14447, 1981.
20. Boron epoxy used to provide vertical bending stiffness for CH-54A helicopter tailcone. Contract NASA NAS 1-10459, 1980.
21. Card, M. F., and M. D. Rhodes. 1980. Graphite-epoxy panel compression strength reduction due to local impact. *AGARD Conf. Proc. No. 288: Effect Serv. Environ. Compos. Mater.,* pp. 11-1 to 11-13, August 1980, Athens, Greece.

22. Rhodes, M. D., and J. G. Williams. Concepts for improving the damage tolerance of composite compression panels. *5th DOD/NASA Conf. Fibrous Compos. Struct. Des.* January 1981, New Orleans, LA.
23. McConnell, V. P. 1989. In need of repair. *Adv. Compos.* May/June:60–70.
24. Jones, J. S., and S. R. Graves. 1984. Repair techniques for Celion/LARC-160 graphite/polyimide composite structures. NASA CR-3794, Contract NAS1-16448, 1984.
25. The advanced composite repair process. *Aerosp. Eng.* December 1992, pp. 7–8.
26. Goering, J., and K. Griffiths. 1991. Design and analysis techniques for aircraft battle damage repair of advanced composite structures. McDonnell Douglas Corporation, Progress Report, November 1989–February 1991. NAWCADWAR-92025-60.
27. Ostrom, R. B., R. H. Stone, and L. D. Fogg. 1983. Field-level repair for composite structures. NADC-79173-60.
28. Bohlmann, R. E., D. A. Glaeser, and J. B. Watson. 1980. Field repair of composite structural components. NADC-78073-60.
29. Labor, J. D., G. M. Button, and N. M. Bhatia. 1985. Depot level repair for composite structures development and validation, vol. I. NADC-79172-60.
30. Hinkle, T., and J. Van Es. 1988. Battle damage repair of composite structures. AFWAL-TR-87-3104.
31. Behrens, R. S. 1986. Structural systems for advanced aircraft. McDonnell Douglas Corporation IR&D Project Description MDC Q0877-7, vol. 2.
32. Hinkle, T. V. 1988. Advanced structural applications. McDonnell Douglas Corporation, IR&D Project Description MDC Q0903-7, vol. 2.
33. Martin, W. 1987. Advanced field level repair materials technology for composite structures. AFWAL-TR-87-4109.
34. Hellard, G. 1986. A.T.R. 42 carbon fiber flap repair—Design and inspection. *AGARD Conf. Proc. No. 402: Repair Aircr. Struct. Involving Compos. Mater.,* Oslo, Norway.
35. Torres, M., and B. Plissonneau. 1986. Repair of helicopter composite structure: Techniques and substantiations. *AGARD Conf. Proc. No. 402: Repair Aircr. Struct. Involving Compos. Mater.,* Oslo, Norway.
36. Armstrong, K. B. 1986. British Airways experience with composite repairs. *AGARD Conf. Proc. No. 402: Repair Aircr. Struct. Involving Compos. Mater.,* Oslo, Norway.
37. Cochran, R. C., T. M. Donnellan, E. L. Rosenzweig, et al. 1986. Composite repair material design development efforts. *AGARD Conf. Proc. No. 402: Repair Aircr. Struct. Involving Compos. Mater.,* Oslo, Norway.
38. Wentworth, S. E., and M. S. Sennett. 1986. Unconventional approaches to field repair. *AGARD Conf. Proc. No. 402: Repair Aircr. Struct. Involving Compos. Mater.,* Oslo, Norway.
39. Cavitt, W. M. 1988. Aircraft battle damage repair of advanced composite structures. Masters Thesis, Naval Postgraduate School, Monterey, CA.
40. Golian, M. A., and T. D. Price. 1986. AV-8B horizontal stabilator composite repair verification. *2nd DOD/NASA Compos. Repair Technol. Workshop,* November 1986.
41. Bhatia, N. M., G. M. Button, and J. D. Labor. 1986. Advanced depot level repair of composite structures. *2nd DOD/NASA Compos. Repair Technol. Workshop,* November 1986.
42. McMillan, D. 1990. Aircraft battle damage repair. McDonnell Douglas Corporation IR&D Project Description MDC Q0952-14, vol. 1.
43. Kovensky, D. 1988. Aircraft battle damage repair, McDonnell Douglas Corporation IR&D Project Description MDC Q0903-7, vol. 3.
44. Dominguez, J. 1987. Design and test of rapid repair of large area damage of graphite/epoxy structure. NADC-87079-60.

45. Swihart, M. S. 1990. Wingbox testing overview. Unpublished correspondence with Tim Wise, Naval Weapons Center, China Lake, CA.
46. Coffenberry, B. S., J. W. O'Neill, M. S. Swihart, et al. 1988. Survivability testing of IM7/8551-7A wingbox at Naval Weapons Center, China Lake, CA. Unpublished.
47. Johnson, A. H., T. W. Hindman, R. L. Nielsen, et al. 1988. Evaluation procedures for aircraft battle damage repair (ABDR) techniques, vol. I: Technical summary. AFWAL-TR-3011.
48. Johnson, A. H., T. W. Hindman, R. L. Nielsen, et al. 1988. Evaluation procedures for aircraft battle damage repair (ABDR) techniques, vol. II: ABDR technique evaluation procedures. AFWAL-TR-3011.
49. Shyprykevich, P. 1986. Design and analysis of field repairs for high-strain composite wing. *2nd DOD/NASA Compos. Repair Technol. Workshop,* San Diego, CA.
50. Mahon, H., H. Moss, P. Shyprykevich, et al. 1989. Repair procedure for skin, plank, and stiffener damage to high strain wing test element. *21st Int. SAMPE Tech. Conf.,* September 1989.
51. Behrens, R. S. 1987. Structural systems for advanced aircraft. McDonnell Douglas Corporation IR&D Project Description MDC Q0885-7, vol. 2.
52. Gunter, G., and L. Lenner. 1986. Composite repair of co-cured J-stiffened panels, design and test verification. *AGARD Conf. Proc. No. 402: Repair Aircr. Struct. Involving Compos. Mater.,* Oslo, Norway.
53. Ledua, K. 1986. Composite repair techniques for J-stiffened composite fuselage structures. *AGARD Conf. Proc. No. 402: Repair Aircr. Struct. Involving Compos. Mater.,* Oslo, Norway.
54. Stone, R. H. 1986. Development of field level repairs for composite structures. *AGARD Conf. Proc. No. 402: Repair Aircr. Struct. Involving Compos. Mater.,* Oslo, Norway.
55. Rosenzweig, E., M. F. DiBerardino, M. D. Poliszuk, et al. 1989. Repair concept development for bismaleimide composite structures. *21st Int. SAMPE Tech. Conf.,* September 1989.
56. Labor, J. D., and S. H. Myhre. 1979. Large area composite structure repair. AFFDL-TR-79-3040.
57. Reisdorfer, D. A. 1992. Evolution of permanent composite repair designs. *24th Int. SAMPE Tech. Conf.,* pp. T791–T801, October 1992, Toronto.
58. Armstrong, K. B. 1986. British Airways experience with composite repairs, IMechE C166/86, pp. 183–90.
59. Dodiuk, H., S. Kenig, and I. Liran. 1991. *Composites* 22(4):319.
60. English, L. K. 1988. Field repair of composite structures. *Mater. Eng.* September:37–39.
61. Cook, L. C. 1989. The repair of aircraft structures using adhesive bonding—A survey of current UK practices, 392/1989. Welding Institute, Abington, England.
62. Hall, S. R., M. D. Raizenne, and D. L. Simpson. 1989. Proposed composite repair methodology for primary structure. *Composites* 20(5):479–83.
63. Reinhart, T. 1989. Composites supportability in the U.S. Air Force. *34th SAMPE Int. Symp. Exhibit.,* May 1989, Reno, NV.
64. Steelman, T. 1989. Repair technology for thermoplastic aircraft structures (REPTAS). *34th SAMPE Int. Symp. Exhibit.,* May 1989, Reno, NV.
65. Purcell, B. 1989. Battlefield repair of bonded honeycomb panels. *34th SAMPE Int. Symp. Exhibit.,* May 1989, Reno, NV.
66. Sivy, G., and P. Briggs. 1989. Rapid low-temperature repair system for field repair. *34th SAMPE Int. Symp. Exhibit.,* May 1989, Reno, NV.
67. Mahon, J. 1989. Induction bonded repair of composite materials in Army service: Planning for the future. MTL, TR-89-45, FR 5/89.

68. John, S. J., A. J. Kinloch, and F. L. Matthews. 1989. In *Bonding and Repair of Composites.* Butterworth, London.

69. Corlet, D. C. 1989. In *Bonding and Repair of Composites, 83,* Butterworth, London.

70. Dodiuk, H., A. Buchman, I. Liran, et al. 1993. Epoxy adhesives for repair of composite structures. Part V. *J. Adhes.* 40(2–4):127–38.

71. Weiss, J. 1983. Development of a stable epoxy resin system for composite repair. NADC-81108-60, Contract N62269-81-C-0153.

72. Kelly, L. G. 1983. Composite structure repair. *AGARD 57th Meet. Struct. Mater. Panel,* October 1983, Vimeiro, Portugal. AGARD 716.

73. Ashizawa, M. 1983. Improving damage tolerance of laminated composites through the use of new tough resins. *6th Conf. Fibrous Compos. Struct. Des.,* January 1983, Lake Tahoe, NV.

74. Hart-Smith, L. J. 1982. Design methodology for bonded-bolted composite joints. Technical Report, Wright-Patterson Air Force Base Off. AFWAL-WTR-81-3154.

75. McCarty, J. E., R. E. Horton, et al. 1978. Repair of bonded primary structures. Technical Report, Army Air Force. AFFDL-TR-78-79.

76. Ong, C.-L., and S. B. Shen. 1992. The reinforcing effect of composite patch repairs on metallic aircraft. *Int. J. Adhes. Adhesiv.* 12(1):19–26.

77. Barr, B. 1994. Repair of nonmetallic laminate and sandwich components-RAH-66 Comanche helicopter. Contract DAAJ09-91-C-A004.

78. Mehrkam, P., and R. Cochran. 1992. Wet layup materials for repair of bismaleimide composites. *24th Int. SAMPE Tech. Conf.,* pp. T1006–16. October 1992, Toronto.

79. Xiao, X. R., S. V. Hoa, and K. N. Street. 1990. Repair of thermoplastic composite structures by fusion bonding. *35th Int. SAMPE Symp.,* pp. 37–43, April 1990, Anaheim, CA.

80. Powers, J. W., and W. J. Trzaskos. 1989. *34th Int. SAMPE Symp.,* pp. 1987–98, Reno, NV.

81. Border, J., R. Salas, and M. Black. 1990. Induction heating development for aircraft repair. *35th Int. SAMPE Symp.,* pp. 1411–20, April 1990, Anaheim, CA.

82. Ong, C.-L., and S. B. Shen. 1989. Adhesive-bonded composite-patching repair of cracked aircraft structure. In D. M. Zakrzewski, ed., *Proc. 34th Int. SAMPE Symp.,* pp 1067–78, Reno, NV.

82. Wegman, R. E., and T. R. Tullos. 1993. Adhesive bonded structural repair. Part I: Materials and processes, damage assessment and repair. *SAMPE J.* 29(4):8–13.

83. Wegman, R. E., and T. R. Tullos. 1993. Adhesive bonded structural repair. Part II: Surface preparation procedures, tools, equipment and facilities. *SAMPE J.* 29(5):8–13.

84. Holcomb, D. H. 1994. Aircraft battle damage repair for the 1990s and beyond. Airpower Research Institute, AU-ARI-93-4. Air University Press, Maxwell Air Force Base, AL.

85. Reinhart, T. J. 1991. *Third DOD/NASA Compos. Repair Technol. Workshop,* May 1991, Wright-Patterson Air Force Base, OH. WLTR-91-4054.

86. Welder, S. M., H. J. Lause, and R. Fountain. 1985. Structural repair systems for thermoplastic composites. *SAMPE Q.* January:33–36.

87. Yamamoto, T., and G. R. Bonnar. 1992. Repair materials and processes for the MD-11 composite tailcone. *24th Int. SAMPE Tech. Conf.,* pp. T1017–T1028, October 1992, Toronto.

88. Wilson, T. A., and M. J. Graves. 1991. Repair of graphite/PEEK APC-2 using thermoforming. *SAMPE Q.* 23(1):51–57.

89. Jones, R., L. Molent, A. A. Baker, et al. 1988. *Bonded Repair of Metallic Components: Thick Sections.* Aeronautical Research Laboratory, Melbourne, Australia. Elsevier, Amsterdam.

90. Wentworth, S. E., M. S. Sennett, and J. W. Gibson. 1985. Some novel approaches to composite field repair. *Proc. Conf. Adhes. Sealants Encapsulants 1985,* pp. 265–71, London, November 1985.

91. Cochran, J. B., T. Christian, and D. O. Hammond. 1988. C-141 repair of metal structures by use of composites, *1987 ASIP/ENSIP Proc.,* ed. J. D. Lincoln. AFWAL-TR-88-4128.

92. U.S. AFLC. 1989. Composite patches for metallic structures. TT-89034.

93. Hammond, D. O., and J. R. Cochran. 1988. Composite repair of primary metallic structure. Contract F09063-87-0741-0035.

94. Baker, A. A. 1987. Fibre composite repair of cracked metallic aircraft components—Practical and basic aspects. *Composites* 18(4).

95. Lynch, T. P. 1991. Composite patches reinforce aircraft structures. *Design News.* April 22:116–117.

BIBLIOGRAPHY

ACS uses adhesives for rapid aircraft repairs. *Adhesiv. Age,* November:44, 1993.

Adhesive Bonding Handbook for Advanced Structural Materials. ESA PSS-03-210 Issue 1. European Space Research and Technology Centre, Noordwijk, Netherlands, 1991. N91-32234.

Aerosp. Eng. July:44, 1993.

ASKINS, R. Characterization of EA9396 epoxy resin for composite repair applications. Dayton University Research Institute, Dayton, OH. UDR-TR-91-77.

Aviat. Week Space Technol. July 19:13, 40, 1993.

BAKER, A. A., R. J. CHESTER, M. J. DAVIS, ET AL. Reinforcement of the F-111 wing pivot fitting with a boron/epoxy doubler system—Materials engineering aspects. *Composites.* 24(6):511–21, 1993.

CHABOT, K. A., AND J. A. BRESCIA. Evaluation of primers for aircraft repair. *25th Int. SAMPE Tech. Conf.,* Philadelphia, October 1993.

CONNOLLY, J. J. A single side access repair solution for highly curved BMI composite structure. *25th Int. SAMPE Tech. Conf.,* Philadelphia, October 1993.

COOK, T. N., AND T. E. CONDON. Inspection and repair of advanced composite airframe structures for helicopters. *J. Am. Helicopter Soc.* 29(10):31–37, 1984.

DAVIS, J. G., J. E. GARDNER, AND M. B. DOW, EDS. *Proc. 4th NASA/DOD Adv. Compos. Technol. Conf.,* Salt Lake City, UT, 1993. NASA-CP-3229-V-1-PT-2.

DIXON, J., AND J. LONERGAN. Repairs for hat stiffened composite skins. *25th Int. SAMPE Tech. Conf.,* Philadelphia, October 1993.

DOMPKA, R. V. V-22 composite repair development program. *25th Int. SAMPE Tech. Conf.,* Philadelphia, October 1993.

DONNELLAN, T. M., E. L. ROSENZWEIG, R. E. TROBOCCO, ET AL. Repair of composites. *58th Meet. Struct. Mater. Panel,* pp. 1-1 to 1-10, Sienna, Italy, April 1984. AGARD-R-716.

ENGLISH, L. K. Field repair of composite structures. *Mater. Eng.* September:37–39, 1988.

HABER, H. S., AND A. QUINN. Making on-orbit structural repairs to space station. *Aerosp. Eng.* August:20–24, 1989.

HOA, S. V., AND X. R. XIAO. Effects of repair processing on the mechanical properties of thermoplastic matrix composites, 1992. DREP-92-02, CTN-92-60594.

JOHNSON, R. Aging aircraft and airworthiness. *Aerosp. Eng.* July:23–30, 1992.

KEARNS, T. M., ET AL. Manufacturing technology for integration of advanced composites repair. Report IR 490-3, III, V, VI, AFWAL/MLTN, November 1984–August 1985.

KREBBEKX, J. A. Repair of composites by means of wet lay-up. Technische Hogeshool, Delft, Netherlands, 1988. LR 551.

MAHON, J. Materials and processes for rapid repair of composites. *25th Int. SAMPE Tech. Conf.,* Philadelphia, October 1993.

MARDOIAN, G. H., AND M. B. EZZO. Flight service evaluation of composite helicopter components. *J. Am. Helicopter Soc.* 39(1):31–40, 1994.

MARSHALL, I. H. *Proc. 6th Int. Conf. Compos. Struct.,* 1991. Elsevier, London, 1991.

MEADE, L. E., AND D. S. BITTAKER. *Manufacturing Technology for Integration of Advanced Repair Bonding Techniques,* vol. II. SA-ALC, OC-ALC, OC-ALC, WR-ALC, and CCAD—Alternate Depots (Honeycomb Repair Center). AFWAL-TR-87-4085, FR September 1982–June 1988.

MEHRKAM, P. A. Cocure of wet layup patch and toughen adhesive for composite repair applications. *25th Int. SAMPE Tech. Conf.,* Philadelphia, October 1993.

MEHRKAM, P., AND R. COCHRAN. Wet lay-up patch repair for elevated temperature composite structures. *26th Int. SAMPE Tech. Conf.,* Atlanta, GA, October 1994.

MIDDLETON, D. H. The first fifty years of composite materials in aircraft construction. *Aeronaut. J.* 96(953):96–104, 1992.

RODGERS, B. A., AND P. J. MALLON. Post-impact repair of thermoplastic composite materials using induction heating. *14th Int. SAMPE Eur. Conf. Exhibit.,* Birmingham, England, October 1993.

RUFFNER, D., AND R. PATEL. Evaluation of low temperature curing repair adhesives. *38th Int. SAMPE Symp. Exhibit.,* Anaheim, CA, May 1993.

SAMPE. 1989. Tomorrow's materials: today. *34th Int. SAMPE Symp. Exhibit.,* pp. 62–74, 75–86, 87–94, 95–99, 185–196, 197–207, 208–222, 223–235, 236–246, Reno, NV.

SIH, G. C., AND S. E. HSU, EDS. *Proc. Int. Conf. Adv. Compos. Mater. Struct.,* Taipei, May 1986. VNU Science Press, Utrecht, Netherlands, 1987.

SUCHAR, K. J. Structural repair of composites, *Adhesives 1990,* Schaumburg, IL, October 1990. SME AD90-452.

24th Int. SAMPE Tech. Conf.: Adv. Mater.—Meet. Econ. Challenge, pp. T536–550, T1006–T1027, October 1992.

UTC-Sikorsky Aircraft. *Composite Repair Training Manual,* 1984.

VILSMEIER, J. W. Composite repair of aircraft structures. *58th Meet. Struct. Mater. Panel,* pp. 2-1 to 2-9, Sienna, Italy, April 1984.

WEGMAN, R. F., AND T. R. TULLOS. Adhesive bonded structural repair. Part III. Repair of composite, honeycomb cored, and solid cored structures. *SAMPE J.* 29(6):8–12, 1993.

WILSON, H. Flight service evaluation of composite components on the Bell Helicopter Model 206L, 1993. NASA Report 191499, Contract NAS1-15279.

8

Nondestructive Evaluation, Testing, and Inspection

8.0 INTRODUCTION

Rapid advances in electronics and computing have produced the new disciplines of nondestructive testing (NDT), nondestructive examination or evaluation (NDE), and nondestructive inspection (NDI) of composites. The subject of NDT and composites has been covered in excellent comprehensive reviews[1–25] which serve to create an outline of the state of the art.

Composite materials must be regarded as very different media from metals when considering which NDT methods are appropriate. Generally, reinforced plastics have poor electric conductivity, low thermal conductivity, high acoustic attenuation, and most importantly significant anisotropy of mechanical and physical properties. The life of a metal component is determined by the nucleation and growth of cracks or damage in the material. The development of linear elastic fracture mechanics is often adequate as a basis for determination of the size of subcritical flaws that must be identified. However, a fiber-reinforced composite is a heterogeneous medium that can contain multiple defect geometries. No single failure model can adequately describe the level of damage that is critical. A multiplicity of models have been developed to describe the various failure modes: interlaminar debonding, matrix degradation, fiber fracture, and fiber-matrix interface separation. These in turn may be caused by improper cure, fiber misalignment, inclusions, poor reinforcement distribution, machining damage, fastener fretting, and environmental degradation.

These voids, void types, causes, and mechanical property influences have been reviewed in the literature by Kardos et al.[26] and Judd and Wright.[27] They have concluded that, regardless of resin, fiber type, or fiber surface treatment, the interlaminar shear strength of a composite decreases by about 7% for each 1% of voids up to a total void content of about 4%.

Chatterjee et al.[19] has also reported on the critical assessment of various types of defects in composites and fracture mechanics-based methodologies for assessing the significance of delamination defects. Walkden and Boutoussov[28] have reported on the feedback mechanisms between NDT and manufacture in atttempts to maximize the efficiency of the overall production process.

NDT technologies applicable to composites (fiber, whisker, and particulate) can be divided into three very broad categories: mechanical, electromagnetic, and spectroscopic (Figure 8.1).

8.0.1 NDT Categories

8.0.1.1 Mechanical vibration

Scanning Acoustic Microscopy (SAM).[29] SAM differs from conventional ultrasonics in two important ways: the frequencies are much higher and an acoustic lens is used to give a focused beam. It differs from conventional optical and electron methods in that the images are formed through the interaction of sound waves with the sample. The contrast therefore depends on the elastic properties of the material. The acoustic waves interact with the sample either by direct reflection of the longitudinal waves back to the coupling medium or by mode conversion to surface acoustic waves which propagate and scatter. SAM can be used to determine the orientation and distribution of fibers in composite materials and their adhesion to the matrix.

Ultrasonics.[30–33] The simplest method of ultrasonic testing is to transmit a short pulse of high frequency (typically 500kKz–10 MHz) energy through the specimen and to measure the transit time or attenuation of the signal. The measurement may be either through-thickness (separate transmit and receive transducers on opposite faces of the sample) or pulse-echo (reflecting the signal from a defect or from the back face of the specimen and using a single transmit-receive transducer). The attenuation is a function of the state of cure or viscoelasticity of the matrix resin and the condition of the fiber-resin interface with dispersion due to fibers, voidage, delaminations, and foreign inclusions. Coupling of the signal to the component may be by gels, water immersion, or water jets. For materials of known, constant constitution the transit time can be used to determine thickness.

With use of the same basic equipment with either water or gel coupling, several types of measurement are possible. The most commonly used display options include

A-scan: pulse and amplitude against time

B-scan: size and position against probe movement on the surface

C-scan: size and position on an area parallel to the surface

P-scan: projection scan on an angled plane (ultrasound tomography)

Acoustography: acoustooptical liquid crystal display.[34]

The most commonly used mode is the A-scan, where the display usually can be seen on the screen of the ultrasonic test set and gives the time history of the echoes received by the receiving transducer. The C-scan presents a plan view of the attenuation as

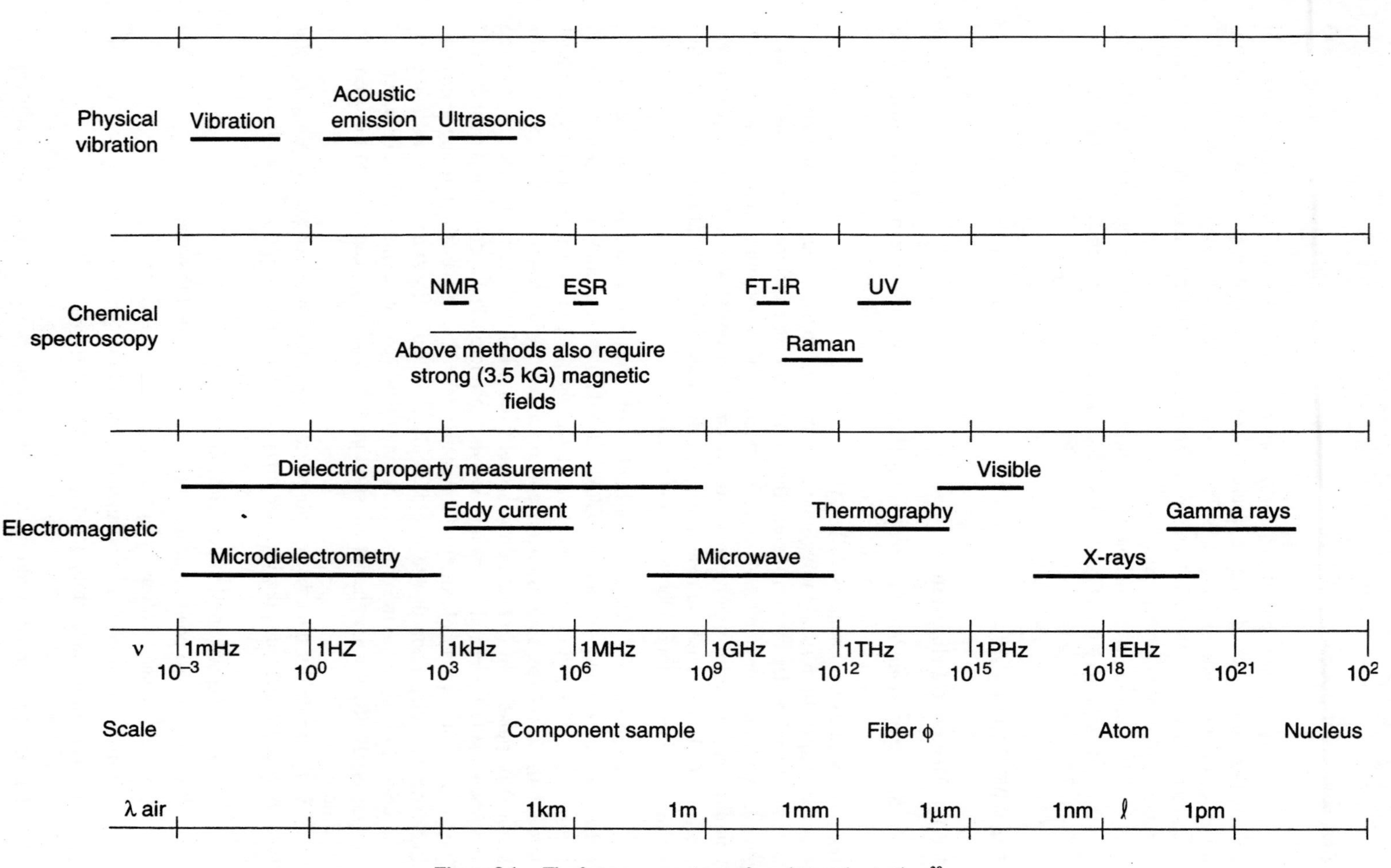

Figure 8.1. The frequency spectrum of nondestructive testing.[20]

a gray-level or color-coded image. The A-scan contains all the data required to establish the position and severity of individual defects, while the C-scan presents an integration of these data throughout the depth of the component (Figure 8.2).

In a comparative study[35] of through-transmission and pulse-echo examinations of a Gr/Ep panel, the scanned panel had varying thicknesses of 16, 32, 48, and 64 plies, increasing from the top to the bottom. Teflon shims and other intentionally implanted flaws of various sizes and depths showed up with either ultrasonic technique. In the through-transmission scans, delaminations appeared as dark areas of lower-than-average intensity; in the pulse-echo scans, they appeared as light areas of higher-than-average intensity. It was also obvious that the through-transmission method was not as dependent on the varying depth as the pulse-echo mode because in the through-transmission mode the waveform traveled through the material only once before reaching the receiver.

Composite materials are usually employed as thin-shell structures, and the introduction of ultrasonic energy can result in plate waves in the component rather than the body waves used for traditional ultrasonic NDT. The propagation of plate (Lamb) waves is more complex than that of body waves. Several modes of propagation exist, being either symmetric (longitudinal) or asymmetric (flexural) with respect to the midplane of the plate.

Lamb waves are dispersive (velocity depends on wave number and hence on frequency), and there is a cutoff frequency below which Lamb waves cannot propagate.

Cohen and Mal[36] recently described their work with an ultrasonic testing system that nondestructively measures elastic properties of, and defects in, a panel of laminated fiber-matrix (e.g., Gr/Ep) materials. The system acquires data on the dispersion of ultrasonic waves at various angles of incidence and reflection and inverts them to obtain the modulus of elasticity of the material.

As shown in Figure 8.3, the specimen panel is immersed in water along with two ultrasonic transducers that operate in a pitch or catch mode. The positions and orientations

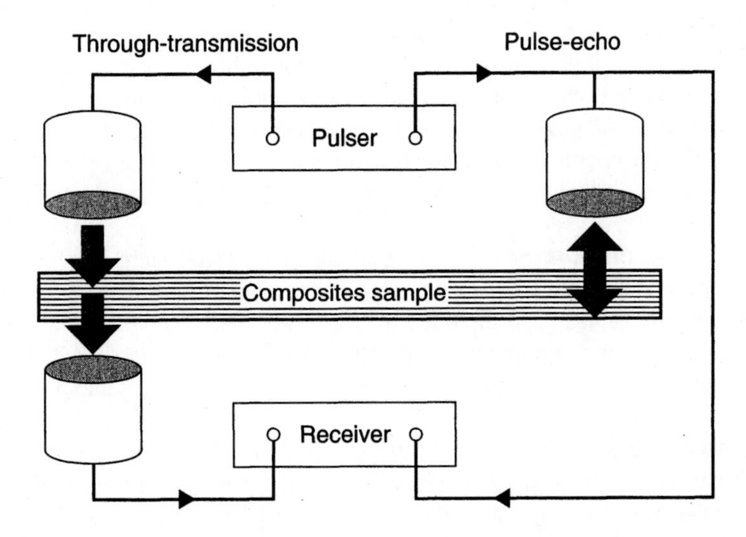

Figure 8.2. Principles of ultrasonic through-transmission and pulse-echo methods.[35]

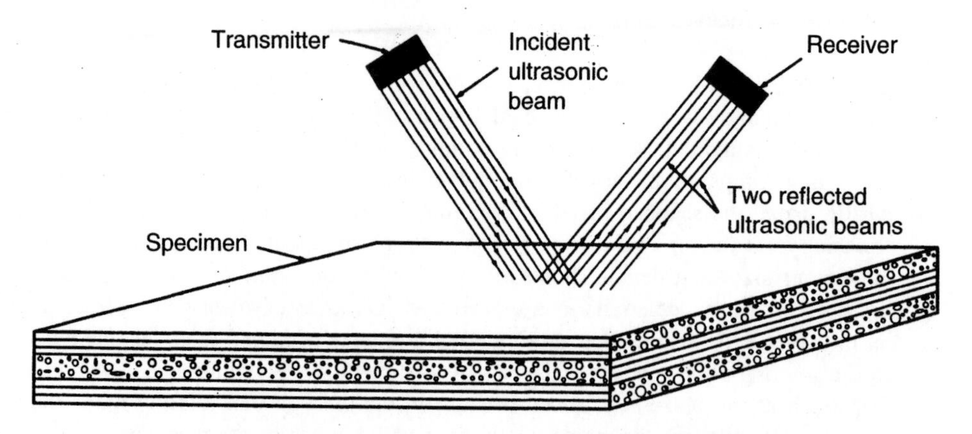

Figure 8.3. Two ultrasonic transducers operating in pitch-catch mode excite and detect leaky Lamb waves in the specimen. Elastic properties of the specimen and defects within it can be characterized from the dispersion curves of the leaky Lamb waves.[36]

of the transducers and specimen are controlled simultaneously so that the fibers in the specimen are oriented at a desired angle with respect to the plane of incidence and the angle of reflection at the receiving transducer equals the angle of incidence.

The angle of incidence and the frequency or frequencies of the ultrasonic waves are chosen to excite leaky Lamb waves in the specimen. (Lamb waves denote guided elastic waves in a plate of finite thickness having free surfaces. Leaky Lamb waves are guided waves in a plate immersed in fluid.) The leaky component of the waves interferes with specular reflection of the incident acoustic wave, giving rise to two reflected beams with a null between them. The receiving transducer is positioned in this null.

The frequencies at which Lamb waves can be excited in the specimen depend on the angle of incidence, the orientation of the fibers in the specimen, the thickness of the specimen, and the elastic properties of the specimen material. The frequency at which two reflected beams with an intervening null are excited at a given angle of incidence can be used to identify a leaky Lamb wave mode. The dispersion curves (phase velocities versus frequencies) of leaky Lamb wave modes at various fiber orientations and at bonds between the specimen and other materials (e.g., metals) provide significant information about defects and about the properties of the specimen composite material. An inversion algorithm extracts the information from the dispersion curves and computes the five stiffness coefficients needed to characterize the composite laminate as an orthotropic material.

Acoustoultrasonics. The acoustoultrasonic (AU) technique involves the use of two ultrasonic transducers placed a fixed distance apart on the test structure surface, one acting as a transmitter and the other as a receiver. The received signal is then processed using the techniques commonly used with acoustic emission testing. There has been considerable interest in a technique for the inspection of composites and bonded joints.[37]

Acoustic emission.[38–43] Acoustic emission (AE), sometimes referred to as stress wave emission (SWE), is a collective term for the generation, propagation, and detection of stress waves resulting from rapid changes in the local stress state of a structure.

Unlike most NDT techniques applied to composites, AE equipment listens for active defects generating stress waves within the volume of the material. Clustering of signals from a small area of the structure indicates the likely site of failure.[44–45] Studies on the practicality of using AE source location for composites have been reported by Glennie and Summerscales,[46] Ibitolu and Summerscales,[47] and Buttle and Scruby.[48]

Vibration. The "quality" of a composite material can be assessed by tapping a succession of locations on the composite with a coin or a tapping hammer. The sounds of good and bad areas are qualitatively different to the human ear, with a clear ring at good positions and a dull sound in poor areas. The technique has been automated in an instrument known as a Tapometer.[49]

Another method, wheel tapping, has been used for checking filament-wound tubes in a production environment. The use of a single tap to excite the vibration of the global structure is known as *wheel-tapping* and is quite distinct from *coin tapping* in which only local vibrations are generated.

Acoustic microscopy (AM). AM resembles conventional ultrasonic scanning except that it uses sharply focused ultrasound waves. Conventional ultrasound units use either unfocused or slightly focused beams, resulting in relatively poor resolution. An acoustic microscope, on the other hand, offers resolution on the order of a wavelength of ultrasound. It permits individual layers or fibers to be examined inside the composite material.

Acoustic microscopy works especially well with quasi-isotropic media, such as Gr/Ep lay-ups commonly used in aircraft. If material is isotropic or quasi-isotropic, relatively thick sections can be examined. The method works poorly with highly anisotropic media because the focused beam is readily refracted. Here, it is limited to near-surface inspection. Resin-poor sections are also difficult to analyze, although this condition poses a problem for any ultrasonic method.

Present systems use mechanical scanning similar to conventional ultrasonics. Because resolution is much finer, scanning an area takes considerably longer. Developments include attachments that allow rapid scanning of areas, with precise rescanning of flaws with the acoustic microscope.

Laser-based ultrasonics. A promising method for remote, high-speed inspection is the laser. With this technique, a laser pulse striking the surface of an object creates an ultrasonic pulse. This ultrasound pulse travels through the object and is reflected by the object's back surface or by flaws that are present. Sound waves returning to the surface produce minute vibrations, which are detected with an interferometer using a laser beam as a probe. Signals are then interpreted to determine the presence of flaws.

Laser-based ultrasonics use ultrabroad bandwidth frequencies, making it better at determining the size, shape, and orientation of flaws. Envisioned uses include in-process inspection in autoclaves. The method also lends itself to rapid scanning at rates two to three times that of conventional ultrasonics (Figure 8.4).

8.0.1.2 Image analysis. Most nondestructive testing technologies can benefit from the use of quantitative image processing. The ability to reliably inspect and interpret images from a composite has been dramatically enhanced through the use of image-

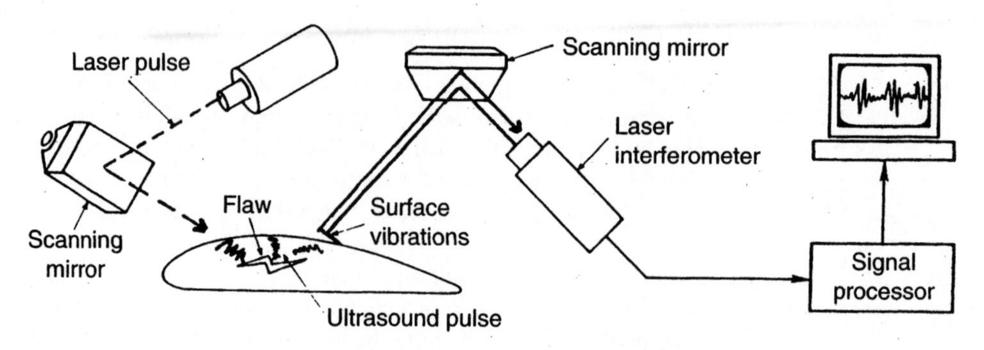

Figure 8.4. Laser-based ultrasonics.[49]

processing techniques. The incorporation of signal pattern recognition techniques allows for the classification of material anomalies by type as well as by location and thus aids in the evaluation of structural integrity and lifetime prediction.[5]

Computer Tomography (CT). CT is a method of imaging an arbitrary slice through a component by measurement of the absorption in a large number of different directions. The measured data are digitized, and a computer program then calculates and displays the chosen image plane. The image produced is as if a thin slice has been cut through the specimen along the plane of the beam and this slice imaged. Most commercial systems use x-rays to obtain the image.

Medical scanners have been used on a nonintervention basis for the imaging of composite components. The requirements for the imaging of materials differ from those for live animals. The usual technique in industrial x-ray tomography is to translate and rotate the component in a series of steps, measuring the absorption at each position.

The CT technique is not limited to radiographic methods. Rothwell et al.[50] used nuclear magnetic resonance CT for nondestructutive imaging of the distribution of water in Gl_f/Ep composites. Other technologies for CT imaging include ultrasound P-scan, eddy current, electron spin resonance, photon emission, microwave, radar, and neutrons[51,52] (Figure 8.5).

Collimated conical anode radiography has been used in detecting flaws in composite materials that are not particularly amenable to ordinary x-ray inspection methods. It is most effective when the constituents of the part have similar densities and attenuation characteristics.

X-ray Tomographic Microscopy (XTM). A high-resolution variant of industrial CT, this technique provides nondestructive, high-resolution "sectioning" of samples and allows three-dimensional mapping of x-ray absorptivity. Stock et al.[53,54] displayed the capabilities of XTM in damage accumulation studies involving five continuous-fiber MMCs (aligned SiC/Al, $\{0_2/\pm 45\}$s SiC/Al, aligned SiC/Ti-6Al-4V, and aligned Al_2O_3/NiAl). The work must be performed and is required in order for a fundamental understanding of damage evolution and failure processes to be accepted and for high-strength MMCs to be used with confidence in critical structures.

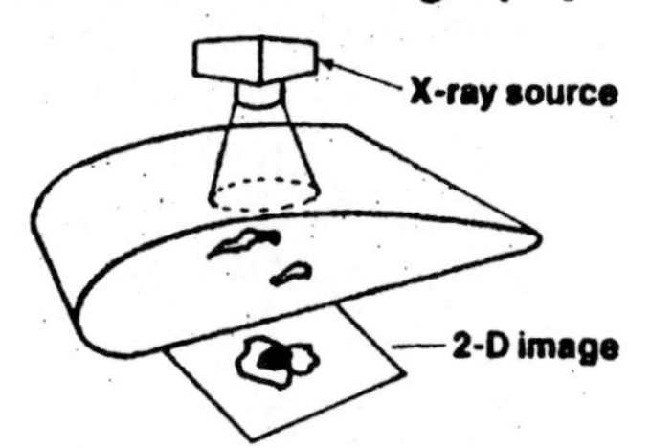

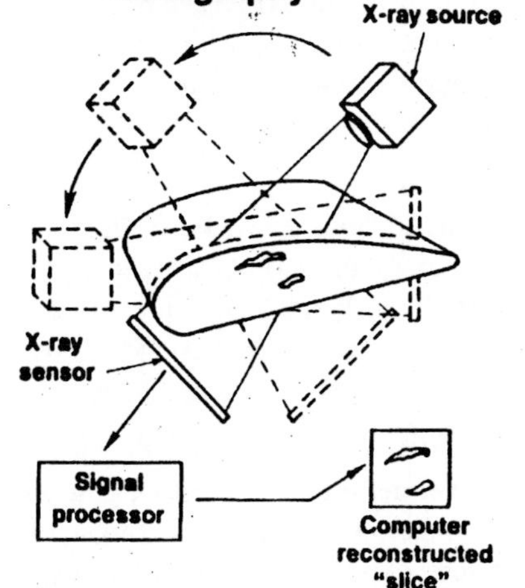

Figure 8.5. Comparison of conventional radiography and computer tomography radiography.[49]

8.0.1.3 Electromagnetic

Radiography. The term radiography encompasses several methods of using high-energy radiation for the examination of opaque objects. The beam energy can be generated electrically (e.g., x-rays with a broad spread of wavelengths), by radioactive isotopes (e.g., gamma rays with discrete wavelengths), and by nuclear reactors. The output is typically recorded as a shadowgraph on photographic film. The molecules in fiber-reinforced composite materials usually have low-atomic-weight nuclei, and hence absorption of x- and gamma-rays is low and contrast is poor especially for thin laminates. Sufficient contrast may be developed with "soft" low-voltage x-rays. The use of penetrant (sulfur, tetrabromoethane, silver iodide) enhancement produces improved images of surface-breaking cracks, but care must be taken to avoid penetrant systems that degrade the performance of the composite. Glass fibers are often added as tracers in Gr/Ep for x-ray examination.

In general, x-ray radiography is capable of locating defects in a plane parallel to the incident photon path. Therefore, factors such as density variations, porosity variations, and cracks are readily apparent. In addition, because the fibers have a density and chemical composition different from those of the matrix, fiber distributions are usually readily apparent.

Cracks or delaminations can also be located with the aid of a high-density penetrant. The high-density penetrant is weeped into the crack, and the specimen is then subjected to x-rays. As the x-rays pass through the material, the penetrant absorbs the x-rays and a light region appears on the film corresponding to the location and shape of the delamination or crack. Although penetrant-enhanced radiography is a valuable tool, caution should be exercised in applying penetrants to crack planes parallel to the incident photon path. In many cases, such cracks are lost in the details of the microstructure. To circumvent this problem, the voltage levels used in penetrant-enhanced radiography are usually higher than the voltage levels used in nonenhanced radiography. Unfortunately, the higher voltage levels then mask all the microstructural details. As a result, radiographs at two different voltage levels may be required.

Overall, x-ray radiography produces important information that strongly correlates with the manufactured condition and failure surface of the specimen. This type of information is invaluable in developing an understanding of the relationships among manufacturing, damage development, and failure in composites, especially CMCs.[55,56]

Newer radiographic systems automatically position the part, and control exposure and power settings. Only interpretation of x-rays is left to the operator. Fully automatic systems that scan and interpret without human intervention are technically feasible today with implementation in the future.

Neutron Radiography.[57] This technique has been identified as one of the most promising candidates for use in detecting corrosion; researchers feel there is also a potential for its use in determining the status of bonded structures. It is based on the principle that hydrogen atoms, which are present in all adhesives and resins, are rendered opaque and thus radiographically visible after they absorb thermal neutrons (Figure 8.6).

Visible Light (Optical Methods)[58]

Visual. Visual testing can be one of the simplest NDT technique for locating defects in transparent and translucent materials and surface errors in opaque materials. Various aids (magnifying glasses, microscopes, high-level illumination, and optical fiber boroscopes) can enhance visual tests.

Penetrants. Liquid penetrants rely on capillary action causing the liquid to enter surface-breaking openings (cracks, delamination, or exposed porosity), according to Vipond and Daniels,[59] who have described specific applications of penetrant testing of composites.

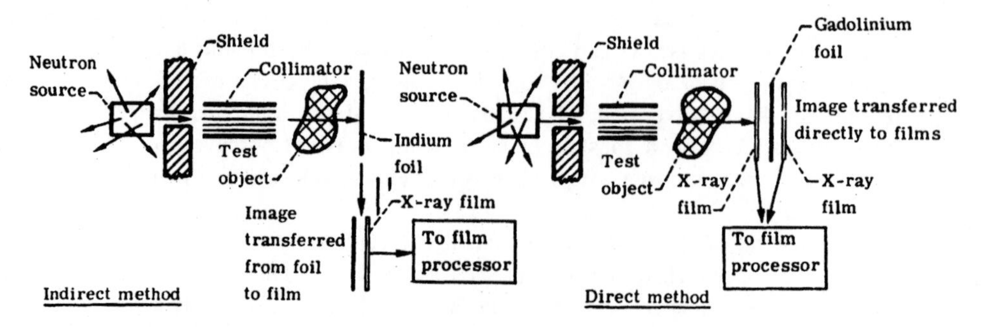

Figure 8.6. Principles of indirect and direct neutron radiography methods.

Edge replication and the deply technique. Harris and Morris[60] used edge replication with enhanced x-rays for NDT of fatigue specimens where edge effects dominated.

Liu et al.[61] extended the use of edge replication from the identification of matrix cracks in damaged composites to the study of delamination results from C_f/Ep plates which were in good agreement with other NDT techniques.

A destructive procedure that can complement NDT methods is a deplying technique that provides a method for sizing and precise location of interlaminar separations and fiber bundle fractures. The procedures have been described by Freeman.[62] Jamison[63] has conducted a systematic examination of microdamage development in Gr/Ep composites using penetrant-enhanced radiography and stereoradiography, edge replication, and laminate deplying.

Photoelasticity. The velocity of polarized light in a material is altered by the stress state of the material. In isotropic materials, the isoclinic fringes are known to give the principal stress or strain directions. In combination with the principal stress difference, from the isochromatics, it is possible to separate the principal stresses.

Pih and Knight[64] reported the use of transmission photoelasticity as a whole-field method of stress analysis for glass-fiber composites.

Coherent Light.[65–71] Coherent light techniques are equally as applicable to conventional materials as to composites, as they monitor surface deformations. Surface finishes (smoothness and reflectivity) may affect the usefulness of these techniques, and the anisotropy may affect subsequent analysis.

Holography. Holography detects defects by observing cancelations and interferences between laser beams. Inhomogeneities or discontinuities affect the way in which the material surface deforms and causes an interference pattern different from that in a "good" composite. The technique is very sensitive and hence only minimal stresses, typically 1 or 2°C or a few millibars pressure change, are necessary to indicate defects. Detectable defects include cracks, impact damage, delaminations, water absorption, and core or skin bonding.[65] See Tables 8.1 and 8.2.

Laser holography. Holographic inspection of composite parts entails taking an image of a part while it is at rest, and a second image when it is slightly stressed.

A problem with laser holography had been that even the slightest vibration of the laser or the object also creates fringe patterns, rendering the image useless. Vibration-isolated optical systems are required, severely limiting the utility of this technique. A system from Laser Technology appears to have overcome this problem by using phase-locked holography. This technique uses a reference laser beam to compensate for movement of the test object with respect to the laser camera. Even though the object and the camera may be moving slightly, vibration isolation is not needed.

The part can be stressed in a number of different ways. Delamination and unbonded regions are detected by heating the part slightly with a flash lamp or heat gun, whereas thicker sections require vacuum stressing. This method can also detect water in a composite structure. Most composite materials are transparent to microwave radiation. However, subjecting a sample to microwave energy heats water entrapped in a composite

TABLE 8.1. Rating of Nondestructive Inspection Techniques With Respect to Flaw Sensitivity and Resolution

Factor or Consideration	Ultrasonics: Through-Transmission	Ultrasonics: Pulse-Echo	Ultrasonics: Resonance	Penetrating Radiation	Microwave	Thermal Imaging	Acoustic Emission	Laser Holography	Acoustic Impact	Visual or Optical
Flaw type sensitivity										
Surface defects	Very poor	Very poor	None	None	None	Very poor	Very poor	Very poor	None	Very good
Delaminations	Good	Good	Good	Very poor	Very poor	Poor	Poor	Fair	Fair	Very poor[a]
Voids and debonds	Fair	Good	Good	Very poor	Poor	Poor	Poor	Fair	Fair	Very poor[a]
Internal fractures	Very poor	Very poor	Very poor	Poor	Poor	Very poor	Poor	Poor	Very poor	Very poor[a]
Flaw size sensitivity										
Area	Good[b]	Good[b]	Fair	Good	Fair	Fair	Very poor	Poor	Fair	Very good[c]
Depth[d]	Very poor	Good[e]	Very poor	Very poor	Very poor	Poor	Very poor	Very poor	Very poor	Fair[c]
Flaw location sensitivity										
Planform location	Good[b]	Good[b]	Fair	Good	Fair	Fair	Very poor	Fair	Poor	Very good[c]
Distance from surface	Poor	Good[e]	Very poor	Very poor	Very poor	Poor	Very poor	Very poor	Very poor	Fair[c]
Flaw resolution capability										
Near surface	Good	Very poor	Good	Good	Poor	Fair	Very poor	Fair	Very good	Good[c]
In planform	Good[b]	Good[b]	Fair	Good	Poor	Fair	Very poor	Fair	Fair	Very good[c]
In section	Very poor	Good	Very poor	Very poor	Very poor	Poor	Very poor	Very poor	Very poor	Fair[c]

[a] Only if surface evidence of internal defect (crazing, chipped paint, etc.).
[b] C-scan presentation.
[c] Surface defects.
[d] Length of the flaw normal to the beam.
[e] A-scan presentation.

TABLE 8.2. Rating of Nondestructive Inspection Techniques With Respect to Adaptability to Advance Composite Airframe Structures

Factor or Consideration	Ultrasonics Through-Transmission	Ultrasonics Pulse-Echo	Ultrasonics Resonance	Penetrating Radiation	Microwave	Thermal Imaging	Acoustic Emission	Laser Holography	Acoustic Impact	Visual or Optical
Adaptability type material										
Fiberglass	Good	Fair	Good	Fair	Fair	Poor	Fair	Good	Good	Good[a]
Kevlar	Good	Fair	Good	Fair	Fair	Poor	Fair	Fair	Good	Good[a]
Graphite	Good	Fair	Good	Fair	Fair	Poor	Fair	Good	Good	Good[a]
Hybrids	Good	Poor	Good	Poor	Poor	Poor	Poor	Good	Good	Good[a]
Core materials	Good	Very poor	Poor	Poor	Very poor	Poor	Poor	Poor	Poor	Good[a]
Adaptability material gage										
Thin gage	Good	Poor	Poor	Very poor	Very poor	Poor	Poor	Poor	Good	Good[a]
Heavy gage	Good	Good	Good	Good	Good	Poor	Fair	Good	Poor	Good[a]
Adaptability-type construction										
Laminates	Good	Good	Fair	Fair	Fair	Poor	Fair	Good	Good	Good[a]
Sandwich construction	Good	Fair	Good	Poor	Poor	Poor	Poor	Poor	Good	Good[a]
Adhesive bonds	Good	Fair	Good	Very poor	Very poor	Poor	Poor	Poor	Good	Good[a]
Adaptability-structural complexity										
Density variations[b]	Poor	Poor	Very poor	Poor	Poor	Poor	Fair	Fair	Very poor	Good
Thickness variations	Good	Fair	Poor	Poor	Poor	Poor	Fair	Fair	Poor	Good
Structural discontinuities[c]	Good[d]	Fair	Poor	Fair	Fair	Poor	Fair	Poor	Poor	Good
Adaptability-structural access										
Inaccessible one side	None	Very good	Very good	None	None	None	None	None	Poor	Poor
Inaccessible to surface contact	None	None	None	Fair	Good	None	None	None	None	Poor

[a] Visible defects.
[b] As might be caused by the introduction and termination of lamina at different points in the structure.
[c] Intersections, cutouts, reinforcements, and so on.
[d] Noncontact methods.

structure, producing slight surface deformations that are easily detected. This system allows real-time viewing of holographic images or permits permanent images to be made on photographic film.

Laser shearography.[66] This technique is an enhanced form of holography and, like holography, requires the part to be under stress. Shearography uses a laser for illumination, and under stress the output takes the form of an image-processed video display. The method has been used effectively in locating defects such as disbonds and delaminations through multiple bond lines. It is capable of showing the size and shape of subsurface anomalies.

Embedded Optical Fibers. Optical fibers embedded in composites have been used in monitoring interfacial corrosion, damage assessment (broken fibers transmit less light), strain measurement, and acoustic emission transduction.

Thermography.[65,67,68] Thermography involves the contour mapping of regions of equal temperature (isotherms) on a surface. There are several classifications of thermographic techniques divided according to contacting or noncontacting measurement system and active or passive heating method.

The technique can be applied to investigation of the state of material of a specimen. As a material deforms, heat is generated, and a variation in the thermal profile indicates variations in the deformation pattern and the state of the material. For example, consider a specimen containing a crack subjected to cyclic loading. Large strains develop in the crack tip region, and these strains produce a local hot spot on the thermal profile that is easily monitored by a thermographic camera. Similarly, any defects or material variations, such as voids, delaminations, fiber orientations, and fiber distributions, cause a variation in local thermal patterns.[69–71]

Thermography has capabilities not found in most NDE methods. The method is noncontacting, and as a result can be used during testing where deformation of the specimen throughout the loading history generates the thermal patterns. The technique may also be used at high temperatures, depending on access to the specimen through the furnace. Although identification of the temperature variations is relatively easy, interpretation of the source of these variations is extremely difficult. For example, surface defects tend to dominate the results. A void near the surface may produce higher thermal gradients than a crack deep within the specimen. However, the crack may dominate the behavior of the specimen. In addition, thermography works best for flat, planer components. Nevertheless, the method can also be used to investigate nonuniform specimens.

In recent years, a form of thermography called stress pattern analysis of thermographic emissions (SPATE) has received a great deal of attention. SPATE is an in situ test procedure that lends itself very well to monitoring damage development in cyclic testing. In short, the SPATE system measures the cyclic variations in the thermal emissions in a specimen and correlates these emissions with the applied cyclic loads. For isotropic materials subjected to fully reversed loading, the measured emissions can be directly converted to stresses, producing a full-field stress pattern in the specimen. The damage development can then be monitored by changes in the stress field. In the case of nonisotropic materials, the conversion from thermal emissions to stresses is not straightforward, and for injection-molded specimens with large variations in fiber orientations, a direct transition from emissions to stresses is impossible. Nevertheless, the evolution of damage can still be monitored by changes in the calculated, albeit inaccurate, stress fields. Therefore, the method is still extremely valuable in locating damaged regions and monitoring how this damage develops until failure. The main limitation of the SPATE system is its cost. A well-equipped system capable of simultaneous data collection and data processing can easily exceed $100,000 in cost.

Defects in composites exposed to either externally applied thermal fields (EATFs) or stress-generated thermal fields (SGTFs) can be imaged by color-change coatings (liquid crystals) or by infrared cameras. The defect acts as a barrier to thermal conduction through the material or the stress concentration generates heat under stress, and hence the thermal field differs from that of undamaged material. Thermographs also clearly show the anisotropy of heat flow in a composite. Vibration is often used to produce SGTFs as part of a technique commonly known as vibrothermography.

Work reported by Reynolds and Wells[72] has indicated that the passive method has been more widely applied than the active technique, and its performance is strongly dependent on the heat source used, a flash gun generally being the most suitable. The conductivity and anisotropy of the composite are also very important; for example, according to Stone and Clark,[73] in CFRP the conductivity in the laminate plane is about nine times that in the through-thickness direction, which tends to obscure defects that are not close to the surface.

Work carried out at Harwell[72,74] by Reynolds has shown that the method is made more convenient by using a video recorder to store the rapidly changing temperature pattern after the structure surface is heated. In this manner, defects in conducting materials that have only a transient effect on the temperature distribution may be detected. The procedure can be used with the heat source and camera on the same side of the structure (pulse-echo) or on opposite sides (through-transmission). The through-transmission method can detect deeper defects than the pulse-echo technique, but for defects close to the surface, the pulse-echo method is superior.

Thermography is attractive as a rapid means of inspecting large areas of a structure. It can be used to find disbonds in adhesive joints, delaminations in composites, and inclusions having conductivity significantly different from that of the base material. Work at British Airways[75] has shown that thermography is also a useful means of detecting moisture ingress in composites. However, equipment costs are high, and the procedure is not as sensitive as ultrasonics, for example, in detecting disbonds and delaminations.

Microwaves.[76] Microwaves are the band of frequencies from 225 MHz to 100 GHz and are capable of penetrating most nonmetallic materials, reflecting and scattering from internal boundaries, and interacting with the molecules. Three factors (shape and dimensions, dielectric constant, and loss tangent) influence the penetration of materials by microwaves.

Sensitivity to the dielectric properties of the material makes microwave testing particularly effective for internal structure (orientation), homogeneity (resin-rich or resin-poor areas), state of cure, moisture content, aging, and porosity. Microwave systems are also capable of detecting internal flaws (delaminations, voids, and inclusions) and of analyzing complex motions, as well as making precise measurements of distance, thickness, and surface waviness. Recent Soviet work has demonstrated the relationship between microwave frequency dielectric properties and deformation strain in composite materials.

Eddy Current.[77] The eddy current method is limited to composites with a high fiber V_f of conductive fiber (B or C) reinforcement. High fiber contents are required for the establishment of continuous conduction paths for the eddy currents. In the presence of these materials, the eddy current coil undergoes a reduction in inductance and an increase in the reactive part of the impedance. Eddy currents can reveal variation in fiber V_f and with the use of a horseshoe probe can indicate fiber orientation. Thickness measurements for nonconducting composites can be achieved by placing a metal item behind the composite.

Eddy current methods are relatively insensitive to porosity, nonconducting inclusions, and delaminations. Therefore, the method is complementary to ultrasonics in materials such as CFRP because it is sensitive to defects that are difficult to find using ultrasonics. Because the resistivity of carbon fibers is much greater than that of metals, lack of penetration depth is not usually a significant problem.[73]

Cure monitoring with eddy currents is based on tracking changes in liftoff caused by variations in composite material thickness. Eddy currents are very sensitive to variations in distance from the surface of conductive materials, and a wide range of thickness testers are based on this principle. To employ the eddy current technique in monitoring changes in the thickness of composite parts during curing, a conductive foil or caul plate is placed on the upper surface of the curing composite to serve as the conductive target. An eddy current probe is positioned above the target material, mounted on the same support tool as the curing composite. Thus, the base of the tool plate serves as a unified reference for thickness measurements.[78]

Dielectric.[79,80] All dielectric measurements involve determination of the electric polarization and conduction properties of a sample, usually in a varying electric field. The dielectric properties of a resin during curing change through a very large range of values. Commercial equipment using prefabricated interdigitated (comblike) electrodes is now available that can follow the curing of resin matrix systems from the liquid state until solidification occurs.

Dielectrometry can also be useful in measuring the moisture content of composites.

Electric.[81] The electric resistivity of commercial fiber-reinforced composites varies in value by 15 decades from the high resistivities of Kv/Ep to the low conductivity of graphite fiber materials. These properties have been used nondestructively to monitor laminate thickness, fiber orientation, moisture content, and crack growth.

8.0.1.4 Spectroscopy[82–84]

Raman.[85,86] The Raman effect is a phenomenon observed in the scattering of light as it passes through a material such that it undergoes a change in frequency and a random alteration in phase. Raman spectroscopy has been used to study curing of the matrix, using optical fibers to monitor the cure in situ. The technique has been applied by Fan and Hsu[87] in monitoring the effects of fiber orientation.

Infrared. All chemical compounds have characteristic intramolecular vibrations that can absorb incident radiation of the correct frequency. The vibrational motions of substituent groups in organic molecules coincide with the infrared spectrum. Infrared spectroscopy has been used primarily for assessing the extent of cure or thermal degradation of RMCs. Other studies have addressed the fiber-matrix interface and fiber stresses.

Infrared imaging is a nondestructive system being evaluated by the U.S. military for composite material components used in its aging fleet of transport aircraft. These systems are able to discriminate temperature differences as small as 0.1°C and produce high-resolution, real-time thermal images.

Several aerospace companies are evaluating the method to determine whether it is economically feasible in a production world. The big drawback is that if a large part needs to be tested, getting a heat source for it will be difficult and expensive.

Electron Spin Resonance (ESR). Any atom or molecule that possesses an unpaired electron is termed a *free radical.* If the molecule is placed in a constant magnetic field with a varying microwave field or is in a constant magnetic field with a varying magnetic field, the energy absorbed or emitted in changing the spin state can be detected electronically. ESR relies on interaction of the magnetic moment of the unpaired electron with the magnetic moment of the nucleus. ESR has been used to monitor the conversion of precursor materials to carbon fibers (preferred orientation, degree of graphitization, nature of unpaired spins, impurities, and defects), impurities and degradation of aramid fibers, aging of glass fiber composites, and radiation-induced damage in epoxy resins.

8.1 NDT PROCESSES FOR PRODUCTION AND AUTOMATION ENVIRONMENTS

Traditional nondestructive evaluation emphasizes finding and sizing flaws. To that end, it has been very successfully applied in improving the safety and reliability of aircraft structures. As the technology matured in the early 1950s, new instrumentation provided relatively easy access to the internal geometry of materials. That insight launched NDE as a profession and created much of the infrastructure present in the technology today. Today NDE is on the threshold of another renaissance, one involving quantitative measurement science combined with materials science. Metamorphosis is normal in most technologies but has been accelerated in NDE by the advent of fast computers, robotics, digitizers, and, of most importance, scientific analysis.

In the progression from isotropic, homogeneous planer structures to anisotropic, inhomogeneous, layered structures, NDE found new challenges requiring reanalysis. For example, are the instruments developed for simple metals appropriate for composites?

Thermal NDE, for example, has been practiced for many years with great success, yet has not achieved its full potential. The early use of thermal NDE involved detecting "defects" by viewing an infrared (IR) image generated after application of heat to the sample. For defects to be detected in this case, the subsurface heat flow must be impeded by the defect in such a way to give a surface temperature contrast of at least 0.3°C.

Currently, high-speed digital systems are in use for real-time enhancements such as averaging of IR digital images. When properly used, these tend to increase the probability of defect detection and lower the minimal amount of contrast required. Often, however, such measurements enhance other artifacts such as uneven heating and emissivity variations which hide the presence of defects or can be mistaken for defects. Therefore, to optimize the use of these new systems, physical models of heat diffusion are required to interpret these IR images and their time evolution.

Such physical models have been developed for NDE that describe the interaction of thermal, ultrasonic, and magnetic energy with structures to permit inversion of the data to give quantitative structural information. By using analyses based on physical models it has been possible to optimize the NDE measurements and determine how to best reduce the data to maximize specificity, resolution, and contrast.[88,89]

Ultrasonic testing is the most widely used NDE method for advanced composites. It is an excellent procedure for detecting delaminations, voids, porosity, and in some cases other flaws such as broken fibers and inclusions.[90,91] In the past ultrasonic waves were required to enter the part within a few degrees of normal and were once limited to flat or gently contoured parts. However, the latest generation systems have no such constraints. The Auss-5 system from McDonnell Aircraft, for example, features nine axes of control and can scan a highly contoured part while maintaining the alignment of two transducers to within 0.76 mm. Computer control allows automatic scanning of one part while the data from another are evaluated.[49]

The Auss (automated ultrasonic scanning system) has been used in the inspection of aircraft composite parts and assemblies and has provided a system that can maintain the highest standard of quality. This specifically designed equipment has been used to inspect adhesively bonded composite assemblies having complex, contoured shapes. It performs through-transmission or pulse-echo tests using water squirters which provide ultrasonic signal coupling between the scanner and the part. The ultrasonic frequencies used are in the 0.4–15 MHz range and inspect for delaminations, porosity, disbonds, voids, ply slippage, and foreign materials.

Using advanced control hardware and software, Auss-5 is able to scan parts having a compound curvature. It acquires data from the three-dimensional real world of structural parts and translates it into data for the two-dimensional world of data image media with minimum surface distortion.

Acoustoultrasonics has been applied principally to assess defects in laminated and filament-wound fiber-reinforced composite materials. The technique can be used to determine variations in such properties as tensile, shear, and flexural strengths and reductions in strength and toughness caused by defects. It can be used to evaluate states of cure, porosities, orientation of fibers, V_f of fibers, bonding between fibers and matrices, and the quality of interlaminar bonds. The term *acoustoultrasonics* is a contraction of "acoustic emission simulation with ultrasonic sources." Conventional acoustic emission testing depends on loading a part to excite spontaneous stress waves like those that accompany plastic deformations and the growth of cracks. Acoustoultrasonics differs mainly in that the ultrasonic waves are benign and are generated externally by pulsed sources (usually piezotransducers).

In a typical apparatus, a transmitting probe and a receiving probe are placed a specified distance apart on the same side of the part under test. The send-receive pair can be moved about as a unit to scan the part. Signal-processing instrumentation analyzes the received sound and generates a map of variations in the properties of the material.

The sensitivity of acoustoultrasonics has already been demonstrated experimentally. The technology has been used to detect and quantify subtle but significant variations in strength and resistance to fracture of fiber-reinforced composites. This achievement is remarkable because it was accomplished with relatively unsophisticated signal processing and signal analysis procedures.

Although acoustoultrasonic technology has been used mainly with RMCs, it is also applicable to other composite materials such as MMCs and CMCs. The use of acoustoultrasonics should be considered whenever it is necessary to quantify damage or degradation of properties after composites have been exposed to hostile environments.

AE, a close relative of ultrasound, has been studied as a potential NDT tool in monitoring curing and viscosity and in predicting ultimate strength.[92,93] Other acoustic-based methods of note include acoustic microscopy, which has been used to measure low-velocity impact damage of Gr/Ep composite plates, and impulse excitation, which can be used to measure elastic properties of materials.

Commercial AE systems such as those available from Lloyd Instruments (Fareham, Hampshire, U.K.) provide microstructure failure information by subjecting the sample to stress in the form of a series of loading and unloading cycles with progressively increasing loads. Under these conditions, the sample produces stress waves, or acoustic emissions, which are caused by a breakdown in the internal microstructure of the material. These waves cover the frequency range of 100 kHz–2 MHz. A microphone attached to the sample transfers the stress or sound waves to an acoustic emission monitor, and the resulting emissions are plotted against the applied stress.

Acoustic microscopy is another interesting spin on acoustic-based NDT procedures. This method actually encompasses three complementary techniques: scanning laser acoustic microscopy (SLAM), scanning acoustic microscopy (SAM), and a modified form of C-scanning ultrasound called C-mode scanning acoustic microscopy (C-SAM). SLAM is a transmission mode method that creates a real-time image of a sample through the entire thickness. In SLAM, a piezoelectric transducer introduces ultrasound to the bottom surface of a sample and the transmitted wave is detected on the top side by a rapid-scanning laser beam.[94,95] SAM and C-SAM, on the other hand, are reflection mode techniques that use a transducer with an acoustic lens to focus the wave at or below the sample surface. The transducer is scanned across the sample to create an image. SAM is designed for high-resolution work at or near the surface, whereas C-SAM is capable of moderate penetration and can be used to analyze at a specific depth.[96,97]

Lawrence, Briggs, and Scruby[98] studied SiC fiber-reinforced composites with a magnesium aluminosilicate (MAS) matrix and a calcium aluminosilicate (CAS) matrix using acoustic microscopy.

The properties of the CAS/SiC and MAS/SiC composites to which acoustic microscopy has shown special sensitivity include the microstructure of both matrix materials, the microstructure of the sigma monofilament, and cracking in the matrix. Perhaps most important of all, it has been possible to accurately calculate the contrast to be expected from interfaces that are bonded (and those that are debonded, corresponding to a crack between them) and to use this value to interpret the contrast from a fiber-matrix interface or debond quantitatively.[98]

The same researchers[99] used SAM to study MMCs reinforced with SiC monofilaments. For most of the specimens the matrix was Ti-6Al-4V, but Ti_3Al intermetallic and 6061 Al matrices were also examined. They also used SAM to study the microstructure of Nicalon-reinforced borosilicate glass.[100] Their study illustrated how acoustic microscopy can provide fresh insights into the microstructural changes accompanying the thermal aging of MMCs. Thus, the radial microcracks that formed in the SiC monofilaments as a result of heat treatment could not be observed by optical microscopy. The extra sensitivity of acoustic images arises because Rayleigh waves still undergo appreciable reflection from tight cracks. Furthermore, degradation of the carbon-rich layer at the fiber-matrix interface could be monitored more sensitively by acoustic microscopy because the microstructural changes caused larger variations in acoustic properties (which control image contrast) than in optical reflectivity.

CT has produced quantitative 3-D data and has been shown to be able to determine material density differences within an accuracy of about 0.1%. It can measure the dimensions of internal structures, and the images are generally free of superimposed structures from outside the area of interest; small flaws can be detected, sized, and located with high precision.

A team at Boeing, for example, has reported significant savings in cost and time in using x-ray CT systems for RMCs.[101] Working on behalf of the U.S. Air Force, these researchers developed a CT system for real-time, on-line analysis of the pultrusion process. The team also successfully investigated a wide range of other composites, including honeycomb structures, basic laminates, filament windings, and braids, and the method was effective for evaluating density variations, voids, delaminations, dimensions, and even wrinkles.[102–104]

Researchers have found the new x-ray backscatter technique to be a solution for the inspection of low-density materials and many types of composites. In comparing conventional x-ray techniques, x-ray computer tomography, and ultrasonic methods to the relatively new Compton backscatter technique "ComScan," it has been found that ComScan will never replace any conventional technique but that there are a number of applications, especially in the field of low-density and composite materials, where conventional techniques fail and where the ComScan technique is the solution.[105–108]

To obtain good contrast sensitivity in the imaging of polymer-based composites, it is necessary to use low-voltage methods. This need is being addressed by such products as the ultralow-voltage, high-current x-ray machine which, with a variable output range of 2–50 kV, is the first system specifically designed for applications requiring extremely low-kilovolt x-ray output. Compared with conventional low-voltage x-ray systems, the IXRS 50/1500 provides lower-energy-level x-rays at a greatly increased current output, up to 10 times that of standard x-ray sources. This allows greatly reduced exposure times for film radiography applications and greatly increased contrast sensitivity for real-time x-ray imaging for composites.[65]

New developments in holography may hold great promise for the composites industry. Advances in vibrational and static mode holographic interferometry, for example, are enabling fast, accurate detection of flaws in composites. Such holographic imaging techniques, in fact, are already being used by some aerospace and defense contractors in the inspection of composite structures.

Some of the most important advances in holographic vibration analysis have resulted in yields with much better picture quality than conventional electronic speckle pattern interferometry.

Laser holographic nondestructive testing (HNDT) has developed rapidly in the last several years into a cost-effective NDI tool.[109] Key to the development of HNDT systems has been the invention of "instant" hologram recording systems and the development of computer-controlled holographic equipment. Rapid film processors can now develop the 35-mm film in less than 20 s. Furthermore, these film processors develop the film in exact register with the object, allowing the techniques of real-time hologram reconstruction to be used for optimizing the NDT technique on a particular object. Controlled by a microprocessor, automation and simplification of the holographic process allows production and development equipment to be operated by personnel with a minimum of technical training.

Electronic shearography has overcome the shortcomings of the holographic technique, providing a method of inspecting complex-shaped composite parts, especially those containing ultrasonic attenuating materials. It can be used for both fabrication and in-service inspections.

Composites engineers have sought methods or sensors to measure pressure, temperature, and degree of cure of composites during a curing cycle.[78] The eddy current (liftoff) cure-monitoring device, developed in 1990, is an inexpensive, reliable method of measuring the thickness of the laminate during curing.

The primary developments of holography have been rapid film processing, real-time holograms, white noise excitation, and phase-locked holography. The development of microwave excitation has allowed the detection of moisture in composites. Holography remains method-dependent, and test methods must be correctly applied to each sample type. However, present data[109] support holography as a powerful tool for NDT of advanced composite materials.

The pulsed video thermography (PVT) technique has been applied to a range of applications including integrity control in aerospace composites. The method has shown great potential for composite quality assurance. Honeycomb composite structure shows up clearly, and damage to a composite part caused by surface impact is clearly revealed at depth, even where it is invisible at the point of impact. With its potential for delamination and internal cell damage, moisture taken up into aircraft structures is easy to detect.

PVT can be applied in two ways: single-sided or double-sided. In single-sided use, the thermal imager is positioned to observe the side of the sample to which the heat pulse is applied. This is the best approach for, say, CRFP up to 5 mm thick. On thicker samples, particularly of composites, this is the best technique. Double-sided PVT, in which the flash tube and thermal imager are on opposite sides of the material, can be used when the sample is relatively thin.[110]

The possibilities and limitations of eddy current methods in the inspection of CFRPs were investigated by De Goeje and Wapenaar.[111] They found that fiber orientations could be detected conveniently with dedicated probes using a polar scan technique. Eddy currents are insensitive to delaminations. Only for unidirectional reinforcements and extensive delamination, over 20%, is the effect large enough to show up in C-scan images. In contrast, fiber fracture is readily detected in unidirectional and weave reinforcements, with a lower limit of about 8% fracture. As a result, eddy current methods are useful in establishing the type of defect when a composite is damaged.

8.2 APPLICATIONS

Production applications for some of the NDT processes and equipment discussed in Sections 8.0 and 8.1 include

1. The AUSS unit has been used to inspect composite honeycomb components 244 cm long and 107 cm wide.

2. CT has been used as an NDT method on thick-walled composite structures in dynamic helicopter components such as main and tail rotors.
3. Cook, Renieri, Cox, et al.[112] studied 10 different NDT methods for helicopter-type composite structures (Table 8.3) and the methods, constraints, and limitations associated with their use.
4. XTM was used to observe the step-by-step fabrication of a SiC/SiC ceramic composite by means of forced chemical vapor infiltration (CVI).
5. Thermography has been introduced and recommended as the standard procedure by Airbus Industries. It has been extensively involved in detecting possible water ingress in composite sandwich structures in their aircraft.[112]
6. Eddy current methods have been used to assess the electric conductivities of SiC_p-reinforced Al alloys—2124, 6061, and 7091 MMC extrusions. The volume percent of SiC_p covered a wide range from 0 to 55%, and the composites demonstrated anisotropic conductivity with the maximum conductivity occurring along the extrusion plane.
7. CT is a powerful evaluation tool for advanced propulsion engine materials and components. CT investigations on MMC ring, rod, and coupon structures were reported by Yancey and Baaklini.[113,114] Moderate-resolution CT data identified density variations within reinforced sections, which were correlated with fiber packing densities. Some composite lay-up features were also imaged. High-resolution CT data provided information on fiber spacing and distribution.
8. Ultrasonics (SAM and acoustic microscopy) have become an integral part of the development of the TMC system, according to Karpur, Stubbs, Matikas, et al.[115]
9. CT has been used to evaluate a C-C two-dimensional fabric mainly used on IUS rocket nozzle exhaust cones. Additionally, Boeing aerospace engineers have looked at 3D C-C with cloth arranged in all directions as well as carbon-phenolic systems with the phenolic group between carbon cloth layers. Finally, they have applied CT techniques in studying the way carbon-phenolic insulation stands up under severe temperature conditions because this composite is used to thermally insulate some IUS parts. The method is also used to provide data on delamination and impact damage.

8.3 FUTURE POTENTIAL OF NDT METHODS

The role of NDE and NDT is changing and will continue to change dramatically. NDT has been used almost exclusively for detection of cracks in objects after they have been manufactured and in service. However, it has become increasingly evident that it is both practical and cost-effective to expand the role of NDE or NDT by introducing it much earlier in the manufacturing cycle. Today, and ever increasingly in the future, the application of advanced sensors, modern measurement technology, signal processing techniques, and artificial intelligence, will permit information on the processing conditions and the characteristics of the materials being processed to be continuously monitored and controlled. Real-time process control resulting in improved product quality and reliability has become a practical reality.

TABLE 8.3. Summary Description of NDT Techniques for Helicopter-type Structures

NDI Technique	Constraints and Limitations
Through-transmission ultrasonics:	
High-frequency sound waves are transmitted into the test article through one surface. The sound energy travels through homogeneous materials with no appreciable loss until a discontinuity is encountered. The discontinuity attenuates the energy through a combination of reflection, refraction, and absorption. The variations in transmitted ultrasound are used to detect the presence of internal flaws. Required equipment includes a pulsed ultrasonic generator, sending and receiving transducers located on opposite sides of the test article, and signal processing and display devices.	Ultrasonic testing requires the sending and receiving transducers to be normally opposed to each other over the test article. A coupling medium is required. For contact testing methods, a thin layer of fluid is generally used. This requires that the surface of the test article be clean and relatively smooth. The requirement for surface contact is a disadvantage when large specimens must be scanned, and a technique involving immersion of the article in a fluid is sometimes used to facilitate large-area inspection.
Ultrasonic test results may be displayed in two basic ways using a cathode ray tube (CRT) or permanent image recorder. The A-scan presentation measures the amplitude of the transmitted and received energy. It reveals the presence of the flaw and provides some indication of its relative size in area but does not locate its depth from the surface of the part. The C-scan presentation requires automated scanning of the test article and the use of signal processing techniques. A persistent image CRT or paper recorder is used to display ultrasonic test responses. The C-scan provides a plan view of the article, revealing the area of the flaw but providing no information relative to its depth or distance from the surface. Successive C-scans with signal gates set to various depths below the surface may be used to locate the flaw in a section.	Ultrasonics is effective in detecting internal flaws in a variety of materials including parts with very thick sections. However, the flaws must be relatively perpendicular to the ultrasonic beam to be detected. A disadvantage of ultrasonics for the detection of flaws in composite materials is the high attenuation caused by absorption in porous resin and scattering by the fibers. Hybrid configurations pose particular problems because the abrupt variations in acoustic impedance from one material to another tend to mask internal flaws. Complex part geometry (variable surface contours, section thicknesses, wall thicknesses, etc.) make ultrasonic inspection difficult. Transmission ultrasonics requires access to both sides of the part, prohibiting its use in many on-aircraft applications.
	Interpretation of ultrasonic test responses requires a reference standard containing flaws of known size and location for comparison purposes. For some types of tests, signal gates and sensitivities must be varied in the course of the test, placing high demands on operator skill.
Pulse-echo ultrasonics:	
Short, rapid pulses of high-frequency sound are transmitted into the test article through one surface. The sound energy travels through homogeneous materials with no appreciable attenuation, scattering, or reflection. When an abrupt change in the conductive medium is encountered, a portion of the sound energy is reflected back toward the source. Reflection is produced at the surfaces of the article, at interfaces between different materials, and at locations where internal discontinuities are present. The amplitude of the reflected sound wave can be used to measure the size of an internal defect and the travel time of the reflected wave to measure its location from the surface of the article.	The pulse-echo technique suffers many of the limitations of the through-transmission ultrasonic technique, namely, the need for coupling to the test article, the insensitivity to flaws that are nonperpendicular to the beam, the difficulty of inspecting parts with hybrid configurations or complex geometries, the need for reliable reference standards, and the high dependency on operator skill. Pulse-echo NDI has the added disadvantage that flaws located close to the surface of the part are usually obscured by near-surface resolution losses. One advantage of the pulse-echo method over through transmission is that the A-scan presentation can be used to reveal both the presence of a defect and its location relative to the surface of the part. It also has the advantage of requiring access to only one side of the test article. The major disadvantage of pulse-echo compared
Equipment required for pulse-echo ultrasonics testing includes a pulsed ultrasonic generator, a transducer which acts alternately as a transmitter and receiver, and a signal processing and display device. The A-scan and C-scan	

TABLE 8.3. *(continued)*

NDI Technique	Constraints and Limitations
methods of presentation covered previously under through-transmission ultrasonics are also used to display pulse-echo information.	with through-transmission testing is that it is more affected by the acoustic attenuation properties of composites. Because the sound must traverse the material twice, it is more likely that a flaw will be obscured by reflection and scattering.
Resonance ultrasonics:	
Continuous, compressional ultrasonic waves are transmitted into the test article from one surface. The sound waves are reflected back toward the source from the opposite surface. The frequency of the sound waves is varied manually or automatically until a frequency is reached that causes the incident and reflected waves to be in phase. This produces standing waves within the material which cause it to vibrate or resonate. The resonant frequency is determined by the acoustic impedance of the material and its thickness. Internal flaws and discontinuities are observed as changes in the strength and location of the resonance indications. Laminar discontinuities and bond separations are particularly sensitive to ultrasonic resonance NDI.	Resonance ultrasonics is useful primarily in the detection of interlaminar defects and debonds. It is quite sensitive to these kinds of defects and can, through the technique of scanning, provide relatively good indications of their size and location in plan view. It is a poor method with respect to locating the depth of the defect below the surface.
Required equipment includes a power supply and oscillator, detection, amplification and indicating circuits, and a transducer. Portable, self-contained battery-operated units are produced for field work. The portable units generally use a meter, and audible alarm and/or headphones to monitor resonance indications. Gating circuits can be used to detect resonant frequencies within a narrow range, thereby providing some discrimination capability with respect to flaw size.	Unlike transmission and pulse-echo ultrasonics, resonance testing is not significantly affected by laminar variations in the material provided the lamina interfaces are intact. Resonance testing is generally not sensitive to fine cracks and other small defects that do not alter the natural frequency of the part. Unlike through-transmission and pulse-echo methods, resonance ultrasonics is able to detect delaminations and debonds close to the surface of the part. Contact with the part is required. The use of a fluid couplant between the transducer and the surface of the part is needed for contact testing. Reference standards are necessary, as they are with the other ultrasonic methods, and the dependency on operator skill is a factor.
Acoustic impact:	
Striking a material with a hard object (hammer, coin, etc.) causes a vibrational response and characteristic sound. Internal flaws and discontinuities produce acoustic damping and an audible change in the sound which is sensed by ear or electronically.	Acoustic impact NDI is useful only when the specimen being investigated produces distinct differences in sound when flaws are present. It has relatively poor resolution capability and is affected by ambient conditions (temperature, humidity, background noise, etc.), changes in material density or thickness, and the presence of structural discontinuities (cutouts, reinforcements, fasteners, etc.).
Visual and optical:	
Surface flaws and internal flaws that produce surface irregularities (wrinkling, crazing, etc.) are detected visually or with the aid of an image intensifier (magnifying glass, microscope, etc.).	Visual inspection is confined to surface defects and some subsurface defects that are manifested at the surface. Its reliability may be degraded by human factors and by surface coatings or contamination which obscure surface features.

TABLE 8.3. *(continued)*

NDI Technique	Constraints and Limitations
Acoustic emission:	
A structure placed under stress experiences plastic deformation or fracture which produces small stress or ultrasonic waves in the material. The rate and intensity of acoustic emissions may be used to detect the initiation and propagation of cracks and delaminations. Acoustic emission NDI has been used to predict static failures and fatigue failures in composites. Required equipment includes a piezoelectric transducer, amplifier, acoustic event counter, and coordinate plotter.	Acoustic emission NDI requires that the test article be statically or cyclically stressed. Acoustic sensing is usually accomplished with piezoelectric transducers placed in contact with the test article. Probe placement and arrangement are critical to test sensitivity. Interpretation of acoustic signals requires a prior history of failure versus acoustic emission rate and intensity. It is not possible to characterize individual failure events with acoustic emission NDI. Triangulation methods with multiple sensors and computer processing have been successful in approximately locating failures in a structure, however.
Holographic interferometry:	
Internal flaws in a structure may create small surface strains or deformations when mechanical, thermal, or acoustic stress is applied to the structure. Holograms are made of the structure before and after the application of a load. When superimposed, the holograms reveal interference fringes in areas where surface deformation has occurred. These "interferons" may be used to detect the presence of subsurface flaws or a change in material stiffness indicative of fatigue damage.	Holographic interferometry requires that the test article be placed under load under precisely controlled conditions. Essentially a laboratory technique, it is useful only when the internal flaw or damage being investigated causes a characteristic displacement of the surface. Quantitative information with regard to the exact location and dimensions of the flaw is not easily obtained with holographic techniques.
Penetrating radiation:	
Low energy x-ray, gamma ray, or neutron emissions penetrate the test article. Variations in material thickness or density and the presence of flaws affect the intensity of the transmitted radiation. An internal image of the test article is recorded using a scintillation detector (radiometry), fluoroscope (fluoroscopy), photographic film (film radiography) or a closed-circuit television camera (video radiography). Voids and other discontinuities attenuate the radiation to different degrees and are revealed as contrasting light and dark areas (shadows) on the recorded image.	The radiograph of a test article is a record of the differential attenuation of radiation as it passes through the material. Subtle variations in material density representative of interlaminar defects in composites may not be detectable unless the flaw is nearly parallel to the direction of the radiation and of significant length. The need to precisely orient the test article to obtain adequate sensitivity to flaws may preclude the use of penetrating radiation for inspection of complex shapes.
Fine-grain film offers the best radiographic sensitivity and provides a permanent record of defects for comparison with reference standards. Recent advancements in video radiography provide radiographic images equivalent to the best fine-grain films. With color television, variations in radiation intensity indicative of internal flaws and discontinuities are displayed as variations in color and color intensity.	Radiometry and fluoroscopy offer the advantage of real-time imaging of the test specimen but are relatively insensitive to small internal flaws and discontinuities, making them suitable only for quick scanning for gross defects and foreign objects. Radiation intensity, focusing, and exposure times are critical, and highly specialized skills are required to set up tests and interpret test results. Penetrating radiation techniques involve equipment that is relatively high in cost, and radiation may pose a safety hazard to personnel. Penetrating radiation methods are most practical for small parts that can be brought to a fixed facility. They are least practical for large, complex structures that require on-site inspection.

TABLE 8.3. *(continued)*

NDI Technique	Constraints and Limitations
Microwave:	
Electromagnetic waves in the frequency band of 300 MHz–300 GHz penetrate the test article. Attenuation, reflection, and scattering are caused by variations in material thickness or density and the presence of flaws and discontinuities. An internal image of the test article obtained from the amplitude and phasing of the transmitted waves is displayed on a meter or plotter. Comparisons with reference standards are used to pinpoint flaws. Microwaves are heavily absorbed and scattered by water molecules, making microwave examination an effective NDI method for determining moisture content. Microwave inspection requires a microwave generator, transmitting and receiving antennas, and a display or recording device.	Microwave NDI has an advantage over ultrasonic methods in that microwaves are able to traverse air space without significant attenuation, avoiding the need for a direct mechanical coupling to the test article. Alignment and orientation of the test specimen are critical, and relatively specialized skills are required to design the test setup and interpret the test results. Microwave NDIs are relatively ineffective for resolving small localized anomalies in a test specimen.
Thermal imaging:	
The test article is heated uniformly and scanned with a heat sensor. Variations in thermal conductivity caused by discontinuities or flaws in the material are revealed as a function of temperature gradients at the surface. With direct contact sensing, a thermally sensitive device or material (paint, liquid crystal, etc.) is placed in physical contact with the test article. With noncontact sensing, thermally generated electromagnetic energy from the test article is measured with an infrared imaging device (scanner, camera, film, etc.). Color thermography displays temperature gradients as changes in color, providing a better definition of discontinuities than black-and-white images.	Locating flaws with thermal NDI involves detecting temperature variations at the surface of the part. Detection of flaws is easiest when the test article has a large impedance to heat flow and the flaw is large and close to the radiating surface. Detection is difficult for small flaws in thermally conductive materials. Designing a thermal NDI requires a knowledge of the thermal conductivity and surface emissivity of the test specimen and a reference standard for comparison purposes. Thermal NDI can be significantly affected by ambient temperature conditions.

Optimization of the processing and properties of composites, the development of synthetically structured materials, the characterization of surfaces and interfaces, and in all cases, the structures, devices, and systems made from these materials demand the innovative application of modern nondestructive materials characterization techniques to monitor and control as many stages of the production process as possible.

Often the various NDT methods complement one another to provide better reliability in detecting the various damage modes that can appear in composite materials. Figure 8.7 summarizes the applicability of the various inspection methods and also gives some information about the reliability or flexibility of the method in detecting the various flaw types. Several new inspection methods or approaches have reached the point where they have been placed in service. As the number of composite components in active service grows and as the incorporation of composites into structural design increases, we will see an even greater demand for rapid and effective nondestructive inspection tools and, with this demand, the incorporation of several new approaches to composite inspection.

Intelligent manufacturing is impossible without integrating modern nondestructive evaluation into the production system. What was once a nonsystematic collection of ex

	Method NDI												
Flaw Type	UT thru trans	UT pulse echo	UT angle	UT reso- nance	X- radiog- raphy	Neutron RT	Holog- raphy	Thermal IR	Tap test	Acoustic emis	Acousto- UT	Eddy current	Visual
Porosity	1	1	2		1	3					2		
Foreign material	2	1			2	3		3					
Shallow delamination	1	1			1		1	1	1	3	3		
Deep delamination	1	1						2		3			
Matrix cracks			1		2					2	1		
Fiber breaks					1					2	2	1	
Impact damage	1	1	2				2	2	2	3	3		3
Skin/skin disbond	1	2		1	2	2	1	1	1				
Skin/core disbond	1	2		3	3	3	1	1	2	2			
Crushed core	1				1		2	2					
Condensed core					1								
Blown core	1				1								
Mode bond failures					1								
Water intrusion	3	3			2	1	3	3		2			
Corroded core	2				2	1				3			
Fatigued core	1				1								
Foam adhesive void					2	2							

[1] Good sensitivity and reliability. Good candidate for primary method.
[2] Less reliability or limited applicability. May be good backup method.
[3] Limited applicability. May provide some useful information.

Figure 8.7. Summary table of NDT methods and flaws.

post facto techniques has become a powerful set of tools, which when used throughout the manufacturing process can help to transform most manufacturing operations and companies and make their products more competitive with one another and with those of outside suppliers.

REFERENCES

1. Start of the composites age? *Int. Reinforced Plast. Ind.* 8(3):4–12.
2. Ashbee, K. 1989. *Fundamental Principles of Fibre Reinforced Composites.* Technomic, Basel.
3. Grayson, M. 1983. *Encyclopedia of Composite Materials and Components.* John Wiley, New York.
4. Halpin, J. C. 1984. *Primer on Composite Materials: Analysis.* Technomic, Lancaster, PA.
5. Harris, B. 1986. *Engineering Composite Materials.* Institute of Metals, London.
6. Kaelble, D. H. 1985. *Computer-Aided Design of Polymers and Composites.* Marcel Dekker, New York.
7. Kelly, A., and Y. N. Rabotnov, eds. 1983–1988. *Handbook of Composites,* vols. I–IV. North-Holland, Oxford.
8. Kelly, A. 1989. *Concise Encyclopedia of Composite Materials.* Pergamon, Oxford.
9. Lee, S. M. 1989–1990. *International Encyclopedia of Composites,* vols. 1–6. VCH, New York.
10. Morley, J. G. 1987. *High Performance Fibre Composites.* Academic Press, London.
11. Partridge, I. K. 1989. *Advanced Composites.* Elsevier, Barking, U.K.
12. Reinhart, T. J. 1987. *Engineered Materials Handbook,* 2nd ed. ASM International, Metals Park, OH.
13. Schwartz, M. M. 1984. *Composite Materials Handbook,* 2nd ed. McGraw-Hill, New York.
14. McMaster, R. C., ed. *Nondestructive Testing Handbook,* 2nd ed. vol. 3, 1985; vol. 4, 1986; vol. 5, 1987; vol. 7, 1990; vol. 9, 1992; vol. 10, 1993. American Society for Nondestructive Testing, Columbus, OH.
15. Halmshaw, R. 1987. *Non-Destructive Testing.* Edward Arnold, London.
16. Hanstead, P. D., ed. 1988–1990. *The Capabilities and Limitations of NDT,* part 3, 1988; part 4, 1990; part 5, 1989; part 7, 1988; part 8, 1988. British Institute of Non-Destructive Testing, Northampton.
17. Jones, L. K., J. M. Kasel, B. B. Risinger, et al., eds. 1985. *Proc. 11th World Conf. Nondestr. Test.,* November 1985, Las Vegas, NV. Taylor, Dallas, TX.
18. Farley, J. M., and R. W. Nichols, eds. 1987–1988. *Non-Destructive Testing, Proc. 4th Eur. Conf. Non-Destructive Test.,* September 1987, London. Pergamon, Oxford.
19. Chatterjee, S. N., K. W. Buesking, B. W. Rosen, et al. 1984. Assessment of significance of defects in laminated composites—A review of the state-of-the-art. *Rev. Prog. Quant. Nondestructive Eval.* 4B:1189–1201.
20. Summerscales, J., ed. 1987. *Non-Destructive Testing of Fibre-Reinforced Plastics Composites,* vol. 1. Elsevier, Barking, U.K.
21. Summerscales, J., ed. 1990. *Non-Destructive Testing of Fibre-Reinforced Plastics Composites,* vol. 2. Elsevier, Barking, U.K.

22. Wegman, R. F. 1989. *Nondestructive Test Methods for Structural Composites,* SAMPE Handbook 1. SAMPE International, Covina, CA.
23. Potapov, A. I., and F. P. Pekker. 1977. *Nondestructive Testing of Composite Materials.* Mashinostroenie, Leningrad. In Russian.
24. Reynolds, W. N. 1988. Nondestructive testing techniques for metal matrix composites. Harwell Laboratory Report AERE R 13040.
25. Snyder, J., and A. Vary, eds. 1987. *Nondestructive Testing of High-Performance Ceramics.* American Ceramic Society, Westerville, OH.
26. Kardos, J. L., R. Dave, and M. P. Dudukovic. 1988. Voids in composites. *Proc. Manuf. Int. 1988,* pp. 41–48. ASME, Atlanta, GA.
27. Judd, N. C. W., and W. W. Wright. 1978. Voids and their effects on the mechanical properties of composites—An appraisal. *SAMPE J.* 14(1):10–14.
28. Walkden, P., and M. Boutoussov. 1990. NDT as a quality control tool. *Proc. 11th Int. Conf. SAMPE Eur.,* pp. 197–207, May 1990, Basel.
29. Smith, G. C. 1986. The scanning acoustic microscope—A new tool for the materials scientist. *Mater. Sci. Technol.* 2(9):881–87.
30. Serabian, S. 1985. Composite characterization techniques: Ultrasonics. *Mantech J.* 10(3):11–23.
31. Bar-Cohen, Y. 1987. Ultrasonic NDE of composites—A review. In *Solid Mechanics for Quantitative Non-Destructive Evaluation,* ed. J. D. Achenbach and Y. Rajapakse, 187–201. Martinus Nijhoff, Dordrecht, Netherlands.
32. Kinra, V. K., and V. Dayal. 1987. Ultrasonic nondestructive evaluation of fibre-reinforced composite materials—A review. *Sadhana* 11(3/4):419–32.
33. Hennecke, E. G. 1990. Ultrasonics. In *Nondestructive Testing of Fibre-Reinforced Plastics Composites,* vol. 2, ed. J. Summerscales, 55–159. Elsevier, Barking, U.K.
34. Sandhu, J. S. 1988. Acoustography: A new imaging technique and its application to nondestructive evaluation. *Mater. Eval.* 46(5):608–13.
35. Steiner, K. V. 1992. *Image Enhancement Techniques for Ultrasonic NDE Applications,* Composite Materials Testing and Design, vol. 10, ed. G. C. Grimes, 330–43, STP 1120. ASTM, Philadelphia.
36. NASA. Technical Brief, November 1993, p. 76.
37. Talreja, R. 1987. Application of acousto-ultrasonics to quality control and damage assessment of composites. *4th Int. Conf. Compos. Struct.,* Paisley College of Technology, U.K.
38. Hamstad, M. A. 1985. Composite characterization techniques: Acoustic emission. *Mantech J.* 10(3):24–32.
39. Hamstad, M. A. 1986. Review: Acoustic emission—A tool for composite material studies. *Exp. Mech.* 26(1):7–13.
40. Arrington, M. 1987. Acoustic emission. In *Non-Destructive Testing of Fibre-Reinforced Plastics Composites,* vol. 2, ed. J. Summerscales, 25–63. Elsevier, Barking, U.K.
41. Drouillard, T. F., and M. A. Hamstad. 1983. A comprehensive guide to literature on AE from composites. *Proc. 1st Int. Symp. Acoust. Emiss. Reinforced Compos.,* July 1983, Paper 23, RFP-3463. AEWG, San Francisco, CA.
42. Drouillard, T. F. 1986. A comprehensive guide to literature on AE from composites. Supplement I. *Proc. 2nd Int. Symp. Acoust. Emiss. Reinforced Compos.,* July 1986, RFP-3918. AEWG, Montreal.
43. Drouillard, T. F. 1989. A comprehensive guide to literature on AE from composites. Supplement II. *Proc. 3rd Int. Symp. Acoust. Emiss. Reinforced Compos.,* July 1989, RFP-4297. AEWG, Paris.

44. Hall, R. F. C. 1985. Investigation into the attenuation and velocity of acoustic emissions in carbon fibre composite aircraft structures. British Aerospace (Kingston) Report for MOD (PE) Contract A93b/1362, BAe-KGT-R-GEN-01349.
45. Ibitolu, E. O., and J. Summerscales. 1987. Acoustic emission source location in bidirectionally reinforced composites. Part 2: Waveform profile. *Proc. 2nd Int. Conf. Test. Eval. Qual. Control Compos.*, pp. 128–33. September 1987, Guildford, U.K.
46. Glannie, A. M. G., and J. Summerscales. 1986. Acoustic emission source location in orthotropic materials. *Brit. J. Non-Destruct. Test.* 28(1):17–22.
47. Ibitolu, E. O., and J. Summerscales. 1988. Acoustic emission source location in bidirectionally reinforced composites. Part 1: Source location. *Proc. 4th Eur. Conf. Non-Destruct. Test.*, London, vol. 4, 2881–91. Pergamon Press, Oxford.
48. Buttle, D. J., and C. B. Scruby. 1988. Acoustic emission source location in fibre reinforced plastics composites. Harwell Report AERE R 13039.
49. Korane, K. J. 1987. Spotting flaws in advanced composites. *Mach. Des.*, December 12:99–104.
50. Rothwell, W. P., D. R. Holecek, and J. A. Kershaw. 1984. NMR imaging study of fluid absorption by polymer composites. *J. Polym. Sci. Polymer Lett.* 22(5):241–47.
51. Summerscales, J. 1990. NDT of advanced composites—An overview of the possibilities. *Brit. J. Non-Destructive Test.* 32(11):568–77.
52. Bathias, C., and A. Cagnasso. 1992. Application of x-ray tomography to the non-destructive testing of high-performance polymer composites. In *Damage Detection in Composite Materials,* ed. J. E. Masters, 35–54, STP 1128. ASTM, Philadelphia.
53. Stock, S. R., T. M. Breunig, A. Guvenilir, et al. 1992. Nondestructive x-ray tomographic microscopy of damage in various continuous-fiber metal matrix composites. In *Damage Detection in Composite Materials,* ed. J. E. Masters, 25–34, STP 1128. ASTM, Philadelphia.
54. Moran, T. J. 1990. Development and application of computed tomography (CT) for inspection of aerospace components. *69th AGARD Struct. Mater. Panel Meet.,* Brussels. AGARD CP 462.
55. Dunyak, T. J., W. W. Stinchcomb, and K. L. Reifsnider. 1992. Examination of selected NDE techniques for ceramic composite components. In *Damage Detection in Composite Materials,* ed. J. E. Masters, 3–24, STP 1128. ASTM, Philadelphia.
56. Vary, M. A., and S. J. Klima. 1991. NDE of ceramics and ceramic composites. NASA-TM-104520, E-6390.
57. Mast, H. U., and R. Schutz. 1990. Neutron radiography—Applications and systems. *69th AGARD Struct. Mater. Panel Meet.,* Brussels. AGARD CP 462.
58. Walker, C. 1987. Optical methods. In *Non-Destructive Testing of Fibre-Reinforced Plastics Composites,* vol. 1, ed. J. Summerscales, 105–49. Elsevier, Barking, U.K.
59. Vipond, R., and C. J. Daniels. 1985. Non-destructive examination of short carbon-fibre reinforced injection moulded thermoplastics. *Composites* 16(1):14–18.
60. Harris, C. E., and D. H. Morris. 1985. An evaluation of the effects of stacking sequence and thickness on the fatigue life of quasi isotropic graphite/epoxy laminates. ASTM Special Technical Publication 864, pp. 153–72.
61. Liu, D., L. S. Lillycrop, L. E. Malvern, et al. 1987. Evaluation of delamination: An edge replication study. *Exp. Tech.* 11(5):20–25.
62. Freeman, S. M. 1984. Correlation of X-ray radiographic images with actual damage in graphite/epoxy composites by the de-ply technique. SME Technical Paper EM 84-101.

63. Jamison, R. D. 1986. On the inter-relationship between fibre fracture and ply cracking in graphite/epoxy laminates. ASTM Special Technical Publication 907, pp. 252–73.
64. Pih, H., and C. E. Knight. 1969. Photoelastic analysis of anisotropic fibre reinforced composites. *J. Compos. Mater.* 3(1):94–107.
65. Leonard, L. 1990. Inside story on composites: NDI looks sharp. *Adv. Comp.* 5(1):52–56.
66. Seidl, T. 1992. Inspection of composite structures. *Aerosp. Eng.* May:9–13.
67. Reifsnider, K. L., and E. G. Henneke. 1985. Composite characterization techniques: Thermography. *Mantech J.* 10(3):3–10.
68. Puttick, K. 1987. Thermal NDT methods. In *Non-Destructive Testing of Fibre-Reinforced Plastics Composites,* vol. 1, ed. J. Summerscales, 65–103. Elsevier, Barking, U.K.
69. Stinchcomb, W. W., ed. 1980. *Mechanics of Nondestructive Testing.* Plenum Press, New York.
70. Cawley, P., and R. D. Adams. 1989. Defect types and non-destructive testing techniques for composites and bonded joints. *Mater. Sci. Technol.* 5(5):413–25.
71. *Adhesive Bonding Handbook for Advanced Structural Materials,* ESA PSS-03-210 Issue 1, 1990. European Space Research and Technology Centre, Noordwijk, Netherlands, 1991. N91-32234.
72. Reynolds, W. N., and G. M. Wells. 1984. *Br. J. Non-Destr. Test.* 26:40–44.
73. Stone, D. E. W., and B. Clarke. 1987. Proc. ICCM VI/ECCM II, ed. F. L. Matthews et al., 1.28–1.59. Elsevier, London.
74. Reynolds, W. N. 1988. *Non-Destr. Test. Int.* 21(4):229–32.
75. Matthews, R. 1988. Safer with the heat on. *The Times,* February 9, 1988, London.
76. Summerscales, J. 1990. Microwave techniques. In *Non-Destructive Testing of Fibre-Reinforced Plastics Composites,* vol. 2, ed. J. Summerscales, 361–412. Elsevier, Barking, U.K.
77. Prakash, R. 1990. Eddy current. In *Non-Destructive Testing of Fibre-Reinforced Plastics Composites,* vol. 2, ed. J. Summerscales, 299–325. Elsevier, Barking, U.K.
78. Hagemaier, D. J. 1991. Nondestructive testing developments in the aircraft industry. *Mater. Eval.* 49(12):1470–78.
79. Senturia, S. D., and N. F. Sheppard. 1986. Dielectric analysis of thermoset cure. *Adv. Polym. Sci.* 80:1–47.
80. Summerscales, J. 1990. Dielectrometry. In *Non-Destructive Testing of Fibre-Reinforced Plastics Composites,* vol. 2, ed. J. Summerscales, 327–60. Elsevier, Barking, U.K.
81. Summerscales, J. 1990. Electrical and magnetic testing. In *Non-Destructive Testing of Fibre-Reinforced Plastics Composites,* vol. 2, ed. J. Summerscales, 253–97. Elsevier, Barking, U.K.
82. Mertzel, E., and J. L. Koenig. 1986. Application of FT-IR and NMR to epoxy resins. *Adv. Polym. Sci.* 75:73–112.
83. Mertzel, E., and J. L. Koenig. 1985. Composite characterization techniques: Physicochemical. *Mantech J.* 10(2):27–36.
84. Summerscales, J., and D. Short. 1987. Chemical spectroscopy. In *Non-Destructive Testing of Fibre-Reinforced Plastics Composites,* vol. 2, ed. J. Summerscales, 207–70. Elsevier, Barking, U.K.
85. Gerrard, D. L., and W. F. Maddams. 1986. Polymer characterization by Raman spectroscopy. *Appl. Spectrosc. Rev.* 22(2/3):251–334.
86. Summerscales, J. 1987. Non-destructive testing by Raman spectroscopy. *Proc. 6th Int. Conf. Compos. Mater.,* vol. 1, 321–32, Imperial College, London.

87. Fan, C. F., and S. L. Hsu. 1989. Effects of fibre orientation on stress distribution in model composites. *J. Polym. Sci. B: Polym. Phys.* 27(13):2605–19.
88. Heyman, J. S., and W. P. Winfree. 1990. Advanced NDE techniques for quantitative characterization of aircraft. *69th AGARD Struct. Mater. Panel Meet.*, AGARD CP 462.
89. Scott, I. G., and C. M Scala. 1986. An overview of Australian research on nondestructive testing. *Proc. Nondestr. Test. Eval. Adv. Mater. Compos. Conf.*, pp. 1–11, August 1986, Colorado Springs, CO.
90. Murphy, R. F., R. W. Reed, and R. S. Williams. 1986. Nondestructive evaluation of ceramic matrix composites. *Proc. Nondestr. Test. Eval. Adv. Mater. Compos. Conf.*, 135–47, August 1986, Colorado Springs, CO.
91. Parker, I. 1989. Looking inside a structure. *Aerosp. Compos. Mater.* May/June:14–22.
92. Leonard, L. 1990. Inside story on composites: NDI looks sharp. *Adv. Compos.* January/February:56.
93. Walker, James, II. 1992. Acoustic emissions testing advances. *Adv. Compos.* May/June:12.
94. MacRae, M. 1993. Details without damage: NDT methods "find fault" with composite materials. *Adv. Compos.* 8(4):28–34.
95. Kessler, L. W., and M. G. Oravecz. 1986. Scanning laser acoustic microscope (SLAM) analysis of advanced materials for internal defects and discontinuities. *Proc. Nondestr. Test. Eval. Adv. Mater. Compos. Conf.*, pp. 173–86, August 1986, Colorado Springs, CO.
96. Bittence, J. C. 1989. Greater precision for materials analysis. *Adv. Mater. Process.* 136(5):11–19.
97. Michaels, T. E., and B. D. Davidson. 1993. Ultrasonic inspection detects hidden damage in composites. *Adv. Mater. Process.* 143(3):34–8.
98. Lawrence, C. W., G. A. D. Briggs, and C. B. Scruby. 1993. Acoustic microscopy of ceramic-fibre composites. Part II: Glass-ceramic-matrix composites. *J. Mater. Sci.* 28(13):3645–52.
99. Lawrence, C. W., G. A. D. Briggs, and C. B. Scruby. 1993. Acoustic microscopy of ceramic-fibre composites. Part III: Metal-matrix composites. *J. Mater. Sci.* 28(13):3653–60.
100. Lawrence, C. W., G. A. D. Briggs, C. B. Scruby, et al. 1993. Acoustic microscopy of ceramic-fibre composites. Part I: Glass-matrix composites. *J. Mater. Sci.* 28(13):3635–44.
101. Bossi, R., J. L. Cline, G. E. Georgeson, et al. 1993. X-ray computes tomography. *Aerosp. Eng.* May:17–22.
102. Bossi, R. H., and G. E. Georgeson. 1991. The application of x-ray computed tomography to materials development. *J. Met.* 43(9):8–15.
103. Georgeson, G., and R. Bossi. 1992. X-ray computed tomography for advanced materials and processes. WL-TR-91-4101.
104. Copley, D. A., J. W. Eberhard, and G. A. Mohr. 1994. Computed tomography. Part I: Introduction and industrial applications. *J. Met.* 46(1):14–26.
105. Roye, W. 1991. The reliability of nondestructive techniques within the field of modern composite materials. *Brit. J. Non-Destr. Test.* 33(11):549–550.
106. Hentschel, M. P., W. Harbich, and A. Lange. 1992. New x-ray tomographic approaches to nondestructive evaluation of composites. *Proc. Int. Symp., Adv. Mater. Lightweight Struct.* pp. 229–39. ESTEC, Noordwijk, Netherlands. ESA SP-336.
107. McKinney, W. E. J. 1992. X-rays and NDT. *Aerosp. Eng.* July:20–21.
108. Henderson, B. W. 1989. USAF seeks aerospace applications for innovative x-ray tomography. *Aviat. Week Space Technol.* July 31:93–99.
109. Newman, J. W. 1986. Laser holographic NDT of advanced materials. *Proc. Nondestr. Test. Eval. Adv. Mater. Compos. Conf.*, pp. 187–99, August 1986, Colorado Springs, CO.

110. Marsh, G. 1990 and 1992. Revealing hidden flaws. *Aerosp. Compos. Mater.* 18–21, 1990; 4(2):26–29, 1992.

111. De Goeje, M. P., and K. E. D. Wapenaar. 1992. Non-destructive inspection of carbon fibre-reinforced plastics using eddy current methods. *Composites* 23(3):147–57.

112. Aircraft Maintenance International. 1994. Infrared NDT method for composites, 6(5):42.

113. Yancey, R. N., and G. Y. Baaklini. 1994. Computed tomography evaluation of metal-matrix composites for aeropropulsion engine applications. *J. Eng. Gas Turbines Power* 116(3):635–39. ASME 93-GT-4.

114. Yancey, R. N., G. Y. Baaklini, and S. J. Klima. 1991. NDE of advanced turbine engine components and materials by computed tomography. ASME-91-GT-287, 1991.

115. Karpur, P., D. A. Stubbs, T. E. Matikas, et al. 1993. Ultrasonic nondestructive characterization methods for the development and life prediction of titanium matrix composites. *77th AGARD Struct. Mater. Panel Meet.*, pp. 13-1 to 13-12, September 1993, Bordeaux, France. AGARD 796.

BIBLIOGRAPHY

AGARD Struct. Mater. Panel 69th Meet.: Impact Emerging NDE-NDI—Meth. Aircr. Des. Manuf. Maint., March 1990. AGARD CP 462.

ALBERTI, F. P. Quality control and nondestructive evaluation techniques for composites. Part IV: Radiography—A state-of-the-art review. AVRADCOM Report TR 82-F-4, Contract IPA05-3363, 1982.

ASHLEY, S. Watching how composites grow. *Mech. Eng.* 115(7):70–71, 1993.

BAR-COHEN, Y., AND A. K. MAL. Ultrasonic system measures elastic properties of composites. NASA Technical Brief, November 1993, pp. 76–77.

BOSSI, R. H. Trends in radiography. *Mater. Eval.* 49(9):1177–78, 1991.

BROWN, R., S. CAUFFMAN, AND M. DERSTINE. Digital imaging inspection and process control for 3-D braiding system. *40th Int. SAMPE Exhibit.*, Anaheim, CA, May 1995.

CAHN, R. W., P. HAASEN, AND E. J. KRAMER, EDS. *Materials Science and Technology*, vol. 2B: *Characterization of materials*, Part II, ed. E. Lifshin. VCH, New York, 1993.

Carbon and Graphite Fiber Composites: Nondestructive Testing. NERAC, Tolland, CT, 1993; NTIS, Springfield, VA, PB93-889988/WMS.

CARRIVEAU, G. W. Benchmarking of the state-of-the-art in nondestructive testing/evaluation for applicability in the composite armored vehicle (CAV) advanced technology demonstrator (ATD) program, Phase I, 1993. NTIAC-7304-103, TARDEC-TR-13604.

CHERN, E. J. Modification of ultrasonic C-scan imaging can detect the integrity of second interfaces in laminated composites. *Mater. Technol.* 8(9/10):189–190, 1993.

CHRESANTHAKES, C. Four new techniques enhance nondestructive evaluation. *Res. Dev.* December:23–24, 1994.

CONNOLLY, M., AND D. COPLEY. Thermographic inspection of composite materials. *Mater. Eval.* 48(12):1461–3, 1990.

COOK, T. N., M. RENIERI, R. COX, ET AL. Advanced composite structures R&M design and repair. Sikorsky Corporation, 1986. SER-510217, USAAVSCOM TR-85-D-12.

COPPEE, J. L. Application of the infrared photoacoustic spectroscopy to the study of the cure mechanism of structural epoxy adhesives. *Proc. Int. Symp. Adv. Mater. Lightweight Struct.*, pp. 233–38, ESA SP-336. ESTEC, Noordwijk, Netherlands, 1992.

CREWS, A. R., AND R. H. BOSSI. X-ray computed tomography for whole system evaluation (small jet engines), 1992, WL-TR-91-4109.

DEMEIS, R. NDI: Ready right now. *Aerosp. Am.* November:48–49, 1989.

DEMEIS, R. CAT scanning composites. *Aerosp. Am.* February:44–45, 1991.

DERRA, S. Aging airplanes—Can research make them safer? *Res. Dev.* January:29–34, 1990.

ELBER, G. Infrared inspection in the field and in the factory. Part I. *Adv. Compos.* 8(6):22–5, 1993.

Fifth generation ultrasonics testing systems developed. *Adv. Mater. Process. Inc. Met. Prog.* November:22–23, 1987.

FORTUNKO, C. M., AND D. W. FITTING. Appropriate ultrasonic system components for NDE of thick polymer composites, National Institute of Standards and Technology, Boulder, CO, 1991. *Rev. Prog. Quant. Nondestr. Eval.* 10B:2105–112, 1991.

FOX, R. L., AND J. D. BUCKLEY. Eddy-current monitoring of composite layups. NASA Technical Brief, December 1993, p. 49.

GALIOTIS, C. Laser raman spectroscopy: A new stress/strain measurement technique for the remote and on-line nondestructive inspection of fiber reinforced polymer composites. *Mater. Technol.* 8(9/10):203–9, 1993.

GENERAZIO, E. R., AND D. J. ROTH. Use of video in microscopic and ultrasonic inspection. NASA Technical Brief, August 1994, pp. 48–49.

GEORGESON, G., AND R. BOSSI. X-ray computed tomography for advanced materials and processes. *TMS 1992 Ann. Meet.: Dev. Ceram. Met.-Matrix Compos.*, ed. K. Upadhya, pp. 143–55, San Diego, CA, March 1992.

GEORGIOU, G. A., AND I. A. MACDONALD. Ultrasonic and radiographic NDT of butt fusion joints of polyethylene. Technical Brief. 465. TWI, Abington, Cambridge, U.K., 1993.

GOLOBOROD'KO, M. N., V. K. KAPRALOV, AND V.A. CHELNOKOV. Ultrasonic strength inspection of carbon-carbon composite materials, Leningrad Polytechnic Institute. Zavod. Lab. 49(1):48–9, 1983.

GREEN, R. E., JR. Nondestructive characterization of material properties. *Mech. Eng.* September:66–70, 1987.

GREEN, R. E., JR. Nondestructive evaluation of materials. *Ann. Rev. Mater. Sci.* 20:197–217, 1990.

GUO, N., AND P. CAWLEY. The nondestructive assessment of porosity in composite repairs. *Composites* 25(9):842–50, 1994.

HAMSTAD, M. A. Quality control and nondestructive evaluation techniques for composites. Part VI: Acoustic emission—A state-of-the-art review, 1983. AVRADCOM Report TR 83-F-7, Contract DAAG29-81-D-0100.

HENNEKE, E. G., AND K. L. REIFSNIDER. Quality control and nondestructive evaluation techniques for composites. Part VII: Thermography—A state-of-the-art review. AVRADCOM Report TR 82-F-5, Contract DAAG29-76-D-0100, 1982.

HOBBS, C., AND A. TEMPLE. The inspection of aerospace structures using transient thermography. *Brit. J. Non-Destr. Test.* 35(4):183–89.

IRVING, R. R. High technology erupts on the NDT scene. *Iron Age* May 15:39–42.

JONES, T. S. Damage-assessment nondestructive inspection methods. *SME Fabric. Compos. Conf.*, pp. 259–63, Dearborn, MI, September 1987.

KAUTZ, H. E., AND B. A. LERCH. Preliminary investigation of acoustoultrasonic evaluation of metal-matrix composite specimens. *Mater. Eval.* 49(5):607–12, 1991.

KESSLER, L. W., AND D. E. YUHAS. Listen to structural differences. *Ind. Res. Dev.* April:102–06, 1978.

KINRA, V. K. Ultrasonic nondestructive evaluation of damage in continuous fiber composites. Annual Technical Report, Texas A&M University, College Station, TX, 1985. NTIS HC A05/MF A01.

KLINE, R. A. Quantitative NDE of advanced composites using ultrasonic velocity measurements. *ASME J.* 112(2):218–22, 1990.

KUPPERMAN, D. S., S. MAJUMDAR, AND J. P. SINGH. Neutron diffraction NDE for advanced composites. *ASME J.* 112(2):198–201, 1990.

LIAW, P. K., R. E. SHANNON, W. G. CLARK, JR. ET AL. Determining material properties of metal-matrix composites by NDE. *J. Met.* 44(109):36–40.

LIAW, P. K., R. PITCHUMANI, D. K. HSU, ET AL. Nondestructive eddy current evaluation of anisotropic conductivities of silicon carbide reinforced aluminum metal-matrix composite extrusions. *J. Eng. Gas Turbines Power* 116(3):647–56, 1994.

Looking closer and faster. *Mach. Des.* December:99–104, 1987.

LUCIER, R. D. Trends in infrared thermography. *Mater. Eval.* 49(9):1162, 1991.

MILLER, B. *Plas. World.* August:19, 1989.

MUMMERY, P. M., B. DERBY, AND C. B. SCRUBY. Acoustic emission from particulate-reinforced metal matrix composites. *Acta Metall. Mater.* 41(5):1431–45, 1993.

MURPHY, W. L. Introduction of expert systems for inspection and repair of composites. *CoGSME Compos. Manuf. Conf.,* pp. 266–79, January 1987.

NEWMAN, J. W. Production and field inspection of composite aerospace structures with advanced shearography. *SAMPE 22nd Int. Tech. Conf.,* pp. 1243–50, November 1990.

NEWMAN, J. W. Test composites with lasers. *Mater. Eng.* July:10–11, 1992.

NEWMAN, J. W. Shearography NDI of composite structures. *40th Int. SAMPE Symp. Exhibit.,* Anaheim, CA, May 1995.

NEWMAN, J. W. Shearography non-destructive testing of production composites. *40th Int. SAMPE Exhibit.,* Anaheim, CA, May 1995.

9th Ann. ASM/ESD Adv. Compos. Conf. Expos. Adv. Compos. Technol., Engineering Society of Detroit. Dearborn, MI, November 1993.

NOKES, J. Inspection of fiber wound pressure vessels. *40th Int. SAMPE Exhibit.,* Anaheim, CA, May 1995.

OSTER, R. Computed tomography (CT) as a nondestructive test method used for composite helicopter components. *Eur. Rotorcr. Forum,* Berlin, September 1991. MBB-UD-0603-91-Pub 17.

PALANISAMY, R. Developments in eddy current nondestructive testing. *Mater. Eval.* 49(9):1158–61, 1991.

PAPADAKIS, E. P. Future of ultrasonics. *Mater. Eval.* 49(9):1180–84, 1991.

PEREIRA, J. M., AND E. R. GENERAZIO. Ultrasonic detection of transverse cracks in composites. NASA Technical Brief, December 1992, pp. 64–65.

REYNOLDS, W. N. Nondestructive examination of composite materials—A survey of European literature. AVRADCOM Report TR 81-F-6, AMMRC TR 81-24, Contract DAJA37-79-C-0553.

ROSE, J. L., AND K. BALASUBRAMANIAM. Ultrasonic NDE potential in composite manufacturing. *SME Fabric. Compos. Conf.,* pp. 327–41, Dearborn, MI, September 1987.

SACHSE, W. Acoustic emission: Current status and future directions. *Mater. Eval.* 49(9):1153–56, 1991.

SAFAI, M. Simultaneous NDE with thermography and shearography. *40th Int. SAMPE Exhibit.,* Anaheim, CA, May 1995.

SATTLER, F. J. Nondestructive testing via optical and acoustical methods. *Chem. Eng.* November:191–98, 1989.

SCHAEFER, L. A. Laser shearography reveals hidden "unbonds." NASA Technical Brief, October 1992, pp. 101–02.

SCHRAMM, S. W., AND I. M. DANIEL. Nondestructive evaluation of metal matrix composites. IIT Research Institute, 1982. DAAG46-80C0070, IITRI M 06084-1, AMMRC TR 82-35.

SEIDL, A. L. Inspection of composite structures. Part I. *SAMPE J.* 30(4):38–44, 1994.

SHEPPARD, L. M. Detecting material defects in real time: Ultrasonics can read the resin-fiber ratio. *Adv. Mater. Process. Inc. Met. Prog.* November:53–60, 1987.

SHEPPARD, L. M. NDE heats up round the world. *Adv. Mater. Process. Metal Prog.* December:10–20, 1987.

SICHINA, W., AND D. SHEPARD. DEA takes the guesswork out of composite manufacture. *Mater. Eng.* July:49–53, 1989.

SMITH, B. Seeing the heat. Agema Infrared Systems Ltd., Leighton Buzzards, Beds, U.K. *Aerosp. Mater.* 26–27, 1994.

STANLEY, R. K. Present state of magnetic nondestructive testing techniques. *Mater. Eval.* 49(9):1169–72, 1991.

STEINER, K. V. Defect classification in composites using ultrasonic NDE techniques. University of Delaware, 1991. CCM 91-08.

TANG, S. S., K.-L. CHEN, A.-Y. KUO, ET AL. Monitoring integrity of composite aircraft components. NASA Technical Brief, October 1994, pp. 62–64.

TITTMAN, B. R. NDE at twelve o'clock high. *Mech. Eng.* September:724, 1987.

TRAVIS, H. S. More about stitching for monitoring composite parts. NASA Technical Brief, October 1994, p. 101.

Ultrasonic evaluation for composite material processing. *Ind. Heat.* August:10, 1992.

VAN VALKENBURG, H. E. Retrospective of ultrasonics. *Mater. Eval.* 49(9):1188–1201, 1991.

VARY, M. Quality evaluation by acousto-ultrasonic testing of composites. NASA Technical Brief, June 1989, p. 81.

Vought develops neutron radiography testing unit. *Aviat. Week Space Technol.* May 16:60–61, 1983.

WATERBURY, M. C., T. E. MATIKAS, P. KARPUR, ET AL. In-situ ultrasonic imaging and acoustic emission monitoring of single fiber fragment metal matrix composites. *ASM Int. Meet.*, Pittsburgh, PA, October 1993.

WEHRENBURG R. H., II. New NDE technique finds subtle defects. *Mater. Eng.* September:59–63, 1980.

WEY, A. C. Load damage in graphite/epoxy laminates. *Adv. Mater. Process.* 142(6):43–44, 1992.

WINFREE, W. P., AND J. N. ZALAMEDA. Heat-pulse measurements reveal fiber volume fractions. NASA Technical Brief, November 1994, p. 69.

WONG, B. S., K. S. TAN, AND T. G. TUI. Ultrasonic testing of solid fiber-reinforced composite plate. *SAMPE J.* 30(6):36–40, 1994.

WOOH, S. C., AND I. M. DANIEL. Enhancement techniques for ultrasonic nondestructive evaluation of composite materials. *ASME J.* 112(2):175–82.

WORKMAN, G. L., ED. *Conf. Nondestr. Eval. Aerosp. Requir.* Huntsville, AL, August 1987. Gordon and Breach, New York, 1991.

WORKMAN, G. L., AND M. WANG. Eddy-current probes for inspecting graphite-fiber composites. NASA Technical Brief, May 1992, pp. 20–24.

WU, W.-LI. Acoustic emissions reveal fiber/matrix bond strength. *Adv. Mater. Process.* 140(2):39–40, 1991.

YANG, B. Z. *Advanced Polymer Composites: Principles and Applications.* ASM International, Metals Park, OH, 1994.

ZAYED, A.-N., AND B. LAWTON. Implementation of a computed tomography system by using a photo-diode as a radiation detector. *40th Int. SAMPE Symp. Exhibit.,* Anaheim, CA, May 1995.

Glossary

ABDR: Aircraft battle damage.

Abhesive: A film or coating applied to one solid to prevent (or greatly decrease) the adhesion to another solid with which it is to be placed in intimate contact, e.g., a parting or mold-release agent.

Ablative plastic: A material that absorbs heat (while part of it is being consumed by heat) through a decomposition process (pyrolysis) taking place near the surface exposed to the heat.

ABL bottle: An internal-pressure-test vessel about 18 in. (457 mm) in diameter and 24 in. (610 mm) long, used to determine the quality and properties of the filament-wound material in the vessel.

ABS: Acrylonitrile-butadiene-styrene.

Accelerator: A material mixed with a catalyzed resin to speed up the chemical reaction between the catalyst and resin; used in polymerizing resins and vulcanizing rubbers; also known as a promoter or curing agent.

Acicular powder: Needle-shaped particles.

ACMC: Advanced Composite Materials Corporation (formerly ARCO).

Activated sintering: An enhanced sintering process involving a treatment that reduces the activation energy for atomic motion. The increase in sintering rate allows faster sintering, lower sintering temperatures, or improved properties.

Activation: A process that increases the surface area of a material such as charcoal or alumina. In the case of charcoal, the surface of the material is oxidized and minute cavities are created which are capable of absorbing gas atoms or molecules.

Activator: An additive used to promote the curing of matrix resins and reduce curing time. (See also Accelerator.)

ACTP: Acetylene-terminated polyimide.

Addition: A polymerization reaction in which no by-products are formed.

Addition polymerization: Polymerization in which monomers are joined together without the splitting off of water or other simple molecules.

Additive: Any substance added to another, usually to improve properties.

Adherend: A body held to another body by an adhesive.

Adhesion: The state in which two surfaces are held together at an interface by forces or interlocking action or both.

Adhesion, mechanical: Adhesion between surfaces in which the adhesive holds the parts together by interlocking action.

Adhesive, contact: See Contact adhesive.

Adhesive failure: A rupture of adhesive bond that appears to be a separation at the adhesive-adherend interface.

Adhesive film: A synthetic resin adhesive, usually of the thermosetting type, in the form of a thin, dry film of resin, used under heat or pressure as an interleaf in the production of laminated materials.

Adhesiveness: The property defined by the adhesion stress $A = F/S$, where F = perpendicular force to glue line and S = surface.

Adsorption: The formation of a layer of gas on the surface of a solid (or occasionally a liquid). The two types of adsorption are (1) chemisorption where the bond between the surface and the attached atoms, ions, or molecules is chemical, and (2) physisorption where the bond is due to van der Waals forces.

Advanced composite: High-performance or cost-performance material.

AE: Acoustic emission.

AES: Auger electron spectroscopy.

AFM: Atomic force microscopy.

Agglomeration: A tendency for fine particles to stick together and appear as larger particles.

Aggregate: A hard, fragmented material used with an epoxy binder, as in epoxy tools.

Aging: The process or the effect on materials of exposure to an environment for an interval of time.

AIMS: Advanced integrated-manufacturing system.

Air bubble void: Noninterconnected spherical air entrapment within and between the plies of reinforcement.

Air locks: Surface depressions on a molded part, caused by air trapped between the mold surface and the plastic.

Air vent: Small outlet to prevent entrapment of gases.

Alloy, alloying: A material formed by the mutual solution within a single phase or crystal structure of two or more elements. Even in alloys where several phases coexist, all the phases present are likely to be alloyed, i.e., contain two or more constituent elements.

Alloy powder: A powder in which each particle is composed of the same alloy of two or more metals.

α-Ti_3Al: Alpha titanium aluminide.

AM: Acoustic microscopy.

Ambient: The surrounding environmental conditions, e.g., pressure or temperature.

Amorphous: Refers to a random or nonsymmetric atomic arrangement. In metals and ceramics it means a noncrystalline state; in polymers it means a random arrangement of the atomic chains of the polymer. For alloys, amorphous structures are obtained by rapidly quenching molten droplets or mechanical alloying.

Amorphous plastic: A plastic that has no crystalline component. There is no order or pattern in the distribution of the molecules.

Anisotropic: Exhibiting different properties in response to stresses applied along axes in different directions.

Anisotropic laminate: A laminate in which the strength properties are different in different directions.

Anisotropy of laminates: The difference of the properties along the directions parallel to the length or width into the lamination planes or parallel to the thickness into the planes perpendicular to the lamination.

ANSYS: Analytical system.

Antistatic compounds: Compounds intermediate between insulators (high resistivity) and conductors (low resistivity).

APC: Aromatic polymer composite.

API: Addition-reaction polyimide.

ARALL: Aramid-reinforced aluminum laminate.

Aramid: Aromatic polyamide fibers characterized by excellent high-temperature, flame resistance, and electrical properties. Aramid fibers are used to achieve high-strength, high-modulus reinforcement in plastic composites. More usually found as polyaramid—a synthetic fiber (trade name Kevlar).

Arc resistance: The total time in seconds that an intermittent arc can play across a plastic surface without rendering the surface conductive.

Areal weight: The weight of fiber per unit area (width times length) of tape or fabric.

Aromatic: An unsaturated hydrocarbon with one or more benzene ring structures in the molecule.

ASD: Alternating stress density.

Ash content: The solid residue remaining after a reinforcing substance has been incinerated or strongly heated.

Aspect ratio: The ratio of length to diameter of a fiber.

A stage: An early stage in the polymerization reaction of certain thermosetting resins (especially phenolic) in which the material, after application to the reinforcement, is still soluble in certain liquids and is fusible; sometimes referred to as resole. (See also B stage, C stage.)

ASTM: American Society for Testing and Materials.

ATF: Advanced tactical fighter.

ATLAS: Automated tape lay-up system.

Atomization: The dispersion of molten metal into droplets by a rapidly moving stream of gas or liquid, or by centrifugal force.

Atomized metal powder: Metal powder produced by the disintegration and subsequent solidifaction of a molten metal stream.

Attenuation: The process of making thin and slender, as applied to the formation of fiber from molten glass. Reduction of force or intensity as in the decrease of an electric signal.

AU: Acoustoultrasonic.

Autoclave: A closed vessel for conducting a chemical reaction or other operation under pressure and heat.

Autoclave molding: A process in which, after lay-up, the entire assembly is placed in a steam autoclave at 50–100 lb/in.^{2} (23.4–47.6 Pa); additional pressure achieves higher reinforcement loadings and improved removal of air.

Automatic mold: A mold for injection or compression molding that repeatedly goes through the entire cycle, including ejection, without human assistance.

Automatic press: A hydraulic press for compression molding or an injection machine that operates continuously, being controlled mechanically, electrically, hydraulically, or by a combination of these methods.

Axial winding: In filament-wound reinforced plastics, a winding with the filaments parallel to the axis.

Back draft: An area of interference in an otherwise smooth-drafted encasement; an obstruction in the taper that would interfere with the withdrawal of the model from the mold.

Backpressure: Resistance of a material, because of its viscosity, to continued flow when a mold is closing.

Bag molding: A technique in which consolidation of the material in the mold is affected by the application of fluid pressure through a flexible membrane.

Balanced design: In filament-wound reinforced plastics, a winding pattern so designed that the stresses in all filaments are equal.

Balanced-in-plane contour: In a filament-wound part, a head contour in which the filaments are oriented within a plane and the radii of curvature are adjusted to balance the stresses along the filament with the pressure loading.

Balanced laminate: A laminate in which all lamina except those at 0°/90° are placed in plus/minus pairs (not necessarily adjacent) symmetrically about the lay-up centerline.

Balanced twist: An arrangement of twist in a plied yarn or cord that will not cause twisting on itself when the yarn or cord is held in the form of an open loop.

Barcol hardness: A hardness value obtained by measuring the resistance to penetration of a sharp steel point under a spring load. An instrument, the Barcol impressor, gives a direct reading on a scale of 0–100. The hardness value is often used as a measure of the degree of cure of a plastic.

Bare glass: Glass (yarns, roving, or fabrics) from which the sizing or finish has been removed or before it has been applied.

Base: The reinforcing material (glass fiber, paper, cotton, asbestos, etc.) that is impregnated with resin in the forming of laminates.

Batch: A measured mix of various materials. (See also Lot.)

Batt: Felted fabrics; structures built by the interlocking action of fibers themselves without spinning, weaving, or knitting. (See also Felt.)

Bearing area: The diameter of a hole times the thickness of the material.

Bearing strength: The bearing stress at the point on the stress-strain curve where the tangent is equal to the bearing stress divided by *n*% of the bearing-hole diameter.

Bearing stress: The applied load in pounds divided by the bearing area. (Maximum bearing stress is the maximum load in pounds sustained by the specimen during the test divided by the original bearing area.)

Bias fabric: A fabric in which warp and fill fibers are at an angle to the length.

Biaxial load: (1) A loading condition in which a laminate is stressed in at least two different directions in the plane of the laminate. (2) A loading condition of a pressure vessel under internal pressure and with unrestrained ends.

Biaxial winding: In filament winding, a type of winding in which the helical band is laid in sequence, side by side, with no crossover of fibers.

Bidirectional laminate: A reinforced plastic laminate with the fibers oriented in various directions in the plane of the laminate; a cross-laminate. (See also Unidirectional laminate.)

Binder: The resin or cementing constituent of a plastic compound that holds the other components together; the agent applied to glass mat or preforms to bond the fibers before laminating or molding, or a material added to a powder for the specific purpose of cementing together powder particles that would not otherwise sinter into a strong body.

Bioglass: A fairly complex ceramic, based on silica glass, developed for its strength, compatibility, stability, and function in the form of implant devices in the human body.

Biomaterial: Synthetic material used for implants in the human (and animal) body.

Bismaleimide (BMI): A type of polyimide that cures by an addition reaction, avoiding formation of volatiles, and has temperature capabilities between those of epoxy and polyimide.

Blank: A pressed, presintered, or fully sintered compact, usually in the unfinished condition, requiring cutting, machining, or some other operation to give a final shape.

Blanket: Plies that have been laid up in a complete assembly and placed on or in the mold all at one time (flexible-bag process); also the form of bag in which the edges are sealed against the mold.

Bleeder cloth: A layer of woven or nonwoven material, not part of the composite, that allows excess gas and resin to escape during cure.

Bleedout: In filament winding, the excess liquid resin that migrates to the surface of the winding.

Blend: A polymer material containing two or more constituents within a given phase. Blends are also referred to as polymer alloys.

Blister: Undesirable rounded elevation of the surface of a plastic with boundaries that are more or less sharply defined, resembling in shape a blister on the human skin; the blister may burst and become flattened.

Block copolymer: An essentially linear copolymer in which there are repeated sequences of polymer segments of different chemical structure; some may be crystalline in nature, others may be amorphous.

Blocking: An unintentional adhesion between plastic films or between a film and another surface usually corrected by antistatic additives.

BMC: Bulk-molding compound.

BMI: Bismaleimide.

Bond strength: The amount of adhesion between bonded surfaces; a measure of the stress required to separate a layer of material from the base to which it is bonded. (See also Peel strength.)

Boron fiber: A fiber usually of a tungsten filament core with elemental boron vapor deposited on it to impart strength and stiffness.

Boss: Protuberance on a plastic part designed to add strength, to facilitate alignment during assembly, to provide for fastenings, etc.

Bottom plate: A steel plate fixed to the lower section of a mold, often used to join the lower section of the mold to the platen of the press.

Braiding: Weaving fibers into a tubular shape.

Breather: A loosely woven material that does not come in contact with the resin but serves as a continuous vacuum path over a part in production.

Breathing: (1) Opening and closing a mold to allow gases to escape early in the molding cycle (also see Degassing). (2) Permeability to air of plastic sheeting.

Bridging: A region of a contoured part that has cured without being properly compacted against the mold.

Broad goods: Woven glass, synthetic fiber, or combinations thereof over 18 in. (457 mm) wide. Fibers are woven into fabrics that may or may not be impregnated with resin, usually furnished in rolls.

B stage: An intermediate stage in the reaction of certain thermosetting resins in which the material swells when in contact with certain liquids and softens when heated but may not dissolve or fuse entirely; sometimes referred to as resistol. The resin in an uncured prepreg or premix is usually in this stage. (See also A stage, C stage.)

Bubble: A spherical internal void; a globule of air or other gas trapped in a plastic.

Buckling: Crimping of fibers in a composite material, often occurring in glass-reinforced thermoset as a result of resin shrinkage during cure.

Bulk density: The density of a molding material in loose form (granular, nodular, etc.), expressed as a ratio of weight to volume.

Burn-off: The removal of an additive by preheating prior to sintering.

Burst strength: Hydraulic pressure required to burst a vessel of given thickness; commonly used in testing filament-wound composite structures.

Butt joint: See Joint.

Butt wrap: Tape wrapped around an object in an edge-to-edge fashion.

CAD: Computer-aided design.

CAE: Computer-aided engineering.

CAM: Computer-aided manufacturing.

Carbon-carbon: A composite of carbon fiber in a carbon matrix.

Carbon fiber: An important reinforcing fiber known for its light weight, high strength, and high stiffness that is produced by pyrolysis of an organic precursor fiber such as rayon, polyacrylonitrile, and pitch in an inert environment at temperatures above 1800°F (982°C). The material may also be graphitized by heat treating above 3000°F (1649°C). The term is often used interchangeably with the term graphite; however, carbon fibers and graphite fibers differ. The basic differences lie in the temperature at which the fibers are made and heat-treated and in the structures of the resulting fibers. Graphite fibers have a more "graphitic" structure than carbon fibers.

Carbon-graphite fibers: Usually prepared from mesophase pitch or polacrylonitrile. Used as reinforcements in composites.

Carbonization: In carbon fiber manufacturing, the process step in which the heteroatoms are pyrolyzed in an inert environment at high temperature to leave a very high-carbon-content material (about 95% carbon and higher). The carbonization temperature can range up to 2000°C. Temperatures above 2000°C (up to 3000°C) are often referred to as graphitization temperatures. The range of temperature employed is influenced by precursor, individual manufacturing process, and properties desired.

CAS: Calcium aluminosilicate.

Casting: An object or finished shape obtained by solidifaction of a substance in a mold.

Catalyst: A substance that changes the rate of a chemical reaction without itself undergoing permanent change in its composition; a substance that markedly speeds up the cure of a compound when added in a small quantity compared with the amounts of primary reactants. (See also Accelerator, Curing agent, Hardener, Inhibitor.)

Catenary: A measure of the difference in length of the strands in a specified length of roving resulting from unequal tension; the tendency of some strands in a taut horizontal roving to sag lower than the others.

Caul: A sheet the size of the platens used in hot pressing.

Cavity: (1) Depression in a mold. (2) The space inside a mold into which a resin is poured. (3) The female portion of a mold. (4) The portion of a mold that encloses the molded article. (5) The portion that forms the outer surface of a molded article (often referred to as the die). (6) The space between matched molds; depending on the number of such depressions, molds are designated as single- or multiple-cavity.

CC: Carbonaceous heat-shield composites.

CCC: Carbon-carbon composite.

CD: Current density.

Cemented carbide: A solid composite consisting of a metal carbide and a binder phase, usually cobalt or nickel aluminide. The composite is formed by liquid phase sintering a mixture of the carbide and binder metal powders.

Centerless grinding: A technique for machining parts having a circular cross section, consisting of grinding the rod which is fed without mounting it on centers. Grinding is accomplished by working the material between wheels rotating at different speeds; the faster, abrasive wheel cuts the stock. Variations of the basic principle can be used to grind internal surfaces.

Centrifugal atomization: The formation of spherical particles by combining a melt with a centrifugal force such that the melt is thrown off into droplets which spheroidize prior to solidification.

Centrifugal casting: A high-production technique for cylindrical composites, such as pipe, in which chopped strand mat is positioned inside a hollow mandrel designed to be heated and rotated as resin is added and cured.

Ceramic-matrix composite (CMC): Material consisting of a ceramic or carbon fiber surrounded by a ceramic matrix, usually silicon carbide.

Cermet: A body consisting of ceramic particles bonded with a metal.

Ce-TZP: Toughened zirconia polycrystal with cerium.

CFG iron: Compact-flake-graphite iron.

CFMMC: Continuous-fiber metal matrix composite.

CFRP: Carbon-reinforced plastic.

Charge: The measurement or weight of material (liquid, preformed, or powder) used to load a mold at one time or during one cycle.

Chase: (1) The main body of a mold, which contains the molding cavity or cavities, or cores, the mold pins, the guide pins or the bushings, etc. (2) An enclosure of any shape used to shrink-fit parts of a mold cavity in place to prevent spreading or distortion in hobbing or to enclose an assembly of two or more parts of a split cavity block.

Chemical vapor deposition (CVD): A process in which desired reinforcement material is deposited from vapor phase onto a continuous core; boron on tungsten, for example.

Chill: (1) To cool a mold by circulating water through it. (2) To cool a molding with an air blast or by immersing it in water.

CIM: Computer-integrated manufacturing.

CIP: Cold isostatic pressing.

Circuit: In filament winding (1) one complete traverse of the fiber-feed mechanism of a winding machine; (2) one complete traverse of a winding band from one arbitrary point along the winding path to another point on a plane through the starting point and perpendicular to the axis.

Circumferential ("circ") winding: In filament-wound reinforced plastics a winding with the filaments essentially perpendicular to the axis.

Clamping plate: A mold plate fitted to a mold and used to fasten the mold to the machine.

Clamping pressure: In injection molding and transfer molding the pressure applied to the mold to keep it closed, in opposition to the fluid pressure of the compressed molding material.

CMC: Ceramic matrix composite.

CN: Chevron notch test (a method used to measure fracture toughness).

Coagulation bath: A liquid bath that serves to harden viscous polymer strands into solid fibers after extrusion through a spinnerette. Used in wet spinning processes such as rayon and acrylic fiber manufacture.

Coalescence: The merging of two objects into a larger object, as can be observed in particles during fabrication, and pores and grains during sintering.

Co-curing: Simultaneous bonding and curing of components.

Coefficient of elasticity: The reciprocal of Young's modulus in a tension test.

Coefficient of expansion: The fractional change in dimension of material for a unit change in temperature. Also called coefficient of thermal expansion (CTE).

Coefficient of friction: A measure of the resistance to sliding of one surface in contact with another surface.

Coefficient of thermal expansion (CTE): The change in length per unit length produced by a unit rise in temperature.

Cohesion: (1) The propensity of a single substance to adhere to itself. (2) The internal attraction of molecular particles for each other. (3) The ability to resist partition from the mass. (4) Internal adhesion. (5) The force holding a single substance together.

Cold isostatic pressing (CIP): The compaction of a powder under isostatic pressure conditions at room temperature using a flexible mold and a high hydrostatic pressure in a hydraulic pressure chamber.

Cold pressing: Forming a compact at room temperature low enough to avoid sintering—usually at room temperature.

Cold-setting adhesive: A synthetic resin adhesive capable of hardening at normal room temperature in the presence of a hardener.

Commingled yarn: A hybrid yarn made with two types of materials intermingled in a single yarn, e.g., thermoplastic filaments intermingled with carbon filaments to form a single yarn.

Compact: An object produced by the compression of metal powders.

Compaction: The shaping, deformation, and densification of a powder by the application of pressure through a tool material.

Compatibility: The ability of two or more substances combined with each other to form a homogeneous composition with useful plastic properties.

Composite: A homogeneous material created by the synthetic assembly of two or more materials (a selected filler or reinforcing elements and a compatible matrix binder) to obtain specific characteristics and properties. Composites are subdivided into the following classes on the basis of the form of the structural constituents. Fibrous:

dispersed phase consists of fibers; flake: dispersed phase consists of flat flakes; laminar: composed of layers of laminar constituents; particulate: dispersed phase consists of small particles; skeletal: composed of a continuous skeletal matrix filled by a second material.

Compression mold: A mold that is open when the material is introduced and shapes the material by heat and by pressure of closing.

Compression molding: A technique for molding thermoset plastics in which a part is shaped by placing the fiber and resin into an open mold cavity, closing the mold, and applying heat and pressure until the material has cured or achieved its final form.

Compression molding pressure: The unit pressure applied to molding material in a mold.

Compressive modulus, E_c: Ratio of compressive stress to compressive strain below the proportional limit. Theoretically equal to Young's modulus determined from tensile experiments.

Compressive strength: (1) The ability of a material to resist a force that intends to crush. (2) The crushing load at the failure of a specimen divided by the original sectional area of the specimen.

Compressive stress: The compressive load per unit area of the original cross section carried by the specimen during a compression test.

Condensation: A polymerization reaction in which simple by-products (e.g., water) are formed.

Conductivity: (1) Reciprocal of volume resistivity. (2) The conductance of a unit cube of any material.

Consolidation: A processing step that compresses fiber and matrix to reduce voids and achieve a desired density.

Contact adhesive: An adhesive that for satisfactory bonding requires the surfaces to be joined to be no farther apart than about 0.004 in. (0.1 mm).

Contact molding: A process for molding reinforced plastics in which reinforcement and resin are placed on a mold, cure is at room temperature using a catalyst-promoter system or by heat in an oven, and no additional pressure is used.

Contact-pressure resins: Liquid resins that thicken or polymerize on heating and require little or no pressure when used to bond laminates.

Continuous filament: An individual flexible glass rod of small diameter and of great or indefinite length.

Continuous-filament yarn: Yarn formed by twisting two or more continuous filaments into a single continuous strand.

Continuous roving: Parallel filaments coated with sizing, gathered together into single or multiple strands, and wound into a cylindrical package. It may be used to provide continuous reinforcement in woven roving, filament winding, pultrusion, prepregs, or high-strength molding compounds, or it may be used chopped.

Cooling fixture: A fixture used to maintain the shape or dimensional accuracy of a molding or casting after it has been removed from the mold and until the material is cool enough to hold its shape.

Copolymer: A composite of two or more polymer types.

Core: (1) The central member of a sandwich construction to which the faces of the sandwich are attached. (2) A channel in a mold for circulation of heat transfer media.

Corrosion: A phenomenon in which certain environments, e.g., salt water, attack and (through an electrochemical reaction) break down the thin oxide films that normally protect metals in air. Corrosion-resistant metals such as stainless steel contain sufficient chromium that a hard, tough film of Cr_2O_3 quickly forms at the surface, thereby protecting the metal under normal wear and tear.

Count: (1) For fabric the number of warp and filling yarns per inch in woven cloth. (2) For yarn, the size based on the relation between length and weight. Basic unit is a tex.

Coupling agent: Any chemical substance designed with two different types of fibers in individual yarns, e.g., thermoplastic fibers woven side by side with carbon fibers.

CP: (1) Cross-ply. (2) Resinous heat-shield composites.

CPI: Condensation-reaction polyimide.

Crack: An actual separation of molding material visible on opposite surfaces of the part and extending through the thickness; a fracture.

Crazing: Fine cracks which may extend in a network on or under the surface of plastic material.

Creel: A device for holding the required number of roving balls or supply packages in the desired postion for unwinding onto the next processing step.

Creep: The change in dimension of plastic under a load over a period of time, not including the initial instantaneous elastic deformation; at room temperature it is called cold flow.

Crimp: The waviness of a fiber; it determines the capacity of fibers to cohere under light pressure; measured either by the number of crimps or waves per unit length or by the percent increase in extent of the fiber on removal of the crimp.

Critical length: The minimum length of a fiber necessary for matrix shear loading to develop fiber ultimate strength by a matrix.

Critical loading: The maximum volume fraction of solid particles that can be incorporated in a polymer binder without forming pores.

Critical longitudinal stress (fibers): The longitudinal stress necessary to cause internal slippage and separation of a spun yarn; the stress necessary to overcome the interfiber friction developed as a result of twist.

Critical strain: The strain at the yield point.

Cross-laminated: Laminated so that some of the layers of material are oriented at right angles to the remaining layers with respect to the grain or strongest direction in tension. Balanced construction above the centerline of the thickness of the laminate is normally assumed. (See also Parallel-laminated).

Crosswise direction: Refers to cutting specimens and to application of load. For rods and tubes, crosswise is the direction perpendicular to the long axis. For other shapes or materials that are stronger in one direction than another, crosswise is the direction that is weaker. For materials that are equally strong in both directions, crosswise is an arbitrarily designed direction at right angles to the length.

Crystalline orientation: The term crystalline applies to sections of all chemical fibers consisting of alternate crystalline and amorphous (noncrystalline) regions. These regions are influenced by manufacturing conditions and can be controlled to some extent. Crystalline orientation implies the extent to which these crystalline regions can be aligned parallel to the fiber optics. Crystalline orientation is an important factor in determining the mechanical properties of carbon fibers.

Crystallinity: The quality of having a molecular structure with atoms arranged in an orderly three-dimensional pattern.

CSM: Chopped strand mat.

C stage: The final stage in the reaction of certain thermosetting resins in which the material is relatively insoluble and infusible; sometimes referred to as resite. The resin in a fully cured thermoset molding is in this stage. (See also A stage, B stage.)

CT: Computerized tomography.

CTBN: Carboxyl terminated butadiene acrylonitrile (an elastomer rubber).

CTE: Coefficient of thermal expansion.

Cull: Material remaining in a transfer chamber after the mold has been filled. (Unless there is a slight excess in the charge, the operator cannot be sure the cavity will be filled.)

Cure: To change the properties of a resin by chemical reaction, which may be condensation or addition; usually accomplished by the action of heat or catalyst, or both, and with or without pressure.

Curing agent: Hardener, a catalytic or reactive agent added to a resin to cause polymerization. Curing agents participate in the polymerization process. They may be latent-curable only at elevated temperatures, or they may be activated at room temperature (25°C).

Curing temperature: Temperature at which a cast, molded, or extruded product, a resin-impregnated reinforcement, an adhesive, etc., is subjected to curing.

Curing time: The length of time a part is subjected to heat or pressure, or both, to cure the resin; interval of time between the instant relative movement between the moving parts of a mold ceases and the instant pressure is released. (Further cure may take place after removal of the assembly from the condition of heat or pressure.)

CVD: Chemical vapor deposition.

CVI: Chemical vapor infiltration.

Cycle: The complete, repeating sequence of operations in a process or in part of a process. In molding, the cycle time is the elapsed time between a certain point in one cycle and the same point in the next.

D glass: A high-boron-content glass made especially for laminates requiring a precisely controlled dielectric constant.

Damage tolerance: A measure of the ability of structures to retain load-carrying capability after exposure to sudden loads (e.g., ballistic impact).

Damping (mechanical): The amount of energy dissipated as heat during the deformation of material. Perfectly elastic materials have no mechanical damping. Damping also diminishes the intensity of vibrations.

Daylight: The distance in the open position between the moving and fixed tables (platens) of a hydraulic press. For a multidaylight press, daylight is the distance between adjacent platens.

Debinding: A step between molding and sintering where the majority of the binder used in molding is extracted by heat, solvent, or another technique.

Debond: An unplanned nonadhered or unbonded region in an assembly.

Deep-draw mold: A mold having a core that is long in relation to the wall thickness.

Deflection temperature under load: The temperature at which a simple beam has deflected a given amount under load (formerly called heat distortion temperature).

Deformation under load: The dimensional change in a material under load for a specific time following the instantaneous elastic deformation caused by initial application of the load; also called cold flow or creep.

Degassing: See Breathing.

Delaminate: To split a laminated plastic material along the plane of its layers. (See also Laminate.)

Delamination: Physical separation or loss of bond between laminate plies.

Denier: A yarn and filament numbering system in which the yarn number is equal numerically to the weight in grams of 30,000 ft (9144 m) (used for continuous filaments). The lower the denier the finer the yarn.

Design allowable: A limiting value for a material property that can be used to design a structural or mechanical system to a specified level of success with 95% statistical confidence. B-basis allowable: material property exceeds the design allowable 90 times out of 100. A-basis allowable: material property exceeds the design allowable 99 times out of 100.

Dew point: A temperature that provides a relative measure of the moisture content in an atmosphere.

Die: The part or parts making up the confining form in which a powder is pressed.

Dielectric: A nonconductor of electricity.

Dielectric constant, ε: (1) The ratio of the capacity of a capacitor having a dielectric material between the plates to that of the same capacitor when the dielectric is replaced by a vacuum. (2) A measure of the electric charge stored per unit volume at unit potential.

Dielectric curing: Curing a synthetic thermosetting resin by passing an electric charge from a high-frequency generator through the resin.

Dielectric loss: The energy eventually converted into heat in a dielectric placed in a varying electric field.

Dielectric strength: The ability of material to resist the flow of an electric current.

Differential scanning calorimeter (DSC): Instrumentation for measuring chemical reactions by observing exothermic or endothermic (heat rise or heat input) reactions of materials, usually over a programmed temperature cycle.

Dimensional stability: The ability of a plastic part to retain the precise shape in which it was molded, cast, or otherwise fabricated.

Dislocation: A linear discontinuity in the perfection of a crystal. A sufficiently large external force acting on the crystal may cause the dislocation, or groups of dislocations, to move, and this is the basis of plastic deformation.

Displacement angle: In filament winding the distance of advance of the winding ribbon on the equator after one complete circuit.

DMOX: Direct melt oxidation.

Doctor roll: A device for regulating the amount of liquid material on the rollers of a spreader; also called a doctor bar.

Doily: In filament winding the planar reinforcement applied to a local area between windings to provide extra strength in an area where a cutout is to be made, e.g., port openings.

Dome: In filament winding the portion of a cylindrical container that forms the integral ends of the container.

Dopant, doping: A very small amount of an alloying addition that, when dissolved by a material, may modify its properties. An example is the doping of silicon by phosphorus or boron to make it semiconducting.

Doubler: Localized area of extra layers of reinforcement, usually to provide stiffness or strength for fastening or other abrupt load transfers.

DRA: Discontinuous reinforced aluminum.

Draft: The taper or slope of the vertical surfaces of a mold designed to facilitate removal of molded parts.

Draft angle: The angle between the tangent to the surface at that point and the direction of ejection.

Drape: The ability of preimpregnated broad goods to conform to an irregular shape; textile conformity.

Dry lay-up: Construction of a laminate by layering preimpregnated reinforcement (partly cured resin) in a female or male mold, usually followed by bag molding or autoclave molding.

Dry spot: (1) Of a laminate the area of incomplete surface film on laminated plastics. (2) In laminated glass an area over which the interlayer and the glass have not become bonded. (See also Resin-starved area.)

Dry winding: Filament winding using preimpregnated roving, as differentiated from wet winding. (See also Wet winding.)

DS: Directionally solidified.

DS: Dispersion-strengthened.

DSC: Differential scanning calorimeter.

Dwell: (1) A pause in the application of pressure to a mold, made just before the mold is completely closed, to allow gas to escape from the molding machine. (2) In filament winding the time the traverse mechanism is stationary while the mandrel continues to rotate to the appropriate point for the traverse to begin a new pass.

Dynamic compaction: Explosive or gas gun compaction of powder at shockwave velocities.

EATF: Externally applied thermal field.

Edgewise: Refers to cutting specimens and to the application of load. The load applied edgewise when it is applied to the edge of an original sheet or specimen. For compression-molded specimens of square cross section the edge is the surface parallel to the direction of motion of the molding plunger. For injection-molded specimens of square cross section this surface is selected arbitrarily; for laminates the edge is the surface perpendicular to the laminas. (See also Flatwise.)

EDM: Electron discharge machining.

EDS: Energy-dispersive spectroscopy.

EDX: Energy-dispersive x-ray spectroscopy.

E glass: A borosilicate glass; the type most used for glass fibers for reinforced plastics; suitable for electrical laminates because of its high resistivity. (Also called electric glass.)

Ejection: Removal of a molding from the mold impression by mechanical means, by hand, or by using compressed air.

Ejection ram: A small hydraulic ram fitted to a press to operate the ejection pins.

Elastic deformation: The part of the total strain in a stressed body that disappears upon removal of the stress.

Elasticity: The property of plastics materials by virtue of which they tend to recover their original size and shape after deformation.

Elastic limit: The greatest stress that a material is capable of sustaining without permanent strain remaining upon complete release of the stress. A material is said to have passed its elastic limit when the load is sufficient to initiate plastic (nonrecoverable) deformation.

Elastic recovery: The fraction of a given deformation that behaves elastically.

Elastic recovery = elastic extension/total extension

Elastic recovery = 1 for perfectly elastic material

0 for perfectly plastic material

Electroformed mold: A mold made by electroplating metal on the reverse pattern on the cavity.

Electrolytic powder: Powder produced by electrolytic deposition and subsequent pulverization. The resulting powder has a dendritic or sponge shape.

Elemental powder: Powder of a single chemical species, like iron, nickel, titanium, copper, or cobalt, with no alloying ingredients.

Elongation: Deformation caused by stretching; the fractional increase in length of a material stressed in tension. (When expressed as a percentage of the original gage length, it is called percentage elongation.)

Elutriation: Classification of powder particles by size using a rising stream of gas or liquid acting against the setting due to gravity; similar to air classification.

EMC: Elastomeric molding tooling compound.

EMI: Electromagnetic interference.

End: A strand of roving consisting of a given number of filaments gathered together (the group of filaments is considered an end or strand before twisting and a yarn after twisting has been applied); an individual warp yarn, thread, fiber, or roving.

End count: An exact number of ends supplied on a ball or roving.

Endothermic atmosphere: A reducing gas atmosphere used in sintering. It is produced by the reaction of hydrocarbon fuel gas and air over a catalyst with the aid of an external heat source. The resulting atmosphere is low in carbon dioxide and water vapor with relatively large percentages of hydrogen and carbon monoxide.

Endurance limit: See Fatigue limit.

Epitaxial films: A term identified with semiconductors, involving the build-up, layer by layer, of thin films with molecular dimensions. Applicable to polymers, ceramics, and metals.

Epoxy plastics: Plastics based on resins made by the reaction of epoxides or oxiranes with other materials such as amines, alcohols, phenols, carboxylic acids, acid anhydrides, or unsaturated compounds.

Equator: In filament winding the line in a pressure vessel described by the junction of the cylindrical portion and the end dome.

Equiaxed powder: Particles with approximately the same size in all three (perpendicular) dimensions.

Equilibrium: The stable state of a phase or structure. Stable states are illustrated for most binary and many ternary alloys in the form of equilibrium diagrams.

ERM: Elastic reservoir molding.

Erosion: The dissolution of a metal compact at the surface where a liquid infiltrant flows into the part.

ESR: Electron spin resonance.

Eutectic: An alloy having the composition indicated by the eutectic point on an equilibrium phase diagram.

Even tension: The process whereby each end of a roving is kept at the same degree of tension as the other ends making up that ball of roving. (See also Catenary.)

Exotherm: The liberation or evolution of heat during curing of a plastic product.

Exothermic atmosphere: A reducing gas atmosphere used in sintering, produced by partial or complete combustion of a hydrocarbon fuel gas and air. The maximum combustible content is 25%.

Expansion: The increase in dimensions of a compact due to unbalanced chemical reactions, pore formation, or pore growth during sintering.

Extrusion: The conversion of a billet into lengths of uniform cross section by forcing metal powder, or a metal powder and plastic mixture, through a die orifice of desired cross section.

Exudation: The expulsion of a low-melting-temperature alloying ingredient from a pore structure due to poor wetting.

Fabric: A material constructed of interlaced yarns, fibers, or filaments, usually planar.

Fabric, nonwoven: A material formed from fibers or yarns without interlacing.

Fabric, woven: A material constructed of interlacing yarns, fibers, or filaments.

Fabricating, fabrication: The manufacture of plastic products from molded parts, rods, tubes, sheeting, or extrusions, or another form of appropriate operation such as punching, cutting, drilling, or tapping. Fabrication includes fastening plastic parts together or to other parts by mechanical devices, adhesives, heat sealing, or other means.

Fan: In glass-fiber forming the fan shape made by the filaments between the bushing and the shoe.

Fatigue: The failure or decay of mechanical properties after repeated applications of stress. (Fatigue tests give information on the ability of a material to resist the development of cracks, which eventually bring about failure after a large number of cycles.)

Fatigue life: The number of cycles of deformation required to bring about failure of test specimens under a given set of oscillating conditions.

Fatigue limit: The stress below which a material can be stressed cyclically for an infinite number of times without failure.

Fatigue strength: (1) The maximum cyclic stress a material can withstand for a given number of cycles before failure occurs. (2) The residual strength after being subjected to fatigue.

FEA: Finite element analysis.

Felt: A fibrous material made from interlocked fibers by mechanical or chemical action, moisture, or heat; made from asbestos, cotton, glass, etc. (See also Batt.)

FEM: Finite element model or modeling or finite element method.

FGM: Functionally gradient material.

Fiber: A relatively short length of very small cross section of various materials made by chopping filaments (converting); also called a filament, thread, or bristle. (See also Staple fibers.)

Fiber-composite material: A material consisting of two or more discrete physical phases in which a fibrous phase is dispersed in a continuous-matrix phase. The fibrous phase may be macro-, micro-, or submicroscopic, but it must retain its physical identity so that it could conceivably be removed from the matrix intact.

Fiber diameter: The measurement of the diameter of individual filaments.

Fiber glass: An individual filament made by attenuating molten glass. (See also Continuous filament, Staple fibers.)

Fiber-matrix interface: The region separating the fiber and matrix phases, which differs from them chemically, physically, and mechanically. In most composite materials, the interface has a finite thickness (nanometers to thousands of nanometers) because of diffusion or chemical reactions between the fiber and matrix. Thus the interface can be more properly described by the term interphase or interfacial zone. When coatings are applied to the fibers or several chemical phases have well-defined microscopic thicknesses, the interfacial zone may consist of several interfaces. In this book interface is used to mean both interphase and interfacial zone.

Fiber orientation: Fiber alignment in a nonwoven or a mat laminate where the majority of fibers are in the same direction, resulting in a higher strength in that direction.

Fiber pattern: (1) Visible fibers on the surface of laminates or moldings. (2) The thread size and weave of glass cloth.

Fiber placement: A continuous process for fabricating composite shapes with complex contours and/or cutouts by means of a device that lays preimpregnated fibers (in tow form) onto a nonuniform mandrel or tool. It differs from filament winding in several ways: there is no limit on fiber angles; compaction takes place online via heat, pressure, or both; and fibers can be added and dropped as necessary. The process produces more complex shapes and permits a faster putdown rate than filament winding.

Filament: Any fiber whose aspect ratio (length to effective diameter) is for all practical purposes infinity, i.e., a continuous fiber. For a noncircular cross section, the effective diameter is that of a circle that has the same (numerical) area as the filament cross section. The smallest unit of a fibrous material. The basic units formed during drawing and spinning, which are gathered into strands of fiber (tows) for use in composites. Filaments usually are of extreme length and very small diameter, usually less than 25 μm (a mil). Normally filaments are not used individually.

Filaments: Individual glass fibers of indefinite length, usually as pulled from the stream of molten glass flowing through an orifice of the bushing. In the operation, a number of fibers are gathered together to make a strand or end of roving or yarn.

Filament weight ratio: In a composite material, the ratio of filament weight to the total weight of the composite.

Filament winding: A process for fabricating a composite structure in which continuous reinforcements (filament wire, yarn, tape, or other) impregnated with a matrix material either previously or during the winding are placed over a rotating removable form or mandrel in a prescribed way to meet certain stress conditions. Generally the shape is a surface of revolution, which may or may not include end closures. When the right number of layers has been applied, the wound form is cured and the mandrel removed.

Fill: Yarn oriented at right angles to the warp in a woven fabric.

Filler: A relatively inert material added to a plastic mixture to reduce cost, modify mechanical properties, serve as a base for color effects, or improve the surface texture. (See also Binder, Reinforced plastic.)

Fillet: A rounded filling for the internal angle between two surfaces of a plastic molding.

Filling yarn: The transverse threads or fibers in a woven fabric, i.e., fibers running perpendicular to the warp; also called the weft.

Fillout: See Lack of fillout.

Film adhesive: A synthetic resin adhesive usually of the thermosetting type in the form of a thin, dry film of resin with or without a paper carrier.

Fine-grained microstructure: A polycrystalline material with a fine grain size. Most crystalline materials are polycrystalline, comprising many small crystals separated by crystal or grain boundaries.

Finish: A material applied to the surface of fibers in a fabric used to reinforce plastics and intended to improve the physical properties of the reinforced plastic over those obtained using reinforcement without finish.

Flake powder: A flat or scalelike powder which is relatively thin and has a large aspect ratio.

Flame resistance: Ability of a material to extinguish flame once the source of heat is removed. (See also Self-extinguishing resin.)

Flame retardants: Chemicals used to reduce or eliminate the tendency of a resin to burn. (For polyethylene and similar resins, chemicals such as antimony trioxide and chlorinated paraffins are useful.)

Flame-retarded resin: A resin compounded with certain chemicals to reduce or eliminate its tendency to burn.

Flame spraying: Method of applying a plastic coating in which finely powdered fragments of the plastic, together with suitable fluxes, are projected through a cone of flame onto a surface.

Flammability: Measure of the extent to which a material can support combustion.

Flash: The portion of the charge that flows or is extruded from a mold cavity during molding; extra plastic attached to a molding along the parting line, which may be removed before the part is considered finished.

Flash mold: A mold designed to permit the escape of excess molding material; such a mold relies upon the back pressure to seal the mold and put the piece under pressure.

Flat lay: (1) The property of nonwarping in laminating adhesives. (2) An adhesive material with good noncurling and nondistention characteristics.

Flatwise: Refers to cutting specimens and the application of load. The load is applied flatwise when it is applied to the face of the original sheet or specimen.

Flexural modulus: The ratio, within the elastic limit, of the applied stress on a test specimen in flexure to the corresponding strain in the outermost fibers of the specimen.

Flexural rigidity: (1) For fibers, a measure of the rigidity of individual strands or fibers; the force couple required to bend a specimen to the unit radius of curvature. (2) For plates, the measure of rigidity is $D = EI$, where E is the modulus of elasticity and I is the moment of inertia, or

$$d = Eh^2/12(1 - \nu) \text{ in./lb}$$

where E = modulus of elasticity, h = thickness of plate, and ν = Poisson's ratio.

Flexural strength: (1) The resistance of a material to breakage by bending stresses. (2) the strength of a material in bending expressed as the tensile stress of the outermost fibers of a bent test sample at the instant of failure. For plastics, this value is usually higher than the straight tensile strength. (3) the unit resistance to the maximum load failure by bending, usually in kips per square inch (megapascals).

Flow: The movement of resin under pressure, allowing it to fill all parts of a mold; flow or creep is the gradual but continuous distortion of a material under continued load, usually at high temperatures.

Foamed plastics: Resins in sponge form; may be flexible or rigid; cells may be closed or interconnected and density anywhere from that of the solid parent resin to 2 lb/ft^3 (32 kg/m^3).

Foam in place: Foam deposition requiring the foaming machine to be brought to the work (as opposed to bringing the work to the foaming machine).

FOD: Foreign object damage.

Force: (1) The male half of a mold, which enters the cavity, exerting pressure on the resin and causing it to flow (also called punch). (2) Either part of a compression mold (top force and bottom force).

Forging: The plastic deformation of a metal into a desired shape. The deformation is usually performed hot at high strain rates and may be done in constraining dies.

FP: Polycrystalline alumina fiber.

Fracture: Rupture of the surface without complete separation of a laminate.

Fracture toughness: A measure of the damage tolerance of a material containing initial flaws or cracks.

FRAT: Fiber-reinforced advanced titanium.

FRP: Fibrous glass-reinforced plastic, any type of plastic-reinforced cloth, mat, strands, or any other form of fibrous glass.

Furan resins: Dark-colored thermosetting resins available primarily as liquids ranging from low-viscosity polymers to thick, heavy syrups, which cure to highly cross-linked, brittle substances. Made primarily by polycondensation of furfuryl alcohol in the presence of strong acids, sometimes in combination with formaldehyde or furfuryldehyde.

Gage length: Length over which deformation is measured.

γ-TiAl: Gamma titanium alumide.

Gap: In filament winding the space between successive windings, which are usually intended to lie next to each other.

Gas atomized powder: A rounded or spherical powder formed by the disintegration of a melt stream by the gas expansion nozzle. The particles solidify during free flight after atomization.

Gel: A semisolid system consisting of a network of solid aggregates in which liquid is held; the initial jellylike solid phase that develops during formation of a resin from a liquid.

Gelation time: For synthetic thermosetting resins the interval of time between introduction of a catalyst into a liquid adhesive system and gel formation.

Gel coat: A resin applied to the surface of a mold and gelled before lay-up. (The gel coat becomes an integral part of the finished laminate and is usually used to improve surface appearance, etc.)

Gel point: The stage at which a liquid begins to exhibit pseudoelastic properties, also conveniently observed from the inflection point on a viscosity-time plot. Also called gel time.

Geodesic: The shortest distance between two points on a surface.

Geodesic isotensoid: Constant-stress level in any given filament at all points on its path.

Geodesic-isotensoid contour: In filament-wound reinforced plastic pressure vessels a dome contour in which the filaments are placed on geodesic paths so that they exhibit uniform tension throughout their lengths under pressure loading.

Geodesic ovaloid: A contour for end domes, the fibers forming a geodesic line. The forces exerted by the filaments are proportioned to meet hoop and meridional stresses at any point.

Glass: An inorganic product of fusion that has cooled to a rigid condition without crystallizing. Glass is typically hard and relatively brittle and undergoes conchoidal fracture.

Glass fiber: A glass filament that has been cut to a measurable length. Staple fibers of relatively short length are suitable for spinning into yarn.

Glass filament: A form of glass that has been drawn to a small diameter and extreme length. Most filaments are less than 0.005 in. (0.13 mm) in diameter.

Glass filament bushing: The unit through which molten glass is drawn in making glass filaments.

Glass finish: A material applied to the surface of a glass reinforcement to improve its effect upon the physical properties of the reinforced plastic; also called a bonding agent.

Glass flake: Thin, irregularly shaped flakes of glass typically made by shattering a continuous thin-walled tube of glass.

Glass former: An oxide that forms a glass easily; also one that contributes to the network of silica glass when added to it.

Glass stress: In a filament-wound part, usually a pressure vessel, the stress calculated using only the load and the cross-sectional area of the reinforcement.

Glass transition: The reversible change in an amorphous polymer or in amorphous regions of a partially crystalline polymer from (or to) a viscous or rubbery condition to (or from) a hard and relatively brittle state. The glass transition generally occurs over a relatively narrow temperature region and is similar to the solidifaction of a liquid to a glassy state; it is not a phase transition. The glass transition temperature is the approximate midpoint of the temperature range over which glass transition takes place.

Glass-transition temperature T_g: The approximate temperature at which increased molecular mobility results in significant changes in properties of a cured resin. The measured value of T_g can vary, depending on the test method. T_g = temperature resistance.

Glass volume percent: The product of the specific gravity of a laminate and the percentage of glass by weight divided by the specific gravity of the glass.

GPa: Gigapascal. Young's modulus is expressed in GPa.

Grain: An individual crystal within polycrystalline (multigrain) material, in which adjacent grains have substantially different orientations.

Grain boundary: Each grain in a polycrystalline material is separated by a grain boundary about two atoms wide.

Green density: The powder density after compaction.

Green strength: The strength of an as-pressed powder compact.

Greige: Fabric before finishing; yarn or fiber before bleaching or dyeing. Also called gray goods, greige goods, greige gray.

Growth: The increase in dimensions of a powder compact which may occur during sintering; the opposite dimensional change from shrinkage.

GRP: Graphite-reinforced plastic.

Guide pin: A pin that guides mold halves into alignment on closing.

Guide pin bushing: The bushing through which the guide pin moves when a mold is closed.

Gusset: A piece used to give added size or strength at a particular location on an object; the folded-in portion of a flattened tubular film.

Hand: The softness of a piece of fabric, as determined by touch (individual judgment).

Hand lay-up: The process of placing (and working) successive plies of reinforcing material or resin-impregnated reinforcement in position on a mold by hand, which is then cured to the formed shape.

Hardener: (1) A substance or mixture added to a plastic composition to promote or control the curing action by taking part in it. (2) A substance added to control the degree of hardness of a cured film. (See also Catalyst.)

Hard metal: Another name for cemented carbides which reflects their high hardness after sintering.

Hardness: The resistance to surface indentation, usually measured by the depth of penetration (or arbitrary units related to depth of penetration) of a blunt point under a given load using a particular instrument according to a prescribed procedure. (See also Barcol hardness, Rockwell hardness number.)

Hat: A member in the shape of a hat.

HCD: High ceramic density.

HCM: Hybrid composite material.

Heat-affected zone: When metal is fusion-welded, the areas near the hot part of the weld metal are heated. This heating cycle can cause substantial changes in microstructure (e.g., grain size) and properties.

Heat buildup: The temperature rise in a part resulting from the dissipation of applied strain energy as heat.

Heat-convertible resin: A thermosetting resin convertible by heat to an infusible, insoluble mass.

Heat-distortion temperature: The temperature at which a test bar deflects a certain amount under a specified temperature and a stated load.

Heat resistance: The property or ability of plastics and elastomers to resist the deteriorating effects of elevated temperatures.

Heavy alloy: A range of high-density, machinable alloys nominally based on tungsten with small concentrations of alloying additions such as nickel, iron, manganese, or copper. The alloys are liquid phase-sintered from mixed elemental powders to create a composite material.

HEHR: High-energy high-rate consolidation.

High-modulus organic fibers (HMOFs): A new generation of organic fibers possessing unusually high tensile strength and modulus but generally poor compressive and transverse properties. Unlike conventional fibers that have a chain-folded structure, all HMOF fibers have an oriented chain-extended structure. One approach to producing such a polymer is to make a structurally modified, highly oriented version of a "conventional" polymer such as high-modulus polyethylene. A second approach is to synthesize polymers that have inherently ordered rigid molecular chains in an extended conformation. Examples of the latter are liquid crystalline aramids and liquid crystalline polyesters.

High-pressure laminates: Laminates molded and cured at pressures not lower than 1 kip/in.^{2} (7 MPa) and more commonly at 1.2–2 kips/in.^{2} (8.3–13.8 MPa).

High-pressure molding: A molding process in which the pressure used is greater than 1 kip/in.^{2} (7 MPa).

HIP: Hot isostatic pressing.

HITEMP: High temperature engine and materials program sponsored by NASA Lewis Research Center.

HM: High-modulus.

HMC: High-strength molding compound.

HME: High-vinyl-modified epoxy.

HMOFs: High-modulus organic fibers.

HNDT: Holographic nondestructive testing.

Honeycomb: Manufactured product of resin-impregnated sheet material (paper, glass, fabric, etc.) or sheet metal formed into hexagonal cells; used as a core material in sandwich construction.

Hoop stress: The circumferential stress in a material of cylindrical form subjected to internal and external pressure.

Hot isostatic pressing (HIP): A process for densifying ceramic or metallic powders combining temperature and high-pressure gas to densify a material into a net-shape component using a pressure-tight outer envelope.

Hot pressing: The high-pressure, low-strain-rate forming of a compact at a temperature high enough to induce sintering and creep processes.

Hot working: The plastic deformation of a metal at a temperature and strain rate that minimizes work hardening.

HPZ: Amorphous Si_3N_4.

HSCT: High-speed civil transport.

HTS: High-temperature strength.

HTS: High tensile strength.

Hybrid: The result of attaching a composite body to another material such as aluminum, steel, etc., or two reinforcing agents in the matrix such as graphite and glass.

Hybrid composite: A composite laminate consisting of laminas of two or more composite material systems. A combination of two or more different fibers, such as carbon and glass or carbon and aramid, into a structure. Tapes, fabrics, and other forms may be combined; usually only the fibers differ.

Hydraulic press: A press in which the molding force is created by the pressure exerted on a fluid.

Hydrophilic: Capable of absorbing or absorbing water.

Hydrophobic: Capable of repelling water.

Hydroscopic: Capable of absorbing and retaining atmospheric moisture.

Hymat: Hybrid material.

ICP-AES: Inductively coupled argon plasma atomic emission spectroscopy.

IFAC: Integrated-flexible automation center.

IHPTET: Initiative for high-performance turbine engine technology.

IITRI: Illinois Institute of Technology Research Institute.

ILC: Integrated-laminating center.

ILSS: Interlaminar shear strength.

IM: Ingot metallurgy.

IM: Intermediate modulus.

IMC: Intermetallic matrix composite.

Impact strength: The ability of a material to withstand shock loading; the work done in fracturing a test specimen in a specified manner under shock loading.

Impact toughness: Toughness refers to a material's resistance to crack growth. Impact toughness is its crack growth resistance under conditions of impact or rapid loading.

Impregnate: In reinforced plastics to saturate the reinforcement with a resin.

Impregnated fabric: A fabric impregnated with a synthetic resin. (See also Prepeg.)

Induction furnace: In the carbon fiber process a carbonization furnace that utilizes induction heating to eliminate heteroatoms from the carbon fiber structure. Heat distribution is obtained by a combination of induced heat to a furnace muffle or susceptor within the induction coil and radiation from the interior surfaces of the furnace chamber in temperatures in excess of 3000°C. Sometimes an induction furnace is used in combination with a resistance furnace to achieve the desired pyrolysis temperature (i.e., the resistance furnace may be used to expose the fiber to a temperature as high as 2000°C, and if higher temperatures are required, as may be the case in graphitization, the fiber can then be passed through an induction furnace).

Inert filler: A material added to a plastic to alter the end item properties through physical rather than chemical means.

Infrared: The part of the electromagnetic spectrum between the visible light range and the radar range; radian heat in this range and infrared heaters are much used in sheet thermosetting.

Inhibitor: A substance that retards a chemical reaction; used in certain types of monomers and resins to prolong storage life.

Initial modulus: See Modulus of elasticity.

Injection molding: A technique developed for processing thermoplastics which are heated and forced under pressure into closed molds and cooled.

Inorganic: Designating or pertaining to the chemistry of all elements and compounds not classified as organic; matter other than animal or vegetable, such as earthy or mineral matter. (See also Organic.)

Insert: An integral part of a plastic molding consisting of metal or other material which may be molded into position or pressed into the molding after the molding is complete.

Insert pin: A pin that keeps an inserted part (insert) inside a mold by screwing or friction; it is removed when the object is being withdrawn from the mold.

Instron: An instrument used to determine the tensile and compressive properties of materials.

Insulating resistance: The electric resistance between two conductors or systems of conductors separated by only insulating material.

Insulator: (1) A material of such low electric conductivity that the flow of current through it can usually be neglected. (2) A material of low thermal conductivity.

Interface: The junction point or surface between two different media; on glass fibers, the contact area between glass and sizing or finish; in a laminate, the contact area between the reinforcement and the laminating resin.

Interlaminar: Existing or occurring between two or more adjacent laminas.

Interlaminar shear: The shearing force tending to produce displacement between two laminas along the plane of their interface; usually the weakest element of a composite.

Interlaminar shear strength: The maximum shear stress existing between layers of a laminated material.

Internal stress: Stress created within an adhesive layer by movement of the adherends at differential rates or by contraction or expansion of the adhesive layer. (See also Stress.)

Interply hybrid: A reinforced plastic laminate in which adjacent laminas are composed of different materials.

Interstitial: A nonmetallic atom in an alloy or compound small enough to occupy sites in the lattice between the metal atoms. In a dilute solution of interstitials in a metal, such as carbon in iron, interstitial atoms can migrate (diffuse) without the need for vacancies.

Ionic bond: Atomic bonding due to electrostatic attraction between charged particles (ions) resulting from the transfer of electrons.

Ion implantation: A sophisticated process in which ions are impacted into the surface of samples, causing modifications in properties of the surface layers. Since the process occurs at low temperatures, the implanted ions cause extreme distortion in the lattice structure of the substrate material, greatly affecting its properties.

IPN: Interpenetrating polymer network.

IPS: Integrated process system.

IR: Infrared.

Irreversible: (1) Not capable of redissolving or remelting. (2) Descriptive of chemical reactions that proceed in a single direction and are not capable of reversal (as applied to thermosetting resins).

ISB: Indentation strength-in-bending method (fracture toughness test method).

Isocyanate plastics: Plastics based on resins made by the condensation of organic isocyanates with other compounds. (See also Urethane plastics.)

Isostatic pressing: The compaction of a powder by subjecting it to equal pressure from all directions; hydrostatic compaction.

Isotropic laminate: A laminate in which the strength properties are equal in all directions.

Izod impact test: A destructive test designed to determine the resistance of a plastic to the impact of a suddenly applied force.

Joint: The location at which two adherends are held together with a layer of adhesive; the general area of contact for a bonded structure. Butt joint: the edge faces of the two adherends are at right angles to the other faces of the adherends. Scarf joint: a joint made by cutting away similar angular segments of two adherends and bonding them with the cut areas fitted together. Lap joint: a joint made by placing one adherend partly over another and bonding together the overlapping portions.

K polymers: Thermoplastic PIs.

Lack of fillout: Characteristic of an area, occurring usually at the edge of a laminated plastic, where the reinforcement has not been wetted with resin.

Lacquer: Solution or natural or synthetic resins in readily evaporating solvents, used as a protective coating.

Lamina: A single ply or layer in a laminate, which is made up of a series of layers.

Laminate: (1) To unite sheets of material by a bonding material usually with pressure and heat (normally used with reference to flat sheets). (2) A product made by so bonding. (See also Bidirectional laminate, Unidirectional laminate.) A composite material made up of a single ply or layer or series of layers with each layer consisting of a reinforcing fiber imbedded in a matrix. Each ply or layer is oriented in a predetermined manner in order to maximize the properties of the laminate.

Laminated molding: A molded plastic article produced by bonding together, under heat and pressure in a mold, layers of resin-impregnated laminating reinforcement; also called laminated plastic.

Laminate ply: One layer of a product that is evolved by bonding together two or more layers of materials.

Land: The portion of a mold that provides separation or cutoff of the flash from the molded article.

Lap: In filament winding the amount of overlay between successive windings, usually intended to minimize gapping.

Lap joint: See Joint.

LARC-TPI: Langley Research Center thermoplastic imide.

LAS: Lithium aluminosilicate.

Lattice: The geometrically perfect network structure on which crystalline materials are based.

Lay: (1) In glass fiber the spacing of the roving bands on the roving package expressed as the number of bands per inch. (2) In filament winding the orientation of the ribbon with some reference, usually the axis of rotation.

Lay-flat: See Flat lay.

Lay-up: (1) As used in reinforced plastics, reinforcing material placed in position in a mold. (2) The process of placing reinforcing material in position in a mold. (3) Resin-impregnated reinforcement. (4) The component materials, geometry, etc., of a laminate.

LCP: Liquid crystal polymer.

Lengthwise direction: Refers to cutting specimens and the application of loads. For rods and tubes, lengthwise is the direction of the long axis. For other shapes of materials that are stronger in one direction than in the other, lengthwise is the direction that is stronger. For materials that are equally strong in both directions, lengthwise is an arbitrarily designated direction that may be with the grain, direction of flow in manufacture, longer direction, etc. (See also Crosswise direction.)

LEO: Low earth orbit.

LHS: Low-cost high-strength.

LIM: Liquid injection molding.

Linear expansion: The increase in a planar dimension, measured by linear elongation of a sample in the form of a beam that is exposed to two given temperatures.

Liner: In a filament-wound pressure vessel the contiguous, usually flexible, coating on the inside surface of the vessel used to protect the laminate from chemical attack or to prevent leakage under stress.

Liquid crystal: Liquid crystal materials are usually made up of rigid, rodlike molecules. They can become ordered in either solution or melt phase, which means that the molecules aggregate under certain conditions so that the materials are anisotropic. It is very easy to achieve high orientation with these materials during either liquid or melt extrusion, so the resulting extrudate (e.g., fiber or film) has excellent tensile properties.

Liquid-crystal polymers (LCP): A newer type of thermoplastic, melt-processible, with high orientation in molding, improved tensile strength, and high-temperature capability.

Liquid phase sintering: Sintering at room temperature where a liquid and solid coexist because of chemical reactions, partial melting, or eutectic liquid formation.

Liquidus: The maximum temperature at which equilibrium exists between molten glass and its primary crystalline phase.

LMC: Low-pressure molding compound.

Load deflection curve: A curve on which the increasing flexural loads are plotted on the ordinate and the deflections caused by these loads are plotted on the abscissa.

Longos: Low-angle helical or longitudinal windings.

Loop tenacity: The strength value obtained by pulling two loops, like two links in a chain, against each other in order to test whether a fibrous material will cut or crush itself; also called loop strength.

Loss on ignition: Weight loss, usually expressed as a percentage of the total, after burning off an organic sizing from glass fibers or an organic resin from a glass-fiber laminate.

Lot: A specific amount of material produced at one time and offered for sale as a unit quantity.

Low-pressure laminates: In general, laminates molded and cured at pressures from 0.4 kips/in.^{2} (2.8 MPa) down to and including pressure obtained by the mere contact of plies.

Low-pressure molding: The distribution of relatively uniform low pressure [0.2 kip/in.^{2} (1.4 MPa) or less] over a resin-bearing fibrous assembly of cellulose, glass, asbestos, or other material, with or without application of heat from an external source, to form a structure possessing definite physical properties.

LST: Large space telescope.

LWV: Lightweight vehicle.

MA: Mechanical alloying.

Macerate: (1) To chop or shred fabric for use as a filler for a molding resin. (2) The molding compound obtained when so filled.

Mandrel: (1) The core around which paper-, fabric-, or resin-impregnated glass is wound to form pipes, tubes, or vessels. (2) In extrusion the central finger of a pipe or tubing die.

MAS: Magnesium aluminosilicate.

Mat: A fibrous material for reinforced plastic consisting of randomly oriented chopped filaments or swirled filaments with a binder, available in blankets of various widths, weights, and lengths.

Mat binder: Resin applied to glass fiber and cured during the manufacture of mat, used to hold the fibers in place and maintain the shape of the mat.

Matched metal molding: A reinforced-plastic manufacturing process in which matching male and female metal molds are used (similar to compression molding) to form the part, as opposed to low-presure laminating or spray-up.

Matrix: The dominating lattice structure of a given phase in a material; the essentially homogeneous resin or polymer material in which the fiber system of a composite is imbedded. Both thermoplastic and thermoset resins may be used, as well as metals, ceramics, and glasses.

Matrix metal: The continuous phase in a polyphase alloy or mechanical mixture; the physically continuous metallic constituent in which separate particles of another constituent are embedded.

Matte: A nonspecular surface having diffused reflective powers.

MDA: Methylenediamine.

Mechanical adhesion: Adhesion between surfaces in which the adhesive holds the parts together by interlocking action.

Mechanical alloying (MA): The formation of an alloy powder by milling elemental powders for a prolonged time; frequently used to create amorphous or dispersion-strengthened alloy powders.

Melamine plastics: Plastics based on melamine resins.

Mesh: The screen number representing the number of wires per inch in a square grid. The higher the mesh number, the smaller the opening size.

Mesophase: An intermediate phase in the formation of carbon fiber from a pitch precursor. It is a liquid crystal phase in the form of microspheres, which, upon prolonged heating above 400°C, coalesce, solidify, and form regions of extended order. Heating above 2000°C leads to the formation of a graphitelike structure.

Metallic bond: The interatomic chemical bond in metals characterized by delocalized electrons in the energy bands. The atoms are considered to be ionized, with the positive ions occupying the lattice positions. The valence electrons are free to move. The bonding force is the electrostatic attraction between ions and electrons.

Metallic fiber: Manufactured fiber composed of metal, plastic-coated metal, metal-coated plastic, or a core completely covered by metal.

M glass: A high-beryllia-content glass designed especially for high modulus of elasticity.

Micron, μ: A unit of length replaced by the micrometer (μm); 1 $\mu m = 10^{-6}$ m = 10^{-3} mm = 0.00003937 in. = 39.4 μin.

Microstructure: A structure with heterogeneities that can be seen through a microscope.

Mil: The unit used in measuring the diameter of glass fiber strands, wire, etc. (1 mil = 0.001 in.).

Milled fibers: Continuous glass strands hammer-milled into small modules of filamentized glass. Useful as anticrazing reinforcing fillers for adhesives.

MIMLC: Microinfiltrated microlaminated composite.

MMC: Metal matrix composite; material in which continuous carbon, silicon carbide, or ceramic fibers are embedded in a metallic matrix.

Modulus: A number that expresses a measure of some property of a material, e.g., modulus of elasticity, shear modulus; a coefficient of numerical measurement of a property. Using “modulus” alone without modifying terms is confusing and should be discouraged. For example, Young’s modulus (*E*) refers to the constant of elasticity, or stiffness, of a material.

Modulus in compression: See Compressive modulus.

Modulus in flexure: See Flexural modulus.

Modulus, initial, or Young's modulus: See Modulus of elasticity.

Modulus in shear: See Shear modulus.

Modulus in tension: See Tensile modulus.

Modulus of elasticity: The ratio of the stress or applied load to the strain or deformation produced in a material that is elastically deformed. If a tensile strength of 2 kips/in.^{2} (14 MPa) results in an elongation of 1%, the modulus of elasticity is 2/0.01 = 200 kips/in.^{2} (1379 MPa); also called Young's modulus.

Modulus of elasticity in torsion: The ratio of the torsion stress to the strain in a material over the range for which this value is constant.

Modulus of rigidity: See Flexural rigidity.

Modulus of rupture: See Flexural strength.

MOE: Modulus of elasticity.

Mohs hardness: A measure of the scratch resistance of a material; the higher the number, the greater the scratch resistance (for diamond it is 10).

Moisture absorption: The pickup of water vapor from air by a material; it relates only to vapor withdrawn from the air by a material and must be distinguished from water absorption. (See also Water absorption.)

Mold: (1) The cavity or matrix in or on which a plastic composition is placed and from which it takes form. (2) To shape plastic parts or finished articles by heat or pressure. (3) The assembly of all parts that function collectively in the molding process.

Molding: The shaping of a plastic composition in or on a mold, normally accomplished under heat and pressure; sometimes used to denote the finished part.

Molding compounds: Plastics in a wide range of forms (especially granules or pellets) to meet specific processing requirements.

Molding cycle: (1) The time required for the complete sequence of operations on a molding press to produce one set of moldings. (2) The operations necessary to produce a set of moldings without reference to time.

Molding press: A press used to form compacts from powder.

Molding pressure: The pressure applied to the ram of an injection machine or press to force the softened plastic to fill the mold cavities completely. (See also Compression molding pressure.)

Molding, pressure bag: See Pressure bag molding.

Molding pressure, compression: See Compression molding pressure.

Mold-release agent: A liquid or powder used to prevent sticking of molded articles in a cavity.

Mold seam: Line on a molded or laminated piece, differing in color or appearance from the general surface, caused by the parting line of the mold.

Mold shrinkage: (1) The immediate shrinkage a molded part undergoes when it is removed from the mold and cooled to room temperature. (2) The difference in dimensions, expressed in inches per inch (millimeters per millimeter) between a molding and the mold cavity in which it was molded (at normal temperature measurement). (3) The incremental difference between the dimensions of a molding and the mold from which it was made, expressed as a percentage of the dimension of the mold.

Monofilament: (1) A single fiber or filament of indefinite length generally produced by extrusion. (2) A continuous fiber sufficiently large to serve as yarn in normal textile operations; also called monofil.

Monomer: (1) A simple molecule capable of reacting with like or unlike molecules to form a polymer. (2) The smallest repeating structure of a polymer; also called a mer.

MOR: Modulus of rupture.

Morphology: The overall form of a polymer structure, i.e., crystallinity, branching, molecular weight, etc. Also, the study of the fine structure of a fiber or other material, such as basal plane orientation across a carbon or graphite fiber.

MPa: Megapascal; tensile strength and stress pressure are expressed in these units.

MT: Microstructural-toughened.

Multicircuit winding: In filament winding a winding that requires more than one circuit before the band repeats by lying adjacent to the first band.

Multidimensional: In order to analyze differences in shape precisely, more than one shape aspect must be monitored. Depending on the number of aspects considered using mutually independent shape parameters, two-, three-, or multidimensional representation of shape can be useful for comparison, and cluster analysis may be appropriate to quantify the significance of differences.

Multifilament yarn: A multitude of fine, continuous filaments (often 5 to 100), usually with some twist in the yarn to facilitate handling. Sizes range from 5 to 10 denier up to a few hundred denier. Individual filaments in a multifilament yarn are usually about 1 to 5 denier.

Multiple-cavity mold: A mold with two or more mold impressions, i.e., a mold that produces more than one molding per molding cycle.

Nanoscale: Refers to powders or microstructures with sizes that can be measured in nanometers. Typically the powders are less than 100 nm in size.

NASA: National Aeronautics and Space Administration.

NASAIR 100: Single-crystal superalloy.

NASP: National aerospace plane.

NASTRAN: Numerous FE analysis codes are commercially available, such as NASTRAN, ADINA, ANSYS, and so forth. These codes provide built-in elements for special applications, e.g., truss and beam elements, two-dimensional solid plane stress or plane strain elements, plate and shell elements, axisymmetric shell or solid elements, three-dimensional solid or "brick" elements, and crack-tip elements.

NDI: Nondestructive inspection.

NDT: Nondestructive testing.

Near-net shape (NNS): A compact that has the general shape of the final product but is oversized and requires final machining.

Needles: Elongated, rodlike particles.

Nesting: In reinforced plastics placing plies of fabric so that the yarns of one ply lie in the valleys between the yarns of the adjacent ply (nested cloth).

Net shape: A compact manufactured to the final density and dimensions without the need for machining.

Netting analysis: The analysis of filament-wound structures which assumes that the stresses induced in the structure are carried entirely by the filaments, the strength of the resin being neglected, and that the filaments possess no bending or shearing stiffness, carrying only the axial tensile loads.

NISA: A general-purpose FEA package; also performs static, dynamic, buckling, and heat transfer analysis of laminated composite structures.

NNS: Near-net shape.

Nol ring: A parallel filament-wound test specimen used for measuring various mechanical strength properties of the material by testing the entire ring or segments of it.

Nondestructive inspection (NDI): A process or procedure for determining material or part characteristics without permanently altering the test subject. Nondestructive testing (NDT) is broadly considered synonymous with NDI.

Nonhygroscopic: Not absorbing or retaining an appreciable quantity of moisture from the air (water vapor).

Nonrigid plastic: A plastic that has a stiffness or apparent modulus of elasticity not over 10 kips/in.^{2} (69 MPa) at 73.4°F (23°C).

Nonwoven fabric: A planar structure produced by loosely bonding together yarns, roving, etc. (See also Fabric.)

Notch sensitivity: The extent to which the sensitivity of a material to fracture is increased by the presence of a surface inhomogeneity such as a notch, a sudden change in section, a crack, or a scratch. Low notch sensitivity is usually associated with ductile materials, and high notch sensitivity with brittle materials.

Novolac: A phenolic-aldehyde resin that remains permanently thermoplastic unless a source of methylene groups is added; a linear thermoplastic B-staged phenolic resin. (See also Thermoplastic.)

OEM: Original equipment manufacturer.

Offset yield strength: The stress at which the strain exceeds by a specific amount (the offset) an extension of the initial proportional portion of the stress-strain curve.

OFHC: Oxygen-free high conductivity.

OML: Outer mold line.

Open-cell foamed plastic: A cellular plastic in which there is a predominance of interconnected cells.

Orange peel: An uneven surface resembling that of an orange peel; said of injection moldings with unintentionally rugged surfaces.

Order: Usually refers to the structure of an alloy of stoichiometric composition. For example, compounds such as aluminum oxide (Al_2O_3) or titanium nitride (TiN) are perfectly ordered, each component species forming its own lattice structure and with the combination forming a superlattice structure. It should be noted, however, that in the present text order is sometimes used to describe the symmetric crystalline state of a metal as opposed to an unordered amorphous state.

Organic: Designating or composed of matter originating in plant or animal life or composed of chemicals of hydrocarbon origin, natural or synthetic.

Oriented materials: Materials, particularly amorphous polymers and composites, whose molecules and/or macroconstituents are aligned in a specific way. Oriented materials are anisotropic. Orientation is generally uniaxial or biaxial.

Orthotropic: Having three mutually perpendicular planes of elastic symmetry.

Out life: The period of time a prepreg material remains in handleable form and with properties intact outside the specified storage environment; e.g., out of the freezer in the case of thermoset prepregs.

Overcuring: The beginning of thermal decomposition resulting from too high a temperature or too long a molding time.

Overflow groove: Small groove used in molds to allow material to flow freely to prevent weld lines and low density and to dispose of excess material.

Overlap: A simple adhesive joint in which the surface of one adherend extends past the leading edge of another.

Overlay sheet: A nonwoven fibrous mat (in glass, synthetic fiber, etc.) used as the top layer in a cloth or mat lay-up to provide a smoother finish or minimize the appearance of the fibrous pattern.

PA: Polyamide.

Package: The method of supplying roving or yarn.

PAEK: Polyarylether ketone.

PAES: Polyarylether sulfone.

PAI: Polyamide-imide.

PAN: Polyacrylonitrile.

Parallel-laminated: Laminated so that all the layers of material are oriented approximately parallel with respect to the grain or the strongest direction in tension. (See also Cross-laminated.)

Parameter: An arbitrary constant, as distinguished from a fixed or absolute constant. Any desired numerical value may be given as a parameter.

Particle size: The controlling linear dimension of an individual particle as determined by analysis with screens or other instruments.

Particle size analyzer: An automated device for determining the particle size distribution.

Particulate: Fine particles, as distinct from fibers.

Parting agent: See Release agent.

Parting line: The linear mark on a compact where two separate tool or die pieces mated during shaping. In injection molding it is where the two halves of the die joined together.

PAS: Polyarylsulfone.

PBBI: Polybutadiene bisimide.

PBI: Polybenzimidazole.

PBO: Polyphenylene benzobisoxazole.

PBT: Polybutylene terephtalate; polybenzothiazole.

PBZ: Polybenzazole.

PC: Polycarbonate.

PDIPN: Pseudointerpenetrating network.

PE: Polyethylene.

PEEK: Polyetherether ketone.

Peel ply: The outside layer of a laminate which is removed or sacrificed to achieve improved bonding of additional plies and leaves a clean resin-rich surface ready for bonding.

Peel strength: Bond strength, in pounds per inch of width, obtained by peeling the layer. (See also Bond strength.)

PEI: Polyetherimide.

PEK: Polyether ketone.

PEKEKK: Polyether ketone ether ketone ketone.

PEKK: Polyetherketone ketone.

Permanence: The property of a plastic that descibes its resistance to appreciate change in characteristics with time and environment.

Permanent set: The deformation remaining after a specimen has been stressed in tension a prescribed amount for a definite period and released for a definite period.

PES: Polyethersulfone.

PET: Polyethylene terephthalate.

Phase: A microstructural constituent of a given lattice structure or type. A material from one to several coexisting phases.

Phenolic, phenolic resin: A synthetic resin produced by the condensation of an aromatic alcohol with an aldehyde, particularly of phenol with formaldehyde. (See also A stage, B stage, C stage, Novolac.)

Physical vapor deposition (PVD): A process for depositing extremely thin layers of material onto a substrate by passing a vapor or gaseous phase over it.

PI: Polyimide.

Pick: (1) An individual filling yarn, running the width of a woven fabric at right angles to the warp; also called fill, woof, weft. (2) To experience tack. (3) To transfer unevenly from an adhesive applicator mechanism because of high surface tack.

Pinch-off: In blow molding a raised edge around the cavity in the mold which seals off the part and separates the excess material as the mold closes around the parison.

Pinhole: A tiny hole in the surface of, or through, a plastic material, usually not occurring alone.

Pin, insert: See Insert pin.

Pit: Small regular or irregular crater in the surface of a plastic, usually with a width about the same order of magnitude as the depth.

Pitch: A high-molecular-weight material left as a residue from the destructive distillation of coal and petroleum products. Pitches are used as base materials for the manufacture of certain high-modulus carbon fibers and as matrix precursors of carbon-carbon composites.

Planar helix winding: A winding in which the filament path on each dome lies on a plane intersecting the dome while a helical path over the cylindrical section is connected to the dome paths.

Planar winding: A winding in which the filament path lies on a plane intersecting the winding surface.

Plasma spraying: A process for spraying materials onto a substrate or surface in which the materials, in powder form, are heated in a hot plasma created by an electric arc. The technique is particularly useful in laying down surface coatings on ceramic materials.

Plastic: A material that contains as an essential ingredient an organic substance of high molecular weight, is solid in its finished state, and at some stage in its manufacture or processing into finished articles can be shaped by flow; made of plastic. A rigid plastic is one with a stiffness or apparent modulus of elasticity greater than 100 kips/in.^{2} (690 MPa) at 73.4°F (23°C). A semirigid plastic has a stiffness or apparent modulus of elasticity between 10 and 100 kips/in.^{2} (69 and 690 MPa) at 73.4°F (23°C).

Plasticate: To soften by heating or kneading.

Plastic deformation: A change in the dimensions of an object under load that is not recovered when the load is removed; opposite of elastic deformation. (See also Elastic recovery.)

Plastic flow: Deformation under the action of a sustained force; flow of semisolids in molding plastics.

Plasticize: To make a material moldable by softening it with heat or a plasticizer.

Plastic tooling: Tools constructed of plastics, generally laminates or casting materials.

Platens: The mounting plates of a press, to which the entire mold assembly is bolted.

Plates: Powder particles with a flat shape and considerable thickness in contrast to flakes which are thin with respect to the width.

Plied yarn: A yarn formed by twisting together two or more single yarns in one operation.

Ply: The number of single yarns twisted together to form a plied yarn; one of the layers that make up a stack or laminate.

P/M: Powder metallurgy.

PMC: Polymer matrix composite.

PMR: Polyimide. Polymerization-of-monomer-reactant polyimide.

Poisson's ratio, ν: A constant relating change in cross-sectional area to change in length when a material is stretched;

$$\nu = 1/2 \text{ for rubbery materials, } 1/4 - 1/2 \text{ for crystals and glasses}$$

Polar winding: A winding in which the filament path passes tangent to the polar opening at one end of the chamber and tangent to the opposite side of the polar opening at the other end. A one-circuit pattern is inherent in the system.

Polyacrylonitrile (PAN): Used as a base material or precursor in the manufacture of certain carbon fibers. The fiber-forming acrylic polymers are high in molecular weight and are produced commercially either by solution polymerization or suspension polymerization. Both techniques utilize free-radical-initiated addition polymerization of acrylonitrile and small percentages of other monomers. Commercial precursor fibers are more than 90% acrylonitrile-based.

Polyamide (PA): A polymer in which the structural units are linked by amide or thioamide groupings; many polyamides are fiber-forming.

Polyamide-imide (PAI): A polymer containing both amide and imide (as in polyamide) groups; its properties combine the benefits and disadvantages of both.

Polycrystalline: Refers to a crystalline material containing several (usually very many) grains.

Polyesters: Thermosetting resins produced by dissolving unsaturated, generally linear alkyd resins in a vinyl active monomer, e.g., styrene, methyl styrene, or diallyl phthalate. The two important commercial types are (1) liquid resins that are cross-linked with styrene and used either as impregnates for glass or carbon fiber reinforcements in laminates, filament-wound structures, and other built-up constructions, or as binders for chopped-fiber reinforcements in molding compounds; and (2) liquid or solid resins cross-linked with other esters in chopped-fiber and mineral-filled molding compounds.

Polyether ether ketone (PEEK): A linear aromatic crystalline thermoplastic. A composite with a PEEK matrix may have a continuous-use temperature as high as 250°C.

Polyimide (PI): A polymer produced by heating polyamic acid; a highly heat-resistant resin [> 600°F (> 316°C)] suitable for use as a binder or an adhesive. Similar to a polyamide, differing only in the number of hydrogen molecules contained in the groupings. May be either thermoplastic or thermosetting.

Polymer: A high-molecular-weight organic compound, natural or synthetic, whose structure can be represented by a repeated small unit (mer), e.g., polyethylene, rubber, cellulose. Synthetic polymers are formed by addition or condensation polymerization of monomers. Some polymers are elastomers, some plastics. When two or more monomers are involved, the product is called a copolymer.

Polymerization: A chemical reaction in which the molecules of a monomer are linked together to form large molecules whose molecular weight is a multiple of that of the original substance. When two or more monomers are involved, the process is called copolymerization or heteropolymerization.

Polymerize: To unite molecules of the same kind into a compound having the elements in the same proportion but possessing a much higher molecular weight and different physical properties.

Poly(phenylene): A high-carbon-content (94.7%) polymer that has the monomeric repeat unit $-C_6H_4$.

Polyphenylene sulfide (PPS): A high-temperature thermoplastic useful primarily as a molding compound. Known for its chemical resistance.

Polysulfone (PS): A high-temperature-resistant thermoplastic polymer with the sulfone linkage and having a T_g of 190°C.

Polyvinyl chloride (PVC): A vinyl-type thermoplastic resin formed by the addition reaction of vinyl chloride monomer.

Porosity: The ratio of the volume of air or void contained within the boundaries of a material to the total volume (solid material plus air or void), expressed as a percentage.

Positive mold: A mold designed to apply pressure to a piece being molded with no escape of material.

Postcure: Additional elevated-temperature cure, usually without pressure, to improve final properties and/or complete a cure. Complete cure and ultimate mechanical properties of certain resins are attained only by exposure of the cured resin to higher temperatures than those of curing.

Postforming: The forming, bending, or shaping of fully cured, C-stage thermoset laminates that have been heated to make them flexible. On cooling, the formed laminate retains the contours and shape of the mold over which it has been formed.

Pot life: The length of time a catalyzed resin system retains viscosity low enough to be used in processing; also called working life.

Powder: Particles of matter characterized by a small size, less than 1 mm in size.

Powder metallurgy: The art and science of producing metal powders and of utilizing metal powders for the production of massive materials and shaped objects.

PPQ: Polyphenylquinoxaline.

PPS: Polyphenylene sulfide.

PQ: Polyquinoxaline.

Prealloyed powder: Each particle contains an intimate mixture of two or more elements in a prescribed ratio to form an alloy; examples include brass, bronze, steel, and stainless steel.

Precipitate: A minor phase or particle that has been precipitated by the movement and coalescence of individual atoms from solid solution during heat treatment.

Precipitation-hardened: Refers to a material in which the precipitation reaction has been so finely dispersed that the material becomes hardened. The small particles provide a barrier to dislocation movement during plastic deformation.

Precure: The full or partial setting of a synthetic resin or adhesive in a joint before the clamping operation is complete or before pressure is applied.

Precursor: For carbon fibers, the rayon, PAN, or pitch fibers from which carbon fibers are made.

Preform: (1) A preshaped fibrous reinforcement formed by the distribution of chopped fibers by air, water flotation, or vacuum over the surface of a perforated screen to the approximate contour and the thickness desired in the finished part. (2) A preshaped fibrous reinforcement of mat or cloth formed to the desired shape on a mandrel or mock-up before being placed in a mold press. (3) A compact "pill" formed by compressing premixed material to facilitate handling and control of uniformity of charges for mold loading.

Preform binder: A resin applied to the chopped strands of a preform, usually during its formation, and cured so that the preform will retain its shape and be handleable.

Preimpregnation: The practice of mixing resin and reinforcement and effecting partial cure before use or shipment to the user. (See also Prepreg.)

Premix: A molding compound prepared prior to, and apart from, the molding operations and containing all components required for molding, i.e., resin, reinforcement, fillers, catalysts, release agents, and other compounds.

Prepreg: Ready-to-mold material in sheet form, which may be cloth, mat, or paper impregnated with resin and stored for use. The resin is partially cured to a B stage and supplied to the fabricator, who lays up the finished shape and completes the cure with heat and pressure.

Presintering: The heating of a compact to a temperature lower than the normal sintering temperature to gain strength for subsequent handling, including machining.

Pressure: Force measured per unit area. Absolute pressure is measured with respect to zero. Gage pressure is measured with respect to atmospheric pressure.

Pressure-assisted sintering: Sintering with the application of an external pressure. It is often performed by initially sintering in vacuum and subsequently pressurizing the furnace to densify any remaining closed pores.

Pressure-bag molding: A process for molding reinforced plasics, in which a tailored flexible bag is placed over the contact lay-up on the mold, sealed, and clamped in place. Fluid pressure, usually compressed air, is exerted on the bag, and the part is cured.

Primary structure: A structure critical to flight safety.

Primer: A coating applied to a surface before the application of an adhesive, lacquer, enamel, or the like, to improve the performance of the bond.

Promoter: See Accelerator.

Proportional limit: The greatest stress a material is capable of sustaining without deviation from proportionality of stress or strain (Hooke's law); it is expressed in force per unit area, usually in kips per square inch (megapascals).

Prototype: A model suitable for use in complete evaluation of form, design, and performance.

PS: Polysulfone.

PSZ: Partially stabilized zirconia.

PT: Phenolic triazine.

PU: Polyurethane.

Pultrusion: Reversed extrusion of resin-impregnated roving in the manufacture of rods, tubes, and structural shapes with a permanent cross section. After passing through the resin dip tank the roving is drawn through a die to form the desired cross section.

PVB: Polyvinyl butadiene.

PVC: Polyvinyl chloride.

PVD: Plasma vapor deposition.

PVI: Plasma vapor infiltration.

PVT: Pulsed video thermography.

Quasi-isotropic: Refers to approximating isotropy by orientation of plies in several directions.

Quenching: A process commonly used in the heat treatment of metals, in which a certain microstructure or property is obtained by rapidly cooling the sample from some high-temperature condition. Blacksmiths have used this technique in the past, e.g., when hardening steel by quenching it from glowing-red temperature to oil or salt water and forming martensite (a solution of carbon and iron).

R glass: A modified E glass that is chemically resistant (ECR glass). ECR glass fibers are used in applications that require good electrical properties coupled with better chemical resistance. The commercial composition is a high-strength, high-modulus R glass.

Ramping: A gradual, programmed increase or decrease in temperature or pressure for controlling the cure or cooling of composite parts.

Random pattern: A winding with no fixed pattern. If a large number of circuits are required for the pattern to repeat, a random pattern is approached; a winding in which the filaments do not lie in an even pattern.

Rapid solidification (RS): The extraction of heat from molten material at cooling rates of 10^4°C/s or higher, resulting in new microstructures, compositions, or phases.

Rayon: A synthetic fiber made up primarily of regenerated cellulose. In the process, cellulose derived from wood pulp, cotton linters, or other vegetable matter is dissolved into a viscose spinning solution. The solution is extruded into an acid-salt coagulating bath and drawn into continuous filaments. Rayon fibers were one of the first precursor materials to be used in the manufacture of carbon fibers. They have now been almost completely replaced by PAN and pitch fibers as starting materials for carbon fiber manufacture because of low yields, high processing costs, and limited physical property formation.

RBSN: Reaction-bonded silicon nitride.

Reactive sintering: A novel sintering process in which an exothermic reaction is initiated in a mixture of dissimilar (elemental) powders. The reaction produces a compound (carbide, boride, aluminide, or other compound), and the heat from the reaction is used to simultaneously sinter the product phase.

Recycle: A function of growing importance in today's energy and ecologically conscious world. It is much cheaper to remelt scrap metal than to produce new metal from ore. Recycling of materials today, however, is beginning to involve most materials, including all metals, paper, plastic, and glass.

Refractory: A metal or ceramic having a high melting temperature, usually over 1700°C. Example refractory metals are tungsten, molybdenum, rhenium, and zirconium.

Reinforced molding compound: Compound supplied by a raw material producer in the form of ready-to-use materials, as distinguished from premix. (See also Premix.)

Reinforced plastic: A plastic with strength properties greatly superior to those of the base resin, resulting from the presence of reinforcements embedded in the composition.

Reinforcement: A strong, inert material bonded into a plastic to improve its strength, stiffness, and impact resistance. Reinforcements are usually long fibers of glass, asbestos, sisal, cotton, etc., in woven or nonwoven form. To be effective, the reinforcing material must form a strong adhesive bond with the resin. ("Reinforcement" is not synonymous with "filler.")

Release agent: A material that is applied in a thin film to the surface of a mold to keep the resin from bonding to it.

Release film: An impermeable film layer that does not bond to the composite during cure.

Resilience: (1) The ratio of energy returned on recovery from deformation to the work input required to produce the deformation (usually expressed as a percentage). (2) The ability to regain an original shape quickly after being strained or distorted.

Resin: A solid, semisolid, or pseudosolid organic material that has an indefinite (often high) molecular weight, exhibits a tendency to flow when subjected to stress, usually has a softening or melting range, and usually fractures conchoidally. Most resins are polymers. In reinforced plastics the material used to bind together the reinforcement material, the matrix. (See also Polymer.)

Resin applicator: In filament winding the device that deposits the liquid resin on the reinforcement band.

Resin content: The amount of resin in a laminate expressed as a percentage of the total weight or total volume.

Resin, liquid: An organic polymeric liquid that becomes a solid when converted into its final state for use.

Resin-rich area: Space that is filled with resin and lacks reinforcing material.

Resin-starved area: Area of insufficient resin, usually identified by low gloss, dry spots, or fiber show.

Resistance furnace: In the carbon fiber process a carbonization furnace that utilizes resistance heating to eliminate heteroatoms from the carbon fiber structure. In this furnace, heat distribution is obtained by a combination of direct radiation from the resistors and reradiation from the interior surfaces of the furnace chamber to temperatures in excess of 1000°C.

Resistivity: The ability of a material to resist passage of electric current through its bulk or on a surface.

Retarder: See Inhibitor.

Reverse helical winding: As the fiber delivery arm traverses one circuit, a continuous helix is laid down, reversing direction at the polar ends; contrasts with biaxial, compact, or sequential winding in that the fibers cross each other at definite equators, the number depending on the helix. The minimum crossover is 3.

RFI: Radiofrequency interference.

Rib: A reinforcing member of a fabricated or molded part.

Ribbon: A fiber having essentially a rectangular cross section, where the width-to-thickness ratio is at least 4:1.

Rigid plastic: See Plastic.

RIM: Reaction-injection molding.

RMC: Resin matrix composite.

Rockwell hardness number: A value derived from the increase in depth of an impression as the load on an indenter is increased from a fixed minimum value to a higher value and then returned to the minimum value.

ROM: Rule of mixture.

Room-temperature-curing adhesives: Adhesives that set (to handling strength) within 1 h at 68–86°F (20°–30°C) and later reach full strength without heating.

Roving: In filament winding a collection of bundles of continuous filaments either as untwisted strands or as twisted yarns. Rovings may be lightly twisted, but for filament winding they are generally wound as bands or tapes with as little twist as possible. Glass rovings are predominantly used in filament winding.

Roving cloth: A textile fabric, coarse in nature, woven from rovings.

RPP: Reinforced pyrolyzed plastic.

RRIM: Reinforced reaction-injection molding.

RSR: Rapid solidification rate.

RTM: Resin transfer molding. A molding process in which catalyzed resin is transferred into an enclosed mold into which the fiber reinforcement has been placed; cure normally is accomplished without external heat. RTM combines relatively low tooling and equipment costs with the ability to mold large structural parts.

RVE: Representative volume element.

SAM: Scanning acoustic microscope.

Sandwich constructions: Panels composed of a lightweight core material (honeycomb, foamed plastic, etc.) to which two relatively thin, dense, high-strength faces or skins are adhered.

Sandwich heating: A method of heating a thermoplastic sheet before forming by heating both sides of the sheet simultaneously.

SAP: Sintered-aluminum powder.

Satin: A plastic finish having a satin or velvety appearance.

Scarf joint: See Joint.

Scratch: Shallow mark, groove, furrow, or channel normally caused by improper handling or storage.

Scrim: A low-cost, nonwoven, open-weave reinforcing fabric made from continuous-filament yarn in an open-mesh construction.

Secondary structure: A structure not critical to flight safety.

Segregation: A process in which alloying elements diffuse out of solid solution and collect at defects in the material such as grain boundaries and nonuniform distribution of ingredients, such as powder separation by size, shape, or density, or chemical separation in the microstructure of a solidified material.

Self-extinguishing resin: A formulated resin that burns in the presence of a flame but extinguishes itself within a specified time after the flame is removed.

Self-propagating high-temperature synthesis: An exothermic reaction of mixed, dissimilar powders that forms a compound and liberates heat. The heat further induces a reaction between unreacted particles to give a continual reaction wave once ignited.

Selvage: The edge of a woven fabric finished off so as to prevent the yarns from raveling.

SEM: Scanning electron microscope.

Semirigid plastic: See Plastic.

SENB: Single-edge notch beam.

Sequential winding: See Biaxial winding.

Set: (1) To convert into a fixed or hardened state by chemical or physical action, such as condensation, polymerization, oxidation, vulcanization, gelation, hydration, or evaporation of volatiles. (2) The irrecoverable deformation or creep usually measured by a prescribed test procedure and expressed as a percentage of the original dimension.

Set up: To harden, as in curing.

S glass: A magnesia-alumina-silicate glass especially designed to provide filaments with very high tensile strength.

S-2 glass: S-glass and S-2 glass fibers have the same glass composition but different finishes (coatings). S-glass is made to more demanding specifications, and S-2 is considered the commercial grade.

SGTF: Stress-generated thermal field.

Shear: An action or stress resulting from applied forces and tending to cause two contiguous parts of a body to slide relative to each other in a direction parallel to their plane of contact.

Shear edge: The cutoff edge of a mold.

Shear modulus, *G*: The ratio of shearing stress τ to shearing strain γ within the proportional limits of a material.

Shelf life: The length of time a material can be stored under specified conditions without harmful changes in its properties; also called storage life.

Shoe: A device for gathering filaments into a strand in glass fiber forming. (See Chase.)

Short-beam shear strength: The interlaminar shear strength of a parallel fiber-reinforced plastic material as determined by three-point flexural loading of a short segment cut from a ring specimen.

Shrinkage: The relative change in dimension between the length measured on a mold when it is cold and the length on the molded object 24 h after it has been taken out of the mold.

Sialon: A commercially successful ceramic alloy, comprising silicon (Si), aluminum (Al), oxygen (O), and nitrogen (N). In its alloy form it thus becomes (Si)(Al)ON.

Silicon carbide fiber: A reinforcing fiber with high strength and modulus; its density is equal to that of aluminum. It is used in organic and metal matrix composites.

Silicone plastics: Plastics based on resins in which the main polymer chain consists of alternating silicon and oxygen atoms with carbon-containing side groups; derived from silica (sand) and methyl chloride.

Silicones: Resinous materials derived from organosiloxane polymers, furnished in different molecular weights including liquids and solid resins and elastomers.

Si_3N_4: Silicon nitride.

Single-circuit winding: A winding in which the filament path makes a complete traverse of the chamber, after which the following traverse lies immediately adjacent to the previous one.

Sintering: A thermal process (fusion of metallic or ceramic powders) that increases the strength of a powder mass by bonding adjacent particles via diffusion or related atomic level events. Most of the properties of a powder compact are improved, and density frequently increases with sintering. The process occurs in the solid state by interparticle diffusion.

Size: Any treatment consisting of starch, gelatin, oil, wax, or another suitable ingredient applied to yarn or fibers at the time of formation to protect the surface and facilitate handling and fabrication or to control fiber characteristics. The treatment involves ingredients that provide surface lubricity and binding action but, unlike a finish, no coupling agent. Before final fabrication into a composite, the size is usually removed by heat-cleaning and a finish is applied.

Sizing: (1) Applying a material on a surface in order to fill pores and thus reduce absorption of the subsequently applied adhesive or coating. (2) To modify the surface properties of the substrate to improve adhesion. (3) The material used for this purpose; also called size. A coating put on fiber, usually at the time of manufacture, to protect the surface and aid the process of handling and fabrication or to control fiber characteristics. Most standard sizes used for aerospace-grade carbon fibers are epoxy-based.

Sizing content: The percentage of the total strand weight made up by the sizing, usually determined by burning off the organic sizing (loss on ignition).

Skein: A continuous filament, strand, yarn, roving, etc., wound up to some measurable length and generally used to measure various physical properties.

Skin: The relatively dense material that may form the surface of a cellular plastic or sandwich.

Slip casting: A forming process applicable to small powders where the powder is dispersed in a low-viscosity fluid to allow shaping by casting the slurry into a porous mold which extracts the fluid.

SMC: Sheet-molding compound.

***S-N* curve:** Stress per number of cycles to failure. (See also Stress-strain.)

Soft flow: The behavior of a material that flows freely under conventional conditions of molding and which, under such conditions, fills all the interstices of a deep mold where a considerable distance of flow can be demanded.

Solid solution: Very few practical materials are pure in form or consist of only one element. If one element or more are dissolved completely in a base material in the solid state, then the whole is referred to as a solid solution.

SPATE: Stress pattern analysis of thermograph emissions.

Specification: A detailed description of the characteristics of a product and of the criteria that must be used to determine whether the product is in conformity with the description.

Specific gravity: The ratio of the weight of any volume of a substance to the weight of an equal volume of another substance taken as standard at a constant or stated temperature.

Specific heat: The quantity of heat required to raise the temperature of a unit mass of a substance 1 degree under specified conditions.

Specimen: An individual piece or portion of a sample used to make a specific test; of specific shape and dimensions.

Spectroscopy: When a focused beam of x-rays or electrons interacts with a material, the emitted x-rays or electrons become highly characteristic of the material through which they have passed. These emissions produce well-defined spectra and provide a "fingerprint" of the material being studied.

SPF/DB: Superplastic-forming diffusion bonding.

Spherical powder: Powder with a uniform spherical shape and a size that can be characterized by a diameter. Gas atomized powders are often spherical.

Spinnerette: A metal disk containing numerous minute holes used in yarn extrusion. The spinning solution or molten polymer is forced through the holes to form yarn filaments.

Spiral: In glass fiber forming the device used to make the strand traverse back and forth across the forming tube.

Splice: To join two ends of glass-fiber yarn or strand, usually by means of an air-drying glue.

Spline: (1) To prepare a surface to its desired contour by working a paste material with a flat-edged tool; the procedure is similar to screeding concrete. (2) The tool itself.

Split cavity blocks: Blocks that, when assembled, contain a cavity for molding articles having undercuts.

Split mold: A mold in which the cavity is formed of two or more components, known as splits, held together by an outer chase.

Split-ring mold: A mold in which a split-cavity block is assembled in a chase to permit forming undercuts in a molded piece. The parts are ejected from the mold and then separated from the piece.

SPMC: Solid polyester molding compound.

Spray: A complete set of moldings from a multi-impression injection mold together with the associated molded material.

Sprayed-metal molds: Molds made by spraying molten metal onto a master until a shell of predetermined thickness is achieved. The shell is then removed and backed up with plaster, cement, casting resin, or other suitable material. Used primarily as a mold in sheet forming.

Spray-up: Techniques in which a spray gun is used as the processing tool. In reinforced plastics, for example, fibrous glass and resin can be simultaneously deposited in a mold. In essence, roving is fed through a chopper and ejected into a resin stream, which is directed at the mold by either of two spray systems. In foamed plastics, very fast-reacting urethane foams or epoxy foams are fed to the gun in liquid streams and sprayed onto the surface. On contact, the liquid starts to foam.

Spun roving: A heavy, low-cost glass fiber strand consisting of filaments that are continuous but doubled back on each other.

Stabilization: In carbon fiber manufacture the process used to render the carbon fiber precursor infusible prior to carbonization.

Staple fibers: Fibers of spinnable length manufactured directly or by cutting continuous filaments to relatively short lengths [generally less than 17 in. (432 mm)].

Starved area: An area in a plastic part that has an insufficient amount of resin to wet out the reinforcement completely. This condition may be due to improper wetting or impregnation or excessive molding pressure.

Starved joint: An adhesive joint that has been deprived of the proper film thickness of adhesive as a result of insufficient adhesive spreading or application of excessive pressure during lamination.

Static fatigue: Failure of a part under continued static load; analogous to creep-rupture failure in metals testing but often the result of aging accelerated by stress.

Static modulus: The ratio of stress to strain under static conditions; calculated from static stress-strain tests in shear, compression, or tension.

Stiffness: The relationship of load and deformation; a term often used when the relationship of stress to strain does not conform to the definition of Young's modulus. (See Stress-strain.)

Storage life: See Shelf life.

Strain, ε: Although strain has several definitions, which depend upon the system being considered, for small deformations, engineering strain is applicable and is the most common definition of strain.

Strain relaxation: See Creep.

Strand count: The number of strands in a plied yarn or in a roving.

Strand integrity: The degree to which the individual filaments making up the strand or end are held together by the sizing applied.

Strands: A primary bundle of continuous filaments (or slivers) combined in a single compact unit without twist. These filaments (usually 51,102 or 51,204) are gathered together in the forming operations.

Strength, flexural: See Flexural strength.

Stress, σ: Most commonly defined as engineering stress, the ratio of the applied load P to the original cross-sectional area A_o.

Stress concentration: Magnification of the level of an applied stress in the region of a notch, void, or inclusion.

Stress corrosion: Preferential attack of areas under stress in a corrosive environment where this factor alone would not have caused corrosion.

Stress crack: External or internal cracks in a plastic caused by tensile stresses less than that of its short-time mechanical strength. The stresses that cause cracking may be present internally or externally or may be combinations of these stresses. (See also Crazing.)

Stress relaxation: The decrease in stress under sustained constant strain; also called stress decay.

Stress-strain: Stiffness, expressed in kips per square inch (megapascals), at a given strain.

Stress-strain curve: Simultaneous readings of load and deformation, converted into stress and strain, plotted as ordinates and abscissas, respectively, to obtain a stress-strain diagram.

Structural adhesive: An adhesive used for transferring loads between adherends.

Structural bond: A bond that joins basic load-bearing parts of an assembly; the load may be either static or dynamic.

Substitutional: Refers to atoms of an alloying addition to a metal that occupy, or substitute for, positions or atoms of the matrix material.

Superalloys: Multiphase alloys, usually cobalt- or nickel-based, that are very heat-resistant.

Superplastic: Most metals are plastic in the sense that they can be plastically deformed up to a reduction in area at fracture of about 20%. Superplastic metals can be deformed several hundredfold before fracturing. The mechanism by which this occurs does not involve dislocations or glide planes but occurs instead by a rotation of grains by diffusion at grain boundaries.

Surface resistance (electric): Between two electrodes in contact with a material, the ratio of the voltage applied to the electrodes to that portion of the current between them that flows through the surface layers.

Surface resistivity (electric): The ratio of potential gradient parallel to the current along the surface of a material to the current per unit width of surface.

Surface treatment: A material applied to fibrous glass during the forming operation or in subsequent processes, i.e., sizing or finishing. In carbon fiber manufacturing, surface treatment is the process step whereby the surface of the carbon fiber is oxidized in order to promote wettability and adhesion with the matrix resin in the composite.

Surfacing mat: A very thin mat, usually 0.007–0.020 in. (0.18–0.51 mm) thick, of highly filamentized fiber glass used primarily to produce a smooth surface on a reinforced plastic laminate. (See Overlay sheet.)

SWE: Stress wave emission.

Syntactic foam: A cellular plastic put together by incorporating preformed cells (hollow spheres or microballoons) in a resin matrix; opposite of foamed plastic, in which the cells are formed by gas bubbles released in the liquid phase by chemical or mechanical action.

Synthetic resin: A complex, substantially amorphous, organic semisolid or solid material (usually a mixture) built up by chemical reaction of comparatively simple compounds, approximating natural resins in luster, fracture, comparative brittleness, insolubility in water, fusibility or plasticity, and some degree of rubberlike extensibility but commonly deviating widely from natural resins in chemical constitution and behavior with reagents.

T_m: Thermal transition temperature.

Tack: Stickiness of an adhesive or filament-reinforced resin prepreg material.

Tack range: The length of time an adhesive remains in the tacky-dry condition after application to the adherend and under specified conditions of temperature and humidity.

Tack stage: The length of time deposited adhesive film exhibits stickiness or tack or resists removal or deformation of the cast adhesive.

Tangent line: In a filament-wound bottle, any diameter at the equator.

Tape: A composite ribbon consisting of continuous or discontinuous fibers aligned along the tape axis parallel to each other and bonded together by a continuous matrix phase.

Tape laying: A fabrication process in which prepreg tape is laid side by side or overlapped to form a structure.

TD: Thoria-dispersed.

TDI: Thermoplastic imide.

TEM: Transmission electron microscopy.

Tenacity: The strength of a yarn or of a filament of a given size; equal to breaking strength divided by denier.

Tensile bar: A compression- or injection-molded specimen of specified dimensions used to determine the tensile properties of a material.

Tensile modulus: The ratio of the tension stress to the strain in the material over the range for which this value is constant.

Tensile strength or stress: The maximum tensile load per unit area of original cross section, within the gage boundaries, sustained by the specimen during a tension test. It is expressed as kips per square inch (megapascals). Tensile load is interpreted to mean the maximum tensile load sustained by the specimen during the test, whether this coincides with the tensile load at the moment of rupture or not.

TFRS: Tungsten fiber-reinforced superalloy.

Theoretical density: The true crystal density of a material corresponding to the limit attainable through full density products without pores.

Thermal conductivity: Ability of a material to conduct heat; the physical constant for the quantity of heat that passes through a unit cube of a substance in unit time when the difference in the temperature of two faces is 1 degree.

Thermal expansion, Coefficient of: See Coefficient of thermal expansion.

Thermal oxidative stability: The resistance of a fiber, resin, or composite material to degradation upon exposure to elevated temperature in an oxidizing atmosphere. It is often measured as percentage weight loss after exposure to a specified temperature for a set period of time. It may also be measured as the percentage of retained properties after elevated-temperature exposure.

Thermoplastic: Capable of being repeatedly softened by an increase in temperature and hardened by a decrease in temperature; applicable to materials whose change upon heating is substantially physical rather than chemical and can be shaped by flow into articles by molding and extrusion. Many natural resins may be described as thermoplastic. Polyester and PVC are examples.

Thermoset: A plastic that changes into a substantially infusible and insoluble material when it is cured by application of heat or by chemical means. Prior to becoming infusible, thermosetting polymers such as phenolic and melamine formaldehyde possess thermoplastic qualities that permit processing. Epoxy is an example.

Thixotropic: Gellike at rest but fluid when agitated; having high static shear strength and low dynamic shear strength at the same time.

Thread count: The number of yarns (threads) per inch (millimeter) in either lengthwise (warp) or crosswise (fill) direction of woven fabrics.

3D: Three-dimensional weave.

TLM: Tape-laying machine.

TMC: Thick molding compound.

Toggle action: A mechanism that exerts pressure developed by the application of force on a knee joint, used as a method of closing presses and applying pressure at the same time.

Tolerance: The guaranteed maximum deviation from the specified nominal value of a component characteristic at standard or stated environmental conditions.

Tooling resins: Plastic resins, chiefly epoxy and silicone, that are used as tooling aids.

Torsional rigidity (fibers): The resistance of a fiber to twisting.

Toughness: The energy required to break a material, equal to the area under the stress-strain curve.

Tow: A large bundle of continuous filaments, generally 10,000 or more, not twisted, usually designated by a number followed by "K," indicating multiplication by 1000; for example, 12 K tow has 12,000 filaments.

TPC: Thermoplastic composite.

TPS: Thermal protection system.

Transfer molding: Method of molding thermosetting materials in which the plastic is first softened by heating and pressure in a transfer chamber and then forced by high pressure through suitable sprues, runners, and gates into the closed mold for final curing.

Transfer pot: (1) A heating cylinder. (2) Transfer chamber in a transfer mold.

Transformation: The transition from one crystalline phase to another. A phase transformation can be brought about either by a diffusion-controlled reaction or by a diffusionless (martensitic) shear process. The transformation is brought about essentially by chemical changes in the energy of atomic bonding that occur on changing the temperature.

Transient liquid phase sintering: A sintering cycle characterized by the formation and disappearance of a liquid phase during heating. The initial compact has at least two differing chemistries, and the first liquid must be soluble in the remaining solid.

Transition temperature: The temperature at which the properties of a material change.

Transverse rupture strength: A three-point fracture test applied to brittle materials or green compacts to assess relative strength.

Twist: The turns about its axis per unit of length in a yarn or other textile strand. (See also Balanced twist.)

TYS: Tensile yield strength.

TZM: Trade name of molybdenum alloy wire.

TZP: Toughened zirconia polycrystal.

UD: Unidirectional.

UDC: Unidirectional composites.

UHM: Ultrahigh modulus.

Ultimate elongation: The elongation at rupture.

Ultimate tensile strength: The ultimate or final stress sustained by a specimen in a tension test; the stress at the moment of rupture.

Ultraviolet: Zone of invisible radiations beyond the violet end of the spectrum of visible radiations. Since ultraviolet wavelengths are shorter than the visible, their photons have more energy, enough to initiate some chemical reactions and to degrade most plastics.

UMC: Unidirectional molding compound.

Unbond: Area of a bonded surface in which bonding of adherends has failed to occur or where two prepreg layers of a composite fail to adhere to each other; also denotes areas where bonding is deliberately prevented so as to simulate a defective bond.

Undercut: Having a protuberance or indentation that impedes withdrawal from a two-piece, rigid mold; any such protuberance or indentation, depending on the design of the mold (tilting a model in designing its mold may eliminate an apparent undercut).

Unidirectional (UD): Refers to fibers that are oriented in the same direction, such as unidirectional fabric, tape, or laminate.

Unidirectional laminate: A reinforced plastic laminate in which substantially all the fibers are oriented in the same direction.

Urethane acrylic polymers: Developed for use in fabricating resin-injected composites when traditional urethanes were too high in viscosity or when polyesters either required dilution with styrene or were too slow in cycle times for large-volume production. These polymers are formed by the reaction of two liquid components, an acrylesterol and a modified diphenylmethane-4,4′-diisocyanate. Acrylamate resin systems form strong, high-modulus, fiber-reinforced composites. Exceptional levels of strength enable composites made with acrylamate resins to be used for load-bearing components.

Urethane plastics: Plastics that generally react with polyols, e.g., polyesters or polyethers, when reactants are joined by formation of a urethane linkage.

UTRC: United Technologies Research Center.

UTS: Ultimate tensile strength.

UV: Ultraviolet.

UW: Ultrasonic welding.

V_f: Volume fraction.

Vacancy: A "point defect" in a crystal, an atom missing from the lattice structure. It has the important function of aiding in the process of diffusion.

Vacuum bag molding: A process for molding reinforced plastic in which a sheet of flexible, transparent material is placed over the lay-up on the mold and sealed. A vacuum is applied betwen the sheet and the lay-up. The entrapped air is mechanically worked out of the lay-up and removed by the vacuum, and the part is cured; also called bag molding.

Van der Waals forces: Interatomic and intermolecular forces of electrostatic origin. These forces arise as a result of small, instantaneous dipole movements of the atoms. They are much weaker than valence bond forces and inversely proportional to the seventh power of the distance between the particles (atoms or molecules).

Vapor: A gas at a temperature below the critical temperature that can be liquefied by compression without lowering the temperature.

Veil: An ultrathin mat similar to a surface mat, often composed of organic fibers as well as glass fibers.

VIM: Vibrational microlamination.

Vinyl ester: A class of thermosetting resins containing esters of acrylic and/or methacrylic acids, many of which have been made from epoxy resin. Cure is accomplished, as with unsaturated polyesters, by copolymerization with other vinyl monomers, such as styrene.

Virgin filament: An individual filament that has not been in contact with any other filament or any other hard material.

Viscosity: The property of resistance to flow exhibited within the body of a material, expressed in terms of the relationship between applied shearing stress and the resulting rate of strain in shear.

VLS: Vapor-liquid-solid.

Void content: For the percentage of voids in a laminate, calculated by use of the formula void % = $100 - x$, where x is usually a weight percent.

Voids: Gaseous pockets trapped and cured into a laminate; unfilled spaces in a cellular plastic substantially larger than the characteristic individual cells.

Volatile content: The percentage of volatiles driven off as a vapor from a plastic or an impregnated reinforcement.

Volatile loss: Weight loss by vaporization.

Volatiles: Materials in a sizing or a resin formulation capable of being driven off as a vapor at room temperature or slightly above.

Warp: (1) The yarn running lengthwise in a woven fabric; a group of yarns in long lengths and approximately parallel put on beams or warp reels for further textile processing, including weaving. (2) A change in the dimensions of a cured laminate from its original molded shape.

Warpage: Distortion in a compact that occurs during sintering or heat treatment typically due to nonuniform heating or density gradients in the initial compact.

Water absorption: Ratio of the weight of water absorbed by a material upon immersion to the weight of the dry material. (See also Moisture absorption.)

Water jet: A high-pressure stream of water used for cutting organic composites.

Weathering: The exposure of plastics outdoors. In artificial weathering plastics are exposed to cyclic laboratory conditions of high and low temperatures, high and low relative humidities, and ultraviolet radiant energy, with or without direct water spray, in an attempt to produce changes in their properties similar to those observed on long continuous exposure outdoors. Laboratory exposure conditions are usually intensfied beyond those in actual outdoor exposure to achieve an accelerated effect.

Weave: The particular manner in which a fabric is formed by interlacing yarns and usually assigned a style number. In plain weave, the warp and fill fibers alternate to make both fabric faces identical; in satin weave, the pattern produces a satin appearance, with the warp tow over several fill tows and under the next one (e.g., eight-harness satin would have warp tow over seven fill tows and under the eighth).

Web: A textile fabric, paper, or a thin-metal sheet of continuous length handled in roll form, as contrasted with the same material cut into sheets.

Weft: The transverse threads of fibers in a woven fabric; fibers running perpendicular to the warp; also called filler, filler yarn, woof.

Weldability: A material has good weldability when it can be joined by fusion welding without problems of cracking or excessive porosity occurring.

Wet flexural strength (WFS): The flexural strength after water immersion, usually after boiling the test specimen for 2 h in water.

Wet lay-up: Reinforced plastic that has liquid resin applied as the reinforcement is laid up; the opposite of dry lay-up or prepreg. (See also Dry lay-up, Prepreg.)

Wet-out: The condition of an impregnated roving or yarn wherein substantially all voids between the sized strands and filaments are filled with resin.

Wet-out rate: The time required for a plastic to fill the interstices of a reinforcement material and wet the surface of the reinforcement fibers; usually determined by optical or light transmission means.

Wetting agent: A surface-active agent that promotes wetting by decreasing the cohesion within a liquid.

Wetting angle: The equilibrium angle formed by a liquid-solid-vapor combination; it provides a relative measure of spreading and capillary forces.

Wet winding: In filament winding the process of winding glass on a mandrel where the strand is impregnated with resin just before contact with the mandrel. (See also Dry winding.)

WFS: Wet flexural strength.

Whisker: A very-short-fiber form of reinforcement, usually crystalline and having almost no crystalline defects. Numerous materials, including metals, oxides, carbides, halides, and organic compounds, have been prepared in the form of whiskers. They are often used to reinforce resin and metallic matrix composites.

Winding, biaxial: See Biaxial winding.

Winding pattern: (1) The total number of individual circuits required for a winding path to begin repeating by laying down immediately adjacent to the initial circuit. (2) A regularly recurring pattern of the filament path after a certain number of mandrel revolutions, eventually leading to complete coverage of the mandrel.

Winding tension: In filament winding the amount of tension on the reinforcement as it makes contact with the mandrel.

Wire: A metallic filament.

WOF: Work of fracture.

Working life: The period of time during which a liquid, resin, or adhesive, after mixing with a catalyst, solvent, or other compounding ingredients, remains usable. (See also Gelation time, Pot life.)

Woven fabrics: Fabrics produced by interlacing strands at more-or-less right angles.

Woven roving: A heavy glass fiber fabric made by weaving roving.

WR: Woven roving.

Wrinkle: A surface imperfection in laminated plastics that has the appearance of a crease in one or more outer sheets of the paper, fabric, or other base that has been pressed in.

Wrought: A material fabricated by conventional fusion metallurgy techniques, with a final stage of plastic deformation to improve mechanical properties.

***X* axis:** The axis in the plane of the laminate used as 0° reference; the *y* axis is the axis in the plane of the laminate perpendicular to the *x* axis; the *z* axis is the reference axis normal to the laminate plane in composite laminates.

XD: Exothermic dispersion.

XMC: Directionally reinforced molding compound.

XPS: X-ray photoelectron spectroscopy.

XRD: X-ray diffraction.

XTM: X-ray tomography.

Yarn: An assemblage of twisted fibers or strands, natural or manufactured, to form a continuous yarn suitable for use in weaving or otherwise interweaving into textile materials. (See also Continuous filament.)

Yield point: The first stress in a material, less than the maximum attainable stress, at which an increase in strain occurs without an increase in stress. Only materials that exhibit this unique phenomenon of yielding have a yield point.

Yield strength: The stress at which a material exhibits a specified limiting deviation from the proportionality of stress to strain; the lowest stress at which a material undergoes plastic deformation. Below this stress, the material is elastic; above it, viscous.

Young's modulus: The ratio of the applied load per unit area of cross section to the increase in length of a body obeying Hooke's law. (See Modulus of elasticity.)

Y-TZP: Toughened zirconia polycrystal with yttrium.

ZTA: Zirconia-toughened alumina.

Index

A

B

C

D

E

F

N

P

R

S

T

U

V

W

X

Z